Molecular and Integrative Toxicology

More information about this series at http://www.springer.com/series/8792

Susan Y. Smith • Aurore Varela
Rana Samadfam
Editors

Bone Toxicology

Springer

Editors
Susan Y. Smith
Montreal, QC, Canada

Aurore Varela
Charles River Preclinical Services
Senneville, QC, Canada

Rana Samadfam
Charles River Preclinical Services
Senneville, QC, Canada

ISSN 2168-4219 ISSN 2168-4235 (electronic)
Molecular and Integrative Toxicology
ISBN 978-3-319-56190-5 ISBN 978-3-319-56192-9 (eBook)
DOI 10.1007/978-3-319-56192-9

Library of Congress Control Number: 2017943684

Printed on acid-free paper

This Springer imprint is published by Springer Nature
The registered company is Springer International Publishing AG
The registered company address is: Gewerbestrasse 11, 6330 Cham, Switzerland

Foreword

Toxicology is often defined as the study of the adverse effects of chemical, physical, or biological agents on organisms and ecosystems. Toxicologists employ a variety of in silico, in vitro, and in vivo experimental approaches to aid in the characterization of these effects. Within the discipline of nonclinical safety assessment of drugs, biotherapeutics, and environmental chemicals, in vivo toxicology studies are used to survey a broad range of organs and tissues to determine if there are any effects on the structure and function of an organ system. While the range of systems examined in these studies is quite broad, much of the focus has been on what are often referred to as major organ systems and there are discrete subdisciplines (e.g., hepatotoxicity, neurotoxicity, nephrotoxicity) which focus on a specific organ system. These subdisciplines have a robust literature spanning many decades and often have multiple comprehensive textbooks and journals focused on a single organ or system. There are a variety of other organ systems for which there is a rich literature; however, there are no comprehensive review textbooks available. The skeletal system is certainly one of the systems that has lacked from an expert review, and this volume entitled *Bone Toxicology*, edited by Susan Y. Smith, Rana Samadfam, and Aurore Varela, provides the first authoritative text in this area.

In the training programs of most toxicologists, there is limited or perhaps no training specific to bone biology or toxicology. Most toxicologic pathologists will have a base of knowledge in skeletal anatomy, physiology, and pathology from their training; however, they are typically not trained in some of the quantitative techniques which can be crucial to the study of the skeleton. As such, the casual student often perceives the skeleton, like other calcified tissues, to be a relatively inert and static organ system. This perception is often reinforced by the fact that practicing toxicologists and pathologists may have never or only rarely encountered the bone as a target organ of toxicity in their experiments. As readily demonstrated in this text, bone is a highly dynamic tissue which can be influenced by a variety of factors including both direct effects of a potential therapeutic or indirect factors such effects on body weight, food consumption, or other experimental factors. As with other systems, the bone does not work in isolation and has important connections and inter-relationships with a variety of systems, most notably the endocrine, neural, and immune networks. The complex and dynamic nature of bone, combined with the increasing diverse range of targets and mechanisms of action of both small molecule and biologic therapeutics, has made it increasingly important for practicing toxicologists to have resources to better understand bone biology, have a working knowledge of appropriate endpoints to assess the skeleton, and appreciate how alterations in a variety of organ systems can impact the skeleton. This volume of *Bone Toxicology* provides such a comprehensive resource for the toxicology community.

It is important to remember that the role of nonclinical safety assessment is not just to identify the effects of potential therapeutic, chemical, or environmental agents on the test systems we study (i.e., hazard identification), but ultimately to understand what the relevance of those findings are for the conditions under which humans will be exposed to the therapeutic agent, chemical agent, or environmental agent (i.e., risk assessment). For disease states such as osteoporosis, much is known about the relative strengths and limitations of the various animal models for predicting responses in humans; however, for the types of effects encountered in repeat-dose toxicity models, the relevance to humans is often a very challenging question and there is often not an existing literature to draw on. By drawing on the information provided in this volume, the toxicology community will be better equipped to accomplish the goals of both detecting effects on the skeleton in animal models and assessing potential human risk. By accomplishing these goals, we will be better able to define reasonably safe conditions of use in the patients and consumers we serve.

John L. Vahle
Senior Research Fellow, Toxicology and Pathology,
Eli Lilly and Company, Lilly Corporate Center,
Indianapolis, IN 46285, USA
jvahle@lilly.com http://www.lilly.com

Preface

The skeleton is traditionally considered a hard structure providing support to allow locomotion, to protect sensitive internal organs, and to serve as a major reservoir for the maintenance of serum calcium. However, bone is a dynamic structure, recognized as having a pivotal role as an endocrine organ, and is coming into focus as an important target tissue in the overall development of a new drug, whether bone is the intended target or not. The skeleton is intimately related to other organ systems through paracrine, endocrine, and neural networks. The objective of this book is to provide the toxicologist in preclinical drug development with the necessary tools to identify and characterize a skeletal effect and to present current research on skeletal regulation and its role as an endocrine organ. This book is not intended to list bone toxic agents or detail their effects, unless needed to illustrate a point. The toxicity of many agents to bone is well described in the literature and is beyond the scope of this book.

The book is divided into three parts. Chapters 1, 2, 3, and 4 in Part I of the book introduce the overarching aspects and goals of skeletal evaluations in drug testing, as well as bone biology, regulatory aspects, pediatric applications, and important animal models. A basic knowledge of bone biology is fundamental for an appropriate assessment of effects of a drug treatment on the skeleton. Many lessons can be learned from Chap. 2, "Bone Physiology and Biology," not the least of which is that a growing skeleton is very different from a mature, adult skeleton. Hence, Chap. 3 is dedicated to specific considerations for bone evaluations for pediatric therapeutics. We have learned much of our current understanding of bone biology from the testing of drugs intended for the prevention or treatment of osteoporosis; the use of simple animal models of osteopenia to test the efficacy and pharmacology of various drug classes has been fundamental to the successful approval of current osteoporosis drugs. These models, highlighted in Chap. 4, can be used to add important key data to a drug development program for many other indications.

In Part II, Chaps. 5, 6, 7, and 8 describe the methods used to derive the four primary outcome measures used to evaluate the skeleton: biochemical markers of bone turnover, imaging, histopathology and histomorphometry, and biomechanical strength testing. Each of these end points has been used extensively in bone research for over 20 years. The adaptation of these end points for use in more general safety assessments of the skeleton has led to interesting challenges while broadening their application to encompass numerous species and important investigations of the juvenile skeleton.

Highlighting the importance of a systems biology approach to drug safety testing, the message that appears throughout the chapters is that bone is not an isolated tissue and is considered an endocrine organ with strong evidence accumulating to support cross talk with many other organ systems. Chapters 9, 10, 11, 12, 13, 14, and 15 in Part III of the book are devoted to topics on bone regulation, including interactions with muscle, pituitary hormones, the kidney, the immune system, the central nervous system, intestinal microbiota, and energy metabolism. These chapters were selected as "hot topics" because of important research advances in these areas and the development of new therapeutic targets; it is by no means intended to cover all possible known interactions of bone.

This book is intended to provide the toxicologist in nonclinical drug development with information on skeletal biology, regulatory requirements, and application of the tools, as well as an appreciation for the regulation of bone and its cross talk with other major organ systems, emphasizing the importance of a systems biology and weight of evidence approach to safety assessments.

Montreal, QC, Canada Susan Y. Smith

Contents

Part III Bone as an Endocrine Organ

About the Editors

Rana Samadfam, MSc, PhD, DABT Dr. Rana Samadfam is a Scientific Director of In Vivo Pharmacology Department at Charles River. She is a Diplomate of the American Board of Toxicology. She received her MSc and PhD degrees from the University of Sherbrook in Quebec, Canada, with focus on arthritis and pharmacology. Following her PhD, she accepted a postdoctoral scholar position at the University of McGill, Montreal, Canada, with focus on bone and endocrinology. She joined Charles River 12 years ago, accepting a position in musculoskeletal group. She held the principal scientist position in Musculoskeletal Research Department for over 5 years. She is a winner of several awards including the ASBMR Young Investigator Award and author of numerous peer-reviewed publications (over 110). Her area of expertise includes inflammation, early in vivo pharmacology, toxicology, bone, diabetes, and the endocrine system.

Susan Y. Smith, MSc Susan has over 30 years experience in nonclinical drug development, with over 20 years as Senior Scientific Director at Charles River, Montreal, specializing in musculoskeletal research. She was instrumental in founding the Department of Bone Research at Charles River, leading the team in the testing of drugs to treat osteoporosis and other bone, joint, and muscle diseases; bone healing; and bone toxicology. Susan has coauthored numerous papers in peer-reviewed journals, contributed to several book chapters, and served as session chair at conferences including a continuing education course in bone toxicology. Susan obtained her Masters of Science degree in Pharmacology from McGill University, Montreal, Canada, and her Bachelor of Technology degree in Applied Biology from Brunel University, Uxbridge, England.

Aurore Varela, DVM, MSc, DABT Aurore is a Scientific Director of Musculoskeletal Research & Imaging at Charles River in Montreal, with 15 years of experience in preclinical radiology, imaging, bone biomarkers, biomechanics, and study design and conduct in musculoskeletal research and toxicology. She plays a key role in designing and executing studies, and interpreting data relevant to musculoskeletal research, musculoskeletal and metabolic toxicology, and imaging. She has coauthored several research papers in peer-reviewed scientific journals and contributed to chapters in several books related to musculoskeletal research and bone pathology. She is also a Diplomate of the American Board of Toxicology.

Part I
Bone Toxicology in Preclinical Drug Testing

Chapter 1
Introduction and Considerations in Bone Toxicology

Susan Y. Smith, Nancy Doyle, and Melanie Felx

Abstract This chapter is intended to provide the researcher with information to facilitate in the design, execution, and interpretation of studies with bone end-points. The ability to use our knowledge of bone biology to ensure study conditions are optimal to detect an effect will be presented. This includes study design considerations and the utility of other models to explore the impact new compounds may have on the skeleton in the development of safe and effective drugs.

Keywords Regulatory • Toxicology • Safety • Bone end-points • Markers • Radiography • Bone densitometry • Histomorphometry • Biomechanics • Study design • Bone quality • Adversity

1.1 Introduction

Bone end-points can be included in any safety study design, for any species, at any age, irrespective of duration and route of administration. Additional mechanistic studies can also be designed on a case-by-case basis to meet regulatory demands. The ability to use our knowledge of bone biology to ensure study conditions are optimal to detect an effect is key to a successful outcome. This includes study design considerations, the utility of animal models (chemically, genetically, or surgically modified) in a safety assessment setting, and the potential benefit of including in vitro studies and studies using nonmammalian species (such as zebra fish). These studies cannot only provide invaluable mechanistic data but also respect the three R's initiative and the ethical use of animals.

S.Y. Smith (✉)
SY Smith Consulting, Montreal, QC, Canada
e-mail: sysmithconsulting@gmail.com

N. Doyle • M. Felx
Charles River, 22022, Transcanadienne, Senneville, QC, Canada

S.Y. Smith et al. (eds.), *Bone Toxicology*, Molecular and Integrative Toxicology,
DOI 10.1007/978-3-319-56192-9_1

This chapter will also tackle the issue of skeletal adversity. One of the most difficult aspects of interpreting the relevance of a skeletal effect is to understand what constitutes a direct effect on bone and what might be considered secondary to effects on growth and body weight. This is a considerable challenge in therapeutic areas such as diabetes and obesity, for example, where the intended pharmacological outcome is likely to impact body weight in non-diseased animals. Animal health and welfare is always the primary concern and of relevance in studies which may become confounded by exaggerated pharmacological effects. Therefore, strategies for mitigating effects of exaggerated pharmacology are important considerations in study design.

The term "bone quality" has many definitions and is often used as a "black box" when unknown factors affect bone strength, which is a product of both bone quantity and bone quality. Bone quality encompasses all the geometric and material factors that contribute to fracture resistance (Donnelly 2011). End-points used to assess skeletal integrity therefore need to characterize the bone geometry and microarchitecture, assess the mechanical properties, and evaluate its tissue composition. No single method can address all aspects of bone quality, but a combination of techniques can help to provide a comprehensive evaluation. The toolbox available in a preclinical setting can be extensive; therefore, the methods used to assess bone should be tailored to the study design and the outcome measures of interest.

1.2 Regulatory Considerations

The primary outcome measures used preclinically to identify and characterize the bone quality are biochemical markers, radiography/bone densitometry, histopathology/histomorphometry, and biomechanics. The selection of these end-points has largely been driven by the requirements or guidelines issued by the regulatory agencies to test drugs intended to treat or prevent osteoporosis (osteoporosis testing guidelines: Japanese MHLW 1999; EMA 2007; FDA 2016b). These various techniques have been employed in bone research for over 20 years and most are successfully validated for use in a good laboratory practice (GLP) setting. GLP requirements may be important if data is intended to support a regulatory submission, but may not be required in early phase exploratory studies. Currently for toxicology studies, regulatory guidelines have been issued to assess skeletal growth only for therapeutics intended for pediatric use (pediatric guidelines: FDA 2006; EMA 2008; Japanese MHLW 2012), with requirements to study growth and skeletal development in detail in the pediatric investigational plans (PIP – EMA 2006) and pediatric study plans (PSP – FDA 2016a). In all other cases, a standard toxicology package typically includes the routine qualitative histopathological evaluation of hematoxylin and eosin-stained decalcified bone sampled from the sternum or femoral-tibial joint for large animal species and the distal femur, femoral-tibial joint, and/or proximal tibia for rodents. Standard histopathology uses transmitted light microscopy and is an important diagnostic tool but is likely the least sensitive method to discriminate effects on bone mass or density; polarized light is also used to verify the lamellar organization of the bone matrix and to examine whether woven bone is present. If changes are seen, then they are probably important (major). This would be a trigger to use more sensitive techniques to discern an effect when looking at lower dose levels, to obtain information on the chronology of the change, and to further characterize the change. These specialized techniques include in vivo biomarkers such as biochemical markers of bone turnover or skeletal imaging, or ex vivo special histopathology stains (such as von Kossa or Goldner's trichrome) and/or histomorphometry on undecalcified tissue (see Chap. 8), and biomechanical strength testing (see Chap. 7). Skeletal imaging in most toxicology studies involves the use of radiography and/or bone densitometry. Bone densitometry equipment used extensively in preclinical research is dual energy X-ray absorptiometry (DXA), a 2-dimensional areal assessment with area, bone mineral content (BMC), and bone mineral density (BMD) as the primary output measures. Bone densitometry is also performed using peripheral quantitative computed tomography (pQCT) which provides three-dimensional measures of area, BMC and BMD, as well as separate measures of area, BMC and BMD for the trabecular and cortical bone compartments. Peripheral QCT scan data provides measures of several important geometric parameters: bone diameter (periosteal circumference), endosteal circumference, and cortical thickness. Software algorithms use these data to derive surrogate indices of bone strength including cross-sectional moments of inertia (CSMI). The advantages and disadvantages of these different skeletal imaging techniques will be further discussed later in this chapter and in Chap. 6 (Skeletal Imaging).

To date, qualitative histopathology remains the gold standard for diagnostic pathology and hazard identification in the safety assessment of potential new therapeutics, but it is not sensitive to evaluate changes in bone mass, geometry, and density. For diagnostic terminology used to describe bone changes, the reader is referred to the INHAND initiative publication devoted to skeletal tissues and teeth of laboratory rats and mice (Fossey et al. 2016). In most cases, qualitative histopathology has sufficient sensitivity to detect test article-related effects on bone marrow and growth plates in standard toxicity studies, although it may lack the sensitivity to detect effects of test articles on key physiological processes in bone tissue such as bone formation, mineralization, and resorption. The known limitations of standard qualitative histopathology and our current knowledge of bone signaling pathways related to specific drug classes has led to increasing demands from the regulatory authorities to include specialized end-points in toxicity studies to identify if the skeleton is a potential target tissue and/or to characterize a skeletal effect. Whenever the skeleton is considered potentially at risk, general toxicology studies should make provision to include retention of additional bones at termination to allow a more comprehensive skeletal evaluation, as needed, Table 1.1.

Table 1.1 Recommendations for bone retention in toxicology studies[a]

Bone	Potential use	Storage
Rodent/non-rodent		
Tibia, proximal	Histomorphometry/micro-CT	NBF then alcohol
Lumbar vertebrae L2 (NHP), L3	Histomorphometry	NBF then alcohol
Lumbar vertebrae L4, L5[b]	Biomechanics	Frozen −20 °C
Femur, whole	Biomechanics	Frozen −20 °C

NBF 10% neutral buffered formalin, *NHP* nonhuman primate

[a] Minimum recommendations in addition to standard regulatory requirements; additional backup bones can be added

[b] L4 + L5 (together) for mice

1.3 Tier Approach to Including Specialized Bone End-Points in Preclinical Toxicology Studies

The decision to include specialized bone end-points into toxicity studies is driven by numerous factors. These specialized techniques when used as part of a systems approach to safety assessments are normally sufficiently sensitive to detect perturbations in bone which are needed to address liability, to monitor drug pharmacology, determine the chronology of an effect, address a regulatory requirement, or provide mechanistic data to characterize an effect. In instances where markers show changes consistent with the expected pharmacology of a compound, blood sampling for pharmacokinetics can be combined with sampling for biochemical markers to establish a pharmacokinetic/pharmacodynamic (PK/PD) profile (e.g., see Ominsky et al. 2010). This PK/PD profiling can help guide selection of a specific marker for use in future studies and potentially to establish the optimal timing for sample collection.

Safety assessments ultimately need to provide information regarding how a drug is affecting the skeleton and whether this is a beneficial or adverse effect. When testing a compound belonging to a specific drug class with known class effects, these end-points can be used to confirm the class effect and to further characterize the drug for equivalence, superiority, or adversity. When testing a new drug with no established toxicity, based on what is known of the target or mechanism of action, standard clinical pathology parameters from acute and/or early dose-range finding studies can be monitored for perturbations in calcium/phosphorus and total alkaline phosphatase levels, as a minimum. Biochemical markers can be added to the clinical pathology analysis to facilitate hazard identification. These early data are used to drive the design of the sub-chronic and chronic studies. In studies 28 days or longer, radiography and/or bone densitometry can be added, either in vivo or ex vivo using excised bones. Radiographs are used to monitor skeletal growth (most notably in juvenile toxicology studies), assess skeletal maturity (growth plate closure), identify and/or monitor the progression of abnormalities and bone healing, and as a tool in skeletal phenotyping. Radiography has become an important end-point in rodent lifetime pharmacology or carcinogenicity studies where there is a risk for osteosarcoma; whole body radiographs obtained prior to termination are used to identify bone masses and lesions to facilitate tissue collection at necropsy and subsequent microscopic evaluation and diagnosis (Jolette et al. 2006; Chouinard et al. 2016). In vivo bone densitometry is highly recommended for non-rodent species where group sizes are small (typically 3–5/sex) and bone density can vary considerably between animals. Inclusion of baseline scans for large animals allows any changes to be determined relative to individual animal baseline data. Baseline scans are normally not required for rodent studies which typically are well powered ($n = 8$ to 10/sex) and have relatively homogeneous bone density within a population. For rodent studies, adequate data for interpretation can often be obtained ex vivo. For rodent and non-rodent studies 3 months or longer, in vivo scans can be acquired at several timepoints during the study to provide a chronology of any effects.

To further characterize a skeletal effect, provision can be made in the protocol to retain appropriate bone specimens to perform additional end-points such as histomorphometry and biomechanical strength testing. Histomorphometry provides unique information on bone microstructure and the dynamics of bone turnover. Biomechanical strength testing is considered the ultimate test of bone quality and is normally performed on a long bone such as the femur and vertebrae. In humans, most osteoporotic fractures occur at the spine or hip; therefore, these are key sites to measure bone strength in animal studies.

High-resolution scanning using micro-CT imaging is currently not used routinely in a toxicology setting; however, application of this technology is expanding, particularly in developmental toxicology studies (Johnson et al. 2014). High-resolution images of whole fetuses can be rapidly assessed qualitatively for abnormalities. Current use of this technology is being standardized with respect to equipment, methodology, and terminology. Submission of image files for regulatory review is intended to expedite this aspect of the drug development process and allow the reviewer to verify reported abnormalities directly. Quantitative micro-CT, used to measure bone microarchitecture, provides a high-resolution scan comparable to histomorphometry structural parameters but without the lengthy processing time (plastic embedding, sectioning, and staining). Micro-CT provides a three-dimensional (3-D) evaluation and includes parameters such as tissue mineral density and connectivity index that cannot be derived using histomorphometry (Kazakia et al. 2008; Donnelly 2011). In toxicology studies, tools with a lower resolution (DXA, pQCT) that are less expensive to acquire and operate are normally adequate to detect changes in bone mass (which reflect changes in bone architecture). An important application of high-resolution CT scanning in bone quality studies is the ability to reconstruct the 3-D images to perform finite element analyses (FEA) to model bone strength (Keaveny 2010). Data from animal models are then compared with actual biomechanical strength data. The micro-CT data from animal studies are compared with clinically acquired data and used as a translational tool to identify effects on bone strength without the need to perform biomechanical testing in humans. Baseline (pre-study) in vivo micro-CT scans of the distal radius of cynomolgus monkeys were used to monitor changes in architecture over time using a

Table 1.2 Summary of application of bone end-points to toxicology studies

Study type	Markers	Growth+	X-ray	DXA/pQCT In vivo	DXA/pQCT Ex vivo	Micro-CT	Histo-morphometry	Bio-mechanics
PK/PD	√							
Acute/DRF	√							
28 days	√				√	√		
3 months	√		√	√	√	√	√[a]	√
Chronic	√		√	√	√	√	√	√
Lifetime	√		√		√			
Juvenile	√	√	√	√	√	√	√	√
Neonatal		√	√[b]	√[b]	√[b]			√[b]

DRF dose-range finding, + physical measurements of long bones and crown-rump
[a]Rodent only
[b]Non-rodent only

registration process based on specific anatomical landmarks (Ominsky et al. 2017b). However, these types of applications are normally used in specialized mechanistic studies, not as part of a toxicity study.

Data from biochemical markers, radiography, bone densitometry, histomorphometry, and biomechanical testing are integrated to "tell the story" regarding the mechanism driving any bone changes and whether the outcome was beneficial or adverse to the skeleton. These data are not interpreted in isolation; however, and the body of data from all toxicology study end-points are considered. Not all studies require inclusion of all bone end-points (Table 1.2). All end-points might typically be included in definitive chronic studies, for example, or where changes in vivo were equivocal, and additional end-points such as the biomechanics may be needed to provide a definitive outcome. However, in many instances, data from in vivo end-points (markers and imaging) are sufficient to make a "go/no-go" decision.

The tier approach to the inclusion of bone end-points in preclinical toxicity studies is summarized in Table 1.3. The use of these primary outcome measures in toxicology studies will serve to provide important information regarding skeletal safety. However, regulatory agencies often request additional information to support a submission which may require specialized mechanistic studies or investigation of other aspects of bone quality. Mechanistic studies can be designed to address specific aspects of drug activity on bone and could include techniques such as 5-bromo-2′-deoxyuridine (BrdU) labeling (Mead and Lefebvre 2014) or sophisticated fluorochrome labeling to investigate the chronology of changes (Boyce et al. 2017; Ominsky et al. 2015, Chap. 8). Other tests of bone quality include microscopic, spectroscopic, physical, and chemical techniques to characterize the mineral and collagenous components of bone tissue and are nicely reviewed by Donnelly (2011). These include high-resolution micro-CT scanning (HR-pQCT, quantitative assessment of microarchitecture and measures true tissue mineral density; ionizing radiation), magnetic resonance imaging (MRI, generates 3-D images of bone geometry

Table 1.3 Tier approach to include specialized bone end-points in preclinical toxicology studies

In vivo
Physical measurements, including radiography to assess growth plate closure and abnormalities
Bone densitometry: bone geometry, BMC, BMD
Biochemical markers of bone turnover and/or hormones
Ex vivo
Bone densitometry: bone geometry, BMC, BMD
Histomorphometry: structural and dynamic parameters – mechanistic information
Biomechanical strength testing – ultimate test of bone quality
Other tests to characterize bone quality

and microarchitecture without ionizing radiation), nuclear magnetic resonance imaging (NMR, nonionizing characterization of bone composition including the chemical bonding of bone mineral but no information on the bone matrix), Fourier transform infrared (FTIR) and Raman techniques (characterize bone tissue composition, both mineral and matrix, Boskey and Robey 2013), scanning electron microscopy (SEM, characterizes the morphology and composition of bone surfaces), chemical analyses of collagen cross-links (determines the total amount of collagen and the types of cross-links present), and gravimetric analyses (determines the mineral content of bone tissue based on the ash weight normalized to the dry weight). These techniques have been used to provide valuable information regarding bone quality to further characterize test article effects as part of drug development programs (e.g., see Saito et al. 2015; Ominsky et al. 2017b).

1.4 Considerations for Study Design

1.4.1 Biochemical Markers of Bone Turnover

Biochemical markers are analyzed in blood and urine so can easily be incorporated into any study design for any species, with the exception of neonatal animals (Table 1.2). In a toxicology study, samples for markers can be collected coincident with routine clinical pathology sampling. Optimally, a panel of markers consists of two bone formation markers (osteocalcin (OC), bone-specific alkaline phosphatase (BAP), procollagen type I propeptide (PINP)) and two bone resorption markers (deoxypyridinoline (DPD), C-telopeptides (CTx), N-telopeptides (NTx), tartrate-resistant acid phosphatase 5b (TRACP5b)). Bone formation markers and CTx, NTx, and TRACP5b can be analyzed in serum; DPD, CTx, and NTx can be analyzed in urine. As a minimum, one formation and one resorption marker can be used and may be the only option for test species where validated bone assays may be limited (e.g., pigs). If limited to one of each, then PINP (or osteocalcin if PINP is not available) (formation) and CTx (resorption, either serum or urine) are recommended (total ALP will also be available as part of the clinical pathology data). Since it is important to understand the net effect of changes in bone formation and bone resorption, use of a single marker is not recommended unless a specific PD marker has been identified. For more information on the markers, the reader is referred to Chap. 5.

For biochemical markers, the most significant limitation is sample volume. Blood sampling in a toxicology study can be dominated by the requirements for assays including clinical pathology, hormones, other PD markers, immunology, pharmacokinetics (PK), and/or sampling for anti-drug antibody (ADA) assessments. Most serum bone marker assays require 75 μL or 150 μL if backup samples are collected. While it may be feasible to analyze multiple markers at a single timepoint in a large animal species study, it is often impractical to do so in rodent studies (notably for mice and juvenile rats) which may require separate subpopulations or use of techniques such as micro-sampling (for relevant assays) to provide samples for all assays. When there is a choice of test system, the rat is recommended for use rather than the mouse to optimize sampling procedures not just for blood and urine, but also bone size for ease of in vivo imaging and postmortem end-points. However, use of the mouse as a test system is unavoidable where the drug target is not appropriate for the rat, driving the need to adapt procedures for blood and urine collection and analysis. The development of Luminex-based assays for several bone markers has significantly reduced the sample volume requirements making the mouse a more feasible test system, as well as increasing the scope of assays that can be performed for the rat. Another strategy adopted when sample volume is limited is to collect the minimal volume to perform a single-point determination for each marker and retain a single backup sample for repeat analyses of any assays that fail.

For bone resorption markers, no special collection procedures are required for urine for the rat or large animals. Following an overnight fast, collection in jars from the cage pan, without ice, is adequate if fecal contamination is avoided. Direct collection from the bladder by catheterization or cystocentesis is an option for the monkey and dog but collection requires animals to be sedated which may not be appropriate in a toxicology study setting. The advantage of collecting a urine sample by catheterization is to reduce variability due to potential bacterial growth. For monkeys, the anesthetic used consists of an intramuscular injection of a cocktail of glycopyrrolate, ketamine, and dexmedetomidine. Atipamezole is used to reverse the anesthesia once the urine sample is obtained. The catheterization procedure is performed in the animal room following aseptic-like procedures. For males, the insertion of the catheter is done very slowly using a catheter of adequate length to prevent the creation of loops which would not allow the removal of the catheter from the urethra. Alternatively, following an overnight fast, a relatively clean urine sample can be obtained from the cage pan early morning during a 4-h collection period with water deprivation. Urine collection for the mouse historically has been avoided although collection in metabolic cages (as used for the rat) has shown some success. A mouse study needs to be adequately powered to compensate for (potentially) several missing samples, to provide adequate data for interpretation. Although yet to be validated for use with bone markers, a relatively new technique using LabSand® allows urine collection from group-housed mice over a 5-h period. LabSand® is a hydrophobic, commercially available sand that keeps the urine afloat. The LabSand® replaces the cage bedding, animals are free to move around the cage, and urine drops are collected using a pipette (Doyle et al. 2017).

The conditions for sample collection are important and need to be consistent. Circadian rhythms, hormonal status, stress, and diet can influence bone markers. In a regulated preclinical study, it is relatively easy to control diet and environmental conditions (light cycle, temperature, and humidity) and dose administration. Samples should be collected at the same time of day and under the same conditions (normally fasted) throughout the study. To ensure samples are obtained consistently, an early morning collection is recommended, between 8 and 10 a.m., following an overnight fast. Alternatively, samples can be collected in the afternoon, following several hours of food deprivation, rotating the groups throughout the collection period. Fruits normally provided to monkeys as part of an enrichment program should not be given for approximately 2 days prior to sampling for bone markers. As a non-certified food supplement, the type and amount of fruit provided can vary from one sampling occasion to another and even among animals in the same study population; therefore, this variable is removed in an attempt to control sampling conditions as much as possible.

Establishing a PK/PD profile is not recommended for all test articles but it can provide information regarding the optimal time for sampling for bone markers and may identify a specific PD marker that can be used in subsequent studies (e.g., see Ominsky et al. 2010). Data derived using a strict sampling regime can be used with confidence and compared cross-sectionally with controls or across time. Toxicology studies typically use young animals with skeletal growth ongoing; as the skeleton matures, bone marker levels decline. Age-related decreases in bone markers are evident in most studies and do not interfere with data interpretation. Historical control data are of great value when compiled for each species, sex, and age-range, particularly for non-rodent species. Samples for bone markers are rarely obtained in pediatric studies from very young (neonatal) animals (unless a decrease is expected) because marker levels are high and more variable than "older" animals. Sample volume is often limiting and removal of offspring from their mother is usually not recommended for data that may have limited value. In longer-term rodent juvenile toxicology studies, samples can be obtained from around 21 days of age, or when animals reach ages similar to those at the start of routine toxicology studies, and can provide important data.

Sample collection for bone marker analysis is normally coincident with sampling for routine clinical pathology so that data from clinical pathology parameters such as calcium, phosphorus, and alkaline phosphatase are available for interpretation. Sample collection should be scheduled around the time scan data is acquired so that marker levels can be correlated with any changes in bone mass. Samples for clinical pathology and bone markers should ideally be collected before fluorochrome labels are injected (see Sect. 4.4) or at least 2 days afterward to allow sufficient clearance of the labels. The presence of fluorochrome labels may interfere with some assays. Samples for markers typically are not required pre-study at baseline for rats because studies are normally well powered (n = 8 to 10/sex/group) and rat populations are relatively homogeneous with respect to marker levels. In rat studies with a large population, sampling for marker analysis may be restricted to a portion of the population (usually 10/sex/group). Because marker levels are relatively homogeneous, animals lost during the study can be readily replaced from others within the same group. Similarly, where sample volume is limited as in pediatric and mouse studies, the use of dedicated subpopulations for various activities is an excellent strategy. Data derived at different timepoints from separate subpopulations are then collated for interpretation. Micro-sampling may be an optional technique for some assays. For monkeys or dogs, marker levels can be variable within a population and group sizes are small. The study design therefore should include adequate sampling pre-study (two occasions recommended) to establish baseline levels for individual animals that can be monitored across time. Bone markers do fluctuate despite rigorous control of sampling conditions, so multiple occasions are useful. Deriving the percent change from average baseline levels may be used for interpretation in some instances, but percent values can often be misleading, especially where the working range of values for the assay is numerically low.

1.4.2 Environmental Conditions and Diet

Environmental conditions are recognized as becoming increasingly important as factors that influence skeletal activity. Diurnal variation is controlled using a consistent light cycle. Room temperature has recently come into focus as an important factor affecting bone mass. Female mice housed under thermoneutral conditions do not show the age-related cancellous bone loss that occurs in mice housed at 22 °C (Iwaniec et al. 2016). The beneficial effects of thermoneutral housing on cancellous bone were associated with decreased Ucp1 gene expression in brown adipose tissue, increased bone marrow adiposity, higher rates of bone formation, higher expression levels of osteogenic genes, and locally decreased bone resorption. Diet and the biotome also affect bone mass (Steves et al. 2016). Manipulating the diet, such as use of low/high fat diets or low/high calcium diets, influences bone metabolism. Primate chow also contains phytoestrogens which may influence the skeleton in conditions of estrogen deprivation (as occurs when testing aromatase inhibitors, for example, or in ovariectomized animals). Tight control of diet and environmental conditions for studies with bone assessments is therefore important.

Stress and the effects of ACTH and glucocorticoids on bone are well documented (Kaltsas and Makras 2010). Social housing vs single housing conditions influence stress levels; therefore, differences in housing may influence the study outcome. Stress associated with laboratory conditions is usually well controlled where experienced staff perform the dosing and animal husbandry procedures. The drug-induced toxicity that can result in some studies however cannot be controlled, and special attention should be paid to the pathologist's observations of lesions that may be attributed to stress. Stress can influence hormone levels in general which can have profound effects on bone. For additional perspective, the reader is referred to Chap. 10 which addresses the influence of pituitary hormones and Chap. 14 for the influence of the intestinal microbiota.

1.4.3 Radiography and Bone Densitometry

The reader is referred to Chap. 6 for more in-depth information regarding skeletal imaging and to Chap. 3 for specific information on the use of young animals to test therapeutics intended for pediatric populations. The following section highlights some of the considerations required for inclusion of these techniques in preclinical toxicology studies.

Radiography and bone densitometry require animals to be anesthetized if performed in-life; the risk associated with this needs to be carefully considered for each study. Isoflurane is the anesthetic of choice for most species and is safe to use. Needless to say, animals that are sick or have overt clinical signs that may be compromising should not undergo anesthesia. Evaluation of excised bones may be an option. As mentioned elsewhere, the bone mass (BMD) of rat bones is normally relatively homogeneous within a population eliminating the need for pre-study baseline scanning. Within the context of a short-term study, up to 28 days duration, scanning of excised rat bones is normally adequate (Table 1.2). Studies 28 days or longer benefit from acquisition of in vivo scan data at intervals during the study. Suggested intervals would be around half way through a 3-month rat study and at 3-month intervals for studies 6 months or longer. Similar intervals are suggested for monkeys and dogs but with large animals it is necessary to obtain pre-study baseline scan data. Data obtained during the treatment period are calculated as the percent change from baseline (pretreatment) for each individual animal to adjust for the variability in bone densitometry parameters seen in large animal species. Non-rodent toxicology studies also have small group sizes which limits the power to obtain robust cross-sectional data.

Radiography is an invaluable tool in studies requiring an assessment of bone growth and is normally combined with more frequent physical measurements of limbs or crown rump in neonatal or juvenile toxicology studies (Table 1.2). The need for anesthesia limits the use of radiography in very young animals. In rat neonatal or juvenile toxicology studies, acquisition of X-rays and bone densitometry scans is typically initiated at around 21 days of age, although animals as young as 14 days of age have been used. If dosing was initiated at an earlier age, then this becomes the first timepoint that meaningful data can be acquired. For dogs (Robinson et al. 2012), in vivo radiographs and limited bone densitometry have been performed as early as 1 day of age. Data acquisition is limited at this very young age by inadequate bone consolidation, restricting bone densitometry to DXA scanning of the whole body. In dogs, regional DXA scans of the lumbar spine and femur become feasible by 2–3 months of age. At this age, scanning using pQCT also becomes feasible since animals can endure a longer anesthesia time. Following in utero exposure of a bone drug (denosumab), ex vivo radiography and bone densitometry were used in part to characterize the skeletal phenotype in infant cynomolgus monkeys at birth (Boyce et al. 2014).

Site selection for data acquisition is driven by the need to provide clinically translatable data and the practicality associated with obtaining these data. The sites considered most at risk for fracture in humans are the spine, hip, and forearm (radius). Fracture incidence data is acquired in clinical trials for vertebral and non-vertebral fractures, with a separate analysis for the incidence of hip fracture, a site considered associated with the greatest morbidity. Agency guidelines for the testing of compounds to treat or prevent osteoporosis therefore recommend evaluation of these specific sites for efficacy in appropriate osteopenic animal models, typically the rat and monkey (Japanese MHLW 1999; EMA 2007; FDA 2016). These preclinical bone quality studies are designed to mimic the end-points used in clinical trials as much as possible. Bone densitometry assessments were therefore developed to focus on assessments of the spine, hip, and radius. Use of these established sites has subsequently been adapted for general use in safety studies and applied to other species including the dog and rabbit. In humans, a clinical diagnosis of osteoporosis and osteopenia is made using DXA BMD, and DXA remains the only bone densitometry technique approved by the FDA. In preclinical research DXA is therefore an important clinically translatable tool and is associated with several advantages and limitations (Blake and Fogelman 2008; Bolotin 2007). DXA can provide information on large areas of the skeleton such as the spine, radius, and femur, as well as the whole body, and can provide measures of whole body lean and fat mass. However, DXA provides a two-dimensional areal measurement of bone where bone size and positioning can affect outcome (BMD) measures, and it cannot discriminate between trabecular and cortical bone. In vivo DXA BMD measurements are dependent upon surrounding soft tissue; therefore, changes in body composition can impact BMD values whether bone is directly affected or not. Use of pQCT provides a three-dimensional bone density assessment, and complementary data which can compensate for some of the limitations of DXA, as well as facilitate DXA data interpretation. Peripheral QCT can provide separate analyses of the trabecular and cortical bone compartments and important measures of bone geometry. The main limitation of pQCT is that data is acquired from single or multiple images (scan slices) through a cross-section of bone at a peripheral site, normally the proximal tibia or distal radius, metaphysis, and diaphysis. To acquire in vivo pQCT scans, the limb is placed within a gantry or opening. The limbs of some animals, such as the adult dog thigh and obese rat hind limb, cannot be positioned appropriately for scanning; thus for the dog, data acquisition is normally restricted to the radius. Larger regions can be scanned in vivo, including vertebrae and the hip, but in most laboratory settings the larger equipment required to do this is not available. Body composition measures can also be obtained with pQCT.

To address the frequently asked question of which bone densitometry technique to use in a preclinical safety study, both modalities (DXA and pQCT) are recommended. However, pQCT is often recommended alone because of the ability to derive data from the trabecular and cortical bone compartments and bone geometry parameters which are used to assess bone size or to monitor growth, especially in juvenile studies. In most non-rodent studies, both the tibia and radius are scanned since the response at these sites can vary. Trabecular bone density at the proximal tibia is considered the most sensitive parameter for most species. However, DXA remains an important tool to evaluate the spine and hip, sites which are not as easily scanned using pQCT. For most test systems, both DXA and pQCT scanning can be performed during the same anesthesia occasion. However, older mice, for example, have low bone mass at the proximal tibia limiting the use of pQCT, and DXA data can only be acquired for the whole body with some DXA equipment that lack the resolution to scan smaller regions of interest. Rabbit vertebrae also have an unusual shape making it difficult to acquire reproducible pQCT data, even ex vivo. Except for dogs, which are the most widely used species in neonatal/juvenile toxicology studies, rats and monkeys remain the most widely scanned species at various ages, from newborn to aged, and are the test systems most recommended for skeletal evaluations based on the wealth of data in the literature and the limitations described for other species.

1.4.4 Bone Histomorphometry

While the imaging techniques (DXA, pQCT, micro-CT) provide quantitative measures of bone mass and bone structure, bone histomorphometry is needed to explain the tissue-level mechanisms that led to a bone mass imbalance or altered microstructure. Bone histomorphometry is an important tool that can detect effects of test articles on bone resorption, formation, mineralization, remodeling rates, and growth, including growth plate thickness. The strength of bone histomorphometry is that it can provide mechanistic insight into the effects of a test article on bone to explain structural changes identified by histopathology or imaging. Importantly, it is also the best tool and is considered the gold standard to identify impairments in bone mineralization. The reader is referred to Chap. 8 for in-depth information and discussion of the considerations for use of bone histomorphometry and specialized histology procedures.

The addition of histomorphometry to a toxicology study requires provision be made in the protocol to collect relevant bones for possible future evaluation, normally the proximal tibia and a vertebra, and that these bones be labelled with a fluorochrome label. Using a tier approach to the inclusion of bone end-points provides for the collection and storage of bones for use depending on the outcome of the primary in vivo measures: the biochemical markers and imaging datasets. Provision to retain bones for histomorphometry is normally recommended in repeat dose studies usually 3 months or longer (Table 1.2). Histomorphometry evaluations are of particular value in rat studies for the reasons described previously (and in Chap. 8), most notably because they are statistically powered, site selection is well established, slides can be prepared with relative ease, and there is a large body of histomorphometry data in the literature. However, histomorphometry can be performed in non-rodent species and is most relevant at the termination of chronic toxicology studies. Table 1.1 includes suggested bones for retention. Bone specimens require careful handling during retention, with long bones shaved to allow adequate penetration of fixative, normally 10 % neutral buffered formalin (NBF) followed by 70 % alcohol (see Chap. 8). To perform dynamic histomorphometry evaluations, bones are required to be labeled with a fluorochrome, such as calcein green (slow intravenous or subcutaneous injection) or alizarin complexone, among others (see Chap. 8). Normally two injections of the same label are given several days apart, the second being given typically 2 or more days prior to bone harvesting. A labeling schema, for example, in the rat might be 8 days and again 3 days prior to bone harvesting (i.e., a 5-day labeling interval). In the context of a toxicology study, however, the decision to inject bone labels into study animals needs to be taken after careful consideration, keeping the primary objectives of the study in mind and the welfare of the animals. These labels are essentially calcium chelators and need to be given slowly and with veterinary oversight. Careful review of clinical pathology data from a recent blood sample should be used to ensure there are no signs of hypocalcemia or any other major abnormalities prior to giving any labels. The reported clinical observations should also be reviewed to verify animal health status.

Since there is a risk associated with injecting bone labels (although minimal when given appropriately), the potential benefit from deriving histomorphometry data needs to be carefully considered. A dedicated mechanistic study may be warranted if the integrity of the toxicology study may be compromised by including this end-point. As mentioned previously, labels are injected several days prior to bone harvesting, so measurements will be made based on bone samples obtained at the end of the study (normally) and will therefore reflect the dynamics of bone turnover only at that time. However, histomorphometry provides other important safety endpoints, most notably an assessment of the mineralization process to identify any impairment.

1.4.5 Biomechanical Strength Testing

Making provision in the protocol to retain bones for possible future biomechanical testing is simple to do since bones do not require any special activities such as bone labeling. The main concern is to maintain bone specimens intact at necropsy; soft tissue should be carefully removed but without scraping the bone surface so as not to damage the periosteum. Specimens need to be kept hydrated and protected from freezer burn so should be wrapped in saline-soaked gauze and plastic wrap and then stored frozen at −20 °C (−80 °C is not recommended) in a plastic bag. Freezer space is a consideration when planning for a study with a large rodent population or for non-rodents with large bones. The collection procedures, preparation of the specimens for testing, and testing methods are well described in Chap. 7.

Freezing at −20 °C is considered the optimal method to store bones until they are needed for testing. Bones can be retained in 10 % NBF or in alcohol but these methods are often less optimal since fixative has limited internal penetration in intact specimens. Frozen specimens can also be retained for several months before testing, providing the flexibility to submit drug submission data and postpone testing until needed. Fixation and storage can affect bone properties. The rule of thumb is that if all specimens from a study are retained under the same conditions for the same duration, then the test data is valid for that study. It may not be appropriate to compare biomechanics data across studies if different methods for retention and storage are used. In most cases, specimens are stored for several weeks before being thawed for testing.

Biomechanical testing needs to be performed by a qualified laboratory with the oversight of a biomechanical or biomedical engineer experienced in the testing of biological specimens to ensure high quality data, adequate data storage, and appropriate data interpretation. Loss of specimens or data due to poor methodology and inadequate data storage can severely impact the timeline of a drug development program if these data are essential and the study needs to be repeated. There is an increasing demand by the regulatory authorities for these data to be added as an end-point in toxicology studies likely for two reasons: bone strength is considered the ultimate end-point reflecting bone quality and bones can easily be retained and tested under GLP conditions. Consideration should be given to the relevance of including bone strength determinations with respect to the study type and test system. It takes time for structural changes to occur in the skeleton, so this end-point is most relevant in studies 3 months or longer for rodent studies and often only in chronic studies for non-rodents (Table 1.2). Data are often generated for shorter-term non-rodent studies too, to rule out an adverse effect rather than attempt to discern a subtle effect. Effects on bone quality cannot always be discerned using bone densitometry; there may be bone matrix changes that may affect bone strength independent of any effects on bone density. As an end-point then, biomechanical strength data serves to provide some level of confidence that there are no other safety issues of concern that may require further characterization.

The impact of biological variability on the homogeneity of data in non-rodent studies is somewhat controlled for with biomechanical end-points. The primary strength output measures, including peak load, yield load, stiffness, and work to failure, describe the extrinsic parameters and are dependent on the size and shape of the bone. Bone geometry data (pQCT-derived) is used to normalize these data to derive intrinsic parameters including ultimate stress/strength, yield stress, modulus, and toughness, which are independent of size and shape. Bone mass (BMC) and bone density (BMD) of the specimens are also used to test the bone mass-strength relationship using correlation analyses. Normal bone, irrespective of site, shows strong, positive correlations where bone strength increases in proportion to increases in bone mass (BMC or BMD). It is important to determine if a drug treatment affected this structure-function relationship. Treatment with high dose fluoride, for example, increases bone but the newly formed bone is woven resulting in reduced strength, due to poor bone quality (Turner et al. 1992). New drugs, such as denosumab and romosozumab, have shown the opposite, with increases in bone strength not fully explained by the increases in bone mass, suggesting some other mechanism improved bone quality (Ominsky et al. 2010, 2011, 2017b). The ability to perform these correlation analyses to test the structure-function relationship using data-points from all animals (control and drug-treated) offsets some of the limitation imposed by the small group sizes in non-rodent studies.

Worthy of note is that laboratory animals are rarely observed with spontaneous fractures. Even animal models intended to mimic human disease such as osteoporosis become osteopenic at best with often minimal effects on bone strength parameters. When laboratory animals are observed with spontaneous fractures related to a drug treatment in a toxicology study, it is therefore regarded as adverse. Depending on the intended indication and treatment regime of the drug under test, it may be appropriate to determine a dose level where fractures do not occur or if the effect can be mitigated, but in most cases development of such a drug is usually terminated.

1.4.6 Considerations for Selection of Test System

Test systems used in standard toxicology studies include rodent (rat, mice) and non-rodent (dog, monkey, pig, rabbit) species. When planning to include bone end-points in the toxicology setting, the most toxicologically and pharmacologically relevant species should be used. Where feasible the species used most often in bone research should be given first consideration since end-points are normally readily available and validated to meet GLP requirements, and a large body of literature exists. The species most frequently used to establish pharmacology and bone quality data are the rat and cynomolgus monkey. Historically these were the preferred animal models of postmenopausal osteoporosis and have been extensively used to meet the regulatory requirements for the safety assessment of bone therapeutics (Smith et al. 2009, 2011, Chap. 4). The rat is an important species used primarily to assess trabecular bone, due to limited Haversian remodeling in cortical bone. Cortical bone evaluations are therefore important to assess in a remodeling species such as the monkey or dog. The dog is noteworthy since cortical remodeling involves the whole cortex like human bone; however, the dog skeleton lacks sensitivity to hormonal influences making it a poor model to study drugs that may affect the reproductive system.

Most standard toxicology studies use young skeletally immature animals which are still growing and accruing bone mass. The rapidly growing juvenile rat and monkey offer the opportunity to examine effects of test articles on longitudinal bone growth and endochondral ossification, relevant to pediatric indications. Many compounds that affect growth will show skeletal effects such as reduced bone mass which most often reflect a failure to accrue bone rather than a frank bone loss. To test certain drugs intended for an adult population, a more skeletally mature test system can be justified to evaluate the skeleton independent of effects on growth. For the rat, the growth plate remains open for most of its life, but longitudinal growth slows significantly by 6 months of age (Martin et al. 2003).

It is important to note the difference between skeletally mature animals and sexually mature animals. Sexual maturation occurs at an earlier age than skeletal maturation. The rat is used as a skeletally mature model at 3–6 months of age even though longitudinal bone growth and modeling continue, albeit slowly, until around 12 months of age (Kilborn et al. 2002; Roach et al. 2003). Skeletal maturity is regarded as the age when growth plate closure is complete and occurs at different ages depending on the site and species (see Chap. 6). Cynomolgus monkeys are generally considered skeletally mature around 6 years of age but continue to accrue bone until around 9 years of age when they reach peak bone mass. Skeletally mature and immature animals are mostly used in toxicology studies. Animals which have attained peak bone mass are most frequently used as models of postmenopausal osteoporosis and rarely used in a toxicology setting.

Bone end-points in toxicology studies with different test systems may lead to apparent disparate outcomes which can often be reconciled based on knowledge of the bone biology (see Chap. 2). The young adult rat (6–8 weeks of age) is used in most safety studies while non-rodent studies use young beagle dogs 5 to 7 months of age or cynomolgus monkeys ranging in age from 2 to 4 years, in general. In the context of a 3-month safety study, during this time the rat skeleton will have undergone around three remodeling bone cycles (that is the estimated time to complete the cycle of bone resorption followed by bone formation), while it takes 8–12 weeks to complete one cycle in the monkey (or dog) skeleton. It is therefore evident that there is more opportunity for a compound to affect the rat skeleton than the monkey. There may appear to be a rat-specific effect but it may be detectable in the monkey when the skeleton is exposed to the test article during a comparable number of bone turnover cycles, as in the longer-term studies.

The safety assessment of potent new drugs that target metabolism, as in indications for diabetes and obesity, has presented the toxicologist with novel challenges. For these drug classes, dose level selection is often limited by exaggerated pharmacological effects including excessive weight loss threatening survival. Weight loss has profound effects on the skeleton (Iwaniec and Turner 2016), and malnutrition may lead to confounding effects including osteomalacia. Different strategies to assess safety to allow dosing at multiple levels of the clinically intended dose are sought to mitigate these effects often with variable success. These strategies may include, for example, selecting a heavier test system to allow a longer period of dosing to be tolerated and use of food supplementation where intake is affected due to central effects on satiety. Diet may take the form of treats to stimulate appetite or gavage (normally for a few days only) to ensure adequate nutrients are provided for growth and survival. The use of such strategies normally requires veterinary oversight and although not sustainable for sub-chronic/chronic studies may allow completion of dose level selection studies and generate the required data to justify further development. Other strategies to extend the dosing period may include use of drug holidays, depending on clearance, and use of special diets. Diets can be modified with respect to fat content and levels of calcium/phosphorus with the intent to mitigate pharmacological effects in drug-treated animals. In some cases, however, modifying the diet may have unwanted effects in control animals and possibly confound data interpretation and use of historical data generated from standard toxicology studies.

1.4.7 Summary of Considerations for Study Design: Take Home Messages

- Biochemical markers of bone turnover provide a rapid and sensitive method to identify a potential hazard of a compound on bone at a specific timepoint.
- Markers of bone formation and resorption should be used: PINP (or osteocalcin if PINP unavailable) and C-telopeptides, as a minimum.
- Samples (blood and urine) should be collected early morning following fasting.
- A standard lab chow is recommended.
- Radiography is an important tool to monitor growth and bone abnormalities.
- In vivo bone densitometry using DXA and/or pQCT should be used for rodent and non-rodent studies 3 months or longer.
- Baseline measurements of bone markers and bone densitometry are essential for non-rodent studies.
- Ex vivo bone densitometry is normally adequate for rodent studies up to 28 days duration.
- Peripheral QCT is recommended (especially for juvenile studies) if one modality is selected in order to acquire separate information on trabecular and cortical bone compartments and bone geometry.
- Peripheral QCT-derived trabecular BMD is considered the most sensitive bone densitometry end-point to discern an effect on bone mass.
- At termination of studies 3-month duration or longer, bones should be retained for possible future use for micro-CT analysis, additional densitometry, biomechanics, and/or histomorphometry: recommend collection of two lumbar vertebrae and a whole femur frozen for micro-CT, densitometry or biomechanics, and a vertebra and proximal tibia in 10% neutral buffered formalin followed by alcohol for histomorphometry.
- Performing histomorphometry or biomechanics on small group sizes is challenging since the precision of these techniques is relatively poor; however, these end-points are important to characterize a toxicological effect on bone.
- To perform histomorphometry or biomechanics in a non-rodent study, it is important to adequately power the study and to select animals as homogeneous as possible with respect to age, sexual, and skeletal maturity.
- Histomorphometry or biomechanics data is superior for rodent studies that are well powered and where bone is relatively homogenous among animals
- The rat and monkey are the preferred species used to establish skeletal effects, except for pediatric studies where the dog is often the non-rodent species of choice.

- Rats should be given first consideration as a model to study potential skeletal effects because of convenience and relevance of that model whenever the appropriate (compatibility with humans) criteria are met.
- The young (4 weeks) intact male rat is often more sensitive than the female rat to detect a change due primarily to the greater growth in the male and resulting larger bones.
- Use the most toxicologically and pharmacologically relevant non-rodent species to assess a skeletal effect only after a signal has been established in the rat which is considered the most sensitive species.
- The monkey or dog are important to evaluate cortical bone since the rat is a poor model of Haversian remodeling.

1.5 Animal Models

Pharmacology models, such as the osteopenic (gonadectomized) rat, are a regulatory agency required test system used during the development of bone-targeted drugs and are well described in Chap. 4. Other rodent models are often used as screening tools in drug development to assess the antiresorptive or anabolic potential of a drug. Such models include the Schenk weanling rat model (Schenk et al. 1973) and the thyroparathyroidectomized rat (Lechowska et al. 2001). While infrequently used in a toxicology setting, on occasion these models can be used to address specific issues. The Schenk model is performed in young (21 day old) rapidly growing rats, normally dosed for 7 days, and is typically used to screen antiresorptive agents which show the hallmark abnormal shaping of the long bone metaphyses ("clubbing") and retention of trabeculae reflected in an increased bone volume fraction. The growth plate is also a focus of this model which is sensitive to deficits in bone mineralization, resulting in a widening or thickening of the physis. It is therefore the model of choice to quickly characterize if a molecule causes defective mineralization. The surgical removal of the parathyroid glands and/or thyroid gland in the rat provides an excellent model of hypoparathyroidism, primarily used as a screening model to test agents that inhibit calcium mobilization from bone following parathyroid hormone (PTH) repletion (infusion) or anti-PTH antibodies, for example. It can also be used to compare the effectiveness of PTH analogues. In addition to the standard application as a short-term screening model traditionally used to test antiresorptive agents (Lechowska et al. 2001), this PTH-deficient model can be useful in the safety assessment of potential drugs to treat hypoparathyroidism.

Other surgical models that are frequently used in bone research include the uremic rat model induced by the surgical removal of one whole kidney and 2/3 of the second kidney (5/6 nephrectomy). Animals subsequently develop chronic kidney disease (CKD) and an osteodystrophy characteristic of late stage CKD. This has recently become a favored model to test drugs intended to treat CKD-MBD (mineral and bone disorder) including anti-fibroblast growth factor-23 (FGF23) antibodies (Silver and Naveh-Many 2013). This model of renal impairment and other bone disease models were used to demonstrate the bone-forming effects of sclerostin antibody, to generate data to support the clinical development of sclerostin antibody for treatment of conditions with low bone mass (Ominsky et al. 2017a).

The use of young, intact, healthy animals to test potent molecules that affect basic systems such as growth, metabolism, and calcium homeostasis may have questionable relevance in safety assessments. It may therefore be relevant to include additional toxicology studies using an animal model where higher doses may be tolerated to better assess off-target effects. Depending on the indication, these models may be chemically or genetically manipulated, in addition to the surgical models described above. Disease areas that may benefit from such models include diabetes, obesity, hyper- and hypoparathyroidism, and skeletal diseases such as dwarfism and rare diseases like fibrodysplasia ossificans progressiva (FOP) and hypophosphatasia, among others. Animal models of diabetes and obesity are well documented in the literature, and further discussion of these models is considered beyond the scope of this chapter. Genetically modified rodent models, in many cases initially developed to study specific skeletal diseases, may be useful in a toxicology setting to identify off-target effects and provide perspective to data from standard toxicology studies where pharmacologically mediated findings may be overwhelming and mask off-target effects. To be effective these models require appropriate validation and need to be bred and supplied by an approved vendo, and to be viable and easily maintained in a standard laboratory environment.

Other models gaining ground in bone research include the use of zebra fish as a screening model that can help to characterize an effect of a drug on bone formation (Kwon and Harris 2016). Such models can be a useful tool in early drug development and may provide data useful to designing preclinical studies. Bone cell culture assays to study effects of a drug on osteoblastogenesis and osteoclastogenesis may also be included as an additional tool that may help in the characterization of the effect of a compound on bone and have been used to provide mechanistic information in several safety studies.

1.6 Data Interpretation: Skeletal Adversity

Adverse effects in bone will be reflected by changes in structure and function which impact bone strength. Data from preclinical studies in multiple species are used to assess whether skeletal fragility may be at risk and to what degree. The severity and extent of the changes, as with other organ systems, determine adverse vs non-adverse effects. The adversity of some skeletal effects however is age dependent, and studies may have dramatically different outcomes with respect to adversity if tested in a skeletally immature vs mature test system. Therefore, use of a test system relevant to the intended human population is pivotal. For example, a compound resulting in premature growth plate closure is important in the context of a skeletally immature test system, but may have no adverse consequences in a skeletally mature animal.

Chronic toxicology studies designed to include the primary outcome measures of biochemical markers, bone densitometry, histomorphometry, and biomechanical strength testing normally provide sufficient mechanistic information to assess whether effects are adverse or not. In the context of a 6-month rat study showing a clear reduction in bone strength parameters, the in vivo bone densitometry data will reflect the time-course and extent of the changes in bone mass that led to the reduced biomechanical competency. The histomorphometry dynamic parameters will describe how the bone machinery, the osteoblasts and osteoclasts, were performing at the end of treatment, which will be further supported by the pattern of changes in biochemical markers sampled at intervals throughout the study. Statistical analyses of the relationship between bone mass and strength will further characterize any effects driven by changes in bone quality other than those predicted by BMD or BMC. In this scenario, the in vivo end-points (markers and densitometry) would be used to justify further characterization of the effect in terms of mechanism (histomorphometry) and the ultimate effects on bone quality (biomechanics). Taking all bone data into consideration, a reduction in intrinsic bone strength parameters (e.g., 40–50 %) would normally be considered adverse irrespective of species in the context of a chronic toxicology study.

In early toxicology studies, which often include skeletal end-points that are used in dose level selection for the chronic definitive studies, it can be difficult to define adversity in the absence of bone strength data. Making provision to test bones if needed to facilitate this decision-making therefore becomes important. It should also be evident that including biochemical markers in the dose-range finding or PK studies facilitates the identification of any potential hazard to the skeleton. These data become important if the drug proceeds to phase I trials to drive the study design and inclusion of appropriate markers and bone densitometry.

Conversely, potential benefits to improve bone quality and strength can also be determined and provide powerful data to support superiority with respect to safety when compared to others in the same class or to provide mechanistic proof-of-concept data when developing a new drug class. Our understanding of skeletal pathway-based mechanisms continues to reveal the consequences of exposure of the skeleton to certain drug classes. Antidiabetic drugs in the thiazolidinediones (TZDs) class, PPARγ (peroxisome proliferator-activated receptors) agonists, for example, rosiglitazone, have subsequently been shown to exacerbate OVX-induced bone loss and to increase skeletal fragility in the osteopenic rat skeleton (Kumar et al. 2013). This highlights the relevance of including additional studies in the osteopenic (ovariectomized (OVX)) rat model to assess mechanism. The OVX rat model was used to test the potential outcome of combining pioglitazone (also a PPARγ agonist) with the potentially bone positive effects of fenofibrate, a PPARα agonist. The results showed the combination of α/γ agonism had no effect on the OVX-induced bone loss in this osteopenic rat model, providing important proof-of-concept safety data to support the development of a drug with combined α/γPPAR agonism (Samadfam et al. 2012; Smith et al. 2015).

A call of skeletal adversity cannot be made based on marker data alone, but any perturbation in these parameters suggests the skeleton is a potential target tissue requiring further investigation. Data from bone end-points form a key component to add to the toxicology study data, linking all relevant findings to use the weight of evidence to determine adversity (Kerlin et al. 2016; Palazzi et al. 2016).

1.7 Conclusions

The tools are available for hazard identification and characterization of the effect of a test article on bone. With careful consideration, the integration of these end-points in safety assessment can be at any level, from sample collection for the analysis of bone markers to more advanced bone densitometry measurements and subsequent mechanistic data from histomorphometry. Bone strength data is considered the ultimate test of bone quality and hence bone fragility, a key end-point in translating nonclinical data to assess fracture risk in humans.

References

Blake GM, Fogelman I. How important are BMD accuracy errors for the clinical interpretation of DXA scans? J Bone Miner Res. 2008;23(4):457–62.

Bolotin HH. DXA in vivo BMD methodology: an erroneous and misleading research and clinical gauge of bone mineral status, bone fragility, and bone remodelling. Bone. 2007;41(1):138–54.

Boskey AL, Robey PG. The regulatory role of matrix proteins in mineralization of bone. In: Marcus R, Feldman D, Dempster DW, Luckey M, Cauley JA, editors. Osteoporosis. New York: Elsevier; 2013. p. 235–58.

Boyce RW, Varela A, Chouinard L, Bussiere JL, Chellman GJ, Ominsky MS, Pyrah IT. Infant cynomolgus monkeys exposed to denosumab in utero exhibit an osteoclast-poor osteopetrotic like skeletal phenotype at birth. Bone. 2014;64:314–25.

Boyce RW, Niu QT, Ominsky MS. Kinetic reconstruction reveals time-dependent effects of romosozumab on bone formation and osteoblast function in vertebral cancellous and cortical bone in cynomolgus monkeys. Bone. 2017;101:77–87.

Chouinard L, Felx M, Mellal N, Varela A, Mann P, Jolette J, Samadfam R, Smith SY, Locher K, Buntich S, Ominsky MS, Pyrah I, Boyce RW. Carcinogenicity risk assessment of romosozumab: a review of scientific weight-of-evidence and findings in a rat lifetime pharmacology study. Regul Toxicol Pharmacol. 2016;81:212–22.

Donnelly E. Methods for assessing bone quality: a review. Clin Orthop Relat Res. 2011;469:2128–38.

Doyle N, Germain C, Barma M-C, Larocque K, Allegre V, Varela A, Parente C, Poitout F. Urine collection for group-housed mice in toxicology studies using the LabSand® techniques – an alternative method. Toxicologist (Suppl Toxicol Sci). 2017;156:169. Abstract 1717

European Medicines Agency (EMA). Regulation (EC) No 1901/2006 of the European Parliament and of the Council on medicinal products for paediatric use, amended by Regulation (EC) No 1902/2006. (Pediatric Investigational Plans – PIP). 2006

European Medicines Agency (EMA). Committee for medicinal products for human use (CHMP). Guideline on the evaluation of medicinal products in the treatment of primary osteoporosis. 2007. http://www.ema.europa.eu/docs/en_GB/document_library/Scientific_guideline/2009/09/WC500003405.pdf.

European Medicines Agency (EMA). Committee for human medicinal products (CHMP). Guideline on the need for nonclinical testing in juvenile animals on human pharmaceuticals for paediatric indications. 2008.

Food and Drug Administration (FDA). Guidance document. nonclinical safety evaluation of pediatric drug products. 2006.
Food and Drug Administration (FDA). Guidance for industry pediatric study plans: content of and process for submitting initial pediatric study plans and amended pediatric study plans. Draft. 2016a.
Food and Drug Administration (FDA). Draft guidance. Center for Drug Evaluation and Research (CDER). Osteoporosis: nonclinical evaluation of drugs intended for treatment guidance for industry. 2016b. http://www.fda.gov/downloads/Drugs/GuidanceComplianceRegulatoryInformation/Guidances/UCM506366.pdf.
Fossey S, Vahle J, Long P, Schelling S, Ernst H, Boyce RW, Jolette J, Bolon B, Bendele A, Rinke M, Healy L, High W, Roth DR, Boyle M, Leininger J. Nonproliferative and proliferative lesions of the rat and mouse skeletal tissues (bones, joints, and teeth). J Toxicol Pathol. 2016;29(3 Suppl):4.
Iwaniec UT, Turner RT. Influence of body weight on bone mass, architecture and turnover. J Endocrinol. 2016;230:R115–30.
Iwaniec UT, Philbrick KA, Wong CP, Gordon JL, Kahler-Quesada AM, Olson DA, Branscum AJ, Sargent JL, DeMambro VE, Rosen CJ, Turner RT. Room temperature housing results in premature cancellous bone loss in growing female mice: implications for the mouse as a preclinical model for age-related bone loss. Osteoporos Int. 2016;27(10):3091–101.
Japanese Ministry of Health, Labour and Welfare (MHLW). Guideline concerning the clinical evaluation method for anti-osteoporosis agents, pharmaceutical examination No. 742. In: Management Bo H, ed.: issued by the Prefecural Bureau Chief, Bureau of Health Management, Section of Examination and Control, Bureau of Drug Safety, Japanese Ministry of Health and Welfare. 1999.
Japanese Ministry of Health, Labour and Welfare (MHLW). Guideline on the nonclinical safety study in juvenile animals for pediatric drugs. Oct 2012.
Johnson C, Winkelmann CT, Wise LD. Considerations for conducting imaging studies in support of developmental toxicology studies for regulatory submission. Reprod Toxicol. 2014;48:41–3.
Jolette J, Wilker CE, Smith SY, Doyle N, Hardisty JF, Metcalfe AJ, Marriott TB, Fox J, Wells DS. Defining a noncarcinogenic dose of recombinant human parathyroid hormone 1-84 in a 2-year study in Fischer 344 rats. Toxicol Pathol. 2006;34:929–40.
Kaltsas G, Makras P. Skeletal diseases in Cushing's syndrome: osteoporosis versus arthropathy. Neuroendocrinology. 2010;92(suppl 1):60–4.
Kazakia GJ, Burghardt AJ, Cheung S, Majumdar S. Assessment of bone tissue mineralization by conventional x-ray microcomputed tomography: comparison with synchrotron radiation microcomputed tomography and ash measurements. Med Phys. 2008;35(7):3170–9.
Keaveny TM. Biomechanical computed tomography-noninvasive bone strength analysis using clinical computed tomography scans. Ann N Y Acad Sci. 2010;1192:57–65.
Kerlin R, Bolon B, Burkhardt J, Francke S, Greaves P, Meador V, Popp J. Scientific and regulatory policy committee: recommended ("best") practices for determining, communicating, and using adverse effect data from nonclinical studies. Toxicol Pathol. 2016;44:147–62.
Kilborn SH, Trudel G, Uhthoff H. Review of growth plate closure compared with age at sexual maturity and lifespan in laboratory animals. Contemp Top Lab Anim Sci. 2002;41:21–6.
Kumar S, Hoffman SJ, Samadfam R, Mansell P, Jolette J, Smith SY, Guldberg RE, Fitzpatrick LE. The effect of rosiglitazone on bone mass and fragility is reversible and can be attenuated with alendronate. J Bone Miner Res. 2013;28(7):1653–65.
Kwon RY, Harris MP. Fishes swim into view at ASBMR: meeting report on FishBone 2015 – advancing the use of fish for skeletal research. IBMS BoneKey. 2016;13:795.
Lechowska BA, Bhattacharyya A, Vasko-Moser J, Stroup GB. Reduction in animal numbers by long-term implantation of intravenous and intra-arterial catheters in thyroparathyroidectomized rats. Comp Med. 2001;51(6):518–23.
Martin EA, Ritman EL, Turner RT. Time course of epiphyseal growth plate fusion in rat tibiae. Bone. 2003;32(3):261–7.
Mead TJ, Lefebvre V. Proliferation assays (BrdU and EdU) on skeletal tissue sections. Methods Mol Biol. 2014;1130:233–43.

Ominsky MS, Vlasseros F, Jolette J, Smith SY, Stouch B, Doellgast G, Gong J, Gao Y, Cao J, Graham K, Tipton B, Cai J, Deshpande R, Zhou L, Hale MD, Lightwood DJ, Henry AJ, Popplewell AG, Moore AR, Robinson MK, Lacey DL, Simonet WS, Paszty C1. Two doses of sclerostin antibody in cynomolgus monkeys increases bone formation, bone mineral density, and bone strength. J Bone Miner Res. 2010;25:948–59.

Ominsky MS, Stouch B, Schroeder J, Pyrah I, Stolina M, Smith SY, Kostenuik PJ. Denosumab, a fully human RANKL antibody, reduced bone turnover markers and increased trabecular and cortical bone mass, density, and strength in ovariectomized cynomolgus monkeys. Bone. 2011;49:162–73.

Ominsky MS, Libanati C, Niu QT, Boyce RW, Kostenuik PJ, Wagman RB, Baron R, Dempster DW. Sustained modeling-based bone formation during adulthood in Cynomolgus monkeys may contribute to continuous BMD gains with Denosumab. J Bone Miner Res. 2015;30(7):1280–9.

Ominsky MS, Boyce RW, Li X, Ke HZ. Effects of sclerostin antibodies in animal models of osteoporosis. Bone. 2017a;96:63–75.

Ominsky MS, Boyd SK, Varela A, Jolette J, Felx M, Doyle N, Mellal N, Smith SY, Locher K, Buntich S, Pyrah I, Boyce RW. Romosozumab improves bone mass and strength while maintaining bone quality in ovariectomized cynomolgus monkeys. J Bone Miner Res. 2017b;32:1–14.

Palazzi X, Burkhardt JE, Caplain H, Dellarco V, Fant P, Foster JR, Francke S, Germann P, Groters S, Harada T, Harleman J, Inui K, Kaufmann W, Lenz B, Nagai H, Pohlmeyer-Esch G, Schulte A, Skydsgaard M, Tomlinson L, Wood CE, Yoshida M. Characterizing "adversity" of pathology findings in nonclinical toxicity studies: results from the 4th ESTP international expert workshop. Toxicol Pathol. 2016;44:810–24.

Roach HI, Mehta G, Oreffo RO, Clarke NM, Cooper C. Temporal analysis of rat growth plates: cessation of growth with age despite presence of a physis. J Histochem Cytochem. 2003;51:373–83.

Robinson K, Smith SY, Viau A. Dog juvenile toxicology. In: Hoberman A, Lewis E, editors. Juvenile toxicology. Hoboken: Wiley; 2012. p. 183–212.

Saito M, Grynpas MD, Burr DB, Allen MR, Smith SY, Doyle N, Amizuka N, Hasegawa T, Kida Y, Marumo K, Saito H. Treatment with eldecalcitol positively affects mineralization, microdamage, and collagen crosslinks in primate bone. Bone. 2015;73:8–15.

Samadfam R, Awori M, Bénardeau A, Bauss F, Sebokova E, Wright M, Smith SY. Combination treatment with pioglitazone and fenofibrate attenuates pioglitazone-mediated acceleration of bone loss in ovariectomized rats. J Endocrinol. 2012;212(2):179–86.

Schenk R, Merz WA, Mühlbauer R, Russell RG, Fleisch H. Effect of ethane-1-hydroxy-1,1-diphosphonate (EHDP) and dichloromethylene diphosphonate (Cl 2 MDP) on the calcification and resorption of cartilage and bone in the tibial epiphysis and metaphysis of rats. Calcif Tissue Res. 1973;11(3):196–214.

Silver J, Naveh-Many T. FGF-23 and secondary hyperparathyroidism in chronic kidney disease. Nat Rev Nephrol. 2013;9:641–9.

Smith SY, Jolette J, Turner CH. Skeletal health: primate model of postmenopausal osteoporosis. Am J Primatol. 2009;71:1–14.

Smith SY, Varela A, Jolette J. Nonhuman primate models of osteoporosis. In: Duque G, Watanabe K, editors. Osteoporosis research – animal models. London: Springer Verlag; 2011. p. 135–57.

Smith SY, Samadfam R, Chouinard L, Awori M, Bénardeau A, Bauss F, Guldberg RE, Sebokova E. Effects of pioglitazone and fenofibrate co-administration on bone biomechanics and histomorphometry in ovariectomized rat. J Bone Miner Metab. 2015;33(6):625–41.

Steves CJ, Bird S, Williams FMK, Spector TD. The microbiome and musculoskeletal conditions of aging: a review of evidence for impact and potential therapeutics. J Bone Miner Res. 2016;31(2):261–9.

Turner CH, Akhter MP, Heaney RP. The effects of fluoridated water on bone strength. J Orthop Res. 1992;10:581–7.

Chapter 2
Bone Physiology and Biology

Jürg Andreas Gasser and Michaela Kneissel

Abstract This chapter is divided into three subsections: The bone biology part will provide some insight into the matrix composition as well as the origin and basic function attributed to the cellular components, the osteoblasts osteocytes and osteoclasts. In the second section, the interplay between these matrix components and the cells will be discussed in the context of skeletal growth, skeletal adaptation (modeling), and skeletal maintenance (bone remodeling). Finally, the third section will address similarities and differences in bone biology between human, nonhuman primate, rat, and mouse, the most commonly used species in the study of bone metabolic disorders. Other large animals like dogs, sheep, mini-pigs, or rabbits which are being successfully used to study orthopedic conditions (implant ingrowth, fracture healing, and bone augmentation) have not been addressed in this chapter.

Keywords Osteoblast • Osteoclast • Osteocyte • Bone formation • Bone resorption • Remodeling • Modeling • Cytokines • Growth factors • Markers • Intramembranous • Endochondral • Mice • Rats • NHP • Humans

2.1 Bone Biology

Bone is a special form of connective tissue, which unlike most other tissues is physiologically mineralized. On the organ level, bone is made up of the cartilaginous joints, the calcified cartilage of the growth plate (during skeletal growth only), the marrow space, and the mineralized cortical and trabecular bone structures (Weiner and Wagner 1998; Seeman 2008; Burr and Akkus 2014) (Fig. 2.1).

J.A. Gasser (✉) • M. Kneissel
Novartis Institutes for BioMedical Research, Department of Musculoskeletal Diseases, 4002 Basel, Switzerland
e-mail: juerg.gasser@novartis.com

S.Y. Smith et al. (eds.), *Bone Toxicology*, Molecular and Integrative Toxicology,
DOI 10.1007/978-3-319-56192-9_2

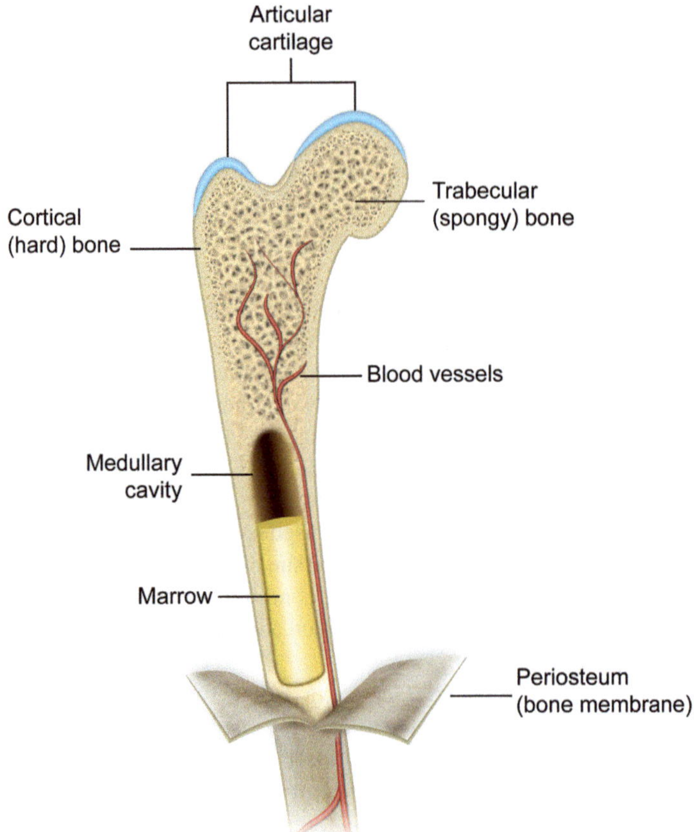

Fig. 2.1 Bone envelopes: On the macroscopic level, long bones are composed of the cartilaginous joints, porous cancellous bone, the more compact, denser cortical bone, and the marrow cavity. The subperiosteal surface in long bones is covered by a thin fibrocellular highly vascularized and innervated membrane, the periosteum

On the tissue level, bone consists of the mineralized and non-mineralized matrix (osteoid) and the three principal cellular components, the bone-forming osteoblasts, some of which become entrapped into the mineralizing bone matrix they lay down to become osteocytes, and the bone-resorbing osteoclasts. The origin of these cell types as well as the communication between them and their involvement in vital processes such as skeletal growth, bone adaptation (modeling), and bone maintenance (remodeling) will be discussed in more detail below.

Bone is a multifunctional tissue which serves as mechanical support and protection, is an essential part of hematopoiesis and mineral metabolism, and has a role as an endocrine organ. Bone is able to resist deformation from impact loading, but at the same time, it is also able to absorb or dissipate energy by changing shape without cracking (Currey 2002). The elastic properties of bone allow it to absorb energy by deforming reversibly when loaded (elastic deformation) (Lanyon and Baggott 1976; Turner 2006). If the load exceeds bones' ability for elastic deformation, plastic deformation occurs, accompanied by permanent shape change and accumulation of microcracks, which helps to release energy (Currey 2002). Microdamage is a defense mechanism against a more serious event, namely, a complete fracture. Repetitive loading at physiologic levels may lead to fatigue damage, which requires a damage detection mechanism (osteocyte) and a repair function (remodeling) capable of preventing its accumulation thereby preventing failure in fatigue (Burr et al. 1998). To allow movement, bone also has to be light, and this lightness is achieved by constructing a version of the mineralized material consisting of trabecular plates and rods and bone marrow cavities, which are sites of hematopoiesis (Seeman 2008).

On the organ level, based on their anatomical location, bones can be divided into four distinct envelopes, namely, subperiosteal, endocortical, cancellous, and intracortical. This division reflects that these envelopes are morphologically distinct (Fig. 2.1). They play different roles in bone biology in health and disease and differ in their response to changes in the mechanical environment.

The subperiosteal surface in long bones is covered by a thin fibrocellular membrane, the periosteum, which is well vascularized and innervated by both sympathetic and sensory fibers. It is composed of an outer fibrous layer containing fibroblast-like cells and a deep cambium layer populated with highly osteogenic cells including mesenchymal stem cells. The osteogenic cells in the cambium layer drive the periosteal expansion during growth, the adaptive response to mechanical stimulation, and the process of fracture repair. Under normal circumstances the cells will produce highly organized lamellar bone, while under pathologic situations and in the initial phases of bone repair, disorganized woven bone may be produced. The mesenchymal cells contained in the cambium contribute to fracture repair, where they can differentiate into chondrocytes to form cartilage.

The endocortical (or endosteal) surface that surrounds the marrow cavity is not lined by a membrane but by a fenestrated layer of osteoprogenitor cells. These so-called lining cells are associated with capillaries near the bone surface and sinusoids in the bone marrow and may act as a "membrane." Lining cells are important for the regulation of the rapid calcium exchange and fluxes of hydrophilic ions between bone and the extracellular fluid.

Cancellous bone surfaces are morphologically and functionally similar to the endocortical surface. They are also covered by a fenestrated layer of lining cells which control ion fluxes (particularly calcium) and respond to mechanical and biological signals to induce bone formation or resorption.

Finally, the intracortical envelope is comprised of the surfaces of the Haversian canals, which are covered by a fenestrated layer of resting osteoprogenitor cells. The neurovascular bundle contained in the canal serves to regulate nutrient exchange between the blood vessels and the extracellular fluid, from where they pass through the canalicular system to be absorbed interstitially. The extracellular fluid is not only important for nutrient and calcium exchange, but it also appears to participate substantially in the transmission of signals between osteocytes embedded in the mineralized matrix and from osteocytes to the cells on the bone surfaces and/or osteoprogenitor cells in the bone marrow.

Calcified tissue presents a considerable barrier for the diffusion of oxygen and nutrients and the removal of cellular waste products. For this reason, cells contained in a mineralized tissue must be localized within a distance of roughly 250 μm of their blood supply. This may be the reason why larger mammals like humans, dogs, primates, and rabbits, with a thicker cortical structure, require osteons. It may also explain why the average diameter of trabeculae in larger mammals is around 200 μm. Interestingly, longitudinal tunneling of substantially thickened trabeculae was observed in primates after bone building treatment with PTH(1–34) (Jerome et al. 2001). Vascular channels were always present in trabeculae thicker than 450 μm and were thought to improve metabolic exchange of deep-seated osteocytes by reducing the distance between cells and filtering surfaces in the thicker trabeculae (Lozupone and Favia 1990). Vascular structures in osteons show typical characteristics of capillaries that are 15 μm in diameter and are often paired. Essentially they are fenestrated, lack smooth muscle cells, are lined by an incomplete layer of endothelial cells, and surrounded by a continuous 40–60 nm thick basement membrane. The basement membrane acts to limit ion transport across the capillary.

Long bones are supplied by a nutrient artery entering the marrow cavity, which then fans out to run aligned with the longitudinal axis of the bone (Marenzana and Arnett 2013; McCarthy 2006; Brandi and Collin-Osdoby 2006). Blood flow in the medullary cavity is regulated by adrenergic and cholinergic sympathetic nerve fibers that either reduce or increase blood flow as required. Branches direct blood from the longitudinally oriented blood vessels to the endosteum, where it enters the intracortical capillary network flowing outward toward the periosteum. Roughly two-thirds of the cortical blood supply enters through the endocortical surface while the remainder is supplied from the highly vascularized subperiosteal side.

Cortical bone comprises most of the bone mass and mechanical function, while the less dense trabecular bone, among other functions, serves to redirect stresses originating in the proximity of the joints from movements to the stronger cortical shell. Throughout skeletal growth and in adult life, bones have to adapt to changes in the mechanical environment to prevent them from fracturing.

A protective function of bone is provided especially for organs situated in vital areas such as the thorax and head, where injuries could be fatal. Developmentally, protective bones like the calvarium and rib, which belong to the axial skeleton, are formed at least in part by a process called intramembranous ossification. This is different from the endochondral ossification (Fig. 2.12) seen in the long bones of the appendicular skeleton. Areas rich in cancellous bone like the iliac crest, vertebrae, and proximal femur continue to contribute significantly to hematopoiesis throughout life, while in adult life, in the marrow cavity of the long bones, red marrow is increasingly replaced with yellow fat. Yellow marrow appears to serve as energy store, and it might contribute to lipid metabolism and the regulation of triglycerides. Both cortical and trabecular bone are sites of long-term storage for calcium which is required for muscle contractions, blood clotting, the transmission of nerve impulses, and many enzymatic reactions. Calcium can be mobilized through two

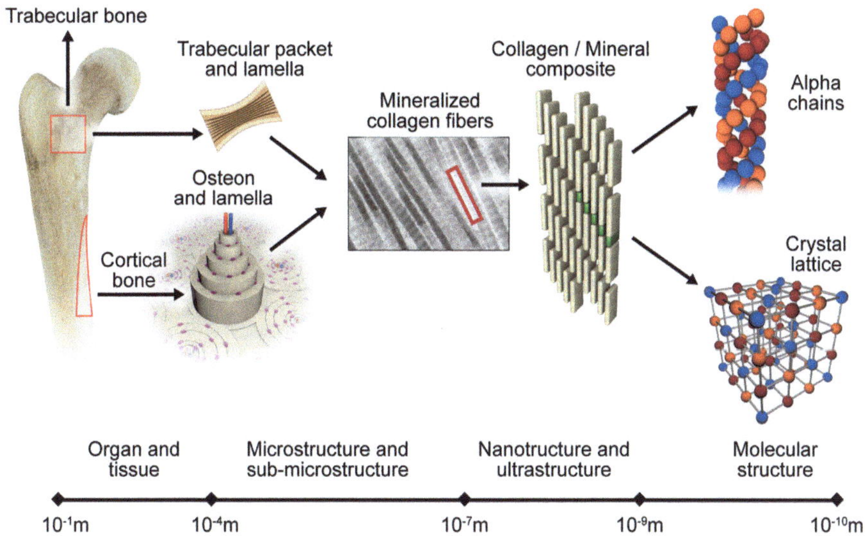

Fig. 2.2 Hierarchical organization of bone: At the organ level, bone is seen as a composite with dense cortical bone forming an outside shell and cancellous (trabecular) bone within the marrow cavity. At the microscopic level, cortical bone is composed of many Haversian systems (osteons). Osteons are composed of a central canal carrying a blood vessel, nerves, and lymphatics surrounded by layers of concentric lamellae. Although trabecular bone is also lamellar, its structure comprises of a combination of lamellae that run largely parallel to the trabecular surface and the remnants of older, remodeled bone that can appear osteon-like in some areas. At the ultra- and nanostructural levels, bone is a composite of collagen fibers with plates of mineral interspersed within the collagen fibrils as well as between the collagen fibers. The collagen fibrils are composed of molecules which form a triple helix out of two α_1 and a single α_2 chain

major processes. The extensive surface area which is formed by the canalicular system and the osteocyte lacunae allows for a rapid mobilization and redeposition of calcium to meet immediate demands, while the release of calcium generated by the more time-demanding process of bone remodeling can address changes in calcium demand over time scales spanning from several days to weeks.

To achieve its multiple diverse functions, bone is organized in a hierarchical manner (Burr and Akkus 2014) (Fig. 2.2). On a nanoscale level, the collagen and mineral form a composite material, with mineral providing stiffness and structure while type I collagen provides resilience and ductility. Collagen also serves as a template structure for the deposition of the mineral. At the microstructural or microscopic level, the individual collagen fibers and the mineral are organized primarily as defined by local functional requirements. Depending on the specific mechanical and biological needs, bone can be dense (cortical bone) or a meshwork (cancellous/trabecular bone). As eluded to previously, cortical bone is composed of many secondary Haversian systems or osteons (Fig. 2.3), which consist of a central canal carrying a blood vessel, nerves and lymphatics, and are surrounded by multiple layers of concentric lamellae. In contrast, while trabecular bone is also lamellar in

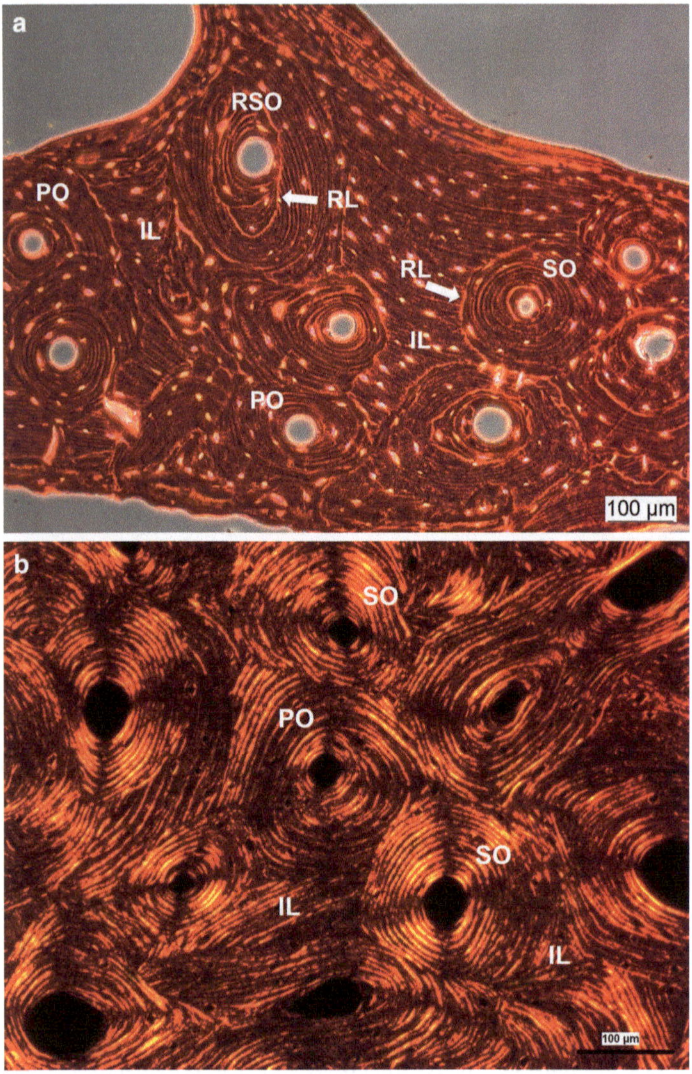

Fig. 2.3 Primary/secondary osteons. (**a**) decalcified human rib showing primary osteons (PO), intact secondary osteons (SO), and remodeled secondary osteons (RSO). Secondary osteons form longitudinally oriented fibers embedded in a matrix of interstitial lamellae (IL), but are separated from the matrix by a cement line (reversal line, ↑RL), a remnant formed during the creation of the secondary osteon at the location where osteoclastic bone resorption stops and bone formation begins (transverse section, 6 µm section, picrosirius staining, circularly polarized light). (**b**) decalcified human rib showing primary osteons (PO) and secondary osteons (SO). Variations in collagen orientation can be viewed using linearly polarized light in human cortical bone. Fibers that are oriented transversely to the direction of light appear bright, whereas those oriented along the path of the light are dark. In a cross section of cortical bone, dark osteons are composed of longitudinally oriented fibers (transverse section, 6 µm section, picrosirius staining, linearly polarized light)

structure, some of these lamellae run approximately in parallel to the trabecular surface, while some of the older remnants of remodeled bone have an almost osteon-like appearance. Finally at the macroscopic level, bone is composed of a dense cortical shell that is wrapped around the cancellous (trabecular) bone which is localized within the marrow cavity. Trabecular bone is made up mostly of primary lamellae, but in areas where bone has been extensively remodeled, hemiosteons which essentially look like half osteons may be present. Woven bone on the other hand represents a highly disorganized repair tissue where randomly arranged collagen I fibers are rapidly mineralized, thus lacking the typical lamellar pattern. It may form in non-pathological situations as a consequence of unusually high loads experienced by bones and in the region of the growth plate during endochondral ossification, or under pathological situations in response to inflammation or bone injury. Following a fracture, woven bone helps forming the callus that bridges the fracture gap providing initial stability during the early healing process. Eventually woven bone is always remodeled and replaced by normal lamellar bone, which provides better mechanical stability.

Primary bone is defined as new bone formed in a space where previously no bone existed, or formed on an existing surface of bone or mineralized cartilage (Burr and Akkus 2014). Thus, primary bone requires only bone formation, whereas secondary bone is defined as bone which is produced by the resorption of previously deposited bone followed by the formation of new bone in its place, a process called remodeling.

Primary lamellar bone is formed on the periosteal surface and is characterized by a series of parallel laminar sheets with few vascular canals. This dense, strong subperiosteal compartment serves primarily a mechanical function. Primary lamellar bone is also deposited on trabecular bone surfaces in the marrow cavity, where it serves a metabolic function in support of calcium metabolism. Plexiform, or fibrolamellar bone is a special variant of primary bone which is found in rapidly growing animals (cows and sheep) and has been reported in humans but limited to the time of major growth spurts. In plexiform bone, primary lamellar bone is deposited on a de novo-formed nonlamellar core substrate within the periosteal membrane (intramembranous ossification). The special arrangement allows increasing bone strength rapidly, as small amounts of bone deposited on the outer surface will contribute significantly to overall bone strength.

Primary osteons are formed by concentric deposition of osteonal lamellae around large vascular canals (Burr and Akkus, 2014) (Fig. 2.3). They consist of less than ten lamellae and measure around 50–100 μm in diameter. In contrast, secondary osteons which form the Haversian system are larger (100–250 μm in diameter) consisting of 20–25 lamellae, and they have a clearly visible cement line at their outer boundary. The cement line (reversal line) is a remnant of the reversal phase of bone remodeling, the process whereby secondary osteons are formed, and it indicates the point where osteoclastic bone resorption stops and bone formation begins. The central Haversian canal carries a neurovascular bundle. Haversian systems run at an average 11–17° angle with respect to the long axis of bone, closely following the dominant loading direction of the bone. Haversian vessels branch extensively and are connected to transversely oriented blood vessels called Volkmann canals, which extend from the marrow vasculature to the periosteal membrane. Cement- or reversal lines serve a very important mechanical function in fatigue and fracture processes. Their ability to absorb energy inhibits crack propagation through the bone matrix, and they provide viscous damping in compact bone. High local shear stresses result in the deformation of the viscous cement line which will diminish the stress and hinder transmission of the energy to the growing crack. The remodeling of cortical bone will inevitably leave behind traces of non-remodeled primary or secondary osteons. These remnants of older bone are referred to as interstitial bone. Since the average tissue age of interstitial bone is older, it is more highly mineralized and more susceptible to microcrack accumulation.

Cortical bone forms the primary component of the diaphysis in the appendicular skeleton. In healthy bone with "normal" bone turnover, Haversian and Volkmann canals create a porosity of around 3–5%. Compact bone is also found in the metaphysis of long bones, surrounding the trabecular bone of vertebral bodies, in the iliac crest, ribs, and the skull, where it provides mechanical support and protection.

Cancellous bone is composed of plate- and rod-like structures of around 200 μm thickness (humans) and smaller dimensions in laboratory animals (Fig. 2.4). Cancellous bone is found in the vertebral bodies of the axial skeleton, the ribs, the iliac crest, and the metaphysis of the long bones in the appendicular skeleton. Due to its position, blood supply to trabecular bone is provided by the blood vessels embedded in the marrow cavity, rendering the need for an osteonal structure in general obsolete. Trabecular bone derives its contribution to mechanical strength

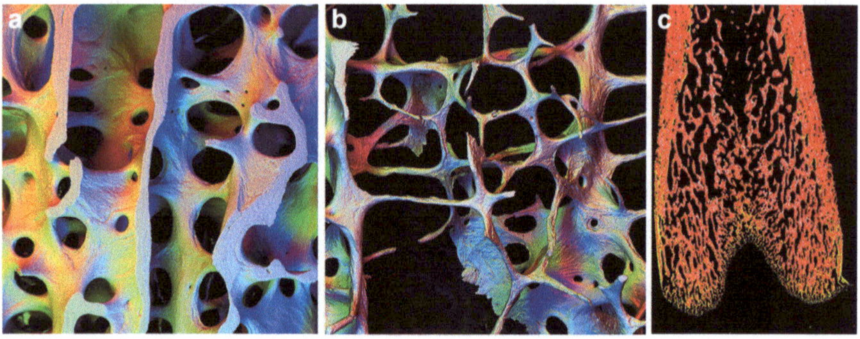

Fig. 2.4 Cancellous bone – rods and plates. (**a**) and (**b**) 20-kV BSE images of macerated, carbon coated, 3-mm-thick section of a human L4 vertebral body. Architecturally, cancellous bone is composed of plates and rods running at various angles to reflect the orientation of the major stresses within the bone. (**a**) shows normal healthy bone, whereas (**b**) shows classical signs of osteoporosis including trabecular thinning, loss of connectivity, and conversion of plate-like into rod-like structures in human bone (Both images courtesy of Prof. Alan Boyde, Queen Mary University of London, UK). (**c**) Trabeculae act as struts to support the outer cortical structure and to funnel the stresses experienced by them to the stronger, more massive cortical bone surrounding them. This important mechanical function is nicely apparent when looking at a longitudinal section through the distal femur metaphysis of a rat (**c**)

essentially from its architecture. Given its location in the metaphysis of long bones, it provides an efficient way of transmitting loads generated in the joints to the adjacent cortical bone. Because it is highly interconnected, vertical struts can be stabilized efficiently and prevented from buckling by the horizontal struts in the structure. Trabecular architecture and especially connectivity is thus an important determinant for the mechanical stability of cancellous bone, and this can be measured effectively by microCT analysis. Studies have shown that a decrease in trabecular number has a much greater negative impact on bone strength than the loss of an equal amount of bone through trabecular thinning (Van den Linden et al. 2001; Hernandez et al. 2006). In healthy human bone, most trabecular structures tend to be shaped in the load-bearing direction as plates rather than rods. Plates provide greater strength against bending in the load-bearing direction of long bones than rods. In conditions of bone loss, plate-like structures will be affected assuming a more elliptical rod-like shape and connectivity decreases. In the axial skeleton, this process seems to occur preferentially outside the primary loading axis, and the remaining struts remain aligned with the loading direction, which enable the structure to preserve some of its strength.

In addition to the mechanical function of metaphyseal cancellous bone for the transmission of loads from the joint to compact bone, a special compartment of trabecular bone is localized beneath the cartilage of the articular surface in the epiphysis, which may function as a cushion by attenuating forces generated by movement. Immediately beneath the joint, the subchondral plate, a layer of condensed cancellous bone is visible. Although the precise role of the subchondral plate and cancellous bone adjacent to the articular cartilage is still debated, it is generally assumed that it may be implicated in the development of osteoarthritis (Goldring and Goldring 2010).

2.1.1 Bone Composition

Bone is a composite material whose extracellular matrix consists of mineral (65%), water (10%), lipids (1%), and organic material (25%), the latter being composed predominantly of type I collagen (90%), noncollagenous proteins (10%). These components have both mechanical and metabolic functions, and the composition and architectural features vary with age (Boskey and Coleman 2010), gender, species, and the site which is studied (Donnelly et al. 2012), and it can be affected by disease and treatment (Boskey 2013; Mandair and Morris 2015).

The mechanical properties and thus the ability of bone to resist fracture are defined by the amount, proper arrangement, and characteristics of each of these components (quantity and quality) (Seeman 2008). In this context, the term "quality" usually refers to factors including composition (weight percent of each component), mineralization (organization of the mineral and its crystallite size), collagen content and collagen crosslinks, morphology (Jepsen 2011), microarchitecture (Patsch et al. 2011), and the presence of microcracks. The physical and chemical properties of the mineral phase have been determined by various methods including chemical analysis, X-ray diffraction, transmission and atomic force microscopy, small angle scattering, and nuclear magnetic resonance (Boskey 2006a). Water can be bound to the mineral-collagen composite, or it can flow freely through canalicular and vascular channels in bone. Unbound water can be redistributed in response to skeletal loading, and it likely contributes to the signals detected by cells such as the osteocytes informing them of the local loading condition.

2.1.2 The Mineral

The largest proportion is occupied by the mineral phase consisting predominately of a nanocrystalline, highly substituted, poorly crystalline analog of the hydroxylapatite [$Ca_{10}(PO_4)_6(OH)_2$], as demonstrated by X-ray diffraction over 60 years ago (Weiner and Wagner 1998; Boskey 2007). In primary mineralization, a process that in humans occurs within the first 3 weeks after osteoid deposition by the osteoblast, mineral is deposited rapidly as an amorphous calcium phosphate along with large amounts of calcium carbonate. During this process about 60–70% of the total mineralization is achieved. As bone matures during the subsequent process of secondary mineralization, crystals grow in size to an eventual size of 40 × 3 × 7.5 nm, become more plate like and crystalline, and orient themselves parallel to one another and to the collagen fibrils. During the final bone maturation which in humans may take more than a year, the calcium carbonate content is reduced, but it continues to serve an important role in the maintenance of acid-base homeostasis. During episodes of acid load, systemic bicarbonate (HCO_3^-) is consumed to buffer the blood pH. The HCO_3^- deficiency is offset by carbonate and phosphate ions present in the bone matrix. In chronic acidosis, this mechanism helps to maintain blood pH, but it also results in bone loss.

The amount of mineral in bone can be quantified by a variety of techniques including gravimetric analyses (determination of ash weight), analysis of calcium and phosphate contents, and spectroscopic and densitometric analyses. Human bone, nonhuman primates, rats, and mice contain ~60–70% mineral/dry weight, depending upon site, species, and stage of development.

Variations in the distribution of mineral and its properties in bone can be illustrated by a variety of imaging techniques, including bone mineralization density distribution (BMDD), Raman (Kazanci et al. 2006; Morris and Mandair 2011; Mandair and Morris 2015), and infrared spectroscopic imaging (Boskey 2006b). Quantitative backscattered electron imaging is used for mapping the calcium concentrations and for the determination of BMDD (frequency distribution of calcium concentrations within the bone sample). Summarized by Boskey (2013), several studies have reported deviations from normal calcium distribution in diseases like osteomalacia, osteoporosis (peak shifted to the left of normal), classical and new forms of osteogenesis imperfecta, and following treatment with bisphosphonates (peak shifted to the right of normal).

2.1.3 The Organic Phase

The organic phase of bone consists primarily of type I collagen (~90%), noncollagenous proteins (NCPs, ~5%), or lipids (~2%). The proteins in the extracellular matrix of bone are often classified into two main groups, structural proteins (collagen and fibronectin) and proteins with specialized functions. The latter are involved in the regulation of collagen fibril diameter; serve as signaling molecules, growth factors, or enzymes; or have other functions.

Structural proteins: The most abundant protein in the bone matrix is type I collagen, a unique triple helical molecule consisting of two identical amino-acid α1-chains and one structurally similar but genetically different α2-chain (Rossert and de Crombrugghe 2002). Each chain is about 1,000 amino acids in length. Collagen fibrils are constructed from collagen molecules which are 300 nm long and 1.5 nm thick. Five of these molecules are stacked in a quarter-staggered array to form a microfibril structure, such that there are 67 nm hole zones between the ends of the molecules and spaces between the laterally contiguous molecules known as pores. The holes and pores serve as spaces which enable the deposition of plates of bone mineral during primary and secondary mineralization. Functionally, collagen serves as a template for mineral deposition, provides elasticity to the tissue, stabilizes the extracellular matrix, and binds other macromolecules. Collagen molecules consist of repeating glycine-X-Y residues which are often hydroxylated and glycosylated, giving rise to some of collagen's unique crosslinking ability. In bone, two types of crosslinks have been demonstrated which increase with age and are altered by disease or treatments. They are either formed enzymatically or by glycation (non-enzymatically mediated collagen cross-linking). Since the number of cross-links is an indicator of collagen maturity, it is not surprising that drugs which reduce bone remodeling (rate of bone renewal) have been shown to increase the number of cross-links. Pyridinoline and deoxypyridinoline are two mature cross-links formed by enzymatic collagen cross-linking near the end of the collagen molecule, the C and N-termini. The content of these mature cross-links increases up to the age of 10–15 years, to remain stable thereafter, possibly declining with age. An increased pyridinoline/deoxypyridinoline ratio appears to enhance compressive strength and stiffness. In contrast, the non-enzymatically formed advanced glycosylation end-products (AGEs), such as pentosidine, furosine, vesperlysine, imidazolone, and Nε-carboxymethyllysine, which are elevated in uncontrolled diabetics and in oxidative stress and are formed randomly, increase the brittleness of bone. Distribution of total cross-links in the bone matrix can be visualized in Fourier transform infrared microscopic imaging (FTIRI) (Boskey 2006b; Mandair and Morris 2015), and pentosidine is often used as a marker since it is the only AGE that can be quantified accurately. AGEs accumulating in the extracellular matrix can influence the proliferation and differentiation of bone-forming cells by interacting with a specific receptor called RAGE. Binding of AGEs to RAGE activates NF-κB in osteoblasts and stimulates cytokine production. AGEs also appear to regulate osteoclastogenesis and osteoclast activity. The presence of AGEs may reduce the solubility of collagen and of the mineralized matrix, thereby slowing down bone resorption.

Observations by microscopical or by X-ray diffraction analysis suggest that collagen fibers are preferentially oriented according to the predominant stress in the bone (Burr and Akkus 2014). Longitudinally oriented fibers are found in portions of bone that are under tension, whereas transversely oriented fibers are more abundant in regions which are usually exposed to compressive loads. It has been shown experimentally that altering the loading direction will change the fiber orientation in newly formed bone. This has been debated however as being an artifact due to characteristics of optics and light transmission. As an alternative, it has been suggested that collagen is organized in a twisted plywood configuration that is continuously rotated through 180° cycles (Weiner and Wagner 1998). Whether the collagen is twisted or not and how it becomes oriented is still under debate.

Noncollagenous proteins (NCPs) contribute 10–15% of the total bone protein content and approximately 2% of the total bone weight. They have important roles in the organization of the extracellular matrix, coordinate mineral-matrix and cell-matrix interactions, and regulate the process of bone mineralization. For a more in-depth review of the various NCPs and their role, the reader is referred to the review written by Boskey and Robey (2013). In brief, NCPs can be subdivided in several groups, namely, the serum-derived proteins, proteoglycans, and glycosylated proteins. In addition, at least 12 glycoproteins which appear to be involved in the cell attachment mediated by transient or stable focal adhesions to extracellular macromolecules are found among the NCPs. Finally, four gamma-carboxy-glutamic acid-containing proteins which appear to be involved in the regulation of cartilage, bone, or vascular calcification, namely, matrix gla protein (MGP), bone gla protein (BGP or osteocalcin), periostin, and protein S, are found in the bone matrix. Most of these NCPs including their origin and potential contribution to bone biology, which is largely based on results from transgenic mouse models, have been described in detail by Boskey and Robey (2013). To date, relatively little information is available for most of the NCPs on how they change in distribution in osteoporosis or other bone diseases, or in response to anabolic or antiresorptive therapies. However, a number of NCPs can interact with collagen fibrils. Osteopontin (OPN) may function as "glue" that provides fiber matrix bonding, thereby enhancing bone's resistance to fracture, as well as crack bridging entity in the case of microcrack formation where it inhibits crack propagation (Sroga and Vashishth 2012). Also, recent gene-wide association studies suggest that osteocalcin (OC) and OPN are important for fracture resistance. Both OPN and OC have an important role in bone mineralization (Mandair and Morris 2015), as demonstrated by the reduction in cortical hardness and elastic modulus in $OPN^{-/-}$ mice (Kavukcuoglu et al. 2007) and the finding that OC appears to regulate the growth of apatite crystals in $OC^{-/-}$ mice, even though genotype mineral/matrix ratios were not different between $OC^{-/-}$ and $OC^{+/+}$ mice (Kavukcuoglu et al. 2009). OPN also binds to osteoclasts and promotes their adherence to the mineral during the process of bone resorption.

A minor but potentially interesting fraction of NCPs in the mineralized matrix of bone is the enzymes, mainly phosphatases (alkaline phosphatase (ALP), tartrate-resistant acid phosphatase (TRAP)), FAM210A, and also matrix metalloproteases (MMP2, MMP9, MMP13, and ADAMTS18) and other signaling factors. While their biological role has been under intense investigation and considerable knowledge of their function is available, only very limited published evidence is available regarding changes in the expression of enzymes and signaling factors in the matrix of patients with metabolic bone disease or under drug treatment (Boskey 2013).

The glycoprotein ALP is used as a biomarker of bone formation, since it hydrolyzes pyrophosphates, which inhibit mineral deposition by binding to mineral crystals. Since ALP is produced by many different organs like the liver and kidney, it is not a specific marker for bone formation. In contrast, the bone-specific isoform of alkaline phosphatase (BSAP) which is found on the surface of osteoblasts has been shown to be a sensitive and reliable indicator of bone metabolism.

Lipids, which make up ~3% of the total bone matrix, are known to be important for cell function regulating the flux of ions and signaling molecules into and out of the cells. The distribution of lipids in the matrix can be observed from histology, based on sudanophilia, from FTIR and Raman analysis, or by nuclear magnetic resonance (NMR) (Reid et al. 2012). However, lipid composition and its association with mechanical properties have not been studied in the context of fragility fractures or other bone diseases in humans and animal models of human disease.

2.1.4 Osteoblasts

Osteoblasts are engaged in bone formation. The extracellular matrix synthesized by osteoblasts is known as osteoid when first deposited and not yet mineralized, but it is subsequently mineralized through the accumulation of calcium phosphate in the form of hydroxyapatite (Fig. 2.5). This process results in the hard but lightweight composite material (with both organic and inorganic components) that is the major constituent of bone. After active bone formation, osteoblasts have three alternative fates: they either undergo apoptosis, or become buried in the mineralizing bone matrix where they undergo terminal differentiation to become osteocytes, or become "inactive" lining cells that continue to exist on quiescent bone surfaces. Bone lining cells show a flattened shape containing only a few cell organelles, indicative of their low metabolic activity. However, these apparently dormant cells are in close contact with matrix-embedded osteocytes through gap junctions and participate in the calcium exchange between the mineralized matrix and the bone marrow compartment, and they can be reactivated within a few hours in response to PTH to contribute to local bone formation processes (Dobnig and Turner 1995; Kim et al. 2012).

Cytologically, actively bone-forming osteoblasts have a strongly basophilic cytoplasm and abundant mitochondria and show typical secretory characteristics, possessing well-developed rough endoplasmic reticulum with dilated cisterna. All these features are consistent with their capacity to produce a large amount of extracellular proteins. Osteoblasts are polarized, in the sense that the part of the cell membrane which is in direct contact with the bone surface possesses many cytoplasmic processes that extend into the newly deposited osteoid.

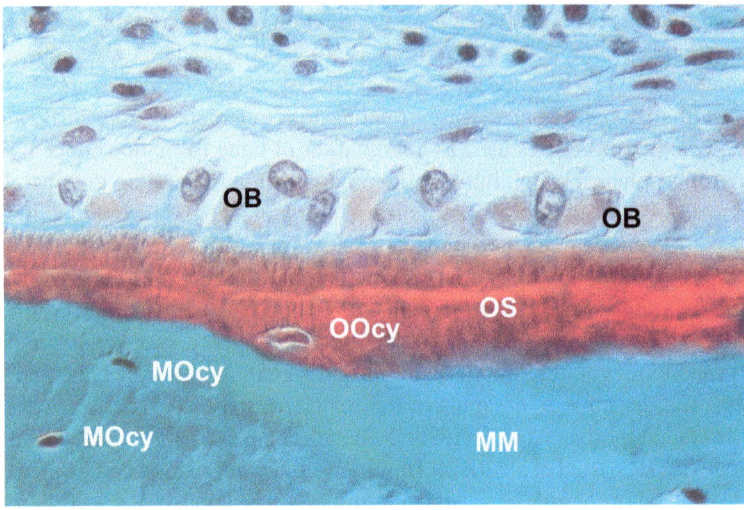

Fig. 2.5 Osteoblasts: matrix synthesizing osteoblasts (OB) are lined up on the endocortical bone-forming surface covered by osteoid (OS, *red*) in rat bone (Goldner Trichrome stain). An osteoblast undergoing differentiation into an osteocyte (OOcy, osteoid osteocyte) has remained completely embedded in the osteoid. Mature osteocytes (MOcy) are visible in the mineralized matrix (MM)

Osteoblasts originate from mesenchymal stem cells (MSCs) which are capable of differentiating into chondrocytes, osteoblasts, myoblasts, and adipocytes (Caplan and Bruder 2001; Jiang et al. 2002). Their differentiation is regulated by specific transcription factors. Sox-5, Sox-6, and Sox-9 regulate chondrocytic differentiation, while their differentiation to adipocytes and myoblasts is determined by PPARγ and Myo D, respectively.

The osteoblast-lineage cells form a group that includes their mesenchymal progenitors, preosteoblasts, osteoblasts, bone lining cells, and osteocytes (Fig. 2.6). Among the cytokines involved in osteoblast differentiation are the Hedgehogs, BMPs, TGF-β, PTH, and WNTs (see reviews by de Gorter and ten Dijke 2013; Nakamura 2007; Long 2012). Runx2/Cbfa (Komori et al. 1997; Ducy et al. 1997) and Osterix/Sp7 (Nakashima et al. 2002) are essential regulators, and they are indispensable for both endochondral and intramembranous ossification. Runx2-deficient mice completely lack osteoblasts, fail to form hypertrophic chondrocytes, and produce a cartilaginous skeleton that is completely devoid of a mineralized matrix (Otto et al. 1997). Runx2 interacts with transcriptional activators and repressors and other coregulatory proteins, to either positively or negatively regulate expression of osteoblast-specific genes including Col1, ALP, OPN, osteonectin (ON), and OC (Harada et al. 1999; Kern et al. 2001). Runx2 appears to directly regulate the expression of the zinc finger-containing transcription factor Osterix (Osx) (Nishio et al. 2006). Similar to mice deficient in Runx2, $Osx^{-/-}$ mice lack osteoblasts, showing the critical role of this transcription factor in bone formation. Osx can interact with the nuclear factor for activated T cells 2 (NFAT2), which cooperates with Osx in

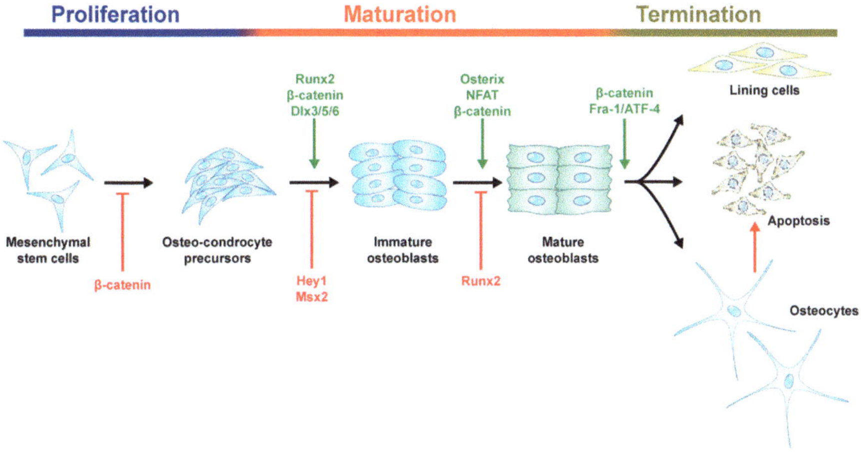

Fig. 2.6 Osteoblastogensis and fate: scheme illustrating osteoblastogenesis and the main transcription factors governing proliferation and differentiation of osteoblast precursors. Upon termination of bone formation, mature osteoblasts can flatten to cover the quiescent bone surface as lining cells, die by apoptosis, or surround themselves by bone matrix while differentiating to become osteocytes. *ATF-4* activating transcription factor-4, *Dlx* distalless homeobox, *Fra* Fos-related antigen, *Osx* Osterix, *Runx2* runt-related transcription factor 2

controlling the transcription of target genes such as OC, OPN, ON, and collagen1 (Nakashima et al. 2002; Koga et al. 2005).

In cell culture, osteoblasts are morphologically nearly indistinguishable from fibroblasts. However, in contrast to fibroblasts, they express Runx2 (Komori et al. 1997; Ducy et al. 1997) and osteocalcin (Ducy et al. 1996), a secreted molecule that inhibits osteoblast function.

Active matrix synthesizing osteoblasts produce a unique combination of extracellular proteins, including osteocalcin, osteopontin, osteonectin, bone sialoprotein, alkaline phosphatase, and a large amount of type I collagen (de Gorter and ten Dijke 2013). They also synthesize and secrete proteoglycans such as decorin and biglycan, and due to their ability to bind calcium, these glycoproteins and proteoglycans are considered to be involved in two functions: the storage of calcium ions for calcification and the regulation of hydroxyapatite growth by preventing excess calcification. Osteoblasts also produce cytokines including insulin-like growth factor I and II, transforming growth factor β (TGF-β), and bone morphogenetic proteins (BMPs) (de Gorter and ten Dijke 2013). These growth factors are stored in calcified bone matrix and play an important role in differentiation and function of osteoblasts. Thus, bone matrix acts as a storage site of growth factors in addition to calcium and phosphates. IGF-1 secreted by osteoblasts is considered an auto- or paracrine regulator of osteoblastic cell function (Canalis 2009). IGF-1 signals via the IGF1 receptor, which in line with other tyrosine kinase receptors activates the phosphatidylinositol 3-kinase (PI3K)-Akt and Ras-ERK MAP kinase pathways.

BMPs belong to the TGF-β superfamily and were originally identified as the active components in bone extracts capable of inducing bone formation at ectopic sites (Urist 1965) and as osteo-inductive factors in decalcified bone and dentin (Urist et al. 1979). They are expressed and stored in skeletal tissue, and are required for skeletal development and maintenance of adult bone homeostasis, and they play an important role in fracture healing (Chen et al. 2004; Gazzerro and Canalis 2006). BMP-2, BMP-4, and BMP-7 trigger endochondral ossification by inducing mesenchymal cells to differentiate into osteogenic cells (Yamaguchi et al. 1991). BMPs bind as dimers to type I and type II serine/threonine receptor kinases, thereby forming an oligomeric complex. Binding of BMPs to their specific receptor complex leads to phosphorylation of Smads (R-Smads) that form heterodimer with Smad4 and regulate gene expression (de Gorter and ten Dijke 2013). BMP activity is also regulated by inhibitory Smads (I-Smads) and antagonists including noggin, chordin, and sclerostin. Although BMPs promote differentiation of osteoblasts by preventing MyoD expression and inducing Runx2 expression, the precise transcriptional mechanism has not been clarified yet.

TGF-β is one of the most abundant cytokines in bone matrix and plays a major role in development and maintenance of the skeleton, where depending on the context and concentration, it can exert both positive and negative effects on bone formation (Janssens et al. 2005; Maeda et al. 2004). TGF-β signals via a similar mechanism as the related BMPs. However, upon binding to its specific receptors, TGF-β induces activation of Smad2 and Smad3 (Feng and Derynck 2005; Massague et al. 2005). Smad3 overexpression in mouse osteoblastic MC3T3-E1 cells enhanced the levels of bone matrix proteins, ALP activity, and mineralization (Sowa et al. 2002). For more details on TGF-b and BMP signaling in osteoblasts, the reader is referred to the article published by de Sánchez-Duffhues et al. (2015).

Numerous markers for osteoblasts have been identified such as cbfa1, Osx, alkaline phosphatase (ALP), and collagen type I. Recent research of tissue nonspecific alkaline phosphatase (TNAP)-deficient mice revealed that TNAP acts as pyrophosphatase hydrolyzing pyrophosphate, an inhibitor of calcification, and increases the concentration of inorganic phosphates required for calcification (Hessle et al. 2002). Although ALP activity is intense in the basolateral plasma membrane of osteoblasts, their membrane facing toward osteoid and the plasma membrane of osteocytes hardly show ALP activity. Histochemical evidence indicates that the distribution of ALP does not always correlate with calcification sites and calcification is not completely disturbed in TNAP-deficient mice. Thus, the exact function of ALP in osteoblasts remains controversial.

2.1.5 Osteocytes

Osteocytes are derived from osteoblast progenitors which become entombed during the process of matrix deposition in small spaces called lacunae (Fig. 2.7). Osteocytes contribute to over 90% of the bone cells as compared to 4–6% osteoblasts and

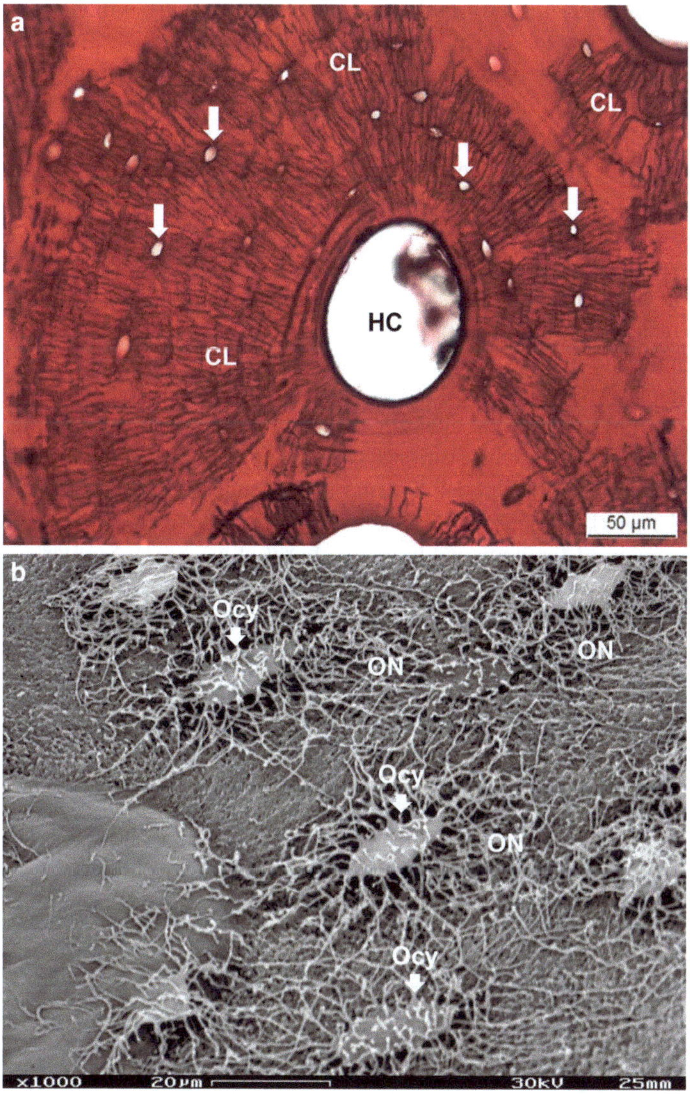

Fig. 2.7 Osteocyte. (**a**) decalcified human femur showing spatial distribution of osteocyte lacunae (*arrows*) and canaliculi (CL) surrounding a Haversian canal (HC). Silver impregnation according to Tripp (Tripp and MacKay 1972) with picrosfuchsin counterstaining. (**b**) SEM image of osteocytes (Ocy, *arrows*) on an acid edged human trabecular bone surface showing the extensive osteocyte network (ON) connecting with each other (SEM Image courtesy of Prof. Alan Boyde, Queen Mary University of London, UK)

1–2% osteoclasts, making them by far the most abundant cell in the bone matrix and on bone surfaces (Bonewald 2013). Osteocytes are regularly distributed throughout the mineralized matrix and connected with each other and the cells on the bone surface and even to cells within the bone marrow through dendrite-like processes contained in fluid-filled micro-canals (canaliculi), which radiate toward the surface and the blood supply (Knothe Tate et al. 2004; Bonewald 2013). Recently developed methods (Schneider et al. 2010; Pacureanu et al. 2012) have allowed to understand the complexity and better define the magnitude of this network, which in terms of size appears to match the neural network of the brain (Buenzli and Sims 2015). The total number of osteocytes within the average adult human skeleton is ~42 billion and that the total number of osteocyte dendritic projections from these cells is ~3.7 trillion.

Their unique location in the bone matrix permits osteocytes to detect local changes in mechanical signals, either through strain or fluid flow; to respond to changes in levels of circulating factors like hormones and ions; and to amplify these signals leading to a coordinated adaptive response of the skeleton to environmental changes (Schaffler et al. 2014). These changes include adaptations in bone shape as well as the microarchitecture through a process called modeling (resorption and/or formation). Observational studies indicate that osteocytes might also detect fatigue-induced microdamage and send signals to the osteoclast progenitor cells on the nearby surface to initiate the complex remodeling process leading to the replacement of the damaged bone (Verborgt et al. 2002). Similarly, apoptotic osteocytes can send out stress signals which will result in their replacement via the process of bone remodeling. Osteocytes thus perform an essential task in the adaptation of bones to changes in mechanical loading as well as the maintenance of "healthy" bone tissue.

Relatively recent evidence suggests that the osteocyte network not only contributes to calcium homeostasis but also that FGF23 is released to regulate phosphate homeostasis in the kidney (Fukumoto and Martin 2009). The role of osteocytes in phosphate metabolism begins with their embedding in osteoid, when molecules known to be involved in phosphate metabolism such as dentin matrix protein 1 (DMP1) (Feng et al. 2006), phosphate-regulating neutral endopeptidase (Phex or PEX) (The HYP Consortium 1995), and matrix extracellular phosphoglycoprotein (MEPE) (Rowe et al. 2004) are elevated.

During the process of bone formation, between 5% and 20% of the osteoblasts become osteocytes. During this transformation the osteoblast cell body is reduced by roughly 30% at the early stage where cytoplasmic processes are formed and by about 70% when osteocyte maturation is complete. It is only recently that osteocyte markers such as E11/gp38, PEX, DMP1, sclerostin, FGF23, and ORP150 have been identified (Guo et al. 2010; Bonewald 2013). Some of these markers overlap in expression with osteoblasts, but others have been identified for specific stages of differentiation. Membrane expression of the membrane-associated proteins E11/PDPN/GP38 and MMP-14 is required for the formation of dendritic processes and canaliculi. The osteocyte-selective promoter, the 8 kb DMP1 driving GFP, showed selective expression in early and late osteocytes (Kalajzic et al. 2004).

A marker for the mature, embedded osteocyte is sclerostin, which is encoded for by the *SOST* (Balemans et al. 2001; Poole et al. 2005). This secreted protein is a negative key regulator of bone mass. It acts in the osteoblastic lineage as an antagonist of canonical WNT signaling, which is critical to bone homeostasis (Baron and Kneissel 2013). It impacts directly bone formation and indirectly bone resorption, and its absence or inhibition results in increased bone mass in several species including humans. Neutralizing antibodies or small molecules are being tested clinically with the aim to increase bone mass. Similar to sclerostin, albeit less selectively, WNT antagonist Dickkopf 1 (Dkk-1) is also robustly expressed by osteocytes. Finally ORP150 which may have a role in the protection of osteocytes from hypoxia is also highly expressed in mature osteocytes (Guo et al. 2010).

Osteocytes are able to modify their microenvironment. The term "osteocytic osteolysis" was introduced over 100 years ago to describe the enlargement of the osteocyte lacunae observed in patients with severe hyperparathyroidism and in immobilized rats (Recklinghausen 1910; Belanger 1969). In addition to the enlargement of the lacunae, changes can take place in the perilacunar matrix (Qin and Bonewald 2009). Perilacunar demineralization (osteocyte halos) was observed in rickets (Heuck 1970) as well as X-linked hypophosphatemic rickets (Marie and Glorieux 1983). Glucocorticoids not only cause enlargement of the osteocyte lacunae but might also be able to reduce mineral in the perilacunar space, thereby generating a zone of hypomineralized bone (Lane et al. 2006). Glucocorticoids may thus not only induce osteocyte death through their proapoptotic effect but in addition compromise the metabolism and function of the osteocyte. Interestingly, tetracycline label uptake, which occurs at sites of active bone mineralization, was observed in the osteocyte lacunae suggesting some ability of osteocytes to impact matrix mineralization. Taken together, this suggests that osteocytes may be capable of adding and removing mineral from their surroundings, which in turn may affect metabolic functions and mechanical properties of bone. Enlargement of the lacunae and canaliculi would reduce bone fluid flow shear stress, thereby reducing mechanical loading on the osteocyte which may trigger an adaptive response. Since the osteocyte lacuno-canalicular system is several orders of magnitudes greater than the bone surface area, the removal of only a few angstroms of mineral could have significant effects on circulating, systemic ion levels.

Osteocytes are long-lived cells, but like osteoblasts and osteoclasts, they die by apoptosis (programmed cell death). Decreased osteocyte viability is associated with bone fragility induced by estrogen and androgen deficiency or withdrawal, glucocorticoid excess, and mechanical disuse or overuse. Apoptosis is also triggered in response to microdamage, and it is proposed that dying osteocytes are targeted for removal by osteoclasts and replaced with new tissue in the process called remodeling. In this context, the expression of antiapoptotic and proapoptotic molecules in osteocytes surrounding microcracks was mapped, and it was found that proapoptotic molecules are elevated in osteocytes immediately at the microcrack locus, whereas antiapoptotic molecules are expressed 1–2 mm from the microcrack (Verborgt et al. 2002).

Given the crucial metabolic, mechanosensory, and tissue maintenance functions, it is clear that osteocyte viability is of crucial for skeletal health. The role of the osteocyte in skeletal growth, in mechanosensation needed for adaptation to mechanical usage (modeling) in the postnatal skeleton, as well as bone tissue renewal and repair of microdamage (remodeling) will be discussed in a separate section of this chapter.

2.1.6 *Osteoclasts*

Osteoclasts are multinucleated giant cells responsible for bone resorption. Ultrastructurally, the multiple nuclei are surrounded by many mitochondria, endoplasmic reticulum, and a well-developed Golgi apparatus (Ross 2013). The presence of vesicles, lysosomes, and vacuoles indicates that these cells are actively involved in energy production and protein synthesis, particularly of lysosomal enzymes.

Osteoclasts are a member of the monocyte-macrophage family derived from the bone marrow macrophage, and they can be generated in vitro from mononuclear phagocyte precursors (Suda et al. 1999) (Fig. 2.8). Osteoclastogenesis critically depends on two cytokines, namely, macrophage-colony stimulating factor (M-CSF or CSF-1) (Pixley and Stanley 2004) and receptor activator of nuclear factor-κB ligand (RANKL) (Suda et al. 1999; Boyle et al. 2003). Both proteins exist as membrane bound and soluble form and are produced by marrow stromal cells and osteoblasts. Thus, recruitment of osteoclasts from their mononuclear precursor is

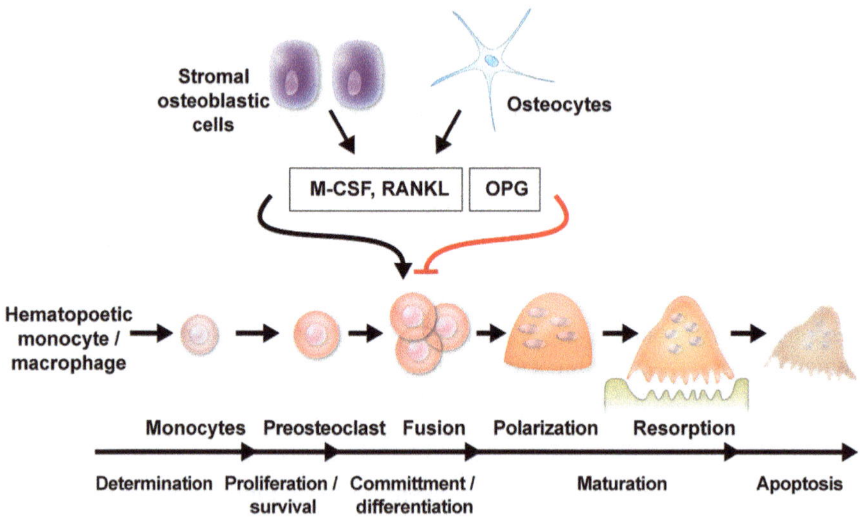

Fig. 2.8 Osteoclastogenesis and apoptosis: Osteoclast differentiation is governed by RANKL and M-CSF, as well as other cytokines secreted by osteoblasts and osteocytes that control various steps of osteoclastogenesis, including precursor proliferation, commitment, differentiation, and maturation. OPG, which is also secreted by osteoblasts and osteocytes, acts as a decoy receptor for RANKL and reduces osteoclast differentiation

critically dependent on the presence of nonhematopoietic cells residing in the bone marrow. RANKL, a member of the tumor necrosis factor (TNF) family, which was discovered originally as a secreted protein of activated T cells (Weitzmann et al. 2000), is the key osteoclastogenic cytokine required for the priming of precursor cells. This has been confirmed by mouse genetic studies showing that mice deficient in either RANK or RANKL show osteopetrosis which is attributable to a defect in osteoclastogenesis. In bone, RANKL produced by osteoblast-lineage cells including osteocytes binds to the RANK receptor, which is expressed in osteoclast progenitors and mature osteoclasts. Stimulators of osteoclastogenesis like PTH, PTH-related protein (PTHrP), prostaglandin E_2 (PGE_2), interleukin 1 (IL-1), and 1,25-$(OH)_2D_3$ act on osteoblast-lineage cells to upregulate RANKL expression. Conversely, osteoprotegerin (OPG), a soluble form of the TNF receptor, which is also expressed by osteoblasts and osteocytes, is acting as a decoy receptor for RANKL to inhibit osteoclastogenesis (Kostenuik and Shalhoub 2001). OPG is upregulated by estrogen, BMP, and TGF-β and suppressed by pro-inflammatory cytokines (Ross 2013). The balance between RANKL and OPG in osteoblast-lineage cells controls differentiation and activation of osteoclasts (Hofbauer et al. 2000; Boyle et al. 2003; Ross 2013). M-CSF appears to contribute to the proliferation, differentiation, and survival of osteoclast precursors, as well as the cytoskeletal rearrangement required for efficient bone resorption.

Active bone-resorbing osteoclasts show cellular polarization with the apical membrane consisting of the clear zone (or sealing zone) and the ruffled border and a basolateral plasma membrane (Nakamura 2007) (Figs. 2.9 and 2.10). The sealing zone (SZ) is the site of bone attachment which delineates the bone-resorbing space where the ruffled border is formed. The capacity of the osteoclast to form a sealed microenvironment between itself and the underlying bone matrix is a prerequisite to the resorptive event. The sealing zone appears as a ring of filamentous actin (F-actin also called actin ring or podosome belt), with the podosomes acting as focal adhesion points. It serves to isolate the bone-resorbing compartment from the extracellular fluid. The attachment of the osteoclast to the bone matrix is mediated primarily by the $\alpha_v\beta_3$ integrin transmembrane protein (vitronectin receptor), which engages in the sealing zone with the recognized amino acid motif Arg-Gly-Asp (RGD)-containing matrix proteins osteopontin and sialoprotein (Ross and Teitelbaum 2005; Teitelbaum 2005). Competitive RGD ligands are able to block osteoclast attachment and bone resorption. Cell matrix interaction stimulates the non-receptor-type tyrosine kinase c-Src, which is crucially involved in the maintenance of cell polarity and activity of osteoclasts.

The actual process of bone resorption is a two-step process, initiating with the dissolution of the mineralized matrix, followed by the enzymatic degradation of the organic matrix. The process takes place under the tightly sealed clear zone where the ruffled border, a surface extension, is formed. The acidification of the enclosed Howship's lacuna is achieved by carbonic anhydrase, which converts CO_2 and H_2O into H^+ and HCO^{3-} (Anderson et al. 1982; Gay and Mueller 1974), and by a vacuolar-type H^+-ATPase (Blair et al. 1989; Teitelbaum and Ross 2003). The pH of

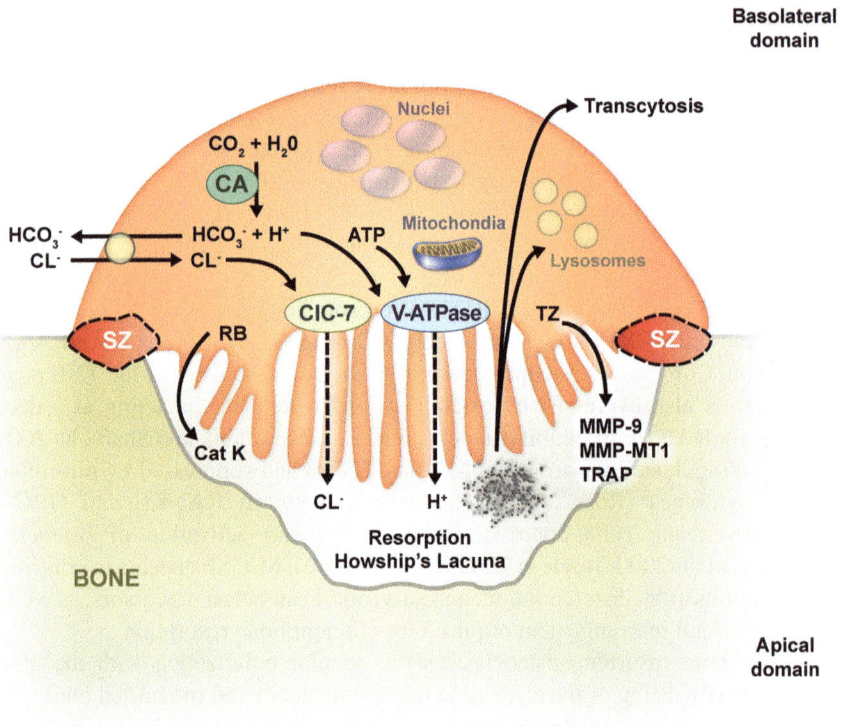

Fig. 2.9 Osteoclast ultrastructure/resorption: schematic representation of a resorbing osteoclast showing the sealing zone (SZ), ruffled border (RB), transition zone (TZ), nuclei, and resorption lacuna. Mature osteoclasts are polarized cells, with the apical domain facing the bone surface. In the process of bone resorption, H^+-protons and chloride ions (Cl^-) generated by carbonic anhydrase (CA) are transported via the V-ATPase and CIC-7 respectively to the RB membrane where they are secreted to acidify the resorption lacuna. Electroneutrality is maintained by a coupled basolateral bicarbonate/chloride exchanger, thus avoiding changes in pH and/or membrane polarization of the cell. Mitochondria generate the necessary ATP. Degradation of the demineralized bone matrix occurs through action of secreted cathepsin K (Cat K), matrix metalloproteinases (MMPs), and tartrate-resistant acid phosphatase (TRAP). Bone degradation products are released into the bone microenvironment, internalized into the cell to be degraded by lysosomes, or secreted at the basolateral membrane via transcytosis

4.5 leads to focal decalcification of hydroxyapatite leading to the exposure of the organic matrix, consisting largely of type I collagen. To prevent intracellular polarization, proton secretion is balanced by parallel extrusion of chloride ions through the ruffled border via a CIC-7-type chloride channel. The degradation of the demineralized extracellular organic components is accomplished by lysosomal enzymes such as cathepsin K (Gelb et al. 1996; Inaoka et al. 1995), MMP-9, and MMP-13 (Reponen et al. 1994; Ross 2013). Cathepsin K belongs to a cysteine protease family, and the acidic conditions in the Howship's lacuna provide an ideal milieu for the

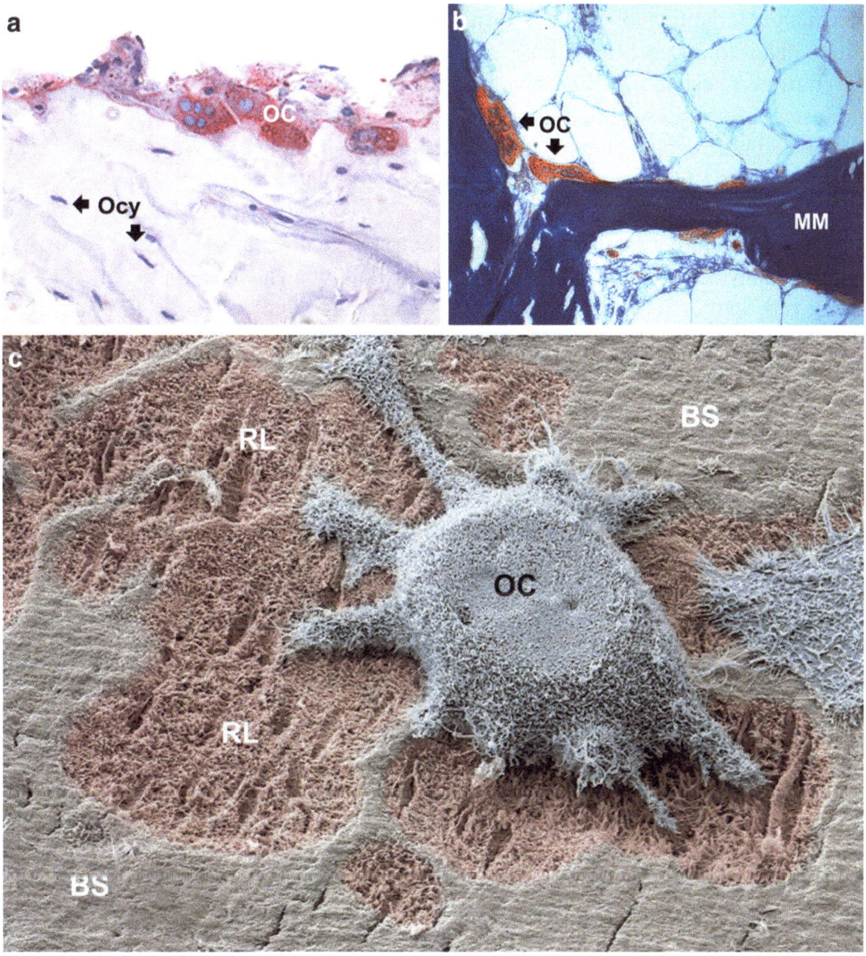

Fig. 2.10 Osteoclast morphology. (**a**) multinucleated osteoclasts (OC) on a subperiosteal bone surface in the rat. TRAP-stain with toluidine blue, (*Ocy* osteocytes). (**b**) TRAP-stained multinucleated rat osteoclasts (OC) on a trabecular surface (*MM* mineralized matrix). (**c**) colored SEM image of an osteoclast (OC). Resorption pit (RL) and 'intact' bone surface (BS) (SEM image courtesy Prof Timothy Arnett, University College London, UK)

degradation of collagen fibers. MMP-9 acts as a gelatinase for further digestion of segmented collagen fibrils. Degraded protein fragments are endocytosed and transported in vesicles to lysosomes where they are either degraded or transported to the basolateral surface where they appear to discharge their content into the surrounding extracellular milieu (Salo et al. 1997; Stenbeck and Horton 2004). The importance of this transcytosis of growth factors released from the mineralized matrix during the step of matrix demineralization and its perceived importance for bone formation (coupling of bone resorption to bone formation) will be discussed in more detail in the section on bone remodeling further below.

Tartrate-resistant acid phosphatase (TRAP) is widely used as a marker enzyme of osteoclasts and secreted in Howship's lacunae. This enzyme appears to be involved in the dephosphorylation of osteopontin and bone sialoprotein, a prerequisite for integrin binding, and for the generation of reactive oxygen species involved in matrix degradation.

The Rho family of small GTPases is central to remodeling of the actin cytoskeleton in many cell types, including osteoclasts. Following attachment to bone, signals from $\alpha_v\beta_3$ and/or receptor tyrosine kinases activate small GTPases Rho and Rac, which bind GTP and translocate to the cytoskeleton to exert their specific effects. Rho signaling appears to affect predominantly cell adhesion mediated by the formation of the actin ring and a constitutively active form of the GTPase which stimulates podosome formation, osteoclast motility, and bone resorption (Jaffe and Hall 2005). In contrast, Rac stimulation in osteoclast precursors prompts appearance of lamellipodia, thus forming the migratory front of the cell to which $\alpha_v\beta_3$ moves when activated (Chellaiah 2005). Bisphosphonates, the potent antiresorptive drugs, inhibit the enzyme farnesyl pyrophosphate synthase in the mevalonate pathway, thereby blocking the addition of hydrophobic moieties onto the GTPases (inhibition of protein prenylation). This in turn prevents their membrane targeting and activation. Since active GTPases also regulate cell viability, bisphosphonates also trigger osteoclast apoptosis.

After completion of bone resorption, osteoclasts undergo programmed cell death or apoptosis. The signals that trigger osteoclast apoptosis are not well understood. High concentrations of extracellular calcium, Fas-ligand secreted by osteoblasts, the process of osteoclast detachment, and a decrease in the production of pro-survival cytokines (M-CSF, and RANKL), all these events have been suggested as potential mechanisms. ERK-activation is crucial for osteoclast survival, and TNF-α and IL-1 delay osteoclast apoptosis via ERK-activation. Finally, M-CSF, RANKL, and TNF-α also activate NF-kB in osteoclasts, a transcription factor which is known to inhibit apoptosis in many cell types. Several steroid hormones are known to regulate osteoclast formation, their activity, and survival. Estrogens and androgens inhibit the production of osteoclastogenic cytokines IL-1 and IL-6, and estrogens also induce osteoclast apoptosis. In mice, glucocorticoids reduce the number of osteoclast progenitors, but they also increase the life span of the mature osteoclast, so that their number does not decrease immediately. This may explain the transient increase in bone resorption in patients exposed to pharmacological doses of glucocorticoids. The inhibitory effect of glucocorticoids on osteoclast progenitors appears to be indirect, and it is likely mediated by the direct inhibitory effect of the steroid hormone on osteoblast numbers. The known effects of glucocorticoids are able to explain the transient increase in bone resorption followed by the low remodeling rate and suppression of bone formation which is typically seen during prolonged exposure in patients.

2.2 Skeletal Growth, Adaptation, and Maintenance

Fetal limb buds removed in utero and grown in vitro develop the shape of the proximal femur implying that bone shape is "imprinted" in the genetic material (Murray and Huxley 1925). This indicates that mechanical strain is required for postnatal, but not for prenatal skeletal development and maintenance. Studies in families and twins support this view (Seeman et al. 1989; Seeman et al. 1996). Both the postnatal and adult skeleton are able to continually adapt their structure where bone is added in response to an increase in loading, or being removed in response to unloading or disuse. The ability of bone to bear these loads is dependent on the type and magnitude of the applied load and the structural properties of the bone that is loaded (Wallace 2014). When loads exceed the structural strength of the bone, damage will occur and the bone may eventually fracture. Bone remodeling is essential for the maintenance of healthy bone tissue. This facilitates the replacement of apoptotic osteocytes, the repair of microdamage in the mineralized matrix, and the remodeling of the fracture callus into normal lamellar bone during the secondary stage of fracture healing.

Cortical bone undergoes both modeling (resorption and formation occurring independently of each other at different locations) and remodeling (coupled and sequential resorption and formation occurring at the same site). The goal in modeling is to either shape bone or change bone mass, whereas the purpose of remodeling is foremost to renew bone. Modeling and remodeling occur on both the endocortical and subperiosteal surfaces of the cortex, whereas within the cortex only remodeling – referred to as Haversian remodeling – takes place. Modeling occurs predominantly during skeletal growth but to a lesser extent also in adults, primarily in response to changes in mechanical loading. Recently, newer drugs such as PTH and anti-sclerostin antibodies used to treat osteoporosis have been shown to be either permissive to modeling or to stimulate it. This has considerably raised the interest in the topic of modeling in the adult skeleton and in its potential for therapeutic manipulation.

2.2.1 *Skeletal Growth and Bone Modeling*

Bone modeling is the process by which bone is either formed on an existing bone surface by osteoblasts without prior resorption (formation modeling) or being removed by osteoclasts (resorption modeling), with the aim to alter the shape of bone or adapt it to a change in mechanical loading (Fig. 2.11). Modeling occurs on subperiosteal, endocortical, and trabecular bone surfaces, and it is the dominant process during skeletal growth and before attainment of peak bone mass, even though it does continue throughout life at low levels. Bone modeling is thus essential for the proper functioning of three major processes, namely, longitudinal growth, radial growth, and cortical and cancellous bone drifts.

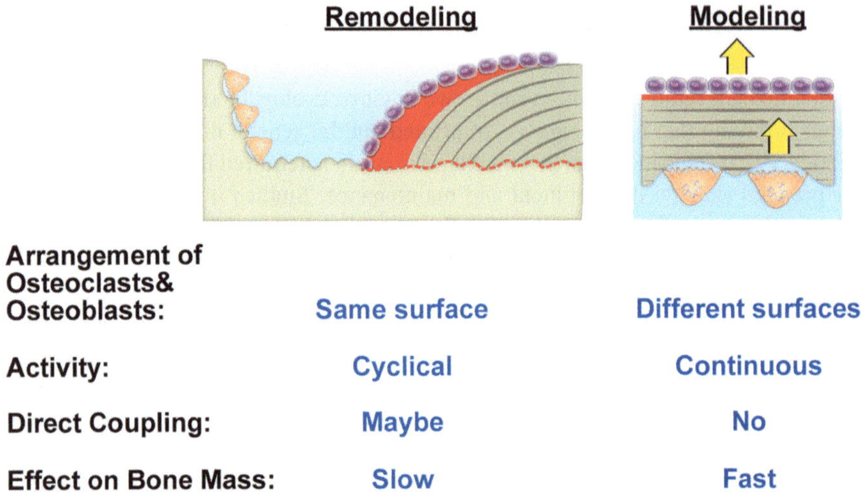

Fig. 2.11 Modeling vs remodeling: Bone remodeling involves sequential osteoclast-mediated bone resorption and bone formation at the same location. Its principal purpose is to repair microdamage and to renew the skeleton. Remodeling occurs on trabecular, endocortical, intracortical, and periosteal envelopes. In contrast, bone modeling occurs predominantly during skeletal growth. Formation modeling and resorption modeling occur on different surfaces. The two processes are either independent of each other (maintenance of local tissue strain) or happen in a coordinated fashion (cortical modeling drift during growth). Modeling can occur on periosteal, endocortical, or trabecular bone surfaces

In humans, skeletal development begins during the first trimester of gestation involving two distinct processes, namely, intramembranous or endochondral ossification. Most parts of the skull, the scapula and clavicula, are formed through intramembranous ossification, whereas the remainder of the bones of the skeleton is formed by endochondral ossification.

Intramembranous ossification is initiated by the formation of the blastema, a consolidation of mesenchymal cells which differentiate into osteoblasts and initiate matrix production, thereby establishing an ossification center (Allen and Burr 2014; Yang 2013). The transcription factor RUNX2 plays a crucial role in driving the cells of the blastema into the osteoblast lineage. The initial collagen matrix produced is disorganized and known as woven bone. Some of the osteoblasts become encapsulated in the mineralizing matrix and differentiate into osteocytes. The primary ossification center forms a small island serving as a template, to which lamellar or woven bone is added on the surface. During growth of these islands, several islands will merge to create a larger structure. When blood supply to the osteocytes becomes insufficient, blood vessels will invade the structure in some bones of the skull and even form a bone marrow space, a process not seen in the clavicula or scapula. Although intramembranous ossification is associated mainly with embryonic development, the process is recapitulated postnatally during bone healing.

In contrast, during *endochondral ossification*, bone is formed on a hyaline cartilage template produced by chondrocytes, which is replaced by mineralized bone tissue

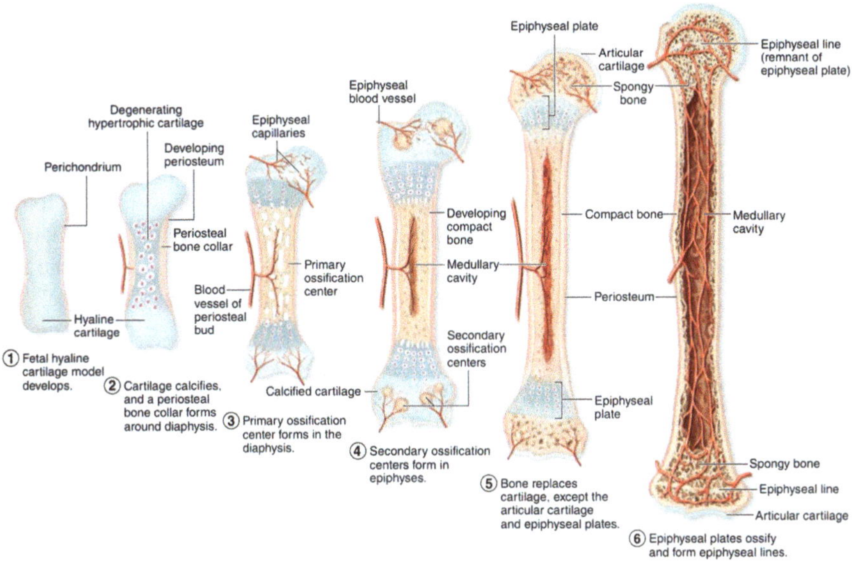

Fig. 2.12 Endochondral ossification forms most bones of the skeleton and occurs in the fetus in models made of hyaline cartilage (1). The process takes many weeks, and major developmental stages include formation of a bone collar around the middle of the cartilage model and degeneration of the underlying cartilage (2), followed by invasion of the resulting ossification center by capillaries and osteoprogenitor cells from the periosteum (3), and osteoid deposition by the new osteoblasts, calcification of woven bone, and its remodeling as compact bone (4). This primary ossification center develops in the diaphysis, along the middle of each developing bone. Secondary ossification centers develop somewhat later by a similar process in the epiphyses. The primary and secondary ossification centers are separated by the epiphyseal plate (5) which provides for continued bone elongation. The two ossification centers do not merge until the epiphyseal plate disappears (6) when full stature is achieved (Reproduced with permission from Junqueira's Basic Histology, 14th Edition)

(Fig. 2.12). In analogy to intramembranous ossification, endochondral ossification begins with a condensation of mesenchymal cells (Allen and Burr 2014; Yang 2013). However, these cells do not develop into osteoblasts, but are differentiated into chondroblasts by the transcription factor SOX-9. Chondroblasts initiate cartilage matrix synthesis with some of the cells becoming engulfed in the matrix where they differentiate into chondrocytes. The hyaline cartilage which is formed in this process is surrounded by a membrane called perichondrium, from where under the control of RUNX2, some of the cells differentiate into osteoblasts and begin forming bone matrix on the cartilage template. This process initiates at the circumference of the diaphysis of long bones (primary or diaphyseal ossification center), thereby creating a lamellar structure referred to as bone collar. The perichondrium is replaced by the periosteum, which provides the source for osteoblasts required for the subperiosteal expansion of the bone collar. The formation and expansion of the bone collar increasingly limit the provision of nutrients to the initial cartilage structure resulting in its calcification and eventually the death of the chondrocytes, followed by the removal of these cartilage remnants by osteoclasts. It is the process of vascular invasion which results in the formation of the bone marrow space.

Over time, through a similar process, secondary ossification centers are being formed at both ends of the long bone in the epiphyses. The central diaphyseal region which contains a marrow cavity remains separated from the two secondary ossification centers by a cartilage layer called the epiphysis. The epiphysis or growth plate which forms at the interface of the two ossification centers is responsible for longitudinal bone growth. In the growth plate, morphologically distinct zones can be identified, namely, the resting zone, the proliferative zone, the prehypertrophic and hypertrophic zone, the calcified cartilage zone, and the ossification zone (Fig. 2.13). The resting zone (or reserve zone) which is comprised of hyaline cartilage matrix with embedded chondrocytes is the most distant region from the primary ossification center. This zone is rich in type II collagen produced by the chondroblasts located near the perichondrium that embed themselves in the matrix and differentiate into chondrocytes. The proliferation zone is readily identifiable by its stacked coin appearance, resulting from chondrocyte mitosis along the longitudinal axis of the bone. Growth hormone (GH), insulin-like growth factors (IGFs), bone morphogenetic proteins (BMPs), Indian hedgehog protein (IHH), and the Wnt-β-catenin signaling pathway all play important roles in stimulation of chondrocyte mitosis, while fibroblast growth factor (FGF) inhibits chondrocyte proliferation in the growth plate (Allen and Burr 2014; Yang 2013).

Longitudinal growth is predominantly driven by the cells located in the prehypertrophic zone. In the hypertrophic zone, chondrocytes begin to enlarge under the control of thyroxine and components of the WNT-β-catenin pathway, followed by their death through apoptosis. In the hypertrophic zone, type II collagen production is switched to type X collagen. This switch provides more stiffness to the region, but it also limits the diffusion of nutrients, and it may act as a trigger for the vascular invasion which is seen further below in the ossification zone. Also, cartilage calcification can be observed around the dying hypertrophic chondrocytes, an active process driven by the release of ALP containing vesicles by chondrocytes into the surrounding matrix, hence the term calcified cartilage zone. Chondroclasts, cells specialized in the removal of mineralized cartilage, are recruited to the site to resorb the calcified cartilage. In the ossification zone, osteoblasts produce woven bone on top of the mineralized cartilage. This woven bone is later removed by osteoclasts and replaced by lamellar bone, thereby forming the trabecular structures. Essentially, the complex process of endochondral ossification results in the production of a large number of parallel-oriented, narrowly spaced, trabecular elements that grow at a 90°

Fig. 2.13 (continued) the matrix is undergoing calcification, and 5. ossification zone, a zone in which blood vessels and osteoblasts have invaded the lacunae of the old cartilage, producing marrow cavities and osteoid for new bone. (**b**) growth plate of a young adult rat. Toluidine blue stain of the epiphysis. (**c**) and (**d**) Movat Pentachrome stain of the growth plate of a skeletally mature, 8-month-old rat. The proliferative zone is "less active," and the epiphysis shows signs of calcification as well as disruption of the pillar-like arrangements

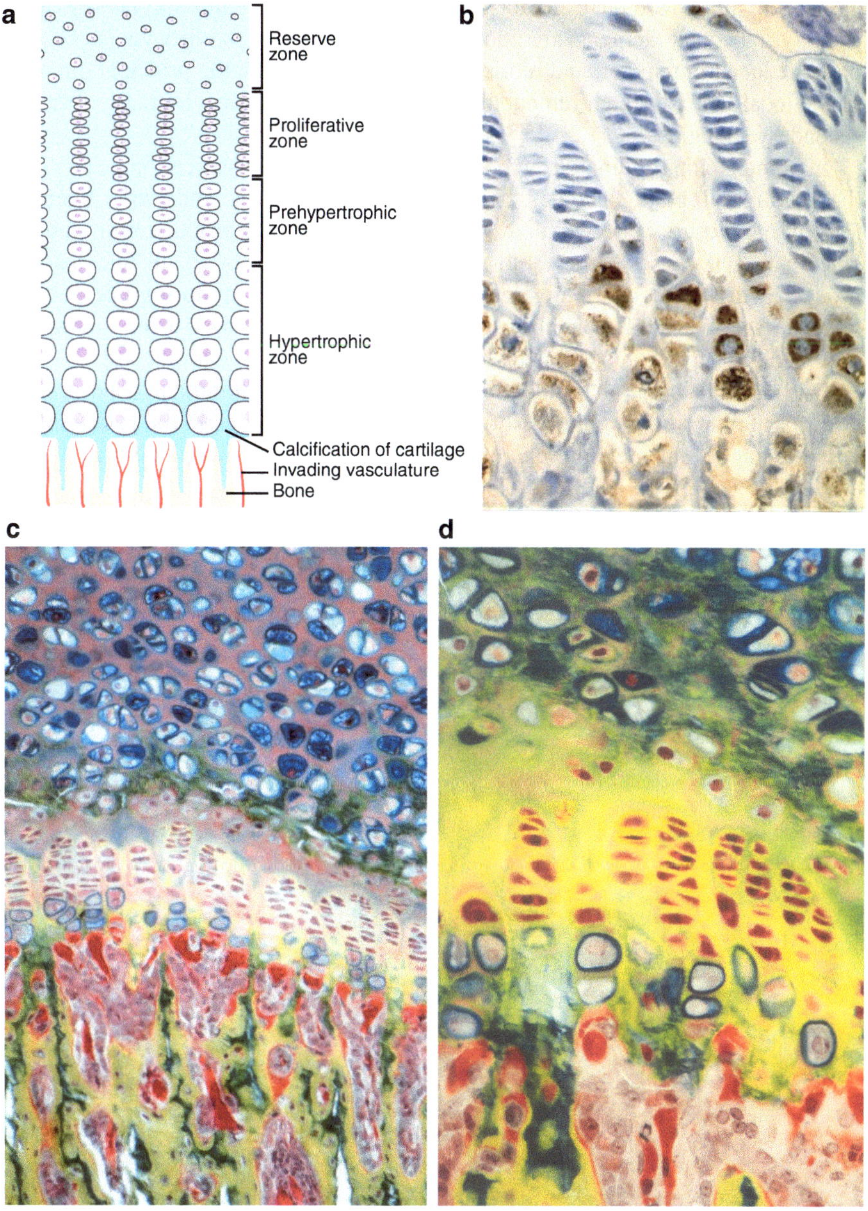

Fig. 2.13 Growth plate: Cells in epiphyseal growth plates are responsible for continued elongation of bones until the body's full size is reached. Based on the histologic appearance of the cells and surrounding matrix, the epiphyseal growth plate is classified into different zones (**a**) 1. resting or reserve zone (hyaline cartilage), 2. proliferative zone consisting of cartilage with proliferating chondroblasts aligned in lacunae as axial aggregates, 3. prehypertrophic- and hypertrophic zone consisting of degenerating cartilage in which the aligned cells are hypertrophic and the matrix condensed, 4. zone of calcified cartilage, an area in which the chondrocytes have disappeared and

angle from the growth plate and form the primary spongiosa. In a second step, these primary trabecular elements are thickened, connected to each other via lateral supports, reduced in number, and reoriented to align them with the principal direction of the local strain through remodeling and modeling activities, to finally form the secondary spongiosa.

In humans, longitudinal growth through endochondral ossification occurs until the epiphyseal plate becomes ossified in the late teens and early twenties. Epiphyseal closure is accelerated by the onset of estrogen production in puberty, since estrogen causes more rapid senescence of chondrocytes. In females, this process is driven by ovarian estrogen production picking up during late puberty, while in the male, different forms of estrogen are synthesized from androgens, specifically testosterone and androstenedione, by the enzyme aromatase.

During endochondral ossification associated with longitudinal bone growth, both formation modeling and resorption modeling play an essential role to maintain the shape in the metaphyseal region (Fig. 2.14). Resorption modeling removes

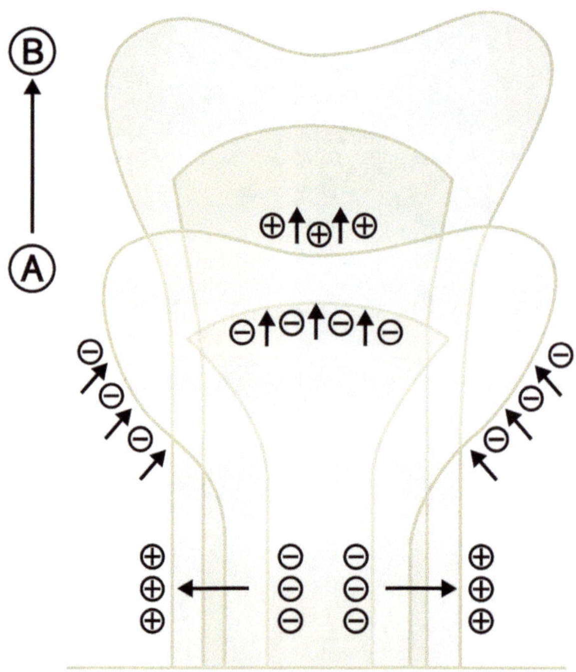

Fig. 2.14 Skeletal growth. Example of classical modeling of ends of long bone: shaping of long bone by modeling drifts during growth. Formation (+) and resorption (−) activities during bone growth from A to B. Resorption drifts (−) in the diaphysis enlarge the marrow cavity, while formation drifts (+) add bone periosteally to maintain the mechanical competence during bone elongation. In the lower metaphysis, an inward drift of cortical bone narrows the cross-sectional diameter of the metaphysis down to the smaller diameter of cortical bone to create the diaphysis and maintain the overall shape of the bone as it grows

bone from the periosteal surface in the metaphysis, while formation modeling adds bone on the endocortical surface which essentially leads to an inward drift of the cortex at this site. This inward drift of cortical bone narrows the cross-sectional diameter of the metaphysis down to the smaller diameter of cortical bone in the diaphysis, thus maintaining the overall shape during bone elongation. In contrast, radial growth is observed at the level of the diaphysis, where bone is added to the periosteal surface (formation modeling) and at the same time being removed at the endocortical surface (resorption modeling) to keep cortical thickness relatively constant, at least during skeletal growth (Fig. 2.14). Subperiosteal expansion is highest during skeletal growth but continues at a much slower pace throughout life. Skeletal growth is sexually dimorphic, and estrogen acts to inhibit periosteal expansion through modeling in the female, while increasing testosterone levels in the male accelerate this process (Garn 1970; Seeman 2008). Gender-specific hormonal differences thus determine the difference with males having larger-sized bones and higher bone mass relative to females at the age where peak bone mass is reached.

During aging, periosteal apposition is believed to increase as an adaptive response to compensate for the loss of strength produced by endocortical bone loss. In a 7-year prospective study of over 600 premenopausal women, Szulc and colleagues report that endocortical bone loss occurred with concurrent periosteal apposition (Szulc et al. 2006). As periosteal apposition was less than endocortical resorption, the cortices were thinned, but there was no net bone loss because the thinner cortex was now distributed around a larger perimeter. As a result of improved spatial distribution, resistance to bending increased despite bone loss and cortical thinning.

Bone modeling is not only important for longitudinal and radial growth during bone elongation, it is also the crucial process driving bone drifts (Fig. 2.15). Bone drifts are related to radial growth, but they can occur in the adult skeleton, although the limited capability for periosteal expansion with increasing age reduces the speed and magnitude at which adaptations can occur. During aging, both increasing endocortical bone resorption and reduced periosteal apposition cause net bone loss, alterations in the distribution of the remaining bone, and the emergence of the bone fragility. In addition, modeling drifts allow changing the position of the cortex relative to its central axis. Bone drifts are essential to allow adaptation of the skeleton to changes in the direction of the principal loading axis. Typical examples are orthodontic tooth movements, the spatial redistribution of bone mass of the humerus in the playing arm of tennis players (Haapasalo et al. 2000; Bass et al. 2002), or the change in the orientation of the hip joint axis during aging. The trabecular network can also undergo modeling drifts for the same reason. A typical example is the conversion of primary trabecular bone growing out of the growth plate at a 90° angle (e.g., independent of the direction of loads), into secondary trabecular bone, which is optimally aligned with the local direction of loads, thus allowing optimal transmission of mechanical forces from the joint to the cortical bone compartment.

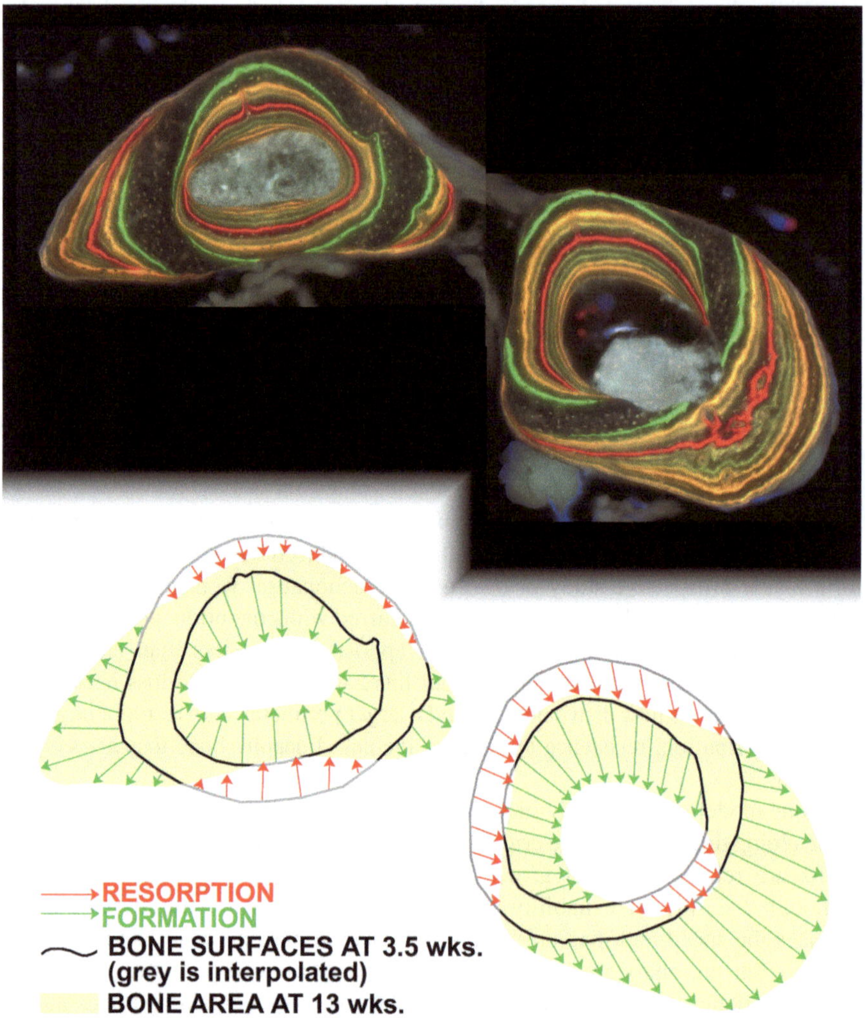

Fig. 2.15 Modeling drifts: Using multiple fluorochrome labels, bone drift can be visualized in a growing mouse. Different fluorochromes were administered roughly 1 week apart to a growing mouse between age 3.5 weeks and 13 weeks. At necropsy, the radius and ulna were sectioned for histological analysis. Radial growth occurred in the ulna (*left bone*) as the periosteal surfaces expanded outward, while there was also some formation modeling on the endocortical surface (modeling drift). The radius (*right bone*) underwent significant drift (*down* and to the *right* in the picture) by formation modeling on one periosteal surface and the opposite endocortical surface, along with resorption modeling on the other two surfaces. The schematic below the photomicrograph conceptualizes what the bone probably looked like at 3.5 weeks of age and how it was transformed through modeling to the eventual geometry (Reproduced with permission from Allen and Burr 2014 (Original image provide by Prof. Matthew Allen, Indiana University School of Medicine, USA))

2.2.2 Bone Remodeling

Bone remodeling involves sequential osteoclast-mediated bone resorption and osteoblast-mediated bone formation at the same location, and it is performed by the so-called basic multicellular unit (BMU) (Frost 1964; Crockett et al. 2011). The process can occur at any of the four envelopes, namely, periosteal, endocortical, intracortical, and trabecular. The final product of bone remodeling in cortical bone is an osteonal structure characterized by concentric layers of bone surrounding a Haversian canal and enclosed by a cement line (Fig. 2.16). In contrast, when remodeling occurs on a cancellous, endocortical, or very rarely the subperiosteal surface, no blood vessel is incorporated, and the resulting structure is called a hemiosteon.

It is commonly believed that two types of remodeling exist, namely, targeted and non-targeted (stochastic) remodeling (Parfitt 2002a). In targeted remodeling, a specific, local signaling event such as osteocyte apoptosis or microdamage would trigger a repair response to restore "healthy" viable bone to the affected site. In contrast, stochastic remodeling is thought to be a random process, which is believed to play a role in calcium homeostasis. The concept of microdamage serving as a signal for initiation of targeted remodeling was first theorized by Harold Frost in the 1960s. Since then, experimental evidence suggests that supraphysiologic mechanical loads can indeed induce microdamage (Mori and Burr 1993). Histological analysis demonstrated that the increased number of resorption cavities found in the overloaded limb was spatially related to the site of microdamage (Burr et al. 1985). Rodent studies showed that microdamage resulted in localized disruption of the osteocyte network via physical breakage of the cytoplasmic connections, which in turn induced osteocyte apoptosis (Verborgt et al. 2000). While the apoptotic osteocytes actively produce RANKL, the key factor in osteoclast development, the nearby healthy osteocytes produce antiapoptotic signals like OPG which may serve to target the remodeling response to the site of damage (Kennedy et al. 2012). A role for osteocyte apoptosis is further suggested by the fact that mechanical disuse, estrogen deficiency, and glucocorticoid excess induce this process and are associated with increased remodeling.

The remodeling cycle is divided into five distinct stages: activation, resorption, reversal, formation, and quiescence (Figs. 2.17 and 2.18) (Allen and Burr 2014). Bone remodeling takes place at various places throughout the skeleton, and these cycles are not coordinated with each other. As will be discussed later in this chapter, the length of the remodeling cycles varies between species, and it can be highly altered by disease. Hauge (Hauge et al. 2001) suggested that cancellous bone remodeling takes place in a special bone remodeling compartment (BRC, Fig. 2.17) (Crockett et al. 2011). Histological evidence suggests that the compartment is lined on its marrow side by flattened cells and on its osseous side by the remodeling bone surface, resembling a roof of flattened cells covering the bone surface (canopy cells). The flat marrow lining cells are in continuity with the bone lining cells at the margins of the BRC. The BRC model has gained support as it may help to explain how coupling of bone resorption and bone formation, two processes in the remodeling

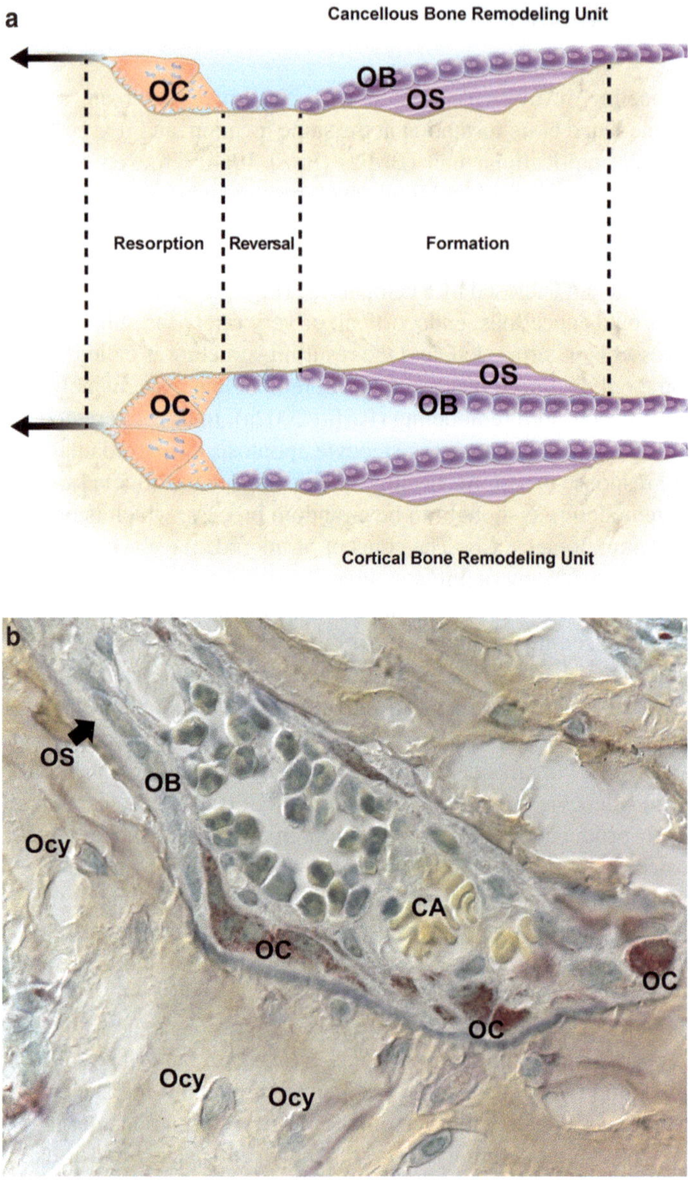

Fig. 2.16 (**a**) Diagram showing schematic of hemi-osteonal and osteonal remodeling carried out by a team of cells referred to as the basic multicellular unit (BMU). In cortical bone, osteonal remodeling in the BMU comprises a cutting cone of osteoclasts (OC) in front, a closing cone lined by osteoblasts (OB) following behind, and connective tissue, blood vessels, and nerves filling the cavity. The BMU maintains its size, shape, and internal organization for many months as it travels through bone in a controlled direction. Individual osteoclast nuclei are short-lived, turning over about 8% per day, and replaced by new pre-osteoclasts that originated in the bone marrow and travel in the circulation to the site of resorption. Refilling of bone at each successive cross-sectional location is accomplished by a team of osteoblasts, probably originating from precursors that reside

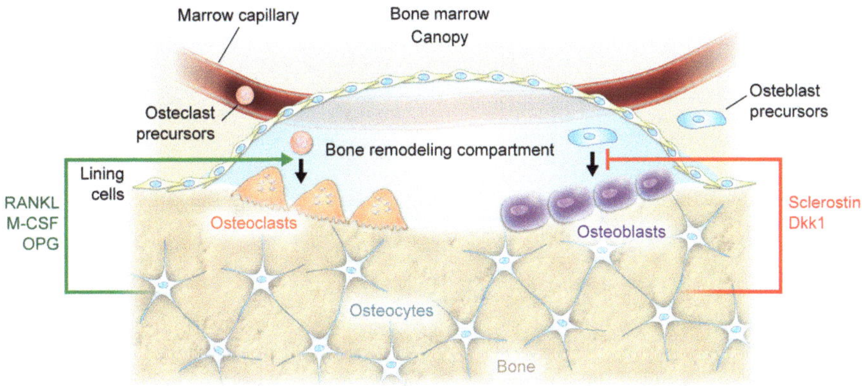

Fig. 2.17 Bone remodeling compartment: When bone remodeling is needed, osteocytes send signals through the canalicular network to the nearby bone lining cells, which retract from the bone surface to form a structure named bone remodeling compartment (BRC). Osteoclast precursors are attracted to the BRC by marrow capillaries where, under the influence of osteocyte-derived pro-osteoclastogenic cytokines RANKL and M-CSF, they differentiate to become mature osteoclasts, leading to the initiation of the bone remodeling cycle. Presumably in response to osteocyte-derived signals (including sclerostin and Dkk-1), and/or bone-derived factors which are released during bone resorption, osteoblast precursors from the bone marrow or the circulation are attracted to the BRC, where they differentiate into mature bone-forming osteoblasts resulting in the refilling of the resorption lacuna

cycle, may be linked, even though they take place at different time windows during the cycle (Sims and Martin 2015).

The activation phase initiates with the lining cells producing collagenase which digests the layer of unmineralized matrix, thereby exposing the bone surface to osteoclastic bone resorption (Chambers and Fuller 1985). Osteoclast precursors are recruited to the bone surface where they differentiate and fuse to become fully functioning osteoclasts. It has been suggested that the signal for the formation of these osteoclasts is provided by the canopy cells of the BRC, which have been shown to express a range of regulators of osteoclastogenesis, including RANKL (Crockett et al. 2011). In humans, the length of the activation phase is estimated to evolve over approximately 10 days (Fig. 2.18). With the presence of mature osteoclasts,

Fig. 2.16 (continued) within the local connective tissue, all assembled within a narrow window of time, at the right location, and in the right orientation to the surface. Hemi-osteonal remodeling, a term reserved for the remodeling of trabecular bone, conforms to the same sequence of surface activation, resorption, and formation as osteonal remodeling. Essentially, the trabecular BMU travels across the surface digging a trench rather than a tunnel, but maintaining its size, shape, and individual identity by the continuous recruitment of new cells. (**b**) Colored SEM of a cutting Cone. A longitudinal section through a cortical BMU, showing the osteoclasts (OC) at the front, followed by osteoblasts (OB) depositing osteoid (OS) at the tunnel wall. The BMU is supported by a capillary (CA). In the bone, osteocytes are visible (Colored SEM image courtesy of Prof. Robert S. Weinstein, University of Arkansas for Medical Sciences, USA)

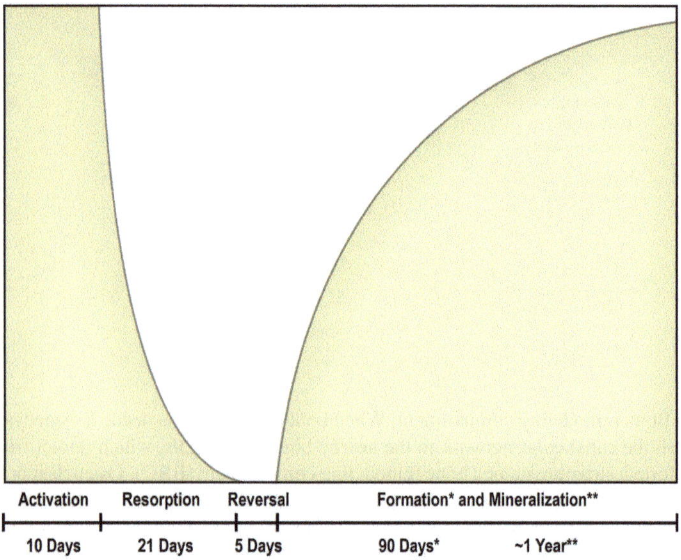

Fig. 2.18 Remodeling cycle: The four stages of bone remodeling, namely, activation, resorption, reversal, and formation, occur over different time frames, with the formation time taking four to five times longer than the resorption phase. Final mineralization of newly formed bone may take up to 1 year. The remodeling period (Rm.P), the time over which these processes occur at a given site, vary greatly between various species, and it may be influenced by disease states or pharmacological interventions. Approximate values reported in the literature for the Rm.P in healthy and estrogen-deficient humans, nonhuman primates, and rats are reported in Table 2.1 later in this chapter. The remodeling cycle is followed by a phase of quiescence of unknown duration. In healthy human bone, the amount of bone formed relative to the amount of bone resorbed by an individual bone remodeling unit (BMU-balance) is slightly negative. In postmenopausal osteoporosis, the BMU balance becomes more negative which together with the observed increase in the number of BMUs (increased activation frequency) results in significant bone loss

the bone lining cells on the targeted surface retract to expose the mineralized matrix to the osteoclasts allowing them to attach. The process of bone resorption, which in humans lasts on average 21 days, has been described earlier in this chapter. In cancellous bone, the size of resorption sites (Howships lacunae) varies considerably, and new osteoclasts can be recruited to the surface undergoing remodeling to extend it or to replace those osteoclasts that die of apoptosis. In contrast, in cortical bone, the size of the cutting cone in osteonal remodeling appears to be relatively constant.

The resorption phase is followed by the reversal phase, which is characterized by the cessation of resorption and initiation of bone formation. During this 5-day phase, remaining collagen fragments which were exposed during the resorption phase are being removed, by little-defined specialized form of cell. This process is essential for the progression of the BMU into the formation phase. These cells also appear to deposit a thin layer of collagen, thus forming the cement or reversal line, a visible delineation of the new hemiosteon or osteon respectively from the surrounding, older matrix. The cement line which is rich in osteopontin (OPN) makes an important contribution to the mechanical properties, with OPN providing fiber matrix bonding which enhances bone's resistance to fracture, as well as crack bridging in the case of microcrack formation and inhibition of crack propagation (Sroga and Vashishth 2012). During the formation phase, osteoblasts deposit the osteoid, which consists primarily of type I collagen fibers and serves as a template for the mineralization. In humans, the bone formation stage (osteoid synthesis), lasts approximately 90 days (see * in Fig 2.18). Mineralization is divided into two phases: primary mineralization accounts for roughly 70% of the mineral deposition and in humans takes place between 2 and 3 weeks. In contrast, secondary mineralization which involves the maturation of mineral crystals to reach the final mineral content may take up to 1 year or more (see ** in Fig 2.18). The difference in time required for resorption (3–6 weeks) and the time needed for bone formation (4–6 months) explain why in histomorphometric analysis of healthy bone, the ratio of bone formation to bone resorption sites is roughly 4:1. The two-step process of bone mineralization has important consequences for the tissue-level mineralization. In general, when bone turnover is high, average tissue mineralization drops and the degree of heterogeneity increases, whereas a low rate of bone remodeling leads to higher average tissue mineralization and greater degree of homogeneity in tissue mineralization. Similarly, nonenzymatic but not enzymatic cross-linking is affected by the rate of bone remodeling. High levels of bone turnover reduce levels of nonenzymatic cross-linking, while low remodeling rates result in increased levels. Changes in tissue-level mineral density and the degree of heterogeneity affect mechanical properties of bones (material stiffness) and as a consequence the way microcracks propagate through the material. On the tissue level, the bone mineral density distribution (BMDD) can be assessed on bone biopsies by quantitative backscattered electron imaging (qBEI) or synchrotron radiation micro computed tomography (SRμCT) (Boyde and Jones 1983; Roschger et al. 2008). For example, transiliacal bone biopsies of postmenopausal osteoporotic women were found to have mostly lower mineralization densities than normal, which were partly associated by an increase of bone turnover, but also caused by calcium and Vit-D deficiency (Roschger et al. 2008). Pharmacological interventions changing bone remodeling and modeling have been shown to influence the BMDD. Antiresorptive treatment for 2–3 years results in a characteristic change in the BMDD profile indicative of an increased degree of homogeneity and a slight shift of the peak position toward higher mineral content. This was observed in postmenopausal osteoporotic women treated with alendronate (Roschger et al. 2001) and in the VertNA-study after 3 years of risedronate (Zoehrer et al. 2006).

Intermittent treatment with PTH is now approved for anabolic treatment worldwide. Given the substantial formation of new bone on trabecular and endocortical envelopes, and the increase in bone turnover under PTH treatment, it is not surprising that the BMDD was shown to shift slightly to lower mineralization densities with an increased peak width (higher degree of heterogeneity in mineral density) (Misof et al. 2003). In an estrogen-depleted rat animal model, comparable effects on BMDD had been observed (Kneissel et al. 2001). It can be assumed that this effect is only transient and if PTH therapy would be followed by an antiresorptive treatment, the matrix would have time for a longer period of secondary mineralization and the BMDD would normalize eventually. In general, antiresorptive therapy causes an increase of degree and homogeneity of mineralization within 3 years of treatment, while normal mineralization levels are not exceeded. In contrast, anabolic therapy like PTH decreases the degree and homogeneity of matrix mineralization, at least transiently (Roschger et al. 2008). Furthermore, BMDD measurements combined with other scanning techniques like nano-indentation, Fourier transform infrared spectroscopy, and small angle X-ray scattering can provide important insights into the structure-function relation of the bone matrix and may ultimately allow for better prediction of fracture risk in diseases and after treatment.

Bone resorption and bone formation in the remodeling cycle are timed in a sequential manner, and under normal conditions, the bone balance after completion of the BMU should be neutral. This stimulation of osteoblast activity in response to resorption is termed "coupling," and it has long been of interest to understand how these two distinct cell types, on the same bone surface but at different times, could be linked so their activities are equal (Sims and Martin 2015; Crockett et al. 2011). There are four main classes of osteoclast-derived signals that are thought to contribute to the process coupling of bone resorption to bone formation in the context of the BMU. They include (1) matrix-derived signals released during bone resorption, (2) factors synthesized and secreted by the mature osteoclast, (3) factors expressed on the osteoclast cell membrane, and (4) topographical changes effected by the osteoclast on the bone surface which are sensed by osteocytes.

During bone formation, osteoblasts deposit a number of growth factors in the mineralizing matrix including TGF-β, platelet-derived growth factor (PDGF), insulin-like growth factors (IGFs), and bone morphogenetic protein 2 (BMP-2), which can be released as a result of the bone resorptive activity of the osteoclast. However, their contribution to the coupling of bone resorption to bone formation appears questionable since it is unlikely that, once released from the matrix, they remain within the BRC during the entire reversal phase to influence mature osteoblasts during the reversal phase and the 90-day formation process. Their principal role may be in the stimulation, recruitment, migration, and differentiation of osteoblast progenitors (Sims and Martin 2015). Similarly, the second class of osteoclast-derived signals, namely, the factors synthesized and secreted by the mature osteoclast, cannot be expected to exist throughout the reversal and bone formation phase of the bone remodeling cycle. Osteoclast-secreted factors, some of which were validated by studies in genetically altered mice, include cardiotrophin-1, sphingosine-1-phosphate, Wnt10b, BMP-6, CTHRC1, and complementfactor3a (C3a) (Sims et al. 2015; Crockett et al. 2011). It is important to note that none of these factors are produced exclusively by osteoclasts making it likely that nearby cells in the vicinity of the BMU are the main sources for the production of these factors which contribute to the coupling. More recently it has been suggested that EphrinB2 (Zhao et al. 2006) and semaphorin (Negishi-Koga et al. 2011), two cell surface regulatory proteins expressed by the osteoclast, may act via direct cell to cell contact with osteoblasts. However, direct cell-cell contact between osteoclasts and osteoblasts at a remodeling site is rarely observed, making it unlikely that they act as coupling factor of bone resorption to bone formation. Direct cell-cell contact between osteoclasts with osteoblast progenitors may exist in the bone marrow or within the constraints of the bone remodeling compartment (Hauge et al. 2001; Kristensen et al. 2014).

Taken together, the question how coupling signals between the osteoclast and osteoblast could be transmitted directly remains elusive, given the fact that bone resorption and bone formation in the BMU are separated in time. The bone remodeling compartment may serve as a way to keep local coupling factors secreted by osteoclasts, osteoblasts, or their progenitors at concentrations sufficiently high to contribute to the coupling. On the other hand, osteocytes located in the vicinity of the Howship lacuna would be in an optimal position to detect the physical changes brought about by the osteoclast-mediated bone resorption (Schaffler et al. 2014). In this model, it is proposed that the increase in mechanical strain caused by the bone resorption process may be detected by the osteocytic network, which may provide the paracrine and endocrine signals required for the initiation as well as completion of the correct amount of matrix production by mature osteoblasts on the bone surface via the extensive canalicular network (McNamara et al. 2006).

It seems hence plausible to propose that the osteocytes could provide the coupling factors between bone resorption and formation and that they provide the final refining control to ensure that sufficient bone is formed by osteoblasts, generated in response to messages from osteoclasts either directly or via other cells residing within the BMC.

In humans, the length of the remodeling cycle, e.g., the time from the initiation of bone resorption until completion of the matrix formation by the osteoblast in the BMU, is roughly 4–6 months. In most laboratory animals, the remodeling cycle is much shorter (see comparative bone biology below). Under pathologic conditions, the length of the remodeling cycle can be altered and the rate of remodeling changed. The rate of bone remodeling is very high during early growth, but continues to slow down until peak bone mass is attained. In adult life, the rate of remodeling is controlled by age and genetics, but also modifiable factors such as nutrition, hormonal status, the level of physical activity, and medications. At menopause, the loss of ovarian estrogen production, which normally acts to suppress osteoclast production and promotes osteoclast apoptosis, results in accelerated bone turnover and bone loss. Men experience a less dramatic increase in bone remodeling and bone loss, and this occurs about a decade later than in women.

In healthy individuals, the bone balance in the remodeling cycle is mildly negative, meaning that the process of bone formation in the BMU does not fully replace the amount of bone that has previously been removed by the osteoclast. As part of the aging process, the mean wall thickness, which is a histomorphometric measure of the bone balance, becomes more negative (Lips et al. 1978). This negative bone balance has serious consequences on the mechanical competence of the skeleton, especially in situations where bone turnover is increased through a disease process (e.g., immobilization, estrogen deficiency, and secondary hyperparathyroidism), leading to accelerated bone loss. Bone remodeling rates double at menopause, triple 13 years later, and remain elevated in osteoporosis (Recker et al. 2004; Seeman 2008).

Over time, the increase in bone turnover will change the material properties by shifting the mineral density profile to a lower level, thereby changing its elastic properties (Roschger et al. 2008). More densely mineralized bone is removed and replaced with younger, less densely mineralized bone, reducing stiffness (Boivin and Meunier 2002; Seeman 2008). Perhaps more importantly, high bone turnover contributes to trabecular thinning and perforation by increasing the likelihood of resorption lacunae coinciding on opposite sides of a trabecular structure (Allen and Burr 2014). But even in the absence of a perforation, an increase in activation frequency will lead to trabecular fragility by causing stress concentrators, which focus stresses to a single point (Hernandez et al. 2006). The negative effects of cavities on bone mechanical performance relative to bone volume are greater when cavities are targeted to regions of high strain and cannot be predicted using standard microarchitecture measures. In computer simulation models in trabecular bone specimens which were run to analyze how apparent level strains are resolved into local tissue-level strains, the importance of trabecular remodeling acting as local stress concentrator was highlighted by the observation that a 10% reduction in relative volume fraction by resorption cavities reduced relative stiffness by approximately 27%, whereas the same loss occurring through trabecular thinning reduced it "only" by approximately 18% (Van der Linden et al. 2001) (Fig. 2.19).

Increased remodeling on the cortical envelope (Haversian canals) resulting from estrogen deficiency, secondary hyperparathyroidism, other pathologies, or aging lead to increased cortical porosity (Martin 1984; Brockstedt et al. 1993; Yeni et al. 1997). Compromised osteoblast function with aging appears to add to this effect by

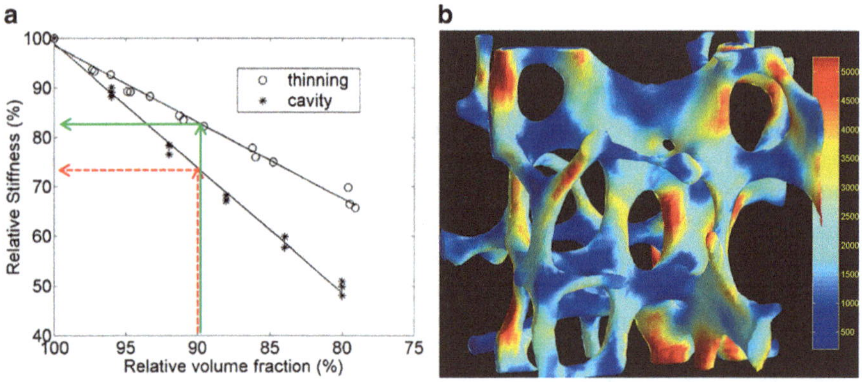

Fig. 2.19 (**a**) Simulation model calculating the change in maximum of apparent Young's modulus resulting from bone loss by resorption cavities (*) and by thinning of trabeculae (o). The reduction in relative stiffness as a consequence of a 10% reduction in relative volume fraction is greater when this deficit is produced by a resorption cavity than by trabecular thinning (Adapted from Van der Linden et al. (2001). With permission from the publisher. (**b**) Original colored SEM image provided by Prof. Harrie Weinans, UMC Utrecht, The Netherlands)

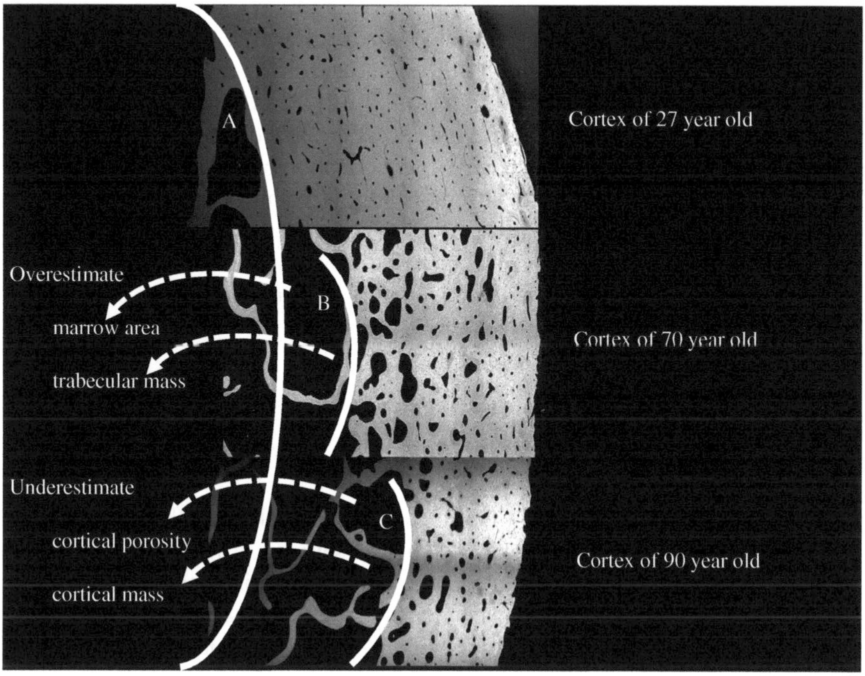

Fig. 2.20 Endocortical trabecularization with aging: Intracortical and endocortical remodeling erode the cortex. The endocortical surface (*white line* A of a specimen from a 27 years old) denotes the true medullary cavity/cortical interface achieved at completion of growth. If the surface of the thinned but still compact-appearing cortex (white line B in a 70 years old or C in a 90 years old) is erroneously described as the endocortical surface, several errors occur by incorrectly apportioning in the cortical fragments and porosity that created them to the seemingly expanded medullary canal (Reproduced with permission from Zebaze and Seeman 2015. Original image provide by Prof. E Seeman, University of Melbourne, Australia)

leading to an increased size of the central canal (larger pores). This effect can be observed particularly in the inner third of the cortical structure leading to the fusion of pores to generate large-sized defects referred to as superosteons and eventually to trabecularization of the endocortical surface (Fig. 2.20) (Bell et al. 2000; Zebaze and Seeman 2015). Clinical evidence suggests that superosteons contribute substantially to cortical porosity and that the elastic modulus of bone decreases approximately as the square root of porosity, suggesting it constitutes an important risk factor for hip fractures (Bell et al. 1999). In addition, from the mechanical point of view, cortical porosity reduces the ability of bone to limit crack propagation so that bone cannot absorb the energy experienced by a fall (Martin 1984; Yeni et al. 1997).

Given the negative effect of increased remodeling on cancellous (trabecular thickness and connectivity, stress raisers, material properties) and cortical bone (cortical thinning and porosity, material properties, endocortical trabecularization), it is not surprising that strategies to reduce the remodeling rate to a more "normal" level have been pursued clinically. For this purpose, various treatment options targeting osteoclast assembly or reducing their function like bisphosphonates and the RANKL-inhibitor denosumab have become available clinically (Kanis et al. 2013; Ferrari et al. 2012; Lekamwasam et al. 2012). Antiresorptive treatments are efficacious in slowing down or arresting cancellous and cortical bone loss, thereby reducing loss of trabecular connectivity, endocortical trabecularization, and trabecular surface-based and intracortical remodeling. Normalization of bone turnover has been shown to lead to a normalization of the mineral density profile. On the other hand, antiresorptive treatment is unable to restore trabecular bone connectivity or normal cortical bone thickness by adding lost bone back at the endocortical envelope.

2.3 Comparative Bone Biology

Animal models provide a more uniform experimental setup and allow for appropriate testing of potential therapies in preclinical settings. Animal models can be divided into three main categories, namely, homologous, isomorphic, and partial, based on how closely they recapitulate human disease processes. Ideally, an animal model would belong to the category of homologous models meaning that it shares the same causes, symptoms, and treatment options. In contrast, isomorphic animal models share the same symptoms but not etiologies, while all other models would be classified as partial models. Due to sometimes limited understanding of disease pathologies and obvious species differences, homologous models are rarely available, and isomorphic models remain the principle research tool for most studies. Animal models can also be divided into spontaneous (naturally occurring) and induced (surgical, genetic, or pharmacological intervention).

Like in other fields of study, genetically modified mice have become the main tool for mechanistic studies of bone metabolism, which have contributed greatly to improve our understanding of the molecular and cellular processes involved in normal bone physiology and disease pathology.

In animal models, bone loss can be induced by a range of interventions: creation of postmenopausal-like state of estrogen deficiency in female animals by surgical bilateral ovariectomy, orchidectomy in male animals, immobilization, or chemically through the administration of high-dose glucocorticoids or of estrogen antagonists such as ICI 182,780 or of aromatase inhibitors (Iwaniec and Turner 2008; Jee and Yao 2001). None of the animal models available to study osteoporosis show the full spectrum of human disease and can be considered a homologous model. Especially the lack of spontaneous fractures, which is perhaps the most important negative consequence of bone loss in humans, is typically only found in few genetically modified mouse strains. However, in animal models, the negative effect of bone loss on bone strength can be evaluated through a number of mechanical testing procedures ex vivo (Turner 2006; Sharir et al. 2008; Wallace 2014). This is relevant since it was realized that in humans, changes in bone mass do not always predict fracture risk. Three-point bending, 4-point bending, and torsion testing are used frequently to assess bone mechanical strength in the diaphysis of the femur, a site where osteoporotic fractures rarely happen in humans. Perhaps more predictive – at least when performed in nonhuman primates – are the vertebral compression tests and femoral head (cantilever testing), sites where fractures are common in humans (Sogaard et al. 1994). Mechanical testing procedures will be discussed in more detail in a separate chapter (see Bone Biomechanics, Chapter 7). In land-dwelling animals, most of the diversity in bone shape and strength is the result of structural diversity, which is obvious at the macroscopic level. This structural diversity is largely due to individual differences in the genetic makeup rather than differences in lifestyle (Pocock et al. 1987; Christian et al. 1989). Also, although variations in material composition of bone exist, the differences are relatively small in land-dwelling animals that are commonly used to study osteoporosis (Keaveney 1998). This chapter will only discuss the most frequently used animals used to study metabolic bone disease, namely, the mouse, rat, and nonhuman primate, and highlight differences and commonalities with humans.

2.3.1 Nonhuman Primates

2.3.1.1 Strengths of the Nonhuman Primate Model

Nonhuman primates (old world monkeys) can be used to study changes in bone remodeling as they occur in human postmenopausal osteoporosis. They are genetically very close to humans and have menstrual cycles and even a natural menopause similar to human females, although their menopause occurs much later chronologically. Some nonhuman primates maintain an upright body posture with a bone biomechanical loading pattern similar to that of humans. At advanced age, they lose bone mass (Jayo et al. 1994; Lees and Ramsay 1999; Colman et al. 1999a, b) and their immune system is similar to humans. In consideration of all these aspects, the nonhuman primate is the most widely used large animal model to evaluate the effects on bone of new drugs designed to treat postmenopausal osteoporosis. Especially with the increase in the need for testing of new biologic agents (Brommage 2001; Smith et al. 2009; Vahle et al. 2015), primates often remain the sole alternative to test therapeutic monoclonal antibodies. Most investigations of bone metabolism have examined rhesus, cynomolgus, or pigtailed macaques, baboons, or African green (vervet) monkeys.

Nonhuman primates have a menstrual cycle very similar to that of women. Ovulatory cycles have been extensively characterized in baboons (Koyama et al. 1977) and in rhesus (Kerber and Reese 1969; Bosu et al. 1973) and cynomolgus macaques (Goodman et al. 1977; Shaikh et al. 1978; Sopelak et al. 1983; Mehta et al. 1986) and involve a follicular phase, ovulation, a luteal phase, and menstruation over 4 weeks. This is different from rats which have a 4- to 5-day estrous cycle that does not involve a true luteal phase unless mating (or pseudopregnancy) occurs.

Also in contrast to rats and similar to humans, a natural menopause occurs in baboons (Chen et al. 1998) and rhesus monkeys (Hodgen et al. 1977; Walker 1995; Gilardi et al. 1997; Colman et al. 1999a; Black et al. 2001) during their third decade of life, which is later than it occurs in humans when taking relative life spans into consideration.

Despite the occurrence of a natural menopause, nonhuman primates must be ovariectomized to induce ovarian hormone deficiency and bone loss of the post-menopausal type. A complete discussion of this model can be found in reviews by Smith et al. (2009) and Brommage (2001). Careful attention to surgical technique is essential, as cynomolgus monkeys continue to cycle normally with as little as 5% of their ovarian tissue present (Danforth et al. 1989). As alternative to surgical induction of the menopause, inhibition of menstrual cyclicity with the use of GnRH agonists and antagonists is also possible (chemical ovariectomy), with the advantage that the loss of ovarian function is reversible (Kenigsberg and Hodgen 1986; Mann et al. 1990; Gordon et al. 1991; Butterstein et al. 1997). Relatively large groups of 15–20 animals are required to provide adequate statistical power to reach valid conclusions which may raise ethical issues (Jerome and Peterson 2001). Special facilities are required resulting in relatively high cost. Following ovariectomy in skeletally mature nonhuman primates, bones exhibit increases in remodeling activity resulting in bone loss reaching a new steady-state bone mass within 8–9 months (Jerome et al. 1997, Smith et al. 2009). Since similar bone changes occur after menopause in women, ovariectomized monkeys provide an excellent model of the early skeletal events following menopause (Brommage 2001; Jerome 2004; Smith et al. 2009).

Another good reason to examine skeletal metabolism and the response to pharmacological treatments in monkeys involve the presence of well-organized osteonal structures in cortical bone. Thus like humans, nonhuman primate cortical bone undergoes osteonal (Haversian) remodeling which in rodents is virtually nonexistent under basal conditions. However, adult human cortical bone is largely composed of Haversian systems occupying about 45% of the total cortical area, whereas in adult macaques the corresponding value is only 5–7% (Schaffler and Burr 1984). This difference may be a reflection of both the longevity and the rate at which cortical bone turnover occurs.

Unlike cortical bone, the volume and structure of vertebral cancellous bone in nonhuman primates are very similar to humans (Vahle et al. 2015). Normative values for fractional bone volume range between 13% and 30% in human iliac crest biopsies and bone volume in intact cynomolgus iliac crest and lumbar vertebrae average about 25% and 27%, respectively (see Table 2.1 below from Vahle et al. 2015). With the exception of trabecular number, which is about 25% higher in nonhuman primates, estimates for other structural cancellous bone parameters in the lumbar vertebrae are consistent with human values. Bone formation rates in macaque lumbar vertebrae tend to be somewhat higher than those in human iliac crest (Vahle et al. 2015).

The fractional intestinal calcium absorption in adolescent pigtailed monkeys has been determined as 37% (Hoffman et al. 1972) and in adult male cynomolgus macaques as 40% of the dietary calcium intake (Lipkin 1998), values which are

Table 2.1 Comparative trabecular bone parameters in humans, macaques, and rats

	Humans		Macaques[a]		Rats[b]	
Parameter	Healthy subjects[c]	Postmenopausal women[d]	Intact	OVX	Intact	OVX
BV/TV (%)	13–28	14–30	26–28	21–25	19–26	~6
Tb.Th (μm)	100–225	93–105	107–127	97–122	40–60	53
Tb.N (mm^{-1})	1.0–1.8	1.2–2.0	2.1–2.6	2.1–2.5	3.89–5.62	1.10
Tb.Sp (μm)	800–1500	480–850	282–359	306–388	133–286	883
MS/BS (%)	5–20	1.0–13.5	11–14	13–22	4.6–15.4	8.2–29.4
MAR (μm/day)	0.83–0.98	0.5–2.0	0.61–0.66	0.65–0.80	0.62–1.56	0.68
BFR/BS ($\mu m^3/\mu m^2$/day)	7–15	1–25	26–31	31–67	103–232	85
Rm.P (days)	120–180	22–131	153	67	28	~20
Ac.f (cycles/year)	1–2	0.15–0.43	0.57–1.43	1.57–2.74	~0.7	0.9–1.6

BV/TV cancellous bone volume, *Tb.Th* trabecular thickness, *Tb.N* trabecular number, *Tb.Sp* trabecular separation, *MS/BS* mineralizing surface, *MAR* mineral apposition rate, *BFR/BS* bone formation rate, tissue level, *Rm.P* remodeling period, *Ac.f* activation frequency

[a]Based on measurements of lumbar vertebrae from Jerome et al. (1994, 1999), Stroup et al. (2009), Jerome CP, 1999, personal communication, and Cabal et al. (2013). Data either are averaged from several sources or presented as ranges.

[b]Based on measures from both the proximal tibia and lumbar vertebra in 3- to 19-month-old female rats. Mori et al. (1992), Jee and Li (1990), Chen et al. (1995a), and Ma et al. (1995)

[c]Based on iliac crest biopsies from 25 healthy subjects, age 19–46 years (for static data); dynamic data are normal values for 13 white adults (Weinstein 1992)

[d]Based on iliac crest biopsies from Recker et al. (1988) and Hauge et al. (1999)

similar to those reported for human adolescents. Interestingly, serum levels of calcitriol, the active metabolite and hormonal form of vitamin D, are five- to tenfold higher in baboons and macaques (Vieth et al. 1987; Knutson et al. 1995) than humans and are higher than in most other animals. The binding affinity for calcidiol, the circulating vitamin D binding protein in rhesus monkeys, appears to be normal, and the reason for this increase remains to be elucidated (Vieth et al. 1990).

2.3.1.2 Limitations of the Nonhuman Primate Model

Pigtailed, cynomolgus, and rhesus macaques raised in colonies attain peak bone mass at the ages of approximately 7, 9, and 10–11 years, respectively (Ott et al. 1997; Jayo et al. 1994; Champ et al. 1996; Cerroni et al. 2000). However, the availability of colony-raised monkeys is clearly limited, and maintaining monkeys in captivity in advance of starting a study increases its expense and raises putative ethical concerns. As an alternative monkeys imported from the wild can be used. A consistent observation is that spine bone mineral density (BMD) values for adult female monkeys imported from the wild increase several percent during the first several years of captivity (Jerome et al. 1995), an effect that can only be mitigated by introducing an extended period of housing prior to the initiation of the study (Jerome et al. 1995; Binkley et al. 1998). These spontaneous gains in spine BMD in control, sham-ovariectomized monkeys are a complication in studies on osteoporosis because, although the ovariectomized monkeys lose spine BMD relative to controls, they may show only small declines or even gains in spine BMD during the experiment. Yet another complication is that the individual ages of wild-caught monkeys are unknown, the only method of verifying maturity is radiographic evidence of closed growth plates. However, even the exclusion of adolescent monkeys with open growth plates from studies does not prevent gains in spinal BMD during the initial period of captivity. The factors responsible for increased spine BMD during early years of captivity have not been identified conclusively, but they result likely from multiple causes. They are likely to include improvement of nutritional status (protein and calcium) and changes in other environmental factors and the recovery from a recent episode of reproduction including weaning.

Monkeys are mostly housed in groups allowing for interactions with peers and formation of hierarchical social groups. In addition, the monkeys have sufficient space for adequate physical exercise. However, group housing, though mimicking the settings in the animals' natural environment, exposes the subordinate individuals to continual psychological stress from losing fights and harassment. In studies involving intact, non-ovariectomized monkeys, the subordinate animals experience ovarian dysfunction and develop more severe atherosclerosis (Kaplan et al. 1991; Shively et al. 1997). Although individual housing eliminates these potential complications, the cages often do not allow sufficient room for movement and weight-bearing exercise, with the possible complication that bone loss due to inactivity might be encountered beside ethical concerns. A reasonable compromise might be pair housing, which allows psychosocial enrichment and a more easily established social hierarchy (Brommage 2001).

2.3.1.3 Methods and Study End-Points Available for Nonhuman Primates

Recent reviews (Vahle et al. 2015; Smith et al. 2009) provide a comprehensive source of information regarding the use of nonhuman primates in skeletal biology.

Ovariectomized monkeys consistently exhibit bone loss, as measured by spine BMD, when compared to sham ovariectomized controls (Jerome 2004). Like in humans, methods used to monitor bone changes associated with disease progression or pharmacological interventions noninvasively include determinations of BMD by dual-energy X-ray absorptiometry (DEXA) and peripheral quantitative computed tomography (QCT) (Jerome et al. 1997; Hotchkiss 1999; Cabal et al. 2013). Bone mineral density is monitored using DEXA bone scanners (e.g., Hologic QDR 2000 or Discovery A densitometers, Bedford, MA) at clinically relevant sites, i.e., the spine, hip, and radius (Smith et al. 2009). PQCT (XCT Research SA, Stratec Medizintechnik, Pforzheim, Germany) is a useful tool in preclinical studies providing complementary bone densitometry data, in conjunction with the DEXA data, and has been previously reported to be a valuable tool for use in nonhuman primates (Dickerson and Hotchkiss 2008; Hotchkiss 1999). In contrast to DEXA, pQCT is not influenced by changes in body composition. It allows separate analysis of trabecular and cortical bone regions to detect BMD changes with greater sensitivity and provides information on geometric parameters such as periosteal circumference, endosteal circumference, and cortical thickness.

In addition, the same serum and urine biomarkers indicative of bone formation or bone resorption that exist in humans are available for nonhuman primates. These include the serum markers of bone formation, serum bone-specific alkaline phosphatase (BAP), osteocalcin and procollagen I intact N-terminal (PINP), and markers of bone resorption N-telopeptides (urinary NTx), serum C-terminal cross-linked peptide of type I collagen (serum CTx), and tartrate-resistant acid phosphatase 5b (TRAP5b).

Ovariectomy-induced bone loss is associated with an elevation of bone remodeling, and markers of bone turnover have been successfully employed including serum activities of TRAP5b and both total and bone-specific alkaline phosphatase, serum levels of osteocalcin, and assays measuring the urinary excretion of various degradation products of bone collagen fragments. Recent experience indicates that bone resorption can be reliably estimated by analyzing serum levels of CTx, thereby avoiding the need to collect urine (Jerome et al. 1994; Register and Jerome 1996; Brommage et al. 1999).

Bone biopsies for chemical and histomorphometric analyses can be taken from monkeys at the start of a study to obtain baseline values or during a study to obtain intermediate data. As is the case in humans, the iliac crest is the easiest and most reliable site to obtain a biopsy with minimal complications, and several technical procedures to obtain such biopsies have been published (Goodwin and Jerome 1987; Nogues and Milhaud 1988; Klein et al. 1991; Inskip et al. 1992). Limited experience is also available for other sites indicating that the rib (Ott et al. 1999; Binkley et al. 1999), vertebral body (Hermann and Smith 1985), and humerus (Ott et al. 1999) are possible. Rib biopsies provide the opportunity to examine osteonal bone metabolism, which is limited in iliac crest biopsies. Histological processing of plastic embedded sections of undecalcified bone (Brommage and Vafai 2000), to conduct static and dynamic histomorphometric measurements of bone structure and activity (Jerome et al. 1994), and bone immunohistochemistry (Carlson et al. 1993; Johnson et al. 2000) are well-established procedures. Other sites which are typically evaluated become available only after sacrifice of the animals including the lumbar vertebral body, femoral neck, tibial shaft, and rib (Smith et al. 2009).

Additional tests to assess bone quality include collagen cross-link analyses, mineral density fractionation to measure the degree of mineralization, and spectrophotometric tests such as FTIR. The use of specimens from nonhuman primate studies to perform these specialized tests provides excellent opportunities to further elucidate the underlying causes of bone fragility and to better understand the effect of pharmacological interventions.

Serum hormone analysis (Hotchkiss et al. 1998) can be performed, and oftentimes, reagents designed for use in human samples can be used successfully in nonhuman primates if no specific test kit is available. Nonetheless, caution is always advised, and all human assays should be adequately validated before being "transferred" for use in monkey samples.

Just like other animal species, nonhuman primates do not develop spontaneous fractures as a consequence of bone loss after ovariectomy. For this reason, bone biomechanical measurements on excised bones (Jerome et al. 1997) have to be performed as a surrogate instead. Measures of biomechanical competency are important end-points in the assessment of bone quality. Strength testing on a materials testing machine is typically performed on a long bone, usually the femur (in 3-point bending), at the femoral neck (shear test) and at the lumbar spine (vertebral compression test) (Smith et al. 2009). Importantly, these basic tests allow evaluation of disease or a treatment at clinically relevant sites. Bone strength in the spine and femoral neck is decreased by ovariectomy (Jerome et al. 1997).

For more complete information on the use of nonhuman primates in bone research including special considerations on experimental design, the reader is referred to Chapter 1.

2.3.2 Rats

Among all animal species, the rat has become the preferred animal for many researchers to study bone disorders (Jee and Yao 2001; Iwaniec and Turner 2008). Its skeleton has been studied extensively, and although there are distinct differences compared to human physiology, these can be mostly overcome through appropriate study design, or at least taken appropriately into account in the interpretation of data generated experimentally. In contrast to humans, rats do not experience natural menopause. Under laboratory conditions, rats will maintain regular estrous cycles of 4–5 days duration. Unlike seasonal breeders (sheep, dogs), the rat skeleton is exposed to regular fluctuations in gonadal steroids and, like the human skeleton, more sensitive to the loss of ovarian hormones. Rats cycle through four distinct stages of the estrous cycle, the pro-estrous (12–18 h), estrous (10–20 h), met-estrous (12 h), and di-estrous (48 h). Not surprisingly, the most common approach for inducing bone loss in the rat is via surgical intervention on the gonadal axis, namely, ovariectomy (Bagi et al. 1993; Jee and Yao 2001; Lelovas et al. 2008) and orchidectomy (Erben et al. 2000; Blouin et al. 2007). Alternative models to surgical ovarie-/orchidectomy are targeting the endocrine system leading to osteoporosis that is reversible after withdrawal. These reversible methods include the administration of gonadotropin-releasing hormone agonists (Goulding and Gold 1989), estrogen receptor antagonists (Gallagher et al. 1993), and aromatase inhibitors (Gasser et al. 2006). The administration of gonadotropin-releasing hormone agonists creates a hypogonadotrophic-hypogonadal model, whereas estrogen receptor antagonists and aromatase inhibitors produce a hypergonadotrophic-hypogonadal model of osteoporosis.

Bone loss can also be achieved through surgical intervention at a higher level of the hypothalamic-pituitary-gonadal axis by performing a hypophysectomy (Chen et al. 1995b; Yeh et al. 1995), a surgical procedure that is more traumatic and complicated, and affects broader and unintended target tissues than needed. More popular are various procedures that cause immobilization/disuse induced and thus regional bone loss (Blouin et al. 2007; Jee and Ma 1999; Bagi et al. 1993). Immobilization models in rats include hind-limb immobilization, limb taping, nerve resection, tenotomy, limb casting, and reversible immobilization through injections of botulinus toxin (Jee and Yao 2001; Lelovas et al. 2008).

2.3.2.1 Strengths and Limitations of the Rat Model

In the past, many scientists were convinced that in the rat skeleton, the prevailing activity is modeling, and consequently the rat was not an appropriate model for human osteoporosis. However, this perception was based largely on data derived from the use of skeletally immature, growing animals, which were inappropriately compared to human adult bones. Although rats reach sexual maturity at the age of 2.5 months, their skeleton is considered fully mature only after the age of 10 months. In female rats, bone growth in the proximal tibia and distal tibia epiphysis stops at the age of 15 and 3 months, respectively, whereas in lumbar vertebrae, it continues for up to 21 months (Jee and Yao 2001). However, in analogy to the human skeleton, the rat skeleton shows a gradual transition from modeling to remodeling that is related to age progression and cessation of longitudinal bone growth in both cancellous and cortical bone. In the cancellous bone of the lumbar vertebrae, this transition is evident from the age of 3 months onward, whereas in the proximal tibia metaphysis, this transition takes place from 6 to 9 months of age (Iwaniec and Turner 2008). After the age of 10 months, the bone growth rate for the proximal tibia epiphysis is less than 3 μm/day, and it stops altogether after the age of 15 months. If experimentation in female rats starts around 10 months, e.g., the time when peak bone mass is achieved, the total longitudinal bone growth adjacent to the epiphyseal plate of the tibia until cessation of growth will be less than 0.5 mm. In contrast in male rats, the closure of the epiphyses of many long bones is delayed with many of them remaining open past 30 months. At the endocortical envelope in vertebral bodies, this gradual transition from modeling to remodeling evolves at the age of 3–6 months and at the endocortex of the proximal tibia metaphysis at 9–12 months (Jee and Yao 2001).

A potential drawback to the use of rat models for osteoporosis is the lack of Haversian remodeling in the rat skeleton. In humans, increased Haversian remodeling is the main cause of cortical porosity, but rats lack a well-developed Haversian remodeling system. In the rat skeleton, cortical bone gain occurs in the periosteum, and cortical bone is lost at the endosteum (Turner 2001). Larger animal models such as rabbits, dogs, and especially primates are considered more appropriate for the study of Haversian remodeling. Despite this drawback, ovariectomy of skeletally mature rats leads to a condition similar to the menopause in humans. Like in humans, cancellous and endocortical bone loss is mediated by an increase in the overall rate of bone remodeling and by altering the balance between bone formation and bone resorption (Jee and Yao 2001).

The ovariectomized skeletally mature rat is the most commonly used animal model to study postmenopausal osteoporosis. After the ovariectomy, accelerated bone remodeling leads to rapid loss of cancellous bone, eventually reaching a steady state, where resorption and formation are balanced (Boyd et al. 2006). Statistically significant bone loss is seen in the proximal tibia metaphysis after 14 days and near steady state is reached from around 90 days onward post-ovariectomy (Wronski et al. 1988; Wronski et al. 1989; Brouwers et al. 2008; Gasser et al. 2006). In lumbar vertebral bodies, significant bone loss is observed after 60 days (Wronski et al. 1990) and in the femoral neck after 30 days (Li et al. 1997), steady state is reached for both sites after 270 days. It should be noted that ovariectomy does not induce bone loss in the epiphyses of long bones, the distal tibia metaphysis, or the caudal vertebrae (Li et al. 1996; Miyakoshi et al. 1999).

Losses of endocortical as well as cancellous bone are the primary causes of postmenopausal osteoporosis in humans, whereas intracortical remodeling-induced bone loss in the Haversian system seems to play a less important role (Iwaniec and Turner 2008). Given the ethical considerations and the high cost of acquisition and maintenance, reduced availability in experimental centers associated with large animal models of osteoporosis, the lack of the Haversian remodeling in the rat skeleton is a shortcoming that can often be accommodated (Turner 2001).

2.3.2.2 Methods and Study End-Points Available for Rats

In general, the methods used in the evaluation of bone mass, architecture, and metabolism in the rat skeleton are the same as those used for humans. Measurements of calcium, phosphorus, and magnesium in blood and urine can be obtained from the rat model. In analogy to humans, the most commonly used biochemical markers representative of "global" bone turnover can be measured in rats including peptides originating from osteoblasts indicative of bone formation (BAP, osteocalcin) or osteoclastic resorption (TRAP5b). In addition, organic compounds released during the synthesis (P1NP) and resorption of bone matrix (urinary or serum CTX-1) (Loeb 1999) can easily be measured.

Like in clinical studies, DEXA and pQCT with small animal software can be used to measure both total and regional bone density and get limited cortical structural parameters in rats and mice (Turner et al. 2001; Gasser and Willnecker 2012; Breen et al. 1998). The recent arrival of high-resolution in vivo microCT provides the ability to noninvasively monitor microarchitectural changes in cancellous bone in individual animals over time (Boyd et al. 2006; Gasser et al. 2006; Bouxsein et al. 2010; van't Hof 2012). Comparison of two-dimensional histomorphometry and three-dimensional microCT demonstrates the superiority of microCT in revealing early changes in bone architecture (Jiang et al. 2005).

Because spontaneous fractures do not occur in aged or ovariectomized rats, data regarding mechanical competence of bones have to be obtained ex vivo on excised bones with mechanical testing procedures. In rats, the most common test procedures are the 3- or 4-point bending of the femur diaphysis, the vertebral compression, and the femoral head test (cantilever testing) (Sogaard et al. 1994). The nature and clinical relevance of the mechanical testing procedures available in rats, which do not reproduce the situation under which spontaneous fractures occur in humans, will be discussed in more detail in a separate chapter (see Bone Biomechanics, Chapter 7). Last but not least, histomorphometric analysis provides a two-dimensional study of bone volume and architecture at a very high resolution unmatched by any of the noninvasive devices. Histomorphometry accurately evaluates bone architecture and indices of bone fragility independently of bone mass. Parameters measured by histomorphometry include the numbers of osteoblasts, osteoclasts, osteocytes, and active osteoblasts relative to bone perimeter; trabecular thickness, number, and separation; and many others (Meunier 1988). The use of fluorochrome labeling is especially valuable in the evaluation of dynamic bone changes (Frost 1964). In rats, this invasive assessment can be carried out ex vivo at every site that is relevant for human disease (Erben and Glösmann 2012). This is in contrast to clinical studies, where the tissue sampling is limited to the iliac crest, a site which may not be fully representative of the actual location where the majority of the fractures occur such as the vertebrae, the hip, and the forearm.

Comparative data for histomorphometric parameters in healthy and ovariectomized rats with human and primate data has been published (Vahle et al. 2015). One of the main differences which is readily apparent when comparing histomorphometric parameters between rat, macaques, and humans is the remodeling period (Table 2.1), e.g., the time that is required for completion of one whole remodeling event. In healthy rats, the remodeling period is on average 28 days, compared to approximately 153 days in macaques and 120–180 days in healthy human subjects. The length of the remodeling cycle is influenced by age, disease, or drug treatment, and some evidence suggests that differences exist between locations where it is measured. Estrogen deficiency, which is known to lead to increased activation of remodeling, also results in shortening of the remodeling period to 20 days in ovariectomized rats, 67 days in ovariectomized macaques, and 22–131 days in postmenopausal women.

Compared to healthy human subjects, the remodeling period in the rat is roughly four to six times shorter which explains the fact that skeletal changes as a consequence of perturbances by disease or pharmacological interventions evolve more rapidly in the rodent. This is convenient since it allows reducing the duration of rat studies under most experimental conditions by a factor of 6.

In summary, after the age of 12 months, remodeling becomes the predominant activity in cancellous and cortical bone in the lumbar vertebral body and proximal tibial metaphysis in the rat skeleton (Jee and Yao 2001; Erben 1996). The modeling-to-remodeling transition is associated with reduction of longitudinal bone growth to very low rates. To conduct research on new potential modalities for the treatment of postmenopausal osteoporosis in rats, the bone site and age of the animal must be such that remodeling is the predominant activity. Despite the limitations discussed above, the skeletally mature rat (>10 months in female rats) is generally considered an appropriate animal model for the research of postmenopausal and immobilization osteoporosis (Turner et al. 2001; Iwaniec and Turner 2008; Jee and Yao 2001).

The broad availability of noninvasive methods to study changes in bone mass, cancellous and cortical bone architecture, and biochemical markers longitudinally, combined with invasive end-points including histomorphometry and bone mechanical testing, makes the rat a very versatile tool for the evaluation of preventive or therapeutic strategies for the treatment of osteoporosis.

2.3.3 Mice

Mice are often employed in bone studies because of their availability, ease of handling, high reproductive rates, and low cost of use (Jilka 2013). Owing to technical feasibility, the mouse is to date the predominant mammalian species for targeted genetic modification, and it has assumed a crucially important role toward dissecting the molecular pathways, which are critical to bone homeostasis. In addition the study of mouse strains has contributed to a better understanding of genetic contributions to peak bone mass and age-related bone loss (Beamer et al. 1996). Mice also successfully replicate the skeletal phenotypes related to several genetic disorders in humans. In fact, the ability to manipulate gene expression in mice in a cell-specific fashion has allowed scientists to address fundamental questions underlying the regulation of bone cells based on observations in the culture dish and to move them into the in vivo situation. The most commonly used technique in mice has been to date genetic modification such as transgenesis, knock-outs and knock-ins, large-scale mutagenesis, and conditional gene modifications (Lyons 2013; Kan 2013). In contrast, the surgical induction of a postmenopausal bone loss by bilateral ovariectomy, as well as immobilization- or glucocorticoid-induced bone loss, is technically more challenging in mice and is thus predominantly performed in rats and primates. This section will focus on mouse biology and the challenges faced when the mouse is used as a preclinical model for postmenopausal- or age-related osteoporosis (Jilka 2013). For a more in-depth discussion of genetically manipulated mice, the reader is referred to the recent reviews written by Lyons (2013) and Kan (2013).

The ovulatory or estrus cycle of female mice has a length of 4–6 days. The estrus cycle can become synchronized in female mice housed together continuously and can be suspended in the absence of exposure to male pheromones. Due to their short 3-week gestation period, high reproductive capacity (five to six animals per litter), and ease of genetic manipulation, standardized outbred and inbred strains of laboratory mice have become very popular in almost every aspect of research, including skeletal biology.

Starting in the early twentieth century, over 34 well-defined outbred strains have been established with the aim to preserve genetic heterogeneity, analogous to the genetic variation in human populations (Chia et al. 2005). They are thus particularly suited for genetic studies and as the starting point for developing models of human disease. As a disadvantage, the genetic variability of outbred strains contributes to differences in phenotype and response to treatment among individuals, just as is observed when studying human populations. As a consequence, larger numbers of animals per treatment group are required to achieve appropriate statistical power.

In contrast, inbred strains of mice were derived from outbred strains by intensive brother-sister mating to produce mice with practically identical genomes. There were 450 known inbred strains in 2000 (Beck et al. 2000). By convention (Committee on Standardized Genetic Nomenclature for Mice 1952), the standard definition of an inbred strain assumes two basic requirements: (i) 20 or more consecutive generations of full-sib mating (or its genetic equivalent in terms of other relationships) and (ii) all members of the strain derived from a single breeding pair of individuals in the 20th or a later generation. These criteria theoretically assure a minimum level of inbreeding of 98.6% or, alternatively, 2% of the genetic variance existing in the base generation. Inbred mice are thus remarkably isogenic, guaranteeing the reproducibility of research experiments in many cases. The mouse phenome database from Jackson Lab contains body weight, body composition, and bone mineral density data assessed by DEXA for 32 strains for ages 6, 12, and 20 months (Ackert-Bicknell et al. 2008). In addition, Beamer and colleagues (1996) applied pQCT, to analyze the femur from 11 inbred strains of female mice to determine the extent of heritable differences at 12 months. In their analysis, C3H/HeJ mice ranked highest in femur density compared to C57BL/6J mice which displayed the lowest density value of the 11 inbred strains analyzed. In addition, bone mineral density in the femur was determined longitudinally in a subset of four strains (C3H/HeJ, DBA/2J, BALB/cByJ, C57BL/6J) at 2, 4, and 8 months. These four strains are representative of high, middle, and low total femur densities. The greatest difference in femur density was found between C57BL/6J and C3H/HeJ females. DBA/2J and BALB/cByJ also had densities significantly lower than C3H/HeJ, although the relative differences were not as great as for C57BL/6J. The difference in femur BMD between strains was visible at 2 weeks and was maintained pretty much at a constant level after peak bone mass was achieved at 4 months of age. With the exception of DBA/2J mice which stopped growing at 8 months, femur length continued to grow after 4 months up to the study end-point at 12 months, but at a much slower rate. Taken together the results indicate that one of the most commonly used mouse strains in skeletal biology, the C57BL/6J strain, exhibits the lowest bone mineral density which may represent a technical challenge when conducting experiments with this mouse strain (see section on limitations below).

In the axial skeleton, mouse strain-related differences in vertebral biomechanics and histomorphometry were assessed in inbred mouse strains C3H/HeJ, C57BL/6J, and DBA/2J (Akhter et al. 2004). Despite the greatest BMC and areal BMD in C3H/HeJ mice, the lack of strain-related differences in vertebral body strength data suggests that the biomechanical properties may be affected by the bone distribution and/or complex combination of cortical and cancellous bone at this site.

With the exception of 2 strains that die before 1 year of age from lymphoma or sarcoma, the median life span among 30 inbred mice ranges from 476 to 964 days

(Yuan et al. 2009). The median life span of C57BL/6 mice is 866 days for females and 901 days for males, similar to the 880-day median life span of UM-HET3 mice (Yuan et al. 2009). Due to the development of age-related pathologies which includes susceptibility to carcinogenesis, metabolic, immunologic, or other abnormalities, short-lived mice are not useful for studies of normal aging. Thus, the relatively long life span of C57BL/6 mice makes them attractive for aging studies.

2.3.3.1 Strengths and Limitations of Mouse Models

Unlike the human skeleton, the murine skeleton continues to grow slowly after puberty (Beamer et al. 1996), osteonal remodeling is absent in younger mice, but cortical porosity increases with age. Cancellous bone loss initiates at approximately 4 months of age soon after peak bone mass is reached and proceeds throughout adult life into senescence (Glatt et al. 2007).

Cancellous bone turnover in the distal femur in mice is approximately 0.7% per day, and each remodeling cycle takes about 2 weeks to complete (Weinstein et al. 1998). In humans, turnover is about 0.1% per day as measured in the iliac crest (Parfitt 2002b), and each remodeling cycle takes 6–9 months to complete (Parfitt et al. 1997). The more rapid pace of events in the murine skeleton is in keeping with the higher metabolic rate of small animals. Metabolic rate among animals varies in proportion to the ¾ power of body mass (Gillooly et al. 2001). This difference may explain at least in part the need to administer higher concentrations of biological agents to mice than to humans to achieve an equivalent physiologic effect.

Size is probably the most limiting factor in the mouse model affecting almost all standard end-points that are routinely measured in larger mammals and humans. As in humans, age-related bone loss in long bones of mice begins shortly after peak bone mass is achieved which in mice is at around 4 months of age (Lazner et al. 1999; Glatt et al. 2007). The cancellous bone volume in long bones in one of the most commonly used mouse strains for genetic manipulations, the C57BL/6, is as low as 3% in 7-month-old females which achieve peak bone mass at roughly 4 months of age (Iwaniec et al. 2006; Glatt et al. 2007). A μCT study on the distal femur of female and male C57BL/6 mice at various ages between 2 and 20 months of age revealed a 94% (female) and 56% drop (male) in trabecular bone volume from their initial value of 17.7% and 24.1%, respectively. This age-related drop of 56% in cancellous bone volume in male C57BL/6 mice is well in line with the paper published by Halloran, reporting a 60% decrease between 6 weeks and 24 months of age (Halloran et al. 2002). Cancellous bone loss appears to result predominantly from a reduction in trabecular numbers and connectivity, but not a change in their thickness (Glatt et al. 2007; Halloran et al. 2002). This is different from humans, where age-related changes in cancellous architecture are characterized by decreased trabecular number, thickness, and connectivity (Seeman 2002). Murine trabecular structures are thinner than human trabeculae (40–50 μm vs. 120–150 μm, respectively), and it has been hypothesized that it is thus more likely that they get disconnected during unbalanced bone remodeling, with a consequent decrease in number and connectivity. A decline in wall thickness contributes to age-related cancellous bone loss in both mice and humans.

In mice of both genders, the majority of the age-related drop in cancellous bone volume had already occurred at 6 months, but cancellous bone loss continued to decline thereafter, albeit at a lower rate. The small amount of cancellous bone present at 4 months when mice are considered adult (especially in female mice), and the rapid age-related decline to 6 months makes a robust detection of changes in cancellous bone with noninvasive X-ray-based technologies (pQCT and microCT) challenging. The same applies to the measurement of cellular and dynamic histomorphometric parameters (osteoclast and osteoblast perimeter and number, mineralizing surface, and mineral apposition rate) that require a minimal cancellous bone perimeter for accurate detection, which makes it difficult in mice to generate a robust readout with a trabecular bone volume of 5% or less.

In male C57BL/6J mice, cortical bone thickness increased between 6 weeks and 6 months of age and then decreased continuously to 24 months (−12%), while cortical bone area remained constant between 6 and 24 months (Halloran et al. 2002).

Like in humans, OVX in the mouse does result in accelerated bone turnover (Beamer et al. 1996), but the magnitude is highly strain dependent and less consistent than in rats (Lazner et al. 1999). Detection of OVX-induced cancellous bone loss can be very challenging, since the trabecular bone volume in the metaphysis of the appendicular skeleton at 4 months, when OVX is carried out, is typically below 10% and the decline in trabecular bone has to be "detected" against the background of the rapid age-related decline.

The limitations discussed above can be partially addressed by investigating lumbar vertebral bodies which have a higher cancellous bone volume, even though the axial skeleton in mice cannot be measured in vivo with noninvasive techniques, thus limiting the readout to a single time-point after necropsy. Trabecular bone volume has been shown to decline by 52% in aging female C57BL/6 mice between 2 and 20 months starting from 28.6% (Glatt et al. 2007). In male mice, the trabecular bone volume at 2 months is 29.1%, declining by 26% at 20 months due to aging (Glatt et al. 2007). Following OVX in skeletally mature C57BL/6 mice, vertebral cancellous bone volume dropped from 12% to 9% in 3 months, representing a 27% change (Iwaniec et al. 2006).

2.3.3.2 Methods and Study End-Points Available for Mice

Several noninvasive methods are available for use in mice. This includes radiological techniques such as plain X-ray (e.g., Faxitron) (Bassett et al. 2012) and DEXA. Although plain x-rays are adequate to detect gross morphological changes in the skeleton of mice, they do not have sufficient resolution and sensitivity to detect subtle changes in bone architecture or to pick up compartment-specific changes in bone density. Since most of the changes occur in cancellous bone, pQCT which allows separate monitoring of cancellous and cortical bone offers greater sensitivity over DEXA (Gasser and Willnecker 2012). In recent years, microCT scanners like the vivaCT40 (Scanco) and the SkyScan 1176 have become the method of choice for skeletal phenotyping and to study skeletal growth and aging in mice (Van't Hof 2012; Bouxsein et al. 2010). In addition to the in vivo microCT scanners, a number of highly versatile desktop systems such as the Skyscan 1275 are available, and nanoCT systems, like the μCT50 from SCANCO, the MicroXCT-200 from Xradia, or the GE Nanotom S nanoCT system, allow nondestructive ex vivo analysis of canalicular structures or osteocyte lacunae in cortical bone at submicron pixel size.

In recent years, optical imaging has evolved as a valuable technique for visualizing and quantifying biological processes in anesthetized mice. Both bioluminescence imaging and fluorescence imaging can be used to study cell- and tissue-specific promoters and to follow trafficking, differentiation, and the fate of reporter gene-expressing cells or biological processes such as apoptosis, protein-protein interactions, angiogenesis, proteolysis, and gene transfer in mice (Snoeks et al. 2012).

Bone histomorphometry is an indispensable tool for assessing the mechanism by which bone diseases occur, studying the effect of pharmacologic interventions, and especially determining the bone phenotype of transgenic mice. The article written by Erben and Glösman provides guidance on the tissue processing and histomorphometric evaluation for rodent bones (Erben and Glösmann 2012). Similarly, Kramer and colleagues described in detail how temporal and spatial patterns of gene expression can be studied in mouse bones by in situ hybridization (Kramer et al. 2012).

Assessment of mouse bones by mechanical testing is a critical step when evaluating the functional effects of an experimental perturbation (Jämsä et al. 1998; Schriefer et al. 2005). Recently, Jepsen and colleagues published detailed practical guidelines for systematically evaluating phenotypic changes in the diaphyses of long bones in mice (Jepsen et al. 2015). Importantly, the recommendation of minimum reportable standards for testing conditions and outcome variables, if adopted, will improve the comparison of data across studies. Typical in vivo loads experienced by mice include bending and torsion. For this reason, whole-bone mechanical tests of mouse long bones are most often performed in bending (Schriefer et al. 2005; Jämsä et al. 1998), but can also be performed in tension, compression (Akhter et al. 2004), and torsion. Descriptions of biomechanical methods, including for vertebral compression testing, that can be used in mice to determine whether a genetic or environmental perturbation affects bone strength have been published recently (Smith et al. 2013). The synthesis of morphological, tissue-level and whole-bone mechanical properties of long bones is a crucial step in the characterization of transgenic mouse strains (Jepsen et al. 2015; Jämsä et al. 1998; Schriefer et al. 2005).

In conclusion, mice will continue to play an important role toward a better understanding of the molecular mechanisms underlying skeletal biology. Their availability, ease of handling, high reproductive rates, and low cost provide additional arguments for their use (Jilka 2013). Full appreciation of the differences in the physiology of the human and murine skeletons is required to distinguish clinically relevant findings from mouse-specific findings. In addition, differences among mouse strains and genders must be considered. Like rats, mice lack osteonal cortical remodeling, but they clearly exhibit an accelerated increase in cortical porosity that may be informative for the situation in humans. As in humans, cancellous bone loss initiates shortly after peak bone mass is achieved; it results from accelerated bone turnover, but in contrast to humans, the predominant feature is a decrease in trabecular number and connectivity with little effect on trabecular thickness. Some mouse strains may continue to show very low rates of longitudinal bone growth even at old age, but the rate is so slow that bone loss due to aging can be detected in the presence of an open but largely inactive growth plate. Despite some species differences, studies using mice will undoubtedly continue to contribute significantly to our understanding of skeletal biology in humans.

Acknowledgements We wish to thank Alan Abrams, Novartis Institutes for BioMedical Research (Cambridge, USA) for contributing the scientific artwork for Figs. 2.1, 2.2, 2.6, 2.8, 2.9, 2.11, 2.14, 2.16a, 2.17, and 2.18. Our sincere thanks go to Nathalie Gris-Accard, Novartis Institutes for BioMedical Research (Basel, Switzerland) for providing histologic images for Figs. 2.3 and 2.7a.

References

Ackert-Bicknell C, Beamer WG, Rosen CJ, Sundberg JP. Aging study: bone mineral density and body composition of 32 inbred strains of mice.MPD;Ackert1. Mouse Phenome Database Web Site. Bar Harbor: The Jackson Laboratory. 2008. http://phenome.jax.org/db/q?rtn=projects/details&id=250.

Akhter MP, Otero JK, Iwaniec UT, et al. Differences in vertebral structure and strength of inbred female mouse strains. J Musculoskelet Neuronal Interact. 2004;4(1):33–40.

Allen MR, Burr DB. Bone modeling and remodeling. In: Burr DB, Allen MR, editors. Basic and applied bone biology. San Diego: Academic, 2014. p. 75–90.

Anderson RE, Schraer H, Gay CV. Ultrastructural immunocytochemical localization of carbonic anhydrase in normal and calcitonin-treated chick osteoclasts. Anat Rec. 1982;204(1):9–20.

Bagi CM, Mecham M, Weis J, et al. Comparative morphometric changes in rat cortical bone following ovariectomy and/or immobilization. Bone. 1993;14:877–83.

Balemans W, Ebeling M, Patel N, et al. Increased bone density in sclerosteosis is due to the deficiency of a novel secreted protein (SOST). Hum Mol Genet. 2001;10(5):537–43.

Baron R, Kneissel M. WNT signaling in bone homeostasis and disease: from human mutations to treatments. Nat Med. 2013;19(2):179–92.

Bass SL, Saxon L, Daly R, et al. The effect of mechanical loading on the size and shape of bone in pre-, peri- and postpubertal girls: a study in tennis players. J Bone Miner Res. 2002;17(12):2274–80.

Bassett JHD, Van der Spek A, Gogakos A et al. Quantitative X-ray imaging of the rodent bone by Faxitron. In: Helfrich, MH and Ralston, SH, editors. Methods in molecular medicine, Vol. 816: bone research protocols. 2nd ed. Totowa: Humana Press; 2012. p. 499–506. 978-1-61779-414-8.

Beamer WG, Donahue LR, Rosen CJ, et al. Genetic variability in adult bone density among inbred strains of mice. Bone. 1996;18(5):397–403.

Beck JA, Lloyd S, Hafezparast M, et al. Genealogies of mouse inbred strains. Nat Genet. 2000;24:23–5.

Belanger LF. Osteocytic osteolysis. Calcif Tissue Res. 1969;4(1):1–12.

Bell KL, Loveridge N, Power J, et al. Regional differences in cortical porosity in the fractured femoral neck. Bone. 1999;24:57–64.

Bell KL, Loveridge N, Jordan GR, et al. A novel mechanism for induction of increased cortical porosity in cases of intracapsular hip fracture. Bone. 2000;27:297–304.

Binkley N, Kimmel D, Bruner J, et al. Zoledronate prevents the development of absolute osteopenia following ovariectomy in adult rhesus monkeys. J Bone Miner Res. 1998;13:1775–82.
Binkley N, Ellison G, O'Rourke C, Hall D, Johnston G, Kimmel D, Keller ET. Rib biopsy technique for cortical bone evaluation in rhesus monkeys (*Macaca mulatta*). Lab Anim Sci. 1999;49:87–9.
Black A, Tilmont EM, Handy AM, et al. A nonhuman primate model of age-related Bone loss: a longitudinal study in male and premenopausal female rhesus monkeys. Bone. 2001;28:295–302.
Blair HC, Teitelbaum SL, Ghiselli R, et al. Osteoclastic bone resorption by a polarized vacuolar proton pump. Science. 1989;245(4920):855–7.
Blouin S, Gallois Y, Moreau MF, et al. Disuse and orchidectomy have additional effects on bone loss in the aged male rat. Osteoporos Int. 2007;18(1):85–92.
Boivin G, Meunier PJ. Changes in bone remodeling rate influence the degree of mineralization of bone. Connect Tissue Res. 2002;43:535–7.
Bonewald LF. Osteocytes. In: Rosen CJ, editor. Primer on the metabolic bone dis-eases and disorders of mineral metabolism. 8th ed. New York: Wiley; 2013. p. 34–41.
Boskey AL. Organic and inorganic matrices. In: Wnek G, Bowlin GL, editors. Encyclopedia of biomaterials and biomedical engineering. London: Dekker Encyclopedias, Taylor & Francis Books; 2006a. p. 1–15.
Boskey AL. Assessment of bone mineral and matrix using backscatter electron imaging and FTIR imaging. Curr Osteoporos Rep. 2006b;4:71–5.
Boskey A. Mineralization of bones and teeth. Elements Mag. 2007;3:385–92.
Boskey AL. Bone composition: relationship to bone fragility and antiosteoporotic drug effects. BoneKEy Reports 2, Article number: 447. 2013. doi:10.1038/bonekey.2013.181.
Boskey AL, Coleman R. Aging and bone. J Dent Res. 2010;89:1333–48.
Boskey AL, Robey PG. The regulatory role of matrix proteins in mineralization of bone. In: Marcus R, Feldman D, Dempster DW, Luckey M, editors. Osteoporosis. New York: Elsevier; 2013. p. 235–58.
Bosu WT, Johansson ED, Gemzell C. Peripheral plasma levels of oestrone, oestradiol-17ß and progesterone during ovulatory menstrual cycles in the rhesus monkey with special reference to the onset of menstruation. Acta Endocrinol. 1973;74:732–42.
Bouxsein ML, Boyd SK, Christiansen BA, et al. Guidelines for assessment of bone microstructure in rodents using micro-computed tomography. J Bone Miner Res. 2010;25:1468–86.
Boyd SK, Davison P, Müller R, et al. Monitoring individual morphological changes over time in ovariectomized rats by in vivo micro-computed tomography. Bone. 2006;39(4):854–62.
Boyde A, Jones SJ. Backscattered electron imaging of skeletal tissues. Metab Bone Dis Rel Res. 1983;5:145–50.
Boyle WJ, Simonet WS, Lacey DL. Osteoclast differentiation and activation. Nature. 2003;423: 337–42.
Brandi ML, Collin-Osdoby P. Vascular biology and the skeleton. J Bone Miner Res. 2006;21:183–92.
Breen SA, Loveday BE, Millest AJ, et al. Stimulation and inhibition of bone formation: use of peripheral quantitative computed tomography in the mouse in vivo. Lab Anim. 1998;32:467–76.
Brockstedt H, Kassem M, Eriksen EF, et al. Age- and sex-related changes in iliac cortical bone mass and remodeling. Bone. 1993;14(4):681–91.
Brommage R. Perspectives on using nonhuman primates to understand the etiology and treatment of postmenopausal osteoporosis. J Musculoskelet Neuronal Interact. 2001;1(4):307–25.
Brommage R, Vafai H. Rapid embedding protocol for visualizing bone mineral and matrix. Calcif Tissue Int. 2000;67:479–80.
Brommage R, Allison C, Stavisky R, et al. Measurement of serum bone-specific alkaline phosphatase activity in cynomolgus macaques. J Med Primatol. 1999;28:329–33.
Brouwers JEM, Lambers FM, Gasser JA, et al. Bone degeneration and recovery after early and late bisphosphonate treatment of Ovariectomized Wistar rats assessed by in vivo micro-computed tomography. Calcif Tissue Int. 2008;82(3):202–11.
Buenzli PR, Sims NA. Quantifying the osteocyte network in the human skeleton. Bone. 2015;75:144–50.

Burr DB, Akkus O. Bone morphology and organization. In: Burr DB, Allen MR, editors. Basic and applied bone biology. San Diego: Academic; 2014. p. 3–25.
Burr DB, Martin RB, Schaffler MB, et al. Bone remodeling in response to in vivo fatigue microdamage. J Biomech. 1985;18:189–200.
Burr DB, Turner CH, Naick P, et al. Does microdamage accumulation affect the mechanical properties of bone? J Biomech. 1998;31:337–45.
Butterstein GM, Mann DR, Gould K, et al. Prolonged inhibition of normal ovarian cycles in the rat and cynomolgus monkeys following a single s.c. injection of danazol. Hum Reprod. 1997;12:1409–15.
Cabal A, Jayakar RY, Sardesai S, et al. High-resolution peripheral quantitative computed tomography and finite element analysis of bone strength at the distal radius in ovariectomized adult rhesus monkey demonstrate efficacy of odanacatib and differentiation from alendronate. Bone. 2013;56:497–505.
Canalis E. Growth factor control of bone mass. J Cell Biochem. 2009;108(4):769–77.
Caplan AI, Bruder SP. Mesenchymal stem cells: building blocks for molecular medicine in the 21st century. Trends Mol Med. 2001;7(6):259–64.
Carlson C, Tulli H, Jayo M, et al. Immunolocalization of noncolla-genous bone matrix proteins in lumbar vertebrae from intact and surgically menopausal cynomolgus monkeys. J Bone Miner Res. 1993;8:71–83.
Cerroni AM, Tomilinson GA, Turnquist JE, et al. Bone mineral density, osteopenia, and osteoporosis in the rhesus macaques of Cayo Santiago. Am J Phys Anthropol. 2000;113:389–410.
Chambers TJ, Fuller K. Bone cells predispose bone surfaces to resorption by exposure of mineral to osteoclastic contact. J Cell Sci. 1985;76:155–65.
Champ JE, Binkley N, Havighurst T, et al. The effect of advancing age on bone mineral content of female rhesus monkeys. Bone. 1996;19:485–92.
Chellaiah MA. Regulation of actin ring formation by rho GTPases in osteoclasts. J Biol Chem. 2005;280:32930–43.
Chen HK, Ke HZ, Jee WS, et al. Droloxifene prevents ovariectomy-induced bone loss in tibiae and femora of aged female rats: a dual-energy X-ray absorptiometric and histomorphometric study. J Bone Miner Res. 1995a;10:1256–62.
Chen MM, Yeh JK, Aloia JF. Effect of ovariectomy on cancellous bone in the hypophysectomized rat. J Bone Miner Res. 1995b;10(9):1334–42.
Chen LD, Kushwaha RS, McGill HC Jr, et al. Effect of naturally reduced ovarian function on plasma lipoprotein and 27-hydroxycholesterol levels in baboons (Papio sp.). Atherosclerosis. 1998;136:89–98.
Chen D, Zhao M, Mundy GR. Bone morphogenetic proteins. Growth Factors. 2004;22(4):233–41.
Chia R, Achilli F, Festing MF, et al. The origins and uses of mouse outbred stocks. Nat Genet. 2005;37:1181–6.
Christian JC, Yu PL, Slemenda CW, et al. Heritability of bone mass: a longitudinal study in aging male twins. Am J Hum Genet. 1989;44(3):429–33.
Colman RJ, Kenmitz JW, Lane MA, et al. Skeletal effects of aging and menopausal status in female rhesus monkeys. J Clin Endocrinol Metab. 1999a;84:4144–8.
Colman RJ, Lane MA, Binkley N, et al. Skeletal effects of aging in male rhesus monkeys. Bone. 1999b;24:17–23.
Committee on Standardized Genetic Nomenclature for Mice. 1952. http://www.informatics.jax.org/mgihome/nomen/strains.shtml
Crockett JC, Rogers MJ, Coxon FP, et al. Bone remodelling at a glance. J Cell Sci. 2011;124:991–8.
Currey JD. Bones. Structure and mechanics. New Jersey: Princeton University Press; 2002. p. 1–380.
Danforth DR, Chillik CF, Hertz R, et al. Effects of ovarian tissue reduction on the menstrual cycle: persistent normalcy after near-total oophorectomy. Biol Reprod. 1989;41:355–60.
De Gorter DJJ, Ten Dijke P. Signal transduction cascades controlling osteoblast differentiation. In: Rosen CJ, editor. Primer on the metabolic Bone diseases and disorders of mineral metabolism. 8th ed. New York: Wiley; 2013. p. 15–24.

Dickerson SS, Hotchkiss CE. Relationships between densitometric and morphological parameters as measured by peripheral computed tomography and the compressive behavior of lumbar vertebral bodies from macaques (*Macaca fascicularis*). Spine. 2008;33:366–72.
Dobnig H, Turner RT. Evidence that intermittent treatment with parathyroid hormone increases bone formation in adult rats by activation of bone lining cells. Endocrinology. 1995;136(8):3632–8.
Donnelly E, Meredith DS, Nguyen JT, et al. Bone tissue composition varies across anatomic sites in the proximal femur and the iliac crest. J Orthop Res. 2012;30:700–6.
Ducy P, Desbois C, Boyce B, et al. Increased bone formation in osteocalcin-deficient mice. Nature. 1996;382(6590):448–52.
Ducy P, Zhang R, Geoffroy V, et al. Osf2/Cbfa1: a transcriptional activator of osteoblast differentiation. Cell. 1997;89(5):747–54.
Erben RG. Trabecular and endocortical bone surfaces in the rat: modeling or remodeling? Anat Rec. 1996;246:39–46.
Erben RG and Glösmann. Histomorphometry in rodents. Helfrich, MH and Ralston, SH, editors. Methods in molecular medicine, Vol. 816: bone research protocols. 2nd ed. Totowa: Humana Press; 2012. p. 279–303. 978-1-61779-414-8.
Erben RG, Eberle J, Stahr K, et al. Androgen deficiency induces high turnover osteopenia in aged male rats: a sequential histomorphometric study. J Bone Miner Res. 2000;15(6):1085–98.
Feng XH, Derynck R. Specificity and versatility in TGF-β signaling through Smads. Annu Rev Cell Dev Biol. 2005;21:659–93.
Feng JQ, Ward LM, Liu S, et al. Loss of DMP1 causes rickets and osteomalacia and identifies a role for osteocytes in mineral metabolism. Nat Genet. 2006;38(11):1310–5.
Ferrari S, Bianchi ML, Eisman JA, et al. IOF Committee of Scientific Advisors Working Group on osteoporosis pathophysiology. Osteoporosis in young adults: pathophysiology, diagnosis, and management. Osteoporos Int. 2012;23(12):2735–48.
Frost HM. Dynamics of bone remodeling. In: Frost HM, editor. BoneBiodynamics. Boston: Little Brown & Co; 1964. p. 315–33.
Fukumoto S, Martin TJ. Bone as an endocrine organ. Trends Endocrinol Metab. 2009;20:230–6.
Gallagher A, Chambers TJ, Tobias JH. The estrogen antagonist ICI 182,780 reduces cancellous bone volume in female rats. Endocrinology. 1993;133:2787–91.
Garn S. The earlier gain and later loss of cortical bone. In: Nutritional perspectives. Springfield: Charles C. Thomas; 1970. p. 3–120.
Gasser JA and Willnecker J. Bone measurements by peripheral quantitative computed tomography in rodents. In: Helfrich, MH and Ralston, SH, editors. Methods in molecular medicine, Vol. 816: bone research protocols. 2nd ed. Totowa: Humana Press Inc.; 2012. 477–98. 978-1-61779-414-8.
Gasser JA, Green JR, Shen V, et al. A single intravenous administration of zoledronic acid prevents the bone loss and mechanical compromise induced by aromatase inhibition in rats. Bone. 2006;39(4):787–95.
Gay CV, Mueller WJ. Carbonic anhydrase and osteoclasts: localization by labeled inhibitor autoradiography. Science. 1974;183(123):432–4.
Gazzerro E, Canalis E. Bone morphogenetic proteins and their antagonists. Rev Endocr Metab Disord. 2006;7(1–2):51–65.
Gelb BD, Shi GP, Chapman HA, et al. Pycnodysostosis, a lysosomal disease caused by cathepsin K deficiency. Science. 1996;273(5279):1236–8.
Gilardi KVK, Shideler SE, Valverde CR, et al. Characterization of the onset of menopause in the rhesus macaque. Biol Reprod. 1997;57:335–40.
Gillooly JF, Brown JH, West GB, et al. Effects of size and temperature on metabolic rate. Science. 2001;293:2248–51.
Glatt V, Canalis E, Stadmeyer L, et al. Age-related changes in trabecular architecture differ in female and male C57BL/6J mice. J Bone Miner Res. 2007;22:1197–207.
Goldring MB, Goldring SR. Articular cartilage and subchondral bone in the pathogenesis of osteoarthritis. Ann N Y Acad Sci. 2010;1192:230–7.

Goodman AL, Descalzi CD, Johnson DK, et al. Composite pattern of circulating LH, FSH, estradiol, and progesterone during the menstrual cycle in cynomolgus monkeys. Proc Soc Exp Biol Med. 1977;155:479–81.
Goodwin BT, Jerome CP. Iliac biopsy for histomorphometric analysis of trabecular bone in cynomolgus monkeys and baboons. Lab Anim Sci. 1987;37:213–6.
Gordon K, Williams RF, Danforth DR, et al. Suppression of ovarian estradiol secretion by a single injection of antide in cynomolgus monkeys during the early follicular phase: immediate, sustained, and reversible actions. J Clin Endocrinol Metab. 1991;73:1262–8.
Goulding A, Gold E. A new way to induce oestrogen-deficiency osteopenia in the rat: comparison of the effect of surgical ovariectomy and administration of the LHRH agonist buserelin on bone resorption and composition. J Endocrinol. 1989;121:293–8.
Guo D, Keightley A, Guthrie J, et al. Identification of osteocyte-selective proteins. Proteomics. 2010;10(20):3688–98.
Haapasalo H, Kontulainen S, Sievanen H, et al. Exercise-induced bone gain is due to enlargement in bone size without a change in volumetric bone density: a peripheral quantitative computed tomography study of the upper arms of male tennis players. Bone. 2000;27(3):351–7.
Halloran BP, Ferguson VL, Simske SJ, et al. Changes in bone structure and mass with advancing age in the male C57BL/6J mouse. J Bone Miner Res. 2002;17:1044–50.
Harada H, Tagashira S, Fujiwara M, et al. Cbfa1 isoforms exert functional differences in osteoblast differentiation. J Biol Chem. 1999;274(11):6972–8.
Hauge E, Mosekilde L, Melsen F. Missing observations in bone histomorphometry on osteoporosis: implications and suggestions for an approach. Bone. 1999;25:389–95.
Hauge EM, Qvesel D, Eriksen EF, et al. Cancellous bone remodeling occurs in specialized compartments lined by cells expressing osteoblastic markers. J Bone Miner Res. 2001;16:1575–82.
Hermann LM, Smith KC. Percutaneous trephine biopsy of the vertebral body in the rhesus monkey (*Macaca mulatta*). Am J Vet Res. 1985;46:1403–7.
Hernandez CJ, Gupt A, Keaveny TM. A biomechanical analysis of the effects of resorption cavities on cancellous bone strength. J Bone Miner Res. 2006;21:1248–55.
Hessle L, Johnson KA, Anderson HC, et al. Tissue-nonspecific alkaline phosphatase and plasma cell membrane glycoprotein 1 are central antagonistic regulators of bone mineralization. Proc Natl Acad Sci U S A. 2002;99(14):9445–9.
Heuck F. Comparative investigations of the function of osteocytes in bone resorption. Calcif Tissue Res Suppl. 1970:148–9.
Hodgen GD, Goodman AL, O'Connor A, et al. Menopause in rhesus monkeys: model for study of disorders in the human climacteric. Am J Obstet Gynecol. 1977;127:581–4.
Hofbauer LC, Khosla S, Dunstan CR, et al. The roles of osteoprotegerin and osteoprotegerin ligand in the paracrine regulation of bone resorption. J Bone Miner Res. 2000;15(1):2–12.
Hoffman RA, Mack PB, Hood WN. Comparison of calcium and phosphorus excretion with bone density changes during restraint in immature *Macaca nemestrina* primates. Aerospace Med. 1972;43:376–83.
Hotchkiss CE. Use of peripheral quantitative computed tomography for densitometry of the femoral neck and spine in cynomolgus monkeys (*Macaca fascicularis*). Bone. 1999;24:101–7.
Hotchkiss CE, Brommage R, Du M, et al. The anesthetic isoflurane decreases ionized calcium and increases parathyroid hormone and osteocalcin in cynomolgus monkeys. Bone. 1998;23:479–84.
Inaoka T, Bilbe G, Ishibashi O, et al. Molecular cloning of human cDNA for cathepsin K: novel cysteine proteinase predominantly expressed in bone. Biochem Biophys Res Commun. 1995;206(1):89–96.
Inskip MJ, Franklin CA, Subramanian KS, et al. Sampling of cortical and trabecular bone for lead analysis: method development in a study of lead mobilization during pregnancy. Neurotoxicology. 1992;13:825–34.
Iwaniec UT, Turner RT. Animal models of osteoporosis. In: Marcus R, Feldman D, Nelson DA, Rosen CJ, editors. Osteoporosis. 3rd ed. Amsterdam: Elsevier; 2008. p. 985–1110.

Iwaniec UT, Yuan D, Power RA, et al. Strain-dependent variations in the response of cancellous bone to ovariectomy in mice. J Bone Miner Res. 2006;21:1068–74.
Jaffe AB, Hall A. Rho GTPases: biochemistry and biology. Annu Rev Cell Dev Biol. 2005;21:247–69.
Jämsä T, Jalovaara P, Peng Z, et al. Comparison of three-point bending test and peripheral quantitative computed tomography analysis in the evaluation of the strength of mouse femur and tibia. Bone. 1998;23:155–61.
Janssens K, ten Dijke P, Janssens S, et al. Transforming growth factor-β1 to the bone. Endocr Rev. 2005;26(6):743–74.
Jayo MJ, Jerome CP, Lees CJ, et al. Bone mass in female cynomolgus macaques: a crosssectional and longitudinal study by age. Calcif Tissue Int. 1994;54:231–6.
Jee WS, Li XJ. Adaptation of cancellous bone to overloading in the adult rat: a single photon absorptiometry and histomorphometry study. Anat Rec. 1990;227:418–26.
Jee WS, Ma Y. Animal models of immobilization osteopenia. Morphologie. 1999;83(261):25–34.
Jee WSS, Yao W. Overview: animal models of osteopenia and osteoporosis. J Musculoskelet Neuronal Interact. 2001;1:193–207.
Jepsen KJ. Functional interactions among morphologic and tissue quality traits define bone quality. Clin Orthop Relat Res. 2011;469:2150–9.
Jepsen KJ, Silva MJ, Vashishth D, et al. Establishing biomechanical mechanisms in mouse models: practical guidelines for systematically evaluating phenotypic changes in the Diaphyses of long bones. J Bone Miner Res. 2015;30(6):951–66.
Jerome C. Hormonal therapies and osteoporosis. ILAR J. 2004;45:170–8.
Jerome CP, Peterson PE. Nonhuman primate models in skeletal research. Bone. 2001;29:1–6.
Jerome CP, Carlson CS, Register TC, et al. Bone functional changes in intact, ovariectomized, hormone-supplemented adult cynomolgus monkeys (*Macaca fascicularis*) evaluated by serum markers and dynamic histomorphometry. J Bone Miner Res. 1994;9:527–40.
Jerome CP, Lees CJ, Weaver DS. Development of osteopenia in ovariectomized cynomolgus monkeys (*Macaca fascicularis*). Bone. 1995;17:403S–8S.
Jerome CP, Turner CH, Lees CJ. Decreased bone mass and strength in ovariectomized cynomolgus monkeys (*Macaca fascicularis*). Calcif Tissue Int. 1997;60:265–70.
Jerome CP, Johnson CS, Vafai HT, et al. Effect of treatment for 6 months with human parathyroid hormone (1–34) peptide in ovariectomized cynomolgus monkeys (*Macaca fascicularis*). Bone. 1999;25:301–9.
Jerome CP, Burr DB, Van Bibber T, et al. Treatment with human parathyroid hormone (1–34) for 18 months increases cancellous Bone volume and improves trabecular architecture in Ovariectomized Cynomolgus monkeys (*Macaca fascicularis*). Bone. 2001;28(2):150–9.
Jiang Y, Jahagirdar BN, Reinhardt RL, et al. Pluripotency of mesenchymal stem cells derived from adult marrow. Nature. 2002;418(6893):41–9.
Jiang Y, Zhao J, Liao EY, et al. Application of μCT-assessment of 3D bone microstructure in preclinical and clinical studies. J Bone Miner Metab. 2005;23:122–31.
Jilka RL. The relevance of mouse models for investigating age-related Bone loss in humans. J Gerontol A Biol Sci Med Sci. 2013; doi:10.1093/gerona/glt046.
Johnson CS, Jerome CP, Brommage R. Unbiased determination of cytokine localization in bone: colocalization of interleukin-6 with osteoblasts in serial sections from monkey vertebrae. Bone. 2000;26:461–7.
Kalajzic I, Braut A, Guo D, et al. Dentin matrix protein 1 expression during osteoblastic differentiation, generation of an osteocyte GFP-transgene. Bone. 2004;35(1):74–82.
Kan L. Animal models of bone diseases – a. In: Conn PM, editor. Animal models for the study of human disease. San Diego: Academic; 2013. p. 353–90.
Kanis JA, McCloskey EV, Johansson H, et al. European guidance for the diagnosis and management of osteoporosis in postmenopausal women. Osteoporos Int. 2013;24:23–57.
Kaplan JR, Adams MR, Clarkson TB, et al. Social behavior and gender in biomedical investigations using monkeys: studies in atherogenesis. Lab Anim Sci. 1991;41:334–43.

Kavukcuoglu NB, Denhardt DT, Guzelsu N, et al. Osteopontin deficiency and aging on nanomechanics of mouse bone. J Biomed Mater Res A. 2007;83(1):136–44.

Kavukcuoglu NB, Patterson-Buckendahl P, Mann AB. Effect of osteocalcin deficiency on the nanomechanics and chemistry of mouse bones. J Mech Behav Biomed Mater. 2009;2:348–54.

Kazanci M, Roschger P, Paschalis EP, et al. Bone osteonal tissues by Raman spectral mapping: orientation-composition. J Struct Biol. 2006;156:489–96.

Keaveney TM. Cancellous bone. In: Black J, Hastings G, editors. Handbook of biomaterials properties. London: Chapman and Hall; 1998.

Kenigsberg D, Hodgen GD. Ovulation inhibition by administration of weekly gonadotropin-releasing hormone antagonist. J Clin Endocrinol Metab. 1986;62:734–8.

Kennedy OD, Herman BC, Laudier DM, et al. Activation of resorption in fatigue-loaded bone involves both apoptosis and active pro-osteoclastogenic signaling by distinct osteocyte populations. Bone. 2012;50:1115–22.

Kerber WT, Reese WH. Comparison of the menstrual cycle of cynomolgus and rhesus monkeys. Fertil Steril. 1969;20:975–9.

Kern B, Shen J, Starbuck M, et al. Cbfa1 contributes to the osteoblast-specific expression of type I collagen genes. J Biol Chem. 2001;276(10):7101–7.

Kim SW, Divieti-Pajevic P, Selig M, et al. Intermittent PTH administration converts quiescent lining cells to active osteoblasts. J Bone Miner Res. 2012;27(10):2075–84.

Klein HJ, Seedor G, Frankenfeld DL, Thompson DD. Method for transiliac bone biopsy in baboons. J Am Vet Med Assoc. 1991;198:1977–9.

Kneissel M, Boyde A, Gasser JA. Bone tissue and its mineralization in aged estrogen-depleted rats after long-term intermittent treatment with parathyroid hormone (PTH) analog SDZ PTS 893 or human PTH(1-34). Bone. 2001;28:237–50.

Knothe Tate ML, Adamson JR, Tami AE, et al. The osteocyte. Int J Biochem Cell Biol. 2004;36:1–8.

Knutson JC, Hollis BW, LeVan LW, et al. Metabolism of 1-hydroxyvitamin D2 to activated dihydroxyvitamin D2 metabolites decreases endogenous 1,25-dihydroxyvitamin D3 in rats and monkeys. Endocrinology. 1995;136:4749–53.

Koga T, Matsui Y, Asagiri M, et al. NFAT and Osterix cooperatively regulate bone formation. Nat Med. 2005;11(8):880–5.

Komori T, Yagi H, Nomura S. Targeted disruption of Cbfa1 results in a complete lack of bone formation owing to maturational arrest of osteoblasts. Cell. 1997;89(5):755–64.

Kostenuik PJ, Shalhoub V. Osteoprotegerin: a physiological and pharmacological inhibitor of bone resorption. Curr Pharm Des. 2001;7:613–35.

Koyama T, de la Pena A, Hagino N. Plasma estrogen, progestin, and luteinizing hormone during the normal menstrual cycle of the baboon: role of luteinizing hormone. Am J Obstet Gynecol. 1977;127:67–72.

Kramer I, Salie R, Susa M et al. Studying gene expression in bone by in situ hybridization. In: Helfrich, MH and Ralston, SH, editors. Methods in molecular medicine, vol. 816: bone research protocols. 2nd ed. Totowa: Humana Press Inc.; 2012. 305–320. 978-1-61779-414-8.

Kristensen HB, Andersen TL, Marcussen N, et al. Osteoblast recruitment routes in human cancellous bone remodeling. Am J Pathol. 2014;184:778–89.

Lane NE, Yao W, Balooch M, et al. Glucocorticoid-treated mice have localized changes in trabecular bone material properties and osteocyte lacunar size that are not observed in placebo-treated or estrogen-deficient mice. J Bone Miner Res. 2006;21(3):466–76.

Lanyon LE, Baggott DG. Mechanical function as an influence on the structure and form of bone. J Bone Joint Surg. 1976;58-B(4):436–43.

Lazner F, Gowen M, Pavasovic D, et al. Osteopetrosis and osteoporosis: two sides of the same coin. Hum Mol Genet. 1999;8(10):1839–46.

Lees CJ, Ramsay H. Histomorphometry and bone biomarkers in cynomolgus females: a study in young, mature, and old monkeys. Bone. 1999;24:25–8.

Lekamwasam S, Adachi JD, Agnusdei D, et al. Joint IOF-ECTS GIO guidelines working group. A framework for the development of guidelines for the management of glucocorticoid-induced osteoporosis. Osteoporos Int. 2012;23(9):2257–76.

Lelovas PP, Xanthos TT, Thoma SE, et al. The laboratory rat as an animal model for osteoporosis research. Comp Med. 2008;58(5):424–30.
Li M, Shen Y, Qi H, et al. Comparison study of skeletal response to estrogen depletion at red and yellow marrow sites in rats. Anat Rec. 1996;245:472–80.
Li M, Shen Y, Wronski TJ. Time course of femoral neck osteopenia in ovariectomizd rats. Bone. 1997;20:55–61.
Lipkin EW. A longitudinal study of calcium regulation in a nonhuman primate model of parenteral nutrition. Am J Clin Nutr. 1998;67:246–54.
Lips P, Courpron P, Meunier PJ. Mean wall thickness of trabecular bone packets in the human iliac crest: changes with age. Calcif Tissue Res. 1978;10:13–7.
Loeb WF. The rat. In: Loeb WF, Quimby FW, editors. The clinical chemistry of laboratory animals. 2nd ed. Ann Arbor: Edwards Brothers; 1999. p. 33–45.
Long F. Building strong bones: molecular regulation of the osteoblast lineage. Mol Cell Biol. 2012;13:27–38.
Lozupone E, Favia A. The structure of the trabeculae of cancellous bone. 2. Long bones and mastoid. Calcif Tissue Int. 1990;46:367–72.
Lyons KM. Animal models: genetic manipulation. In: Rosen CJ, editor. Primer on the metabolic bone diseases and disorders of mineral metabolism. 8th ed. New York: Wiley; 2013. p. 69–75.
Ma YF, Li XJ, Jee WS, et al. Effects of prostaglandin E2 and F2 alpha on the skeleton of osteopenic ovariectomized rats. Bone. 1995;17:549–54.
Maeda S, Hayashi M, Komiya S, et al. Endogenous TGF-β signaling suppresses maturation of osteoblastic mesenchymal cells. EMBO J. 2004;23(3):552–63.
Mandair GS, Morris MD. Contributions of Raman spectroscopy to the understanding of bone strength. BoneKEy Reports 4, Article number: 620. 2015. doi:10.1038/bonekey.2014.115.
Mann DR, Gold KG, Collins DC. A potential primate model for bone loss resulting from medical oophorectomy or menopause. J Clin Endocrinol Metab. 1990;71:105–10.
Marenzana M, Arnett TR. The key role of blood supply in bone. Bone Res. 2013;3:203–15.
Marie PJ, Glorieux FH. Relation between hypomineralized periosteocytic lesions and bone mineralization in vitamin D-resistant rickets. Calcif Tissue Int. 1983;35(4–5):443–8.
Martin RB. Porosity and specific surface of bone. CRC Critical Rev Biomed Eng. 1984;10:179–221.
Massague J, Seoane J, Wotton D. Smad transcription factors. Genes Dev. 2005;19(23):2783–810.
McCarthy I. The physiology of bone blood flow: a review. J Bone Joint Surg Am. 2006;88(Suppl 3):4–9.
McNamara LM, Van Der Linden JC, Weinans H, et al. Stress-concentrating effect of resorption lacunae in trabecular bone. J Biomech. 2006;39:734–41.
Mehta RR, Jenco JM, Gaynor LV, et al. Relationships between ovarian morphology, vaginal cytology, serum progesterone, and urinary immunoreactive pregnanediol during the menstrual cycle of the cynomolgus monkey. Biol Reprod. 1986;35:981–6.
Meunier PJ. Assessment of bone turnover by histomorphometry in osteoporosis. In: Riggs BL, Melton LJ, editors. Osteoporosis: etiology, diagnosis, and management. New York: Raven Press; 1988. p. 317–32.
Misof BM, Roschger P, Cosman F, et al. Effects of intermittent parathyroid hormone administration on bone mineralization density in iliac crest biopsies from patients with osteoporosis: a paired study before and after treatment. J Clin Endocrinol Metab. 2003;8(3):1150–6.
Miyakoshi N, Sato K, Tsuchida T, et al. Histomorphometric evaluation of the effects of ovariectomy on bone turnover in rat caudal vertebrae. Calcif Tissue Int. 1999;64:318–24.
Mori S, Burr DB. Increased intracortical remodeling following fatigue damage. Bone. 1993;14:103–9.
Mori S, Jee WS, Li XJ. Production of new trabecular bone in osteopenic ovariectomized rats by prostaglandin E2. Calcif Tissue Int. 1992;50:80–7.
Morris MD, Mandair GS. Raman assessment of bone quality. Clin Orthop Relat Res. 2011;469: 2160–9.
Murray PDF, Huxley JS. Self-differentiation in the grafted limb bud of the chick. J Anat. 1925;59:379–84.

Nakamura H. Morphology, function, and differentiation of bone cells. J Hard Tissue Biol. 2007;16(1):15–22.
Nakashima K, Zhou X, Kunkel G, et al. The novel zinc finger containing transcription factor osterix is required for osteoblast differentiation and bone formation. Cell. 2002;108(1):17–29.
Negishi-Koga T, Shinohara M, Komatsu N, et al. Suppression of bone formation by osteoclastic expression of semaphoring 4D. Nat Med. 2011;17:1473–80.
Nishio Y, Dong Y, Paris M, et al. Runx2-mediated regulation of the zinc finger Osterix/Sp7 gene. Gene. 2006;372:62–70.
Nogues C, Milhaud C. A new technique for iliac crest biopsy in rhesus monkeys for use in weightlessness experiments: some results of ground studies. Aviat Space Environ Med. 1988;59:374–8.
Ott SM, O'Hanlan M, Lipkin EW, et al. Evaluation of vertebral volumetric vs areal bone mineral density during growth. Bone. 1997;20:553–6.
Ott SM, Lipkin EW, Newell-Morris. Bone physiology during pregnancy and lactation in young macaques. J Bone Miner Res. 1999;14:1779–88.
Otto F, Thornell AP, Crompton T, et al. Cbfa1, a candidate gene for cleidocranial dysplasia syndrome, is essential for osteoblast differentiation and bone development. Cell. 1997;89(5):765–71.
Pacureanu A, Langer M, Boller E, et al. Nanoscale imaging of the bone cell network with synchrotron X-ray tomography: optimization of acquisition setup. Med Phys. 2012;39:2229–38.
Parfitt AM. Targeted and nontargeted bone remodeling: relationship to basic multicellular unit origination and progression. Bone. 2002a;30(1):5–7.
Parfitt AM. Misconceptions (2): turnover is always higher in cancellous than in cortical bone. Bone. 2002b;30:807–9.
Parfitt AM, Han ZH, Palnitkar S, et al. Effects of ethnicity and age or menopause on osteoblast function, bone mineralization, and osteoid accumulation in iliac bone. J Bone Miner Res. 1997;12:1864–73.
Patsch JM, Burghardt AJ, Kazakia G, et al. Noninvasive imaging of bone microarchitecture. Ann N Y Acad Sci. 2011;1240:77–87.
Pixley FJ, Stanley ER. CSF-1 regulation of the wandering macrophage: complexity in action. Trends Cell Biol. 2004;14:628–38.
Pocock NA, Eisman JA, Hopper JL, et al. Genetic determinants of bone mass in adults. A twin study. J Clin Invest. 1987;80(3):706–10.
Poole KE, van Bezooijen RL, Loveridge N, et al. Sclerostin is a delayed secreted product of osteocytes that inhibits bone formation. FASEB J. 2005;19(13):1842–4.
Qing H, Bonewald LF. Osteocyte remodeling of the perilacunar and pericanalicular matrix. Int J Oral Sci. 2009;1(2):59–65.
Recker RR, Kimmel DB, Parfitt AM, et al. Static and tetracycline-based bone histomorphometric data from 34 normal postmenopausal females. J Bone Miner Res. 1988;3:133–44.
Recker R, Lappe J, Davies KM, et al. Bone remodeling increases substantially in the years after menopause and remains increased in older osteoporosis patients. J Bone Miner Res. 2004;19:1628–33.
Recklinghausen FV, editor. Untersuchungen über rachitis and osteomalacia. Jena: Fischer Verlag; 1910.
Register TC, Jerome CP. Increased urinary markers of collagen degradation accompany ovariectomy in skeletally mature cynomolgus macaques. J Bone Miner Res. 1996;11:S196.
Reid DG, Shanahan CM, Duer MJ, et al. Lipids in biocalcification: contrasts and similarities between intimal and medial vascular calcification and bone by NMR. J Lipid Res. 2012;53:1569–75.
Reponen P, Sahlberg C, Munaut C, et al. High expression of 92-kD type IV collagenase (gelatinase B) in the osteoclast lineage during mouse development. J Cell Biol. 1994;124(6):1091–102.
Roschger P, Rinnerthaler S, Yates J, et al. Alendronate increases degree and uniformity of mineralization in cancellous bone and decreases the porosity in cortical bone of osteoporotic women. Bone. 2001;29(2):185–91.
Roschger P, Paschalis EP, Fratzl P, et al. Bone mineralization density distribution in health and disease. Bone. 2008;42:456–66.

Ross FP. Osteoclast biology and bone resorption. In: Rosen CJ, editor. Primer on the metabolic bone dis-eases and disorders of mineral metabolism. 8th ed. New York: Wiley; 2013. p. 25–33.
Ross FP, Teitelbaum SL. alphavbeta3 and macrophage colony-stimulating factor: partners in osteoclast biology. Immunol Rev. 2005;208:88–105.
Rossert J, de Crombrugghe B. Type I collagen: structure, synthesis and regulation. In: Bilezikian JP, Raisz LA, Rodan GA, editors. Principles of bone biology, vol. 1. 2nd ed. San Diego: Academic; 2002. p. 189–210.
Rowe PS, Kumagai Y, Gutierrez G, et al. MEPE has the properties of an osteoblastic phosphatonin and inhibin. Bone. 2004;34(2):303–19.
Salo J, Lehenkari P, Mulari M, et al. Removal of osteoclast bone resorption products by transcytosis. Science. 1997;276:270–3.
Sánchez-Duffhues G, Hiepen C, Knaus P, et al. Bone morphogenetic protein signaling in bone homeostasis. Bone. 2015;80:43–59.
Schaffler MB, Burr DB. Primate cortical bone microstructure: relationship to locomotion. Am J Phys Anthropol. 1984;65:191–7.
Schaffler M, Cheung W-Y, Majeska R, et al. Osteocytes: master orchestrators of bone. Calcif Tissue Int. 2014;94:5–24.
Schneider P, Meier M, Wepf R, et al. Towards quantitative 3D imaging of the osteocyte lacuno-canalicular network. Bone. 2010;47:848–58.
Schriefer JL, Robling AG, Warden SJ, et al. A comparison of mechanical properties derived from multiple skeletal sites in mice. J Biomech. 2005;38(3):467–75.
Seeman E. Pathogenesis of bone fragility in women and men. Lancet. 2002;359:1841–50.
Seeman E. Modeling and remodeling: the cellular machinery responsible for the gain and loss of Bone's material and structural strength. In: Bilezikian JP, Raisz LA, Martin TJ, editors. Principles of bone biology, vol. 1. 3rd ed. San Diego: Academic; 2008. p. 3–28.
Seeman E, Hopper J, Bach L, et al. Reduced bone mass in the daughters of women with osteoporosis. New Engl J Med. 1989;320:554–8.
Seeman E, Hopper JL, Young NR, et al. Do genetic factors explain associations between muscle strength, lean mass, and bone density? A twin study. Am J Phys. 1996;270(2 Pt 1):E320–7.
Shaikh AA, Naqvi RH, Shaikh SA (1978) Concentrations of oestradiol-17, and progesterone in the peripheral plasma of the cynomolgus monkey (*Macaca fascicularis*) in relation to the length of the menstrual cycle and its component phases. J Endocrinol 79:1–7.
Sharir A, Barak MM, Shahar R. Whole bone mechanics and mechanical testing. Vet J. 2008; 177:8–17.
Shively CA, Laber-Laird K, Anton RF. Behavior and physiology of social stress and depression in female cynomolgus monkeys. Biol Psychol. 1997;41:871–82.
Sims NA and Martin TJ. Coupling signals between the osteoclast and osteoblast: how are messages transmitted between these temporary visitors to the bone surface? Frontiers Endocrinol. 2015. http://dx.doi.org/10.3389/fendo.2015.00041 Vol 6 (Article 41) 1–5.
Sims NA, Quinn JM, Martin TJ. Coupling between immune and bone cells. In: Lorenzo JA, Choi Y, Horowitz MC, Takayanagi H, editors. Osteoim- munology: interactions of the immune and skeletal systems. 2nd ed. London: Academic; 2015.
Smith SY, Jolette J, Turner CH. Skeletal health: primate model of postmenopausal osteoporosis. Am J Primatol. 2009;71:752–65.
Smith L, Bigelow EM, Jepsen KJ. Systematic evaluation of skeletal mechanical function. Curr Protoc Mouse Biol. 2013;3:39–67.
Snoeks TJA, Van Beek E, Que I et al (2012) Bioluminescence imaging of bone metastasis in rodents. Helfrich, Miep H and Ralston, Stuart H, editors. Methods in molecular medicine, vol. 816: bone research protocols 2nd ed. Totowa: Humana Press Inc.; 507–15. 978-1-61779-414-8.
Sogaard CH, Wronski TJ, McOsker JE, et al. The positive effect of parathyroid hormone on femoral neck bone strength in ovariectomized rats is more pronounced than that of estrogen or bisphosphonates. Endocrinology. 1994;134:650–7.
Sopelak VM, Lynch A, Williams RF, et al. Maintenance of ovulatory menstrual cycles in chronically cannulated monkeys: a vest and mobile tether assembly. Biol Reprod. 1983;28:703–6.

Sowa H, Kaji H, Yamaguchi T, et al. Smad3 promotes alkaline phosphatase activity and mineralization of osteoblastic MC3T3-E1 cells. J Bone Miner Res. 2002;17(7):1190–9.
Sroga GE, Vashishth D. Effects of bone matrix proteins on fracture and fragility in osteoporosis. Curr Osteoporos Rep. 2012;10:141–50.
Stenbeck G, Horton MA. Endocytic trafficking in actively resorbing osteoclasts. J Cell Sci. 2004;117:827–36.
Stroup GB, Kumar S, Jerome CP. Treatment with a potent cathepsin K inhibitor preserves cortical and trabecular bone mass in ovariectomized monkeys. Calcif Tissue Int. 2009;85:344–55.
Suda T, Takahashi N, Udagawa N, et al. Modulation of osteoclast differentiation and function by the new members of the tumor necrosis factor receptor and ligand families. Endocr Rev. 1999;20:345–57.
Szulc P, Seeman P, Duboeuf F, et al. Bone fragility: failure of periosteal apposition to compensate for increased endocortical resorption in postmenopausal women. J Bone Miner Res. 2006;21:1856–63.
Teitelbaum SL. Osteoporosis and integrins. J Clin Endocrinol Metab. 2005;90:2466–8.
Teitelbaum SL, Ross FP. Genetic regulation of osteoclast development and function. Nat Rev Genet. 2003;4:638–49.
The HYP Consortium. A gene (PEX) with homologies to endopeptidases is mutated in patients with X-linked hypophosphatemic rickets. Nat Genet. 1995;11(2):130–6.
Tripp EJ, Mac Kay EH. Silver staining of bone prior to decalcification for quantitative determination of osteoid in sections. Stain Technol. 1972;47(3):129–36.
Turner AS. Animal models of osteoporosis – necessity and limitations. Eur Cell Mater. 2001;1:66–81.
Turner CH. Bone strength: current concepts. Ann N Y Acad Sci. 2006;1068:429–46.
Turner RT, Lotinun S, Hefferan T, et al. Animal models for osteoporosis. Rev Endocr Metab Disord. 2001;2:117–27.
Urist MR. Bone: formation by autoinduction. Science. 1965;150(698):893–9.
Urist MR, Mikulski A, Lietze A. Solubilized and insolubilized bone morphogenetic protein. Proc Natl Acad Sci U S A. 1979;76(4):1828–32.
Vahle JL, Ma YL and Burr DB. Skeletal assessments in the nonhuman primate. Chapter 32. In: Bluemel J, Korte S, Schenck E, Weinbauer GF, editors. The nonhuman primate in nonclinical drug development and safety assessment. San Diego: Academic; 2015. 605–25. 978–0–12-417144-2.
Van der Linden JC, Homminga J, Verhaar JAN, et al. Mechanical consequences of bone loss in cancellous bone. J Bone Miner Res. 2001;16:457–65.
Van't Hof, R (2012) Analysis of bone architecture in rodents using microcomputed tomography. In: Helfrich, Miep H and Ralston, Stuart H, editors. Methods in molecular medicine, vol. 816: bone research protocols. 2nd ed, Totowa: Humana Press; 461–76. 978-1-61779-414-8.
Verborgt O, Gibson GJ, Schaffler MB. Loss of osteocyte integrity in association with microdamage and bone remodeling after fatigue damage in vivo. J Bone Miner Res. 2000;15:60–7.
Verborgt O, Tatton NA, Majeska RJ, et al. Spatial distribution of Bax and Bcl-2 in osteocytes after bone fatigue: complementary roles in bone remodeling regulation? J Bone Miner Res. 2002;17(5):907–14.
Vieth R, Kessler MJ, Pritzker KPH. Serum concentrations of vitamin D metabolites in Cayo Santiago Rhesus macaques. J Med Primatol. 1987;16:347–57.
Vieth R, Kessler MJ, Pritzker KP. Species differences in the binding kinetics of 25-hydroxyvitamin D3 to vitamin D binding protein. Can J Physiol Pharmacol. 1990;68:1368–71.
Walker ML. Menopause in female rhesus monkeys. Am J Primatol. 1995;35:59–71.
Wallace JM. Skeletal hard tissue biomechanics. In: Burr DB, Allen MR, editors. Basic and applied bone biology. San Diego: Academic; 2014. p. 115–30.
Weiner S, Wagner HD. The material bone: structure-mechanical function relations. Annu Rev Mater Sci. 1998;28:271–98.
Weinstein RS. Clinical use of bone biopsy. In: Coe FL, Favus MJ, editors. Disorders of bone and mineral metabolism. New York: Raven Press; 1992. p. 455–74.

Weinstein RS, Jilka RL, Parfitt AM, et al. Inhibition of osteoblastogenesis and promotion of apoptosis of osteoblasts and osteocytes by glucocorticoids. Potential mechanisms of their deleterious effects on bone. J Clin Invest. 1998;102:274–82.
Weitzmann MN, Cenci S, Rifas L, et al. Interluekin-7 stimulates osteoclast formation by upregulating the T-cell production of soluble osteoclastogenic cytokincs. Blood. 2000;96:1873–8.
Wronski TJ, Cintron M, Dann LM. Temporal relationship between bone loss and increased bone turnover in ovariectomized rats. Calcif Tissue Int. 1988;43:179–83.
Wronski TJ, Dann LM, Scott KS, et al. Long-term effects of ovariectomy and aging on the rat skeleton. Calcif Tissue Int. 1989;45:360–6.
Wronski TJ, Dann LM, Horner SL. Time course of vertebral osteopenia in ovariectomized rats. Bone. 1990;10:295–301.
Yamaguchi A, Katagiri T, Ikeda T, et al. Recombinant human bone morphogenetic protein-2 stimulates osteoblastic maturation and inhibits myogenic differentiation in vitro. J Cell Biol. 1991;113(3):681–7.
Yang Y. Skeletal morphogenesis and embryonic development. In: Rosen CJ, editor. Primer on the metabolic bone diseases and disorders of mineral metabolism. 8th ed. New York: Wiley; 2013. p. 3–14.
Yeh JK, Chen MM, Aloia JF. Skeletal alterations in hypophysectomized rats: I. A histomorphometric study on tibial cancellous bone. Anat Rec. 1995;241(4):505–12.
Yeni YN, Brown CU, Wang Z, et al. The influence of bone morphology on fracture toughness of the human femur and tibia. Bone. 1997;21:453–9.
Yuan R, Tsaih SW, Petkova SB, et al. Aging in inbred strains of mice: study design and interim report on median lifespans and circulating IGF1 levels. Aging Cell. 2009;8:277–87.
Zebaze R, Seeman E. Cortical bone: a challenging geography. J Bone Miner Res. 2015;30(1):24–9.
Zhao C, Irie N, Takada Y, et al. Bidirectional ephrinB2-EphB4 signaling controls bone homeostasis. Cell Metab. 2006;4:111–21.
Zoehrer R, Roschger P, Durchschlag E, et al. Bone mineralization density distribution in triple biopsies of the iliac crest in post-menopausal women. J Bone Miner Res. 2006;21(7):1106–12.

Chapter 3
Specific Considerations for Bone Evaluations for Pediatric Therapeutics

Keith Robinson

Abstract Growth and development of the immature organism can be adversely affected by indirect and direct exposures to xenobiotics. There has been an increase in the requirements for juvenile toxicity testing of pharmaceuticals and biopharmaceuticals since 1998, with regulatory agencies commonly requiring nonclinical studies; particular emphasis is often being placed on evaluations of growth and the development of the skeleton. A number of therapeutic classes have been shown to affect the human pediatric skeleton and various juvenile animal models. A nonclinical approach must be taken for full evaluation of effects on the musculoskeletal system's development in pharmaceutical testing, because the prolonged times for completion of development in the human precludes a clinical approach. A range of techniques can be used in toxicology studies, including biochemical markers and in vivo imaging that can also be incorporated in clinical trials. These techniques along with routine toxicology endpoints, growth measurements, and ex vivo techniques, such as histomorphometry and biomechanical testing, can be incorporated into juvenile toxicology studies, each of which is designed on a case-by-case basis. Key factors in study design of the nonclinical program are the species, age, dose route, duration of dosing, and the post-dosing development period, and which outcome measures to include.

Keywords Juvenile • Toxicology • Growth • Skeleton • Pharmaceuticals

3.1 Introduction

There has long been an understanding that growth and development of the immature organism can be adversely affected by indirect and direct exposures to xenobiotics. One of the features of fetal alcohol syndrome (FAS) is an effect upon growth; this was identified in 1973 (Jones and Smith 1973). Similarly, maternal tobacco use can cause in utero effects on fetal growth (Lowe 1959) and tobacco use among the

K. Robinson (✉)
Reproductive and Juvenile Toxicology, Charles River Laboratories,
22022 Transcanadienne, Senneville H9X 3R3, QC, Canada
e-mail: keith.robinson@crl.com

S.Y. Smith et al. (eds.), *Bone Toxicology*, Molecular and Integrative Toxicology,
DOI 10.1007/978-3-319-56192-9_3

pediatric population results in postnatal effects on growth and the skeleton, as well as an increased risk of bone fractures (O'Loughlin et al. 2008; Jones et al. 2004). To study the nonclinical effects of in utero and lactational exposure to pharmaceuticals, the International Conference on Harmonization (ICH) guideline entitled "Detection of Toxicity to Reproduction for Medicinal Products and Toxicity to Male Fertility (S5[2R])" contains two studies that investigate these times in the reproductive cycle, the first, the embryo-fetal development study (ICH-3), looks only at development in utero during major organogenesis, and the second, the pre- and postnatal study (ICH-2), examines in utero (major organogenesis and the fetal growth period) and lactational exposures up to weaning (ICH 2005). However, these ICH guidelines do not require any direct dosing of juvenile animals to examine pediatric treatments.

There has been an increase in the requirements for juvenile toxicity testing since the US Food and Drug Administration (FDA) pediatric rule (FDA 1998) in 1998 and subsequent nonclinical juvenile toxicology guidances from the FDA (FDA 2006) in 2006. In Europe, the European Medicines Agency (EMA) promulgated the Paediatric Regulation (European Medicines Agency [EMA] 2006) and the nonclinical guidelines (EMA 2008) in 2006 and 2008, respectively. In 2012, Japanese nonclinical guidelines were provided by the Ministry of Health, Labour and Welfare (MHLW 2012). Each of these nonclinical guidelines places particular emphasis on evaluation of growth and the development of the skeleton. The FDA nonclinical guidance "Nonclinical Safety Evaluation of Pediatric Drug Products" (FDA 2006 discusses the importance "that studies include measurement of overall growth (e.g., body weight, growth velocity per unit time, tibial length)." The EMA "Guideline on the Need for Nonclinical Testing in Juvenile Animals on Human Pharmaceuticals for Pediatric Indications" (EMA 2008) states: "When addressing overall effects on growth and development, studies should include, at a minimum, measurement of growth." In the Japanese guideline the MHLW "Guideline on the Nonclinical Safety Study in Juvenile Animals for Pediatric Drugs," there are several comments on required assessments including "organs/functions growth and development…skeletal system" and "the effects on development and its reversibility." All three guidances (FDA, EMA, and MHLW) discuss the skeletal system as an "organ system of concern," others being the cardiovascular, pulmonary, gastrointestinal, neurobehavioral, immune, reproductive, renal, and metabolic. As a result, there are often requirements to study growth and skeletal development in detail in the Pediatric Investigational Plans (PIP) (European Medicines Agency [EMA] 2006) and Pediatric Study Plans (PSP) (United States Food and Drug Agency [FDA] 2016).

Regulatory agencies frequently have specific requests when reviewing the nonclinical components of the PIPs and PSPs. Sometimes these are focused on organ systems of concern like the skeleton and can give detailed requests of species, duration of treatment, and endpoints to be studied. For the skeletal system, requests to study bone density are relatively common along with more detailed requests to study skeletal growth/maturation, muscle/bone growth, bone development with bone density and length, bone quality endpoints of bone mass and density, biomechanical properties, and biochemical markers of bone turnover. Requests have also included the study of specific components of the skeleton (e.g., growth plates), as well as dentition.

Table 3.1 Human organ system development

Organ system	Age of development
Behavior/ cognition	Up to adulthood
Immune system	Up to 12 years old
Reproduction	Up to adulthood
Skeleton (growth)	Up to adulthood
Renal	Up to 1 year old
Pulmonary	Up to 2 years old
Cardiovascular	Up to 5–7 years old (based on ECG)
Gastrointestinal	Up to 1–2 years old

A nonclinical approach must be taken with evaluations of some organ systems, such as the musculoskeletal system, because the prolonged times to achieve full development in the human (up to 25 years for the musculoskeletal system) precludes assessment in clinical trials (Table 3.1).

3.2 Postnatal Skeletal Development

Since the most commonly used laboratory species (rats, mice, rabbits and dogs) are born less developed than humans, these species provide models that can be used to study processes seen in both postnatal development and the development of premature infants (Table 3.2).

A comprehensive review of postnatal bone growth and development was prepared by Zoetis et al. (2003). Critical postnatal milestones in development of the skeleton were identified as appearance of secondary ossification centers, longitudinal bone growth, fusion of secondary ossification centers, diametric bone growth, and bone vascularity.

Primary ossification centers mostly develop during the embryonic and early fetal period; however, in the long bones secondary centers of ossification develop in areas where primary centers have not extended (Zoetis et al. 2003). In many larger mammals (non-rodent), the growth plates close at skeletal maturity and longitudinal bone growth stops. In rats and mice growth continues after skeletal maturity; in rats skeletal maturity is achieved at 11.5–13 weeks; however, bone growth continues at a lower rate until approximately 26 weeks (Roach et al. 2003). The timing of epiphyseal closure is shown in Table 3.3 for various laboratory species. In studying the time course of the tibial growth plate fusion using micro-CT (computed tomography), Martin (Martin et al. 2003) showed 90% of maximal bone bridging having occurred by 7.4 months in male rats and 6.5 months in females; longitudinal growth had effectively ceased after 8 months of age. It is worthy of note that the minipig also shows continued growth after sexual maturity, which occurs at 5–8 months, with epiphyseal closure as late as 21–42 months (Tortereau et al. 2013; Tsutsumi

Table 3.2 Age groups and comparative ages

Age groups (EMA/FDA)	Human	Rat (mouse[a]) (days)	Rabbit (weeks)	NHP (cyno) (months)	Minipig (days)	Dog (days)
Premature	less than term	1–4	0 to 1–2	–	–	1–4/10
Neonate	Birth to 1 month	4–10 (7)	2–3	Birth to 0.5	0–14	5/11–21
Infant	1 month–2 years	10 (7) to 21	3–5	0.5–5	15–28	22–42
Children	2–12 years	21–28F/35M	5–13	Jun-35	29–108	43–140F/170M
Juvenile	12–16 years	28F/35M–49F/70M (28F/35M–42F/56M)	13–21	36–48	120–180	150F/180M–250F/260M

M male, *F* female
[a]Mouse added where different

Table 3.3 Species – postnatal skeletal development

Parameter	Human (years)	Rat (mouse) (days)	Rabbit (weeks)	NHP (months)	Minipig (weeks)	Dog (weeks)
Age compared to human at birth (*estimate*)	–	7–14 (7–10)	0	Precocious (120 gd)	?	1–2
Puberty	12–16	30F to 49M (21F–35M)	13–17	24–36 F to 43–52 M	13–17	26–34F to 34–36M
Sexual maturity	16–18	56F to 70M (49F–56M)	18–32	31–48F to 53–60M	17–32	34–52F to 52–56M
Epiphyseal closure	11–20	105–301 (91–105)	16–32	21–72	78–182	26–56

M male, *F* females, *gd* gestation day

et al. 2004). As bones lengthen they also grow in diameter by appositional bone growth; in this type of growth, new bone is deposited on the periosteal surface.

As cortical bone matures, the immature woven bone and primary osteons are removed by the osteoclasts, which hollow out a channel through the bone, usually following existing blood vessels. Layers of bone-forming cells, or osteoblasts, follow the osteoclasts and lay down new bone on the sides of the channel; the layers of bone built up in this way slowly narrow the channel until a tunnel not much larger than the central blood vessel remains. The blood supply for the osteocytes then passes through these channels, the Haversian canals. Osteons are characteristic of mature bone and occur extensively in cortical bone in humans and large animal (non-rodent) species. Limited Haversian remodeling is seen in rodents such rats, mice, hamsters, and guinea pigs, whereas in the rabbit some Haversian systems are seen (Zoetis et al. 2003). There is clear similarity among mammalian species in the vascularity in long bones. This is discussed in more detail in the preceding chapter (Skeletal Growth, Adaptation, and Maintenance in the Bone Physiology and Biology Chap. 2).

3.3 Growth

3.3.1 Concepts

Growth can be defined as "an increase in cell size and number" and is distinct from development or maturation which "represents a progression through the stages leading to maturity" (Rovner and Zemel 2009). The endpoint of human physical growth is attainment of adult size and the endpoint of development of the skeleton is closure of the epiphyses (Rovner and Zemel 2009) and attainment of peak bone mass. In a favorable environment, the genetic potential of growth may be achieved, and adverse environmental factors may alter this depending on their strength, duration, and frequency (Rovner and Zemel 2009).

The phases of growth in the neonate and infant human (from birth to 2 years) are rapid and then slowing; for childhood (2–12 years) a continued constant growth with lower, decelerating velocity; and then from puberty (12–16 years) to adulthood, a rapid acceleration and then deceleration (Tassinari 1999). There are marked sex differences in both humans and common laboratory species for the timing of puberty and thus growth and development of the skeleton (Fig. 3.1). These timings can be compared to the FDA/EMA pediatric age groups (FDA 2006; EMA 2008) (Table 3.2) and equivalent ages for nonclinical species.

Because rodent animal models do not have the equivalent of the prolonged childhood of humans, the comparative ages are compressed into the first 5–8 weeks of life for rats. A peak of growth can be seen in Fig. 3.1 just before vaginal opening (and first estrous) in females and slightly later before preputial separation for males. Growth velocity for the skeleton can be defined as the change in height (length or width) divided by the time between measurements.

Above normal changes in growth following periods of undernutrition or illness have been reported in animals and humans (Osborne and Mendel 1915). The term "catch-up growth" was introduced by Prader (Prader et al. 1963) who described it as "the phase of rapid linear growth that allowed the child to accelerate toward and, in favorable circumstances, resume his/her pre-illness growth curve." Catch-up growth may be defined as a "height velocity above the statistical limits of normality for age and/or maturity during a defined period of time, following a transient period of growth inhibition" (Boersma and Maarten Wit 1997). Theoretically three types of catch-up growth have been suggested by these authors, although clear distinctions are not always evident.

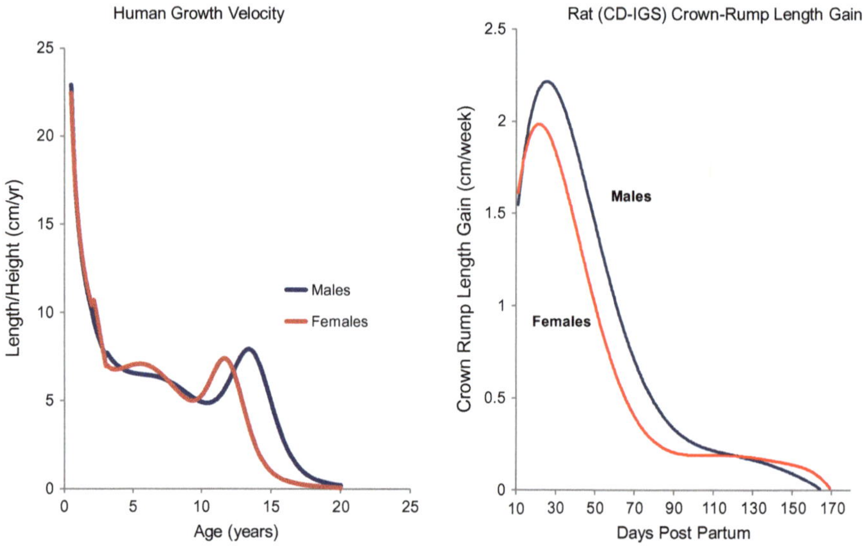

Fig. 3.1 Growth patterns

Type A is where growth restriction ceases and height velocity increases to such an extent that the deficit is swiftly eliminated.

Type B is where when growth restriction ceases a delay in growth and somatic development persists, but growth continues for longer than normal.

Type C is a mixture of the first two types.

3.3.2 Importance

Growth velocity is an excellent indicator of current nutritional status and overall well-being for pediatric patients whose growth is incomplete (Rovner and Zemel 2009).

In presenting a set of principles for teratology, in the fourth principle Wilson defined adverse effects as "death, malformation, growth retardation and functional disorder" (Wilson 1977). Because many of the common laboratory species are born less developed than humans, in some nonclinical studies when treatment is initiated at young ages, both for the animal in general and certain organ systems in particular, this earliest time may represent growth and development that happens in utero in the human. Therefore, adverse effects may be viewed as representing a continuum from teratologic to toxicologic events during the neonatal and juvenile periods.

Growth retardation following in utero exposure to ethanol is followed by a lack of catch-up growth in the postnatal period with fetal alcohol syndrome (FAS). However, growth retardation following in utero exposure to tobacco smoke is followed by full catch-up growth for weight and partial catch-up growth in terms of height, as well as with increased obesity (Power and Jeffris 2002; Strauss 1997). Adolescent smoking affects growth in terms of height in males, but not females, there being no indications of catch-up growth (O'Loughlin et al. 2008).

3.3.3 *Preclinical Measurements*

3.3.3.1 Rodents: In Vivo

Simple techniques that can be used to measure skeletal growth include crown-rump (or snout-rump) length, external tibia length, and tail length measurements (Figs. 3.2 and 3.3). These measurements can be taken frequently and without anesthesia using calipers and/or rulers (Hoberman and Barnett 2012). The variability in these conscious measurements is comparatively small with coefficients of variation comparing favorably with body weight as shown in Table 3.4 (Robinson et al. 2012a), with group sizes of at least ten/sex typically used. In studies starting at young ages, the day of incisor eruption can be assessed.

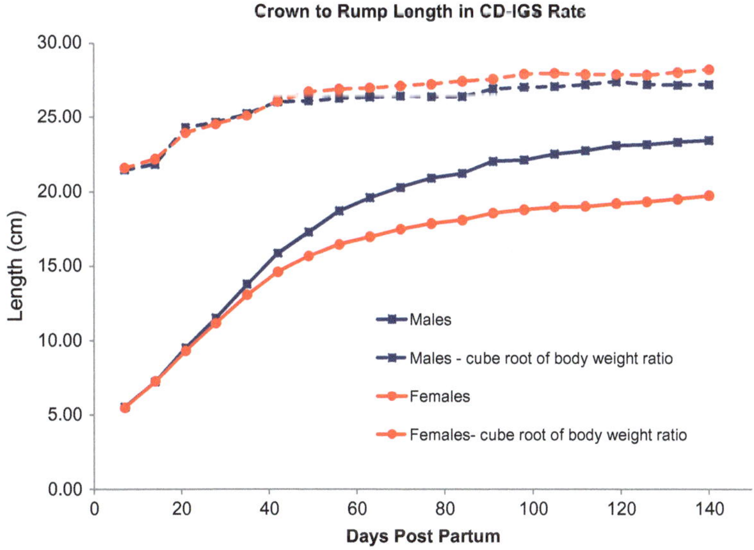

Fig. 3.2 Crown to rump length in CD-IGS rats

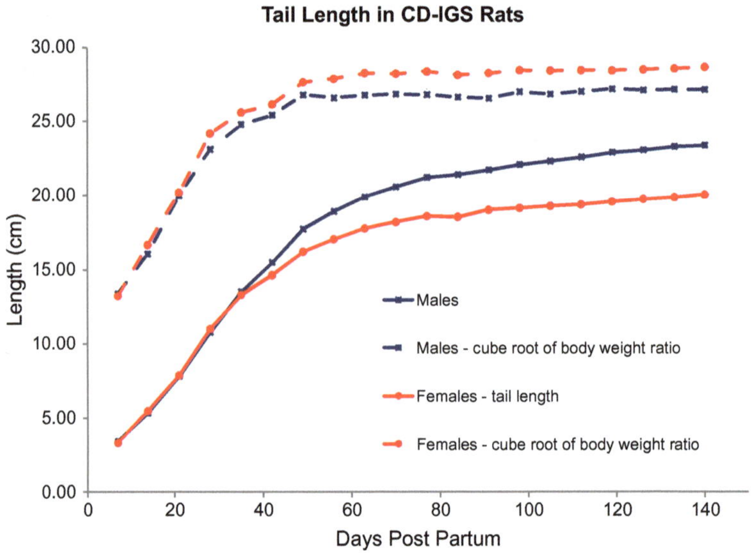

Fig. 3.3 Tail length in CD-IGS rats

Measurements of the length and width of the tibia and femur and the length of distinct regions of the spinal column, for example, lumbar vertebrae 1–6, can be made using radiographs; however, the frequency of these is restricted by the need for anesthesia (Robinson et al. 2012). Data for the tibia and femur are shown in Table 3.4. Other components of the skeleton like the skull and pelvis can also be measured (Hughes and Tanner 1970). In addition to the measurements, the radiographs also allow for the detection of abnormalities of skeletal development (Barbeau et al. 2006). Other imaging techniques such as bone densitometry using pQCT and DXA can be utilized to measure bone mineral density (BMD), bone mineral content (BMC), and some geometry parameters (Tables 3.5 and 3.6) as described elsewhere. Commonly radiographs, pQCT and/or DXA scanning, can be undertaken during one occasion of anesthesia. By having several subsets of animals, these imaging techniques can be conducted at frequent intervals.

Biochemical markers of skeletal growth can be measured at regular intervals during the postnatal period of development of the skeleton; these include markers for bone formation: procollagen type I propeptide (PINP), osteocalcin (serum) and total and/or bone specific alkaline phosphatase (serum) and for bone resorption, N- and/or C-telopeptide (serum/urine) and deoxypyridinoline (urine). Other clinical chemistry parameters that can be measured include serum calcium (total and ionized), serum phosphorus, 1,25-dihydroxyvitamin D, 25-hydroxyvitamin D, parathyroid hormone, and calcitonin. The volume of blood may restrict analyses at the younger ages or require pooling of blood from littermates.

Table 3.4 Variance of in vivo and ex vivo parameters for rats

Parameter	Strain	Sex	Mean (CV)	Min (CV)	Max (CV)
In vivo					
Crown-rump length	CD-IGS	Male	4.01	2.40	7.79
	CD-IGS	Female	4.00	2.59	7.72
	Han Wistar	Male	4.35	2.77	7.44
	Han Wistar	Female	3.36	1.94	7.92
External tibia	CD-IGS	Male	2.75	1.53	4.37
	CD-IGS	Female	3.47	1.80	4.83
	Han Wistar	Male	3.56	2.37	5.57
	Han Wistar	Female	3.66	2.45	5.56
Body weight	CD-IGS	Male	7.97	7.13	9.29
	CD-IGS	Female	9.21	7.05	11.14
	Han Wistar	Male	9.68	7.87	12.85
	Han Wistar	Female	8.75	6.88	10.75
Radiographs					
Femur length	CD-IGS	Male	3.00	2.26	4.29
	CD-IGS	Female	2.74	1.22	3.88
	Han Wistar	Male	2.73	2.09	3.38
	Han Wistar	Female	2.65	1.76	3.53
Femur width	CD-IGS	Male	5.62	4.09	9.55
	CD-IGS	Female	6.00	4.77	8.14
	Han Wistar	Male	3.51	0.48	6.55
	Han Wistar	Female	5.74	5.41	6.07
Ex vivo					
Femur length	CD-IGS	Male	3.17	2.19	5.75
	CD-IGS	Female	2.62	1.88	3.77
	Han Wistar	Male	3.25	1.51	3.94
	Han Wistar	Female	3.16	1.51	6.54
Femur width	CD-IGS	Male	7.22	2.86	10.78
	CD-IGS	Female	7.12	4.12	13.07
	Han Wistar	Male	8.17	5.28	12.92
	Han Wistar	Female	8.82	4.27	13.64

3.3.3.2 Rodents: Ex Vivo

Bones can be measured at necropsy, typically the tibia and femur lengths and widths are recorded (Table 3.7). These and other bones can be preserved for subsequent imaging, histopathology, histomorphometry, and/or biomechanical strength testing. Bone growth dynamics can be measured using fluorochrome labeling and histomorphometry, so that parameters such as bone formation rate (BFR) and mineral apposition rate (MAR) can be calculated (as described later in this book, Application of Histopathology and Bone Histomorphometry for Understanding Test Article-Related Bone Changes and Assessing Potential Bone Liabilities Chap. 8).

Table 3.5 In vivo densitometry in CD®-IGS rats – pQCT

		Days postpartum									
Parameter		20	20	46	46	105	105	193	193	231	231
pQCT – proximal tibia		M	F	M	F	M	F	M	F	M	F
Metaphysis – cortical/subcortical											
Area (mm^2)	Mean	3.41	2.77	6.65	4.73	8.13	4.38	8.60	4.28	8.78	4.21
	CV	9.96	11.97	7.01	10.23	7.34	10.17	8.33	9.30	9.80	5.44
BMC (mg/mm)	Mean	1.35	1.14	3.45	2.94	7.02	4.45	8.36	4.73	8.54	4.62
	CV	10.48	13.02	5.68	7.50	6.25	10.69	8.34	9.85	10.54	6.13
BMD (mg/cm^3)	Mean	394.33	411.30	520.01	622.38	865.06	1017.80	973.21	1105.42	971.82	1097.66
	CV	4.15	6.11	4.65	4.50	3.79	2.47	3.70	2.08	3.14	1.34
Diaphysis – cortical											
Area (mm^2)	Mean	1.40	1.39	3.49	3.23	6.64	4.95	8.28	5.65	8.47	5.87
	CV	10.30	10.67	8.54	6.88	8.72	7.11	7.85	7.20	9.60	8.74
BMC (mg/mm)	Mean	0.86	0.86	3.31	3.07	7.49	5.53	9.73	6.60	9.99	6.85
	CV	17.17	14.10	9.25	6.95	8.68	8.45	8.21	8.26	9.64	9.14
BMD (mg/cm^3)	Mean	612.66	618.29	948.31	950.27	1126.78	1115.29	1175.09	1167.42	1178.95	1166.54
	CV	7.53	5.00	2.24	2.68	1.40	1.84	1.25	2.14	1.83	1.11
Thick (mm^2)	Mean	0.34	0.34	0.65	0.65	0.99	0.86	1.10	0.95	1.09	0.96
	CV	16.61	14.01	6.00	6.12	4.71	5.70	5.49	5.91	1983.78	4.18
Metaphysis – trabecular											
Area (mm^2)	Mean	5.13	4.16	10.04	7.13	12.22	6.58	12.94	6.42	13.18	6.36
	CV	9.18	11.62	6.83	10.31	7.38	10.07	8.35	9.20	9.96	6.12
BMC (mg/mm)	Mean	1.48	0.97	2.30	1.86	2.84	3.45	4.35	4.36	4.28	4.54
	CV	14.63	23.77	13.71	15.09	15.22	15.56	10.98	12.50	8.42	11.97
BMD (mg/cm^3)	Mean	288.23	231.28	229.13	261.00	233.09	528.24	338.26	683.21	327.13	715.58
	CV	6.29	13.98	12.69	12.08	16.05	17.87	14.41	3.37	12.46	10.81

Table 3.6 In vivo bone densitometry in CD®-IGS rats – DXA

		Days postpartum									
Parameter		20	20	46	46	105	105	193	193	231	231
DXA		M	F	M	F	M	F	M	F	M	F
Whole body area (cm²)	Mean	16.68	16.29	47.10	40.55	83.96	61.58	99.59	69.67	102.20	72.87
	CV	8.34	9.77	4.33	6.85	5.46	5.84	5.55	5.87	4.24	6.26
Whole body BMC (g)	Mean	1.38	1.31	5.31	4.64	14.06	9.80	18.95	12.09	19.67	12.96
	CV	9.72	13.17	6.97	8.48	8.86	9.12	8.81	8.89	10.10	8.07
Whole body BMD (g/cm²)	Mean	0.08	0.08	0.11	0.11	0.20	0.16	0.19	0.17	0.19	0.18
	CV	5.68	5.96	4.61	4.55	3.65	3.71	4.11	3.46	6.19	2.81
Lumbar vertebrae area (cm²)	Mean	0.52	0.50	1.29	1.20	2.52	1.93	2.87	2.11	2.93	2.16
	CV	9.51	8.40	5.19	8.89	4.94	5.55	6.42	5.75	4.35	5.94
Lumbar vertebrae BMC (g)	Mean	0.020	0.019	0.174	0.172	0.618	0.459	0.811	0.558	0.839	0.584
	CV	15.00	15.79	6.61	12.21	8.41	6.49	11.96	11.48	8.71	12.50
Lumbar vertebrae BMD (g/cm²)	Mean	0.040	0.038	0.134	0.144	0.250	0.237	0.281	0.263	0.285	0.269
	CV	16.58	11.46	4.84	4.74	5.48	6.32	4.21	6.41	5.92	7.19
Whole femur area (cm²)	Mean	0.343	0.340	0.996	0.933	1.966	1.504	2.214	1.626	2.297	1.669
	CV	10.79	10.88	6.12	5.79	5.47	5.15	4.63	4.18	6.38	4.91
Whole femur BMC (g)	Mean	0.014	0.015	0.154	0.157	0.614	0.441	0.805	0.526	0.852	0.556
	CV	14.29	20.00	8.79	8.92	7.98	9.53	6.96	8.85	9.57	10.43
Whole femur BMD (g/cm²)	Mean	0.042	0.043	0.154	0.168	0.311	0.292	0.362	0.323	0.369	0.332
	CV	9.98	16.32	3.91	4.82	3.95	5.09	4.04	5.38	3.99	5.99

M male, *F* female

Table 3.7 Terminal femur and tibia measurements (cm) in rats

Weeks post partum	7		10/11		15/16		21–23	
Strain	CD-IGS	HW	CD-IGS	HW	CD-IGS	HW	CD-IGS	HW
Femur								
Male -length	3.38	3.00	3.85	3.41	3.97	3.53	4.26	3.89
Cube root propn bwt	4.99	5.05	4.79	4.96	4.68	4.91	4.79	5.08
Male – width	0.44	0.36	0.48	0.41	0.51	0.39	0.52	0.48
Cube root propn bwt	0.64	0.60	0.60	0.59	0.60	0.54	0.58	0.62
Female – length	3.07	2.81	3.40	3.04	3.49	3.16	3.67	3.35
Cube root propn bwt	5.12	5.24	5.06	5.27	5.03	5.21	5.07	5.23
Female – width	3.89	0.35	4.03	0.35	4.20	0.33	4.23	0.40
Cube root propn bwt	0.65	0.65	0.60	0.60	0.60	0.54	0.59	0.62
Tibia								
Male – length	3.82	3.49	4.31	3.86	4.51	4.01	4.75	4.33
Cube root propn bwt	5.63	5.89	5.37	5.61	5.32	5.57	5.34	5.65
Male – width	0.32	0.25	0.32	0.31	0.39	0.30	0.41	0.37
Cube root propn bwt	0.47	0.42	0.39	0.44	0.46	0.41	0.46	0.49
Female – length	3.48	3.28	3.84	3.46	3.96	3.62	4.10	3.79
Cube root propn bwt	5.81	6.11	5.72	6.00	5.71	5.95	5.67	5.91
Female – width	0.31	0.24	0.27	0.27	0.31	0.27	0.32	0.31
Cube root propn bwt	0.52	0.44	0.40	0.46	0.44	0.45	0.45	0.48

3.3.3.3 Non-rodents: In Vivo

For rabbits, crown-rump length can be readily measured (Fig. 3.4). Radiographic measures of the femur, tibia, and lumbar spine can be made (Robinson et al. 2015). Also, data from pQCT and DXA measurements are shown in Table 3.8 (Primakova et al. 2013).

For dogs in-life measures of length and height (Figs. 3.5 and 3.6) can be readily made using a simple apparatus (Robinson et al. 2012b). As for rodents, these data show low CVs (4–14%) compared to body weight (12–30%) with typical data sets of eight or more pups per sex. Radiographic measures can also be made for the spine and long bones (Fig. 3.7). Other imaging techniques mentioned above can be employed to study skeletal development. Data from pQCT and DXA are shown in Table 3.9 (Adamo et al. 2013). For dogs, the ages of eruption of the incisors, molars, and canines can be recorded.

For neonatal/juvenile nonhuman primates, many external measurements can be made, including long bone (tibia and femur), crown-rump length, and head circumference, Charles River Historical Control Data (Chellman 2016). Radiographic measurements are shown in Fig. 3.8, which typically need to be taken with anesthesia at younger ages using simple techniques (Weinbauer et al. 2012). As for the dog, bone densitometry assessments (DXA and pQCT) can also be made.

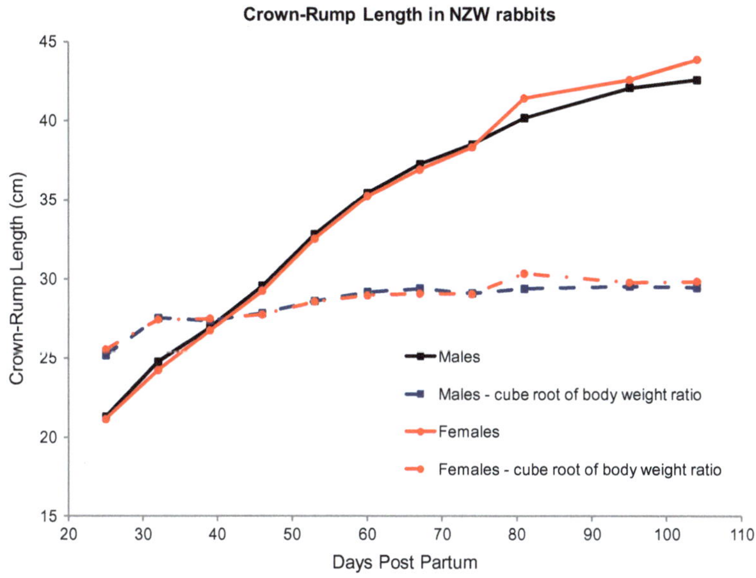

Fig. 3.4 Crown to rump length in NZW rabbits

3.3.3.4 Non-rodents: Ex Vivo

Although less commonly conducted, various bones can be measured terminally at necropsy; the tibia and femur lengths and widths can be readily recorded. These and other bones can be preserved for subsequent imaging, histopathology, histomorphometry, and/or biomechanical strength testing, as described in more detail in other chapters of this book.

3.3.3.5 Data Interpretation

The coefficients of variation (CV) for the various techniques for measurement of bone size (length and geometry) were generally less than 5% across a full range of ages for two commonly used rat strains (Sprague Dawley and Han Wistar) (Table 3.4). The widths of the femur and tibia, whether measured as excised bones or in vivo with radiographs, tended to show more variance than lengths, with averages up to 9%. In comparison, body weight values for these subsets of animals showed average CVs of 8–10%. Measurements of crown to rump length, external tibia, and tail length can be made during routine handling of the animals. Body weight changes are expected in juvenile toxicology studies, and so data analysis techniques that allow for evaluation of the growth measurements relationship to body weight may be needed. Calculation of the ratio of the cube root of the body weight (Wise et al. 1991), rather than just a simple percentage of body weight, was

Table 3.8 In vivo bone densitometry in Crl:KBL[NZW] rabbits

		Days 24/25 postpartum				Days 52/53/54 postpartum			
		Male		Female		Male		Female	
Parameter	Units	Mean	CV	Mean	CV	Mean	CV	Mean	CV
Radiography									
Femur length	mm	45.5	4.4	44.6	5.1	71.8	1.9	70.6	3.2
Tibia length	mm	49.3	4.2	48.4	4.7	77.3	2.2	76.3	3.7
Lumbar spine (L1–L6) length	mm	52.0	4.9	51.3	5.3	87.3	2.4	87.1	3.7
pQCT proximal tibia metaphysis									
Total area	mm^2	38.0	16.1	36.5	13.2	74.0	9.0	70.2	10.0
Total BMC	mg/mm	11.4	22.1	11.1	20.5	27.6	8.2	26.7	13.8
Total BMD	mg/cm^3	296.8	9.4	301.0	9.7	374.2	6.7	380.1	7.0
Trabecular area	mm	23.1	13.1	22.1	11.1	37.0	13.9	35.1	9.4
Trabecular BMC	mg/mm	3.7	20.9	3.6	20.2	5.8	18.5	5.5	21.1
Trabecular BMD	mg/cm^3	159.5	12.3	164.1	14.5	156.8	14.8	156.5	20.3
Cortical/subcortical area	mm^2	15.0	26.7	14.4	24.8	37.0	12.1	35.1	17.2
Cortical/subcortical BMC	mg/mm	7.7	25.9	7.4	24.7	21.8	10.0	21.2	14.1
Cortical/subcortical BMD	mg/cm^3	512.7	4.9	516.2	5.1	592.2	5.9	609.0	6.4
***pQCT** proximal tibia diaphysis*									
Total area	mm^2	12.7	11.2	12.1	12.8	26.5	7.2	25.2	8.2
Cortical area	mm^2	5.9	12.6	5.7	11.6	13.1	6.8	13.1	8.1
Cortical BMC	mg/mm	5.1	15.2	4.9	14.6	13.5	7.9	13.5	9.1
Cortical BMD	mg/cm^3	864.7	3.6	863.6	4.1	1030	1.8	1033	2.1
Cortical thickness	mm	0.5	11.0	0.5	10.5	0.8	7.4	0.9	7.3
Periosteal circumference	mm	12.6	5.6	12.3	6.4	18.2	3.6	17.8	4.1
Endosteal circumference	mm	9.2	7.6	9.0	9.6	13.0	6.3	12.3	6.4
DXA whole femur									
Area	cm^2	2.6	8.9	2.5	9.3	6.8	6.1	6.5	6.1
BMC	g	0.5	18.4	0.4	18.4	1.9	9.7	1.8	12.1
BMD	g/cm^2	0.2	11.6	0.2	12.4	0.3	8.0	0.3	7.3
DXA whole body									
Area	cm^2	89.3	8.7	86.7	10.2	199.6	4.5	198.0	5.6
BMC	g	12.7	13.2	12.2	0.1	29.2	8.6	28.8	9.0
BMD	g/cm^2	0.1	5.3	0.1	6.1	0.1	5.1	0.1	4.1
DXA lumbar spine (L1–L4)									
Area	cm^2	2.4	7.7	2.4	9.2	4.9	9.7	5.0	9.9
BMC	g	0.3	16.5	0.3	16.1	1.1	14.1	1.1	13.1
BMD	g/cm^2	0.1	9.5	0.1	8.2	0.2	5.0	0.2	4.9

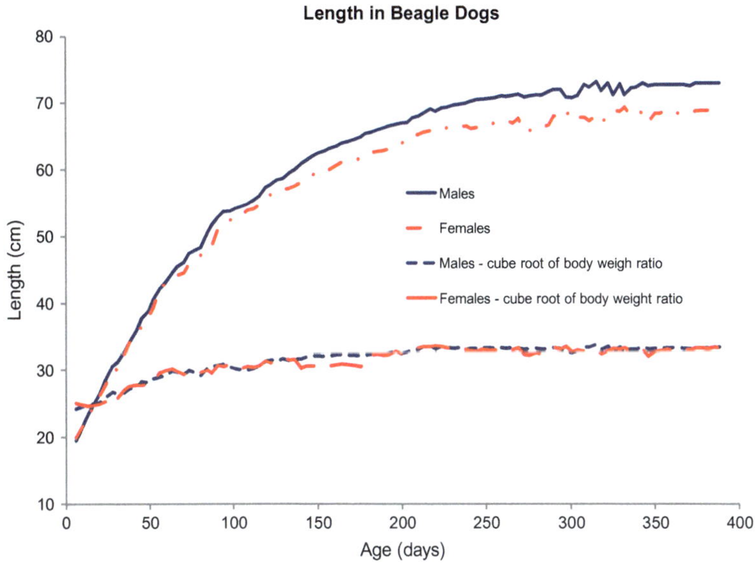

Fig. 3.5 Length in beagle dogs

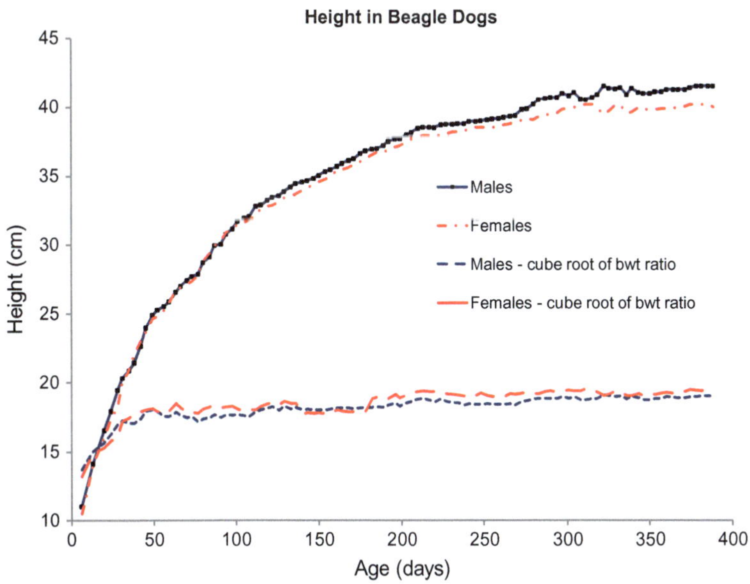

Fig. 3.6 Height in beagle dogs

conducted for several data sets and the utility of this approach to normalize the data can be seen for both in vivo and ex vivo measures (Figs. 3.2, 3.3, 3.4, 3.5, 3.6, 3.7, and 3.8 and Table 3.7). Analyses of these ratios can be helpful to discern generalized effects on growth from specific effects on skeletal development.

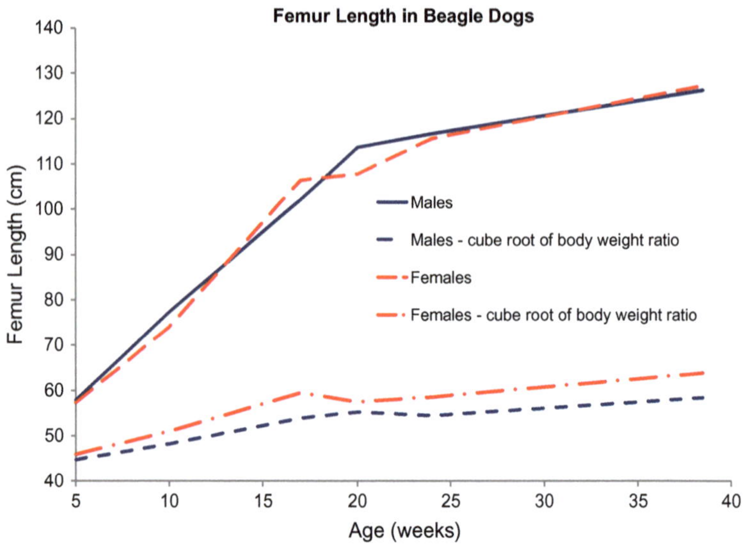

Fig. 3.7 Femur length in beagle dogs

Skeletal growth in the rat continues at a low rate for a number of months after sexual maturity. The dosing of juvenile toxicology studies may be completed prior to this time, and so growth may be continuing in the post-dosing development (or recovery) period. Catch-up growth, following treatment, can occur as a result of either an increased velocity or an extended growth period or as a result of a mix of velocity and extended growth. The aforementioned length of the post-dosing development period should consider that catch-up growth may occur after the cessation of dosing and that the normal growth period may need to be extended.

3.4 Postnatal Therapies

3.4.1 Anticancer Agents

Therapies for pediatric cancers have produced a population of survivors who may have problems resulting from their treatment. These effects can be delayed; cancer therapies may cause subclinical changes that may only be of clinical relevance as the subject ages (Wasilewski-Masker et al. 2008). Peak bone mass may be adversely affected by bone metabolism perturbations (Wasilewski-Masker et al. 2008). For example, treatment of leukemia with prednisone (corticosteroid) and methotrexate (cytotoxic agent) results in decreased bone mineral accretion (Wasilewski-Masker et al. 2008).

Table 3.9 DXA and pQCT data for dog pups

Parameter	Units		Postpartum day or week							
			Day 10/11	Day 15/16	Day 18/19	Week 5	Week 9/10	Week 13	Week 17	Week 20/21
DXA										
Whole body BMC	g	M + F	11.5	18.4	15.9	52.4	87.9	–	–	239.6
Lumbar vertebrae area	cm²	M + F	2.65	3.82	2.48	5.64	8.54	8.73	11.2	14.7
Lumbar vertebrae BMC	g	M + F	0.38	0.49	0.40	1.43	2.80	3.07	4.82	7.36
Lumbar vertebrae BMD	g/cm²	M + F	0.14	0.13	0.16	0.27	0.33	0.35	0.42	0.48
Femur area	cm²	M + F	0.18	0.28	2.25	4.64	7.57	8.75	11.27	15.78
Femur BMC	g	M + F	0.31	0.49	0.50	1.59	2.80	3.15	4.64	6.98
Femur BMD	g/cm²	M + F	0.17	0.18	0.22	0.28	0.33	0.36	0.40	0.47
pQCT tibia metaphysis										
Total area	mm²	M + F	19.9	34.1	27.4	37.6	99.78	129.7	149.7	167.9
Total BMC	mg/mm	M + F	10.7	14.0	13.0	18.3	34.72	42.9	57.9	68.5
Total BMD	mg/cm³	M + F	543.6	413.7	477.3	489.4	349.1	334.2	389.2	412.5
Trabecular area	mm²	M + F	6.62	20.6	13.9	19.7	57.46	79.1	76.0	77.3
Trabecular BMC	mg/mm	M + F	1.96	5.79	4.36	5.11	12.20	17.8	19.8	20.8
Trabecular BMD	mg/cm³	M + F	305.5	283.0	316.6	261.9	209.4	223.8	261.0	272.9
Cortical/subcortical area	mm²	M + F	13.3	13.5	13.5	17.9	42.33	50.6	73.6	90.6
Cortical/subcortical BMC	mg/mm	M + F	8.74	8.26	8.67	13.2	22.53	25.1	38.2	47.9
Cortical/subcortical BMD	mg/cm³	M + F	657.9	612.5	643.0	737.7	531.8	496.0	524.9	534.9
pQCT tibia diaphysis										
Total area	mm²	M + F	15.7	22.5	18.5	23.8	47.7	68.8	92.2	105.6
Cortical area	mm²	M + F	9.93	14.2	12.1	14.6	21.7	26.9	34.8	41.7
Cortical BMC	mg/mm	M + F	7.41	10.2	9.16	13.0	20.4	24.7	26.6	37.5
Cortical BMD	mg/cm³	M + F	747.5	716.9	760.5	890.0	817.4	839.6	837.2	850.5
Cortical thickness	mm	M + F	0.89	1.05	0.99	1.04	1.02	1.15	1.17	1.32
Periosteal circumference	mm	M + F	14.0	16.8	15.2	17.3	24.4	29.2	33.5	35.9
Endosteal circumference	mm	M + F	8.48	10.2	8.98	10.7	18.0	22.0	26.5	28.1

M male, *F* female

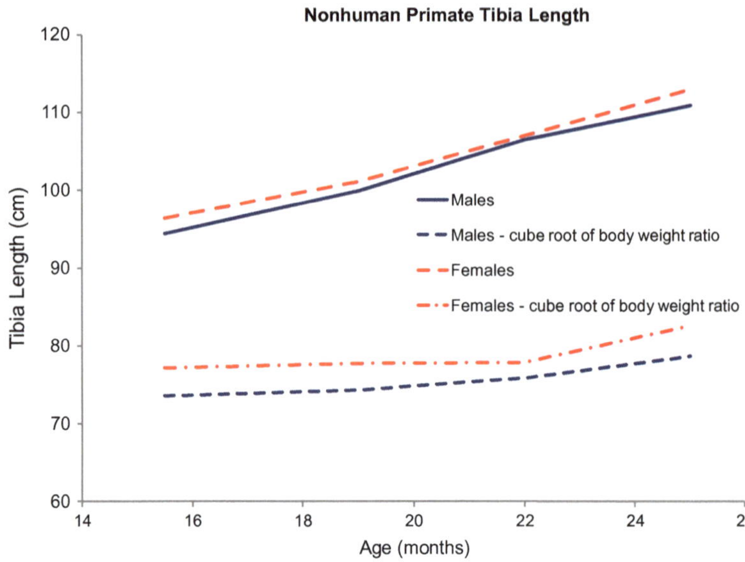

Fig. 3.8 Nonhuman primate tibia length

Methotrexate therapy can cause reduced bone volume and a decrease in bone formation due to its cytotoxic effect on osteoblasts. Adverse effects on the skeleton include osteopenia, fractures, and growth plate changes (Roebuck 1999). Corticosteroids such as prednisone and dexamethasone can have adverse effects on bone mineral density (BMD) and are associated with a higher incidence of fractures, resulting from effects on bone that include lower osteoblast activity (Wasilewski-Masker et al. 2008; Roebuck 1999).

Radiation can cause growth hormone deficiency associated with hypothalamic-pituitary injury resulting in BMD impairment. Also, radiation can damage the gonads and the consequent lower hormone levels (estrogen and testosterone) can cause lower BMD (Wasilewski-Masker et al. 2008). Alkylating agents can have effects on the gonads similar to radiation, with which combination therapy can have synergistic effects on hormones (Wasilewski-Masker et al. 2008).

Bortezomib, a proteasome inhibitor, is used to treat multiple myelomas, and it was noted to have beneficial effects on the bone in adult patients (Mohty et al. 2014). In myeloma, adverse effects on bone are related to an imbalance in remodeling of the bone with increased osteolytic bone destruction and no compensatory bone formation (Terpos and Christoulas 2013; Mohty et al. 2014). At the time of writing, pediatric studies are ongoing. In a mouse model, Eriksson et al. 2012 showed that bortezomib caused growth retardation by effects on growth plate chondrocytes. In vitro studies showed effects on rat metatarsal bones, by causing chondrocyte apoptosis and effects in human growth plate cartilage.

Monoclonal antibodies targeting vascular endothelial growth factor (VEGF), a cytokine that is pro-angiogenic, have been shown to cause growth plate abnormalities in pediatric patients (Voss et al. 2015); these abnormalities included growth plate widening and physeal cartilage hypertrophy.

Schunior studied the effects of irradiation, prednisolone, and methotrexate on growth in juvenile rats (Schunior et al. 1990). Exposures/doses were given to Sprague Dawley rats on postnatal days 17 and 18. Measurements included body length, femur length, and a range of craniofacial dimensions. Effects seen included lower body weight, a lack of catch-up growth and/or an adolescent growth spurt, and microcephaly. Also, ratios of body length to body weight were affected. Radiation alone caused microcephaly and with chemotherapy craniofacial proportions were altered. Chemotherapy also affected growth.

An anti-angiogenic estrogen metabolite, 2-methoxyestradiol, a potential antitumor therapy, caused suppression of long bone elongation in juvenile rats by decreasing the number of proliferating chondrocytes in the growth plate and increasing apoptosis in differentiated chondrocytes (Sibonga et al. 2002).

Ryan and coworkers reported 4 and 13 week studies in cynomolgus monkeys with rhuMAbVEGF (bevacizumab), an anti-angiogenic humanized monoclonal antibody, in which physeal dysplasia was seen in the proximal and distal femoral and humeral growth plates (Ryan et al. 1999). These changes were described as "being characterized by markedly thickened growth plate cartilage, degeneration of the growth plate cartilage, disorganization of chondrocyte columns, which formed packets or clusters; decreased numbers of primary bony trabeculae with inappropriate retention of hypertrophied chondrocytes in trabeculae; absence of metaphyseal capillary invasion into the zone of hypertrophied chondrocytes; and formation of a transverse subchondral bony plate" (Ryan et al. 1999). It was thought the more severe effects in males as compared to females were related to the males being less sexually mature. After 4 weeks, partial recovery of the physeal dysplasia was seen by resumption of capillary invasion into the growth plate.

Doxorubicin, an anthracycline chemotherapeutic agent, given to 10-day-old rat pups by intraperitoneal injection, caused vertebral and femoral growth deficits with decreased BMD and bone mineral content (BMC) in the lumbar vertebrae in the females (Mwale et al. 2005). Coadministration of the cardioprotective agent dexrazoxane did not prevent these effects. There were also changes noted histologically in the growth plate and intervertebral discs; for the tibial growth plate, the dexrazoxane did have some partial rescue effects.

3.4.2 Retinoids

Pennes (Pennes et al. 1984) described observations of hyperostosis in mostly pediatric patients of 5–26 years of age following treatment with 13-cis-retinoic acid. Doses of 1–3 mg/kg/day given for a year resulted in six of the eight patients showing hyperostoses as identified by x-rays in the spine and ankles. Lawson and McGuire (1987) reported osteophyte formation and other ossification of ligaments, proliferative enthesopathies, decreased bone density, premature fusion of epiphyses, and modeling abnormalities in pediatric patients 13–16 years old receiving 13-cis-retinoic acid for 16–87 months, at doses of 1.4–2.9 mg/kg.

Premature epiphyseal closure was described in two pediatric patients treated with etretinate (an aromatic retinoid) for multiyear periods (Prendiville et al. 1986).

A range of skeletal changes related to etretinate in children were noted, including periosteal thickening and bone resorption, osteoporosis, disc narrowing, and premature epiphyseal closure (White and MacKie 1989).

There is a dearth of published studies on the effect of therapeutic retinoids on the skeleton of juvenile animals; however, hypervitaminosis A studies have been conducted. Early studies of hypervitaminosis A included assessments in weanling rats which showed fractures with 7–10 days of dosing (Moore and Wang 1945). In an early study of hypervitaminosis A in dogs (Greyhounds), skeletal lesions, described as an "increased tempo of growth processes" and "an acceleration of skeletal aging," were seen (Maddock et al. 1949). Also, studies in young adult rats (11–12 weeks of age) have shown marked effects of 13-cis-retinoic acid and/or all trans-retinoic acid on the skeleton (Hotchkiss et al. 2006). Spontaneous bone fractures occurred along with decreased bone diameter and cortical thickness in the femur, shown by histomorphometry, and DXA scanning revealed BMD and BMC effects; the lumbar spine was not affected (Hotchkiss et al. 2006).

3.4.3 *Corticosteroids*

Oral corticosteroids have been known to adversely affect growth as described above (Wasilewski-Masker et al. 2008; Roebuck 1999); however, some inhaled corticosteroids also have adverse effects. Inhaled budesonide has been shown to decrease growth velocity in one of three studies when compared to conventional asthma therapy (Skoner et al. 2000). The same authors also showed decrements in overall growth rates with beclomethasone dipropionate for 6–9 year old patients, effects were evident within 1 month suggesting the effects on growth occurred with the initial treatments (Skoner et al. 2000). A comparison of beclomethasone and fluticasone showed no effects with fluticasone, while beclomethasone resulted in slower rates of growth and decreased serum cortisol (Rao et al. 1999). Vertebral aseptic osteonecrosis, although rare, occurs most commonly with glucocorticoid treatment and occurs in the pediatric population.

Blood flow to the vertebrae is thought to be the cause of vertebral aseptic osteonecrosis, and this was studied in juvenile Danish Landrace pigs (Drescher et al. 2000) treated orally with methylprednisolone. Blood flow to cervical, thoracic, and lumbar vertebrae was decreased in the cancellous vertebral body bone and end plates of the vertebrae, as well as in regions of other bones such as the humerus. Bone mineral density, measured by DXA, was shown to be decreased in the lumbar vertebrae. There was also marked growth retardation for these animals.

Methylprednisolone administered to weanling rats for a week caused decreases in growth and increases in cortical area but no significant effect on trabecular BMD

(McHugh et al. 2003). Neonatal rat pups treated with cortisone acetate at a dose of 1.25 mg/pup (Sobel 1978) resulted in some effects on growth and smaller bones, but also indications of accelerated skeletal maturation were seen at day 21 postpartum, but not by day 84 postpartum.

3.4.4 Anti-infectives

One of the most commonly studied class of pharmaceutical compounds for effects on the skeleton is the quinolone antibiotics. Following a clinical report (Bailey et al. 1972) of arthropathy, initial preclinical studies were conducted in beagle dogs and showed cartilage damage in younger dogs (Ingham et al. 1977). These dogs also showed gait abnormalities (the only laboratory species to do so) including lameness. Fluid-filled vesicles or "blisters" were seen macroscopically, particularly on the distal epiphyses of the femur and humerus, and there were also changes in synovial fluid volumes (Gough et al. 1992; Davis and Mckenzie 1989; Stahlmann and Schwabe 1997; Takayama et al. 1995). The histopathological observations showed synovial joints were affected, including necrosis of chondrocytes and loss of extracellular matrix (Gough et al. 1992; Takayama et al. 1995). Other species including the rat and marmoset also show similar pathologic lesions; however, the dog has remained the species of choice for testing quinolones because of similarities in both the toxicity and toxicokinetics to the human (Gough et al. 1992; Davis and Mckenzie 1989).

3.4.5 CNS Therapies: Selective Serotonin Reuptake Inhibitors (SSRIs)

Spencer (Spencer et al. 2007) monitored the growth of children (6–17 years of age) treated with atomoxetine, a selective serotonin reuptake inhibitor (SSRI) for 5 years. There was a transitory effect on height after 12 months which had recovered by 24 months, weight effects were seen from 1 to 15 months, and in patients larger than average at the start of treatment, there were indications of a persistent effect.

A study of both various SSRIs and risperidone (an antipsychotic; see below) in boys 7–17 years of age showed hyperprolactinemia and effects on the BMD of the trabecular bone at the ultra-distal radius related to both SSRIs and risperidone, with effects on the lumbar spine with SSRIs (Calarge et al. 2010). Hyperprolactinemia, without effects on testosterone, were seen with risperidone.

Investigations of the effects of inhibition of the serotonin 5-hydroxytryptamine (5-HT) transporter by SSRIs were studied in mice with a null mutation for the gene for the 5-hydroxytryptamine transporter and normal juvenile mice (Warden et al. 2005). In the null mutation mice, there were effects on the skeleton of reduced bone mass (lower BMC overall and in the cranium, spine, and femur), changes in bone architecture (femoral mid-shaft cortical area), and decreased mechanical strength measured using 3 point bending (Warden et al. 2005). In normal (C57Bl/6) juvenile mice effects of fluoxetine upon bone mineral content (BMC) were seen at a dose of 20 mg/kg.

3.4.6 CNS Therapies: Antipsychotics

Male adolescents with autism spectrum disorders treated with antipsychotics, mainly risperidone, that resulted in hyperprolactinemia (49%), also had decreases in lumbar spine BMD (Roke et al. 2012). There were also lower levels of the biochemical bone resorption marker, C-telopeptide of bone collagen type I, and higher body fat. As mentioned above, Calarge and colleagues found effects in boys including effects on BMD of the trabecular bone with risperidone and also hyperprolactinemia, without effects on testosterone (Calarge et al. 2010).

Treatment of juvenile pigtail macaques, of 13–20 months of age, with risperidone and quetiapine resulted in no effects on body weight or skeletal dimensions, but there were some differences in BMC and BMD with risperidone, measured by DXA between 13 and 16 months of age (Sackett et al. 2010).

Weanling mice were treated orally with risperidone for 5–8 weeks from post weaning by the oral (dietary) route and by subcutaneous infusion for 7 weeks of age (Motyl et al. 2012). Evaluations including histomorphometry measurements of bone formation rates, using labeling techniques, showed increased resorption parameters, but no change in osteoblast number and function; subcutaneous infusion caused lower levels of biochemical markers of bone formation (Motyl et al. 2012). An in vitro study, however, showed some increases in osteoclasts (Motyl et al. 2012).

3.4.7 Bisphosphonates

Bisphosphonate treatment results in a greater response in children as compared to adults in terms of BMD (Ward et al. 2009), accompanied by reshaping of vertebrae. Osteogenesis imperfecta is commonly treated with bisphosphonates with most of the benefit being seen in the first 2–4 years of therapy (Ward et al. 2009). In one study alendronate increased BMD, but did not decrease fractures in patients with osteogenesis imperfecta aged 4–19 years (Ward et al. 2011). Bisphosphonates are also sometimes used to treat some forms of pediatric osteoporosis, resulting from chronic conditions like cerebral palsy and cystic fibrosis (Henwood and Binkovitz 2009). Treatment of idiopathic osteoporosis with bisphosphonates was not recommended by Ward and colleagues for asymptomatic reductions in bone parameters (Ward et al. 2009).

The *oim* mouse model of osteogenesis imperfecta was used to study the effects of pamidronate (an aminobisphosphonate) on the growth plate. Homozygous and heterozygous mice were treated from 4 weeks of age intraperitoneally with pamidronate weekly for 9 weeks (Evans et al. 2009). In both homozygous and heterozygous females, reduced bone length was seen, also indications of reduced growth plate cell turnover were seen, and there were indications of reduced resorptive capacity.

Smith (Smith et al. 2005) studied the effects of zoledronic acid, a nitrogen-containing bisphosphonate, on the physis and bone length in rabbits. New Zealand White rabbits of 8–10 weeks of age underwent right-sided tibial distraction osteogenesis, and the effects on the non-operated limb were studied after treatments with zoledronic acid at 8, or 8 and 10 weeks of age. Tibial length was lower in the zoledronic acid-treated animals at 26 weeks of age and this persisted until 52 weeks. Transient changes in physeal morphology were seen; resorption was noted but remnants of cartilaginous matrix remained in the cortical bone.

3.4.8 Therapies to Affect Growth

Several conditions including Turners syndrome and short stature homeobox-containing (SHOX) gene deficiency can be treated with growth hormone (Blum et al. 2009). There have also been recommendations to use growth hormone to treat idiopathic short stature (Cohen et al. 2008).

In a study of the safety of the use of growth hormone in children with growth disorders, a risk of secondary neoplasms following irradiation was apparent, notably after bilateral retinoblastomas; potential skeletal changes evaluated included slipped capital femoral epiphysis and scoliosis (Bell et al. 2010).

Antiestrogens like aromatase inhibitors and selective estrogen receptor modulators (SERMs) have been studied for use to decrease the rate of skeletal maturation and increase adult height (Kreher et al. 2005) in the pediatric population with short stature. Tamoxifen, a SERM, was shown to decrease the rate of skeletal maturation, resulting in increased height without effects on sexual maturation (Kreher et al. 2005). Shulman and colleagues (2008) concluded third-generation aromatase inhibitors combined with glucocorticoids for the treatment of the condition congenital hyperplasia may provide a better treatment regimen. Further, concerns regarding long-term adverse effects related to bone health, reproductive functional development, and lipid metabolism need to be addressed.

A juvenile toxicology study in rats with letrozole given orally from days 7 to 91 of age showed decreased growth in males and increased growth in females, relative to gender-respective controls, with delays in both preputial separation and vaginal opening (Pouliot et al. 2013). Decreased bone growth measurements for males were confirmed by radiographs, DXA and pQCT, with generally the opposite effect in females. BMC values were increased for females and decreased for males; however, BMD values were comparable when the diaphysis site of the proximal tibia was viewed. The authors indicated that letrozole caused delayed bone maturation and pubertal progression in both sexes (Pouliot et al. 2013), highlighting the importance of sex steroid hormones on bone development.

In a study of exemestane, an aromatase inhibitor, 26-day-old male rats were treated for 6 weeks (Van Gool et al. 2010a). Decreased body weight and tail length were seen but nose-anus length, and measures of bone growth plate thickness and radial width were unaffected. At the highest dose (100 mg/kg/day), effects included decreased trabecular bone thickness in the epiphysis and metaphysis of the distal femur measured using micro-CT. For females treated similarly, increased weight gain and growth plate width were seen. At the highest dose (100 mg/kg/day) effects included decreased trabecular thickness in the epiphysis and metaphysis of the femur, as well as polycystic ovaries (Van Gool et al. 2010b).

3.4.9 Bone Morphogenetic Protein

In reviewing the pediatric literature, use of recombinant bone morphogenetic protein-2 (BMP-2) in spinal fusion was reported to be successful (Poon et al. 2016).

Juvenile rabbits were used to study the effects of recombinant bone morphogenetic protein-2 on an absorbable sponge in treatment of a skull defect (Kinsella et al. 2011). Effects of the treatment included premature suture fusion and skull development. It was concluded that there was a need to further limit the effect to the targeted defect before clinical use in the pediatric population.

3.4.10 Review of Agents

A broad range of pharmaceuticals and biopharmaceuticals can have effects upon the developing skeleton. In some cases these may be the desired therapeutic effects, such as for bisphosphonates for treatment of osteogenesis imperfecta and bone morphogenetic protein-2 (BMP-2) for spinal fusion; in other cases potential beneficial effects may be found serendipitously as for bortezomib (Terpos and Christoulas 2013; Mohty et al. 2014). As in the case of BMP-2, the potential for unwanted effects (i.e., premature suture closure) on the developing skeleton that are not present in the adult may occur (Kinsella et al. 2011). In other cases, transitory or persistent adverse effects may occur. In some cases these effects may be treatable by life style changes and/or with therapies to alleviate the adverse effects. For example, after treatment with cancer therapies, effects on the skeleton may be ameliorated with appropriately timed life style changes, including exercise and nutritional changes and supplements, like calcium and vitamin D (Wasilewski-Masker et al. 2008). As described above, clinical trials have been conducted to assess the use of various therapies like bisphosphonates, aromatase inhibitors, BMPs, and SERMs for skeletal growth disorders, including for iatrogenic diseases (Henwood and Binkovitz 2009; Shulman et al. 2008; Kreher et al. 2005; Poon et al. 2016).

For the therapies discussed above, nonclinical data studying skeletal development in juvenile animals was available in the scientific literature. The techniques utilized to assess the skeleton and growth, as well as routine nonclinical tests such as body weight, clinical condition and gross and histopathology, included specialized skeletal assessments such as biochemical biomarkers of bone growth, imaging techniques including radiography, DXA and pQCT, histomorphometry, and biomechanical testing terminally. Some of these techniques such as biochemical assays and imaging can be utilized in clinical trials as biomarkers of skeletal development with the knowledge of the effects in the nonclinical studies.

Prenatal exposures have not been discussed in detail here; however, there are some prenatal exposures that can have adverse effects on the skeleton that persist into the postnatal period. For example, first trimester exposure to warfarin, and other related coumarin anti-coagulants, can cause facial abnormalities with nasal hyperplasia, growth retardation, and radiological stippling, particularly in the lumbar-sacral region (Schardein 2000).

An animal model in the rat in which pups were treated with warfarin and vitamin K1 from day 1 postpartum for up to 12 weeks showed craniofacial abnormalities with shortened nasal bones (Howe and Webster 1992). The timing of the dosing regimen was chosen for the timing of nasal bone development in the rat as compared to the human. Interestingly treatment of adult rats with warfarin resulted in adverse effects on the skeleton including bone strength reductions in the femur as measured by biomechanical parameters and decreased femoral cancellous bone (Simon et al. 2002). For denosumab, a fully human IgG2 monoclonal antibody treatment for osteoporosis and other skeletal problems, bone resorption is inhibited by affecting the activity of RANKL (Boyce et al. 2014). Effects of in utero treatment were seen on skeletal development early in the postpartum period and included increased long bone fractures, decreased long bone length, effects on dentition, and a range of changes detected by imaging and histopathology (Boyce et al. 2014). By 6 months of age there was evidence of recovery of bone-related changes but effects on long bone length, dentition, and strength remained (Boyce et al. 2014). In juvenile (adolescent) nonhuman primates (cynomolgus), changes noted included inhibition of bone resorption giving widened growth plates with retention of spongiosa (Boyce et al. 2014). It was concluded that denosumab's skeletal effects were related to its pharmacological effects and impaired bone modeling and remodeling.

3.5 Designs of Juvenile Toxicology Testing Programs for Pharmaceuticals and Biopharmaceuticals Where the Skeleton Is an Organ System of Concern

3.5.1 Overview

The first consideration should be to determine if there is a need for any nonclinical juvenile studies. The comparatively young ages of many species at the start of routine subchronic and chronic toxicology testing may obviate the need for testing of therapies for the juvenile pediatric (regulatory definition of 12–16 years) population (Table 3.1). If it is concluded, there is a need for nonclinical juvenile testing and then it is best to integrate these studies into the toxicology testing program early in drug development. There may be a need for interaction with regulatory bodies to ensure acceptance of the plans comparatively early in the drug development process. The possibility of conducting combined juvenile toxicology and adult toxicology studies when the drug is primarily for a pediatric indication is included in the ICH M3(R2) guidance. This approach can be particularly helpful when skeletal development is a concern, since the lengthy period of time for full skeletal development can then be fully studied in many of the standard laboratory species, except for the nonhuman primate and minipig. Further, it can allow for the testing of two species, since for most small molecules a rodent and non-rodent species are evaluated in the nonclinical toxicology.

3.5.2 Program Design

In the design of juvenile toxicity testing programs, consideration should be given as to whether both rodent and non-rodent species are required to be tested and, if so, which species should be selected. With particular relevance in consideration of the use of the non-rodent as the potential test species is the need to compare the adult and neonatal/juvenile toxicities, especially if the adult target organ is only noted in one species or there are specific characteristics of development of the test species similar to the human that may be important to study (Hurtt et al. 2004; Cappon et al. 2009). For the study of the developing skeleton, the selection of a species showing Haversian remodeling may be of importance.

Practical considerations include the age of weaning, i.e., 3 weeks of age for most rat and mouse strains, 4–6 weeks for rabbits, and typically 7–9 weeks for dogs, which can impact the study design since more flexibility in design is possible when dosing starts post weaning, without the restrictions of the litter.

For study design purposes, overall estimates of comparable ages of development are helpful (Table 3.1), but these must be used with caution since individual tissues, regions of complex organs, organs, and organ systems develop at different rates in different species; this is especially true for the musculoskeletal system which is varied and complex (Zoetis et al. 2003).

It is worthy of consideration that humans have a prolonged childhood compared to the non-primate laboratory species; this is particularly important for rodents where the first ovulation can occur within days of weaning and male puberty can occur a few weeks later. For a chronic therapy, a child could have an exposure of 10 years (from 2 to 12 years) that is difficult to model in a rodent species where only comparatively few doses can be given in the relevant time period. Some species like the dog and rabbit show a period of 3–8 months between weaning (at 4–5 and 7–9 weeks, respectively) and puberty (at 4–6 and 6–9 months, respectively) which allows for the study of chronic exposures during the juvenile period.

For drugs being developed primarily for a pediatric population, the conduct of chronic studies starting at juvenile ages may be an appropriate design where there are concerns for organ systems with development lasting up to adulthood. In the ICH M3(R2) guidance for "Non-Clinical Safety Studies for the Conduct of Human Clinical Trials and Marketing Authorization for Pharmaceuticals" (ICH 2009), there is an example of a 12-month dog study that covers the full period of postnatal development of the dog. Dosing for such a study would start at an age relevant for the clinical use.

There have been a number of reviews published of postnatal organ system development of laboratory animals, many of which contain detailed information on the commonly used species (Hurtt and Sadler 2003a, b). Since these series of reviews, there have been further publications so that additional literature reviews will generally provide sufficient information for the design of studies. In cases where this is insufficient, additional data can be generated. For example, examination of relevant tissue or biological fluid samples from animals over a range of ages of the putative test species and strain, in combination with the knowledge of the equivalent development in the human, can provide more relevant data.

3.5.3 Species and Strain Selection

To allow for comparison to a species tested in the adult animal, it is typical to consider the rodent and non-rodent species being studied in the general toxicology program. For the EMA guidance (EMA 2008), the first species considered should be a rodent. In practice the rat is the most commonly used species for juvenile toxicology tests followed by the dog. Routinely, two albino strains of rats are most commonly used: a Sprague Dawley (e.g., the CD® IGS (Crl:CD[SD]) or a Han Wistar (e.g., Crl:WI [Han]). For mice, the Swiss (e.g., Crl:CD1(ICR)) or one of the pigmented C57/Bl6 strains (e.g., C57BL/6NCrl) are typically used. While the postnatal development of the rat, and sometimes the mouse, is routinely studied in pharmaceutical toxicology testing laboratories in the pre- and postnatal (ICH-2) study, the development of other species has been assessed far less frequently. As a result, there have been efforts to adapt techniques used in areas such as veterinary medicine, physiology studies, and adult toxicity testing for use in these juvenile toxicology studies (Hoberman and Lewis 2012). In some instances, a non-rodent species may be the most suitable model due to the toxicokinetics, toxicology in the adult animals, or development of an organ system. An example of this included selection of the dog to study quinolone arthropathy based upon similarity to the human in the toxicology and the toxicokinetic parameters (Gough et al. 1992; Davis and Mckenzie 1989).

As mentioned earlier, the dog is the most commonly used non-rodent species for the toxicology testing of pharmaceuticals, and as a result it is often the species of choice when non-rodent juvenile toxicology studies are indicated. This may be the result of findings in the aforementioned adult toxicity testing. Other factors that may be of importance include similarities in organ system development and the relative age of development compared to the human. This relative similarity to the human, as compared to some other laboratory species of lung development, makes the dog a species commonly selected for inhalation juvenile toxicology studies. Therefore, when studying skeletal development of an inhaled therapy, the dog is often a suitable species. The dog, in common with rodent species like rats and mice, and non-rodents such as rabbits, is born comparatively less developed than the human, resulting in its possible use as a nonclinical model for the premature infant. Depending upon the organ system of concern, the first week or two of the dog's life can be considered a potential time for such assessments.

Other potential non-rodent species include the rabbit, where skeletal development has been studied extensively with some classes of compounds such as the bisphosphonates (Smith et al. 2005). Where there is a potential continuum of effects on the skeleton from the prenatal to postnatal exposures, the rabbit has the advantage of the prenatal exposures having been evaluated in the embryo-fetal development (ICH-3) studies for many small molecules and some biologics. Strains of rabbits typically used are the albino New Zealand White (e.g., Crl: KBL [NZW]) or the pigmented Dutch-Belted rabbit. Since ocular toxicology studies are often conducted in pigmented animals, sometimes the rabbit, this strain (Dutch-Belted) can be used for juvenile toxicology studies. To date there have been relatively few juvenile toxicology studies in the minipig focused on the skeleton, in part because of the prolonged period between sexual maturity and epiphyseal closure, from 22 to 42 months (Tsutsumi et al. 2004) and also because few compounds to date have used the minipig as the non-rodent species in the adult toxicology studies. The strain of minipig most commonly used in toxicology studies is the Gotttingen.

The nonhuman primate species used most commonly for nonclinical testing and reproductive and juvenile toxicology studies is the cynomolgus monkey. For many biopharmaceuticals the cynomolgus monkey may be the only relevant laboratory species and therefore may be the only species tested in the nonclinical studies. The ICH S6(R1) guidance (ICH 2011), for biotechnology-derived pharmaceuticals, allows for this; further, the use of homologous molecules to test a second species is generally not encouraged.

There are several difficulties in using the cynomolgus for juvenile toxicology studies, firstly practical considerations of animal supply make studies starting with animals less than 12–15 months difficult and secondly the length of time to achieve sexual maturity and full development of the skeleton being up to approximately 5 years in the female and 6 years in the male. Therefore, even in a chronic study of 6–12 months, only a small part of postnatal development can be studied. Nevertheless studies can be undertaken and provide useful data. For example, as discussed above the studies in cynomolgus monkeys of the monoclonal antibodies like bevacizumab and denosumab showed skeletal changes (Ryan et al. 1999; Boyce et al. 2014).

3.5.4 Study Design

Various designs can be used in neonatal and juvenile toxicology testing. Split litter (or within litter) and whole litter (or between litters) designs can be utilized with studies that start dosing preweaning. For initial studies, split litter designs, where pups within the same litter are administered different dosages, can decrease the number of animals used. For later, more definitive studies where toxicokinetic sampling is included, whole litter designs are generally used for both rodents and non-rodents when dosing starts preweaning to avoid contamination. When dosing starts at weaning or shortly thereafter, it is possible to balance litter mates across groups in many species.

In assessing growth and skeletal development, consideration should always be given to including a post-dosing development period (often termed recovery or regression), of sufficient length to allow for catch-up growth to be studied and the possible regression of other changes notes at the end of dosing. In some instances, not just growth can be affected by treatment but also development of other systems such as reproductive function; in such cases a post-dose development (recovery) period is included to assess the potential for "catch-up development."

Group sizes for the types of evaluations discussed earlier (preclinical measurements 3.3.3.) do not have to exceed those for routine toxicology endpoints like body weight given the favorable CVs noted. However, in rodents, if frequent assessments are required that necessitate anesthesia or blood sampling, then several subsets may be needed. Micro-sampling techniques can be used to decrease the numbers of rodents needed but few compounds reaching juvenile toxicology testing at the time of writing have valid assays for micro-sampling volumes (Powes-Glover et al. 2014).

There should be consideration of the inclusion of biomarkers that can be measured in the clinical trials, for example, if there is a focus on the skeleton, then measurements of growth, imaging/bone densitometry, and biochemical markers of bone turnover are suitable parameters (Biochemical Markers of Bone Turnover Chap. 5).

Dose selection for these juvenile toxicology studies can be quite difficult because the metabolism of the drug may change dramatically as the animals develop and because of the range of endpoints that will be assessed. It is not uncommon for large differences in toxicokinetic parameters (sometimes greater than 100-fold) between young pups and adult animals to be observed. The range of endpoints can be from typical toxicology parameters like mortality, clinical condition, and body weight to subtle biochemical and/or functional parameters. Partly as a result of the aforementioned difficulties, dose-range-finding studies, either long enough to cover a reasonable age range or over several age ranges, are often required, though these studies do not have to be over elaborate in terms of the endpoints discussed above. When preweaning animals are being tested, small numbers of animals can be tested initially in a staggered fashion, in a "tolerability" phase, to minimize animal use in the dose-range-finding studies.

One of the advantages of non-rodents over rodents is that typically the non-rodent can be used for many evaluations and the sampling for bioanalysis is conducted on the same animals, allowing for correlation of the toxicology and toxicokinetics on an individual basis.

3.6 Conclusions

Nonclinical juvenile toxicology studies of skeletal growth and development show concordance with clinical findings and have been performed in a range of laboratory species. Therefore, careful design of the juvenile toxicology testing program, since each study is designed on a "case-by-case" basis, can be expected to produce a suitable nonclinical study. For the evaluation of new chemical entities, various biomarkers (biochemical and imaging) can be included in the nonclinical studies and potentially in the clinical studies.

References

Adamo M, Pouliot L, Primakova I, Martin A, Doyle N, Samadfam R, Varela A, Smith SY, Robinson K. Post natal skeletal development assessments in dogs and rabbits in non-clinical pediatric studies. Reprod Toxicol. 2013;41:26.

Bailey RR, Natale R, Linton AL. Nalidixic acid arthralgia. Can Med Assoc J. 1972;107:604.

Barbeau S, Varela A, Moreau IA, Pouliot L, Pinsonneault L, Smith SY, Robinson K. Growth and post natal skeletal development assessments in non-clinical neonatal and juvenile toxicity studies. Birth Defects Res Part A. 2006;76(5):399.

Bell J, Parker KL, Swinford RD, Hoffman AR, Maneatis T, Lippe B. Long-term safety of recombinant human growth hormone in children. J Clin Endocrinol Metab. 2010;95(1):167–77.

Blum WF, Cao D, Hesse V, Fricke-Otto S, Ross JL, Jones C, Quigley CA, Binder G. Height gains in response to growth hormone treatment to final height are similar in patients with SHOX deficiency and Turner syndrome. Horm Res. 2009;71(3):167–72.

Boersma B, Maarten Wit J. Catch-up growth. Endocr Rev. 1997;18(5):646–61.

Boyce RW, Varela A, Chouinard L, Bussiere JL, Chellman GE, Ominsky MS, Pyrah IT. Infant cynomolgus monkeys exposed to denosumab in utero exhibit an osteoclast-poor osteopetrotic-like skeletal phenotype at birth and in early postnatal life. Bone. 2014;64:314–25.

Calarge CA, Zimmerman B, Xie D, Kuperman S, Schlechte JA. A cross-sectional evaluation of the effect of risperidone and selective serotonin reuptake inhibitors on bone mineral densityin boys. J Clin Psychiatry. 2010;71(3):338–47.

Cappon GD, Bailey GP, Buschmann J, Feuston MH, Fisher JE, Hew KW, Hoberman AM, Ooshima Y, Stump DG, Hurtt ME. Juvenile animal toxicity study designs to support pediatric drug development. Birth Defects Res (Part B) Dev Reprod Toxicol. 2009;86:463–9.

Chellman GJ. Charles River, Nevada, Historical control data in non human primates. 2016.

Cohen P, Rogol AD, Deal CL, Saenger P, Reiter EO, Ross JL, Chernausek SD, Savage MO, Wit JM, on behalf of the 2007 ISS Consensus Workshop participants. Consensus statement on the diagnosis and treatment of children with idiopathic short stature: a summary of the Growth Hormone Research Society, the Lawson Wilkins Pediatric Endocrine Society, and the European Society for Paediatric Endocrinology Workshop. J Clin Endocrinol Metab. 2008;93(11):4210–7.

Davis GJ, Mckenzie BE. Toxicology evaluation of ofloxatin. Am J Med. 1989;87(suppl 6C):43–6.

Drescher W, Li H, Ovesel D, Jensen SD, Flo C, Hansen E, Bunger C. Vertebral blood flow and bone mineral density during long term cortocisteroid treatment. Spine. 2000;25(23):3021–5.

Eriksson E, Zaman F, Chrysis D, Wehtje H, Heino TJ, Savendahl. Bortezomib is cytotoxic to the human growth plate and permanently impairs bone growth in young mice. PLOS One. 2012;7(11):e50523.

European Medicines Agency (EMA). Pediatric Regulation (EC) No 1901/2006 of the European Parliament and of the Council on medicinal products for paediatric use, amended by Regulation (EC) No 1902/2006.

European Medicines Agency (EMA). Committee for Human Medicinal Products (CHMP). Guideline on the need for nonclinical testing in Juvenile Animals on Human Pharmaceuticals for Paediatric Indications. Jan 2008.

Evans KD, Sheppard LE, Rao SH, Martin RB, Oberbauer AM. Pamidronate alters the growth plate in the *Oim* mouse model for osteogenesis imperfecta. Int J Biomed Sci. 2009;5(4):345–52.

Gough AW, Olajide BK, Sigler RE, Baragi V. Quinolone arthropathy – acute toxicity to immature articular cartilage. Toxicol Pathol. 1992;30(3):436–49.

Henwood MJ, Binkovitz L. Update on pediatric bone health. Am Osteopath Assoc. 2009;109(1):5–12.

Hoberman A, Barnett JF. Juvenile toxicology study design in the rodent and rabbit. In: Hoberman A, Lewis E, editors. Pediatric non-clinical drug testing. New Jersey: Wiley; 2012. p. 141–82.

Hoberman A, Lewis E. Pediatric non-clinical drug testing. New Jersey: Wiley; 2012.

Hotchkiss CE, Latendresse J, Ferguson SA. Oral treatment with retinoic acid decreases bone mass in rats. Comp Med. 2006;56(2):502–11.

Howe AM, Webster WS. The warfarin embryopathy: a rat model showing maxillonasal hyoplasia and other skeletal disturbances. Teratology. 1992;46:379–90.

Hughes PCR, Tanner JM. A longitudinal study of the growth of the black-hooded rat: methods of measurement and rates of growth for skull, limbs, pelvis nose-rump and tail lengths. J Anat. 1970;2:349–70.

Hurtt ME, Sadler JD. Comparative organ system development: introduction. Defects Res (Part B). 2003a;68:85.

Hurtt ME, Sadler JD. Comparative organ system development: continuing the series. Birth Defects Res B Dev Reprod Toxicol. 2003b;68:307–8.

Hurtt ME, Daston G, Davis-Bruno K, Feuston M, Silva Lima B, Makris S, McNerney ME, Sandler JD, Whitby K, Wier P, Cappon GD. Juvenile animal studies: testing strategies and design. Birth Defects Res B Dev Reprod Toxicol. 2004;71:281–8.

ICH Tripartite Guideline M3(2R). Non-clinical safety studies for the conduct of human clinical trials and marketing authorization for pharmaceuticals. International conference on harmonization of technical requirements for the registration of pharmaceuticals for human use. ICH steering committee. 2009.

ICH Tripartite Guideline S5(2R). Detection of toxicity to reproduction for medicinal products and toxicity to male fertility. International conference on harmonization of technical requirements for the registration of pharmaceuticals for human use. ICH steering committee. 2005.

ICH Tripartite Guideline S6(R1). Preclinical safety evalaution of biotechnology derived pharmaceuticals. International conference on harmonization of technical requirements for the registration of pharmaceuticals for human use. ICH steering committee. 2011.

Ingham B, Brentnall DW, Dale EA, McFadzean JA. Arthropathy induced by antibacterial fused N-alkyl-4-pyridone-3-carboxylic acids. Toxicol Lett. 1977;1:21–6.

Jones KL, Smith DW. Recognition of the fetal alcohol syndrome in early infancy. Lancet. 1973;302(7836):999–1001.

Jones IE, Williams SM, Goulding A. Associations of birth weight and length, childhood size, and smoking with bone fractures: evidence from a birth cohort. Am J Epidemiol. 2004;159:343–59.

Kinsella CR, Cray JJ, Durham EL, Burrows AM, Vecchione L, Smith DM, Mooney MP, Cooper GM, Losee JE. Recombinant human bone morphogenetic protein-2–induced craniosynostosis and growth restriction in the immature skeleton. Plast Reconstr Surg. 2011;127(3):1173–81.

Kreher NC, Eugster EA, Shankar R. The use of tamoxifen to improve height potential in short pubertal boys. Pediatrics. 2005;16(6):1513–5.
Lawson JP, McGuire J. The spectrum of skeletal changes associated with long-term administration of 13-cis-retnoic acid. Skelet Radiol. 1987;16:91–7.
Lowe CR. Effect of mothers smoking on birth weight of their children. Brit Med J. 1959;2:673–6.
Maddock CL, Wolbach SB, Maddock S. Hypervitaminosis a in the dog. J Nutr. 1949;39:117–37.
Martin EA, Ritman EL, Turner RT. Time course of epiphyseal growth plate fusion in rat Tibae. Bone. 2003;32:261–7.
McHugh N, Vercesi HM, Egan RW, Hey JA. In vivo rat assay: bone remodeling and steroid effects on juvenile bone by pQCT quantification in 7 days. Am J Physiol Endocrinol Metab. 2003;284:E70–5.
Ministry of Health, Labour and Welfare (Japan). Guideline on the nonclinical safety study in juvenile animals for pediatric drugs. Oct 2012.
Mohty M, Malard F, Mohty B, Savani B, Moreau P, Terpos E. The effects of Bortezomib on bone disease in patients with multiple myeloma. Cancer. 2014;120:618–23.
Moore T, Wang YL. Hypervitaminosis A. Biochem J. 1945;39(2):222–8.
Motyl KJ, Dick-dePaula MAE, Lotinun S, Bornstein S, dePaula FJE, Baron R, Houseknecht KL, Rosen CL. Trabecular bone loss after administration of the second generation anti-psychotic risperidone is independent of weight loss. Bone. 2012;50:490–8.
Mwale F, Antoniou J, Héon Servant N, Wang C, Kirby GM, Demers CN, Chalifour LE. Gender-dependent reductions in vertebrae length, bone mineral density and content by doxorubicin are not reduced by Dexrazoxane in young rats: effect on growth plate and intervertebral discs. Calcif Tissue Int. 2005;76(3):214–21.
O'Loughlin J, Karp I, Henderson M, Gray-Donald K. Does cigarette use influence adiposity or height in adolescence? Ann Epidemiol. 2008;18(5):395–402.
Osborne TB, Mendel LB. The resumption of growth after long continued failure to grow. J Biol Chem. 1915;23:439–54.
Pennes DR, Ellis CN, Madison KC, Voorheees JJ, Martel W. Early skeletal hyperostoses secondary to 13-cis-retinoic acid. AJR. 1984;141:979–83.
Poon B, Kha T, Tran S, Dass CR. Bone morphogenetic protein-2 and bone therapy: successes and pitfalls. J Pharm Pharmacol. 2016;68:139–47.
Pouliot L, Schneider M, DeCristofaro M, Samadfam R, Smith S, Beckman DA. Assessment of a nonsteroidal aromatase inhibitor, letrozole, in juvenile rats. Birth Defects Res (Part B). 2013;98:374–90.
Power C, Jeffris B. Fetal environment and subsequent obesity: a study of maternal smoking. Int J Epidmol. 2002;31:413–9.
Powes-Glover N, Kirk S, Jardine L, Clubb S, Stewart J. Assessment of haematological and clinical pathology effects of blood microsampling in suckling and weaned juvenile rats. Regul Toxicol Pharmacol. 2014;69(3):425–33.
Prader A, Tanner JM, Harnack V. Catch-up growth following illness or starvation: an example of developmental canalization in man. J Pediatr. 1963;62(5):646–59.
Prendiville J, Bingham EA, Burrows D. Premature Epipyseal closure – a complication of etretinate therapy in children. J Am Acad Dermatol. 1986;15:1259–562.
Primakova I, Doyle N, Martin AI, Pouliot L, Barbeau S, Samadfam R, Varela A, Smith SY, Robinson K. Assessments of skeletal development in rats and rabbits using imaging techniques. Birth Defects Research Part A. 2013;97(5):324.
Rao R, Gregson RK, Jones AC, Miles EA, Campbell MJ, Warner JO. Systemic effects of inhaled cortocosteroids on growth and bone turnover in childhod asthma: comparison of fluticasone with beclomethasone. Eur Respir J. 1999;13(1):87–94.
Roach HI, Mehta G, Oreffo ROC, Clarke NMP, Cooper C. Temporal analysis of rat growth plates: cessation of growth with age despite presence of a physis. J Histochem Cytochem. 2003;51:373.
Robinson K, Adamo M, Martin A, Primakova I, Barbeau S, Pouliot L, Varela A, Samadfam R, Smith SY. Measurements of growth in rodent juvenile toxicology studies. Birth Defects Res Part A. 2012a;94(5):337.

Robinson K, Smith SY, Viau A. Dog juvenile toxicology. In: Hoberman A, Lewis E, editors. Pediatric nonclinical drug testing: principles, requirements, and practices. Hoboken: John Wiley & Sons, Inc.; 2012b. p. 183–212.
Robinson K, Adamo M, Martin A, Samadafan R. Measurements of growth in nonrodent juvenile toxicology studies. Birth Defects Res Part A. 2015;103(5):435.
Roebuck DJ. Skeletal complications in pediatric oncology patients. Radiographics. 1999;19:873–85.
Roke Y, van Harten PN, Jan K, Buitelaar JK, Tenback DE, Quekel LGBA, de Rijke YB, Boot AM. Bone mineral density in male adolescents with autism spectrum disorders and disruptive behavior disorder with or without antipsychotic treatment. Eur J Endocrinol. 2012;167:855–63.
Rovner AJ, Zemel BZ. Growth and physical maturation. In: Mulberg AE, Silber SA, Van den Anker JN, editors. Pediatric drug development: concepts and applications, vol. 29. Hoboken: John Wiley; 2009. p. 363–83.
Ryan AM, Bates Eppler D, Hagler KL, Bruner RH, Thomford PJ, Hall RL, Shopp GM, O'Neill CA. Preclinical safety evaluation of rhuMAbVEGF, an antiangiogenic humanized monoclonal antibody. Toxicol Pathol. 1999;27(1):78–96.
Sackett G, Unis A, Crouthamel B. Some effects of risperidone and quetiapine on growth parameters and hormone levels in young pigtail macaques. J Child Adolesc Psychopharmacol. 2010;20(6):489–93.
Schardein JL. Drugs affecting blood. In: Schardein JL, editor. Chemically induced birth defects, vol. 4. New York: Marcel Dekker; 2000. p. 125.
Schunior A, Zengel AE, Mullenix PJ, Tarbell NJ, Howes A, Tassinari MS. An animal model to study toxicity of central nervous system therapy for childhood acute lymphoblastic leukemia: effects on growth and craniofacial proportion. Cancer Res. 1990;50:6455–60.
Shulman DI, Francis GL, Palmert MR, Eugster EA. Use of aromatase inhibitors in children and adolescents with disorders of growth and adolescent development. Pediatrics. 2008;121(4):975–83.
Sibonga JD, Sommer U, Turner RT. Evidence that 2-methoxyestradiol suppresses proliferation and accelerates apoptosis in normal rat growth plate chondrocytes. J Cancer Res Clin Oncol. 2002;128:477–83.
Simon R, Beaudin SM, Johnston M, Walton KJ, Shaughnessy SG. Long term treatment with sodium warfarin results in decreased femoral bone sStrength and cancellous bone volume in rats. Thrombosis. 2002;105:353–8.
Skoner DP, Rachelefsky GS, Meltzer EO, CHevinsky P, Morrris RM, Seltzer JM, Storms WW, Wood RA. Detection of growth suppression in children during treatment with intranasal Beclomethasone dipropoinate. Pediatrics. 2000;105(2):E23.
Smith EJ, Little DG, Briody JN, McEvoy A, Smith NC, Eisman JA, Gardiner EM. Transient disturbance in physeal morphology is associated with long-term effects of nitrogen-containing bisphosphonates in growing rabbits. J Bone Miner Res. 2005;20(10):1731–41.
Sobel EH. Effects of neonatal stunting on the development of rats: early and late effects of neonatal cortisone on physical growth and skeletal maturation. Pediatr Res. 1978;12:945–7.
Spencer TJ, Kratochvil CJ, Sangal RB, Saylor KE, Bailey CE, Dunn DW, Geller DA, Casat CD, Lipetz RS, Jain R, Newcorn JH, Ruff DD, Feldman PD, Furr AJ, Allen AJ. Effects of atomoxetine on growth in children with attention-deficit/hyperactivity disorder following up to five years of treatment. J Child Adolesc Psychopharmacol. 2007;17(5):689–99.
Stahlmann R, Schwabe R. Safety profile of grepafloxin compared with other fluoroquinolones. J Antimicrob Chemother. 1997;40(suppl A):83–92.
Strauss RS. Effects of the intrauterine environment on childhood growth. Br Med Bull. 1997;33(1):81–95.
Takayama S, Hirohashi M, Kato M, Shimada H. Toxicity of quinolone antimicrobial agents. J Toxicol Environ Health. 1995;45:1–45.
Tassinari M. Growth from birth to puberty: dynamics of growth and skeletal development: education course teratology society meeting. 1999.
Terpos E, Christoulas D. Effects of proteasome inhibitors on bone cancer. Bone Key Reports. 2013;2:395. 129

Tortereau A, Howroyd P, Lorentsen H. Onset of puberty and normal histological appearances of the reproductive organs in peripubertal female Göttingen Minipigs. Toxicol Pathol. 2013;41:1116–25.

Tsutsumi H, Katagiri K, Takeda S, Nasu T, Igarashi S, Tanigawa M, Mamba K. Standardized data and relationship between bone growth and bone metabolism in female Gottingen Minipigs. Exp Anim. 2004;53(4):331–7.

United States Food and Drug Administration (FDA). Guidance document. Nonclinical safety evaluation of pediatric drug products. Feb 2006.

United States Food and Drug Administration (FDA). Guidance for industry pediatric study plans: content of and process for submitting initial pediatric study plans and amended pediatric study plans. Draft. Mar 2016.

United States Food and Drug Agency (FDA). Regulations requiring manufacturers to assess the safety and effectiveness of new drugs for pediatric patients. 1998.

Van Gool SA, Wit JM, de Schutter T, de Clerck N, Postnov AA, Hovinga SK, van Doorn J, Veiga SJ, Garcia-Segura LM, Karperien M. Impaired body weight and tail length and altered bone quality after treatment with the aromatase inhibitor exemestane in male rats. Horm Res Paediatr. 2010a;73:376–85.

Van Gool SA, Wit JM, de Schutter T, de Clerck N, Postnov AA, Hovinga SK, van Doorn J, Veiga SJ, Garcia-Segura LM, Karperien M. Marginal growth increase, altered bone quality and polycystic ovaries in female prepubertal rats after treatment with the aromatase inhibitor exemestane. Horm Res Paediatr. 2010b;73:49–60.

Voss SD, Glade-Bender J, Spunt SL, DuBois SG, Widemann BC, Park JR, SES L, Nelson MD, Adamson PC, Blaney SM, Weigel B. Growth plate abnormalities in pediatric cancer patients undergoing phase anti-angiogenic therapy: a report from the children's oncology group phase I consortium. Pediatr Blood Cancer. 2015;62:45–51.

Ward LM, Petryk A, Gordon CM. Use of bisphosphonates in the treatment of pediatric osteoporosis. Int J Clin Rheumatol. 2009;4(6):657–72.

Ward LM, Rauch F, Whytte MP, D'Astous JD, Gates PE, et al. Aledronate for the treatment of pediatric osteogenesis imperfecta: a randomized placebo controlled study. J Clin Endocrinol Metab. 2011;96(2):355–64.

Warden SJ, Robling AG, Sanders MS, Bliziotes MM, Turner CH, Haney EM. Inhibition of serotonin (5-hydroxytryptamine) transporter reduces bone accrual during growth. In vitro and animal-based studies. Endocrinology. 2005;146(2):685–93.

Wasilewski-Masker K, Kaste SC, Hudson MM, Esiashvili N, Mattano LA, Meacham LR. Bone mineral density deficits in survivors of childhood cancer: long-term follow-up guidelines and review of the literature. Pediatrics. 2008;121(3):705–14.

Weinbauer GF, Chellman GJ, Rasmussen AD, Vogelwedde E. Use of primate pediatric model. In: Hoberman A, Lewis E, editors. Pediatric nonclinical drug testing: principles, requirements, and practices. Hoboken: John Wiley & Sons; 2012. p. 213–2.

White SI, MacKie RM. Bone changes associated with retinoid therapy. Pharmac Ther. 1989;40(1):137–44.

Wilson JG. Current status of teratology. General principles and mechanisms derived from animal studies. In: Wilson JG, Fraser FC, editors. Handbook of teratology, vol. 1/2. New York: Plenum Press; 1977. p. 47–74.

Wise LD, Vetter CM, Anderson CA, Antonello JM, Clark RL. Reversible effects of tramcinolone and lack of effects on external genitalia of male Sprague Dawley rats exposed in utero. Teratology. 1991;44:507–20.

Zoetis T, Tassinari MS, Bagi C, Walthall K, Hurtt ME. Species comparison of postnatal bone growth and development. Birth Defects Res B Dev Reprod Toxicol. 2003;68:86–110.

Chapter 4
Animal Models in Bone Research

Donald B. Kimmel

Abstract Many studies of the response of isolated osteoblasts and osteoclasts to treatments exist. While such studies are informative, a reliable path to translating those effects into effects in humans is required, because of the skeletal intermediary organization that only exists in vivo. Bone cells function in vivo within the intermediary organizations of modeling, which controls bone growth and shape changes, and remodeling, which controls bone renewal during adulthood (Frost, Metab Bone Dis Rel Res 4:281–290, 1983). Modeling and remodeling govern how important tissue-level properties of the whole skeleton, such as bone mass, microarchitecture, strength, and calcium homeostasis, respond to treatment. Animal models are the lowest level at which the tissue-level intermediary organizations of modeling and remodeling can be examined. The purpose of this chapter is to give toxicologists an overview of some popular animal models of diseases of low bone strength, which can be used to translate basic research findings to the whole animal and human level. It focuses on osteoporosis and how various models not only provide in vivo skeletal behaviors that reflect tissue-level organizations found in adult humans but also can be applied to assist in understanding ways to prevent and treat various types of osteoporosis.

Keywords Osteopenia • Estrogen deficiency • Glucocorticoids • Murine • Rat • Nonhuman primate

4.1 Regulatory Guidelines for Animal Models of Osteoporosis

The most prevalent type of osteoporosis occurs in postmenopausal women. Thus, regulatory guidelines for preclinical testing of anti-osteoporosis agents recommend doing prevention studies in adult animals that are losing bone and treatment

D.B. Kimmel, DDS, PhD (✉)
Department of Physiological Sciences, JHMHC, University of Florida,
Box 100144, Gainesville, FL 32610, USA
e-mail: kimmeldb@comcast.net

S.Y. Smith et al. (eds.), *Bone Toxicology*, Molecular and Integrative Toxicology,
DOI 10.1007/978-3-319-56192-9_4

(rebuilding) studies in adult animals with established estrogen-deficiency osteopenia. Guidelines strongly suggest acquiring data from one small and one large animal species. They further suggest measuring endpoints similar to those used in human trials that evaluate (a) bone density, (b) biochemical markers of bone turnover, (c) bone strength by biomechanical testing, and (d) mineralization defects and turnover.

To achieve regulatory approval, osteoporosis treatment drugs must reduce fracture risk in humans. Though animal studies provide relevant data on bone density, biochemical markers of turnover, and mineralization defect and turnover assessment, these data do not substitute for human fracture risk data. The bone strength portion of animal studies is best considered safety data. Most current guidelines are based principally on testing specific anti-resorptives that treat osteoporosis by slowing bone turnover and improving the balance of bone resorption and formation (e.g., bisphosphonates, estrogen, selective estrogen receptor modulators [SERMs], anti-RANK ligand antibodies, and calcitonin). Current guidelines thus focus on detecting modest changes in bone mass over extended time periods. Evaluating agents that stimulate bone formation (e.g., teriparatide, sclerostin antibody) often can be done in shorter experiments, due to faster changes in measurable endpoints.

4.2 A Perspective on In Vivo Experimentation

Past in vivo animal investigations of osteoporosis are excellent examples of how animal models can be used to test agents/concepts that improve the understanding and control of human disease. The main reason for the success of animal models of postmenopausal osteoporosis is that the tissue-level features of human osteoporosis are understood well enough to allow identifying animals with skeletons that reflect human tissue-level behaviors. Because of both the variety of measurements that can be applied and their favorable timeframe compression, animal studies are the most appropriate way to predict tissue-level mechanisms of drug actions in humans. The combined results from human Phase III and IV trials whose outcomes were accurately predicted by preclinical studies attest to the appropriateness of the animal models. The ongoing examination of human osteoporosis data further improves the animal models (Cummings et al. 1995; Boggild et al. 2015; Boonen et al. 2011; Garnero et al. 1996; Ross et al. 1993).

The most certain way to learn the effects of a safe agent on the human skeleton is to study it in a large human population with appropriate disease. However, despite improved noninvasive imaging in humans (Genant et al. 2008), when osteoporosis treatment agents need proof of concept, efficacy testing prior to safety assessment, or development of models for clinical adverse events, in vivo animal experimentation is the best choice. In addition, when questions arise about their effects on turnover and cellular activity in humans in osteoporotic fracture sites (e.g., vertebrae, hip, and wrist) or extremely rare adverse events (Khan et al. 2015; Shane et al.

2014), human experiments cannot generate suitable samples. The failure of animal models to predict sodium fluoride adverse effects seen in humans (Kleerekoper et al. 1991; Riggs et al. 1990) and failure of fracture risk reduction to follow bone mineral density (BMD) improvements with etidronate (Chapuy and Meunier 1995) led the USFDA to require that anti-osteoporosis agents demonstrate *both* BMD increases and anti-fracture efficacy in clinical trials to gain approval.

4.3 Criteria for Animal Models in Osteoporosis

Full parallelism of in vivo animal models with human symptoms rarely exists. The criteria are flexible, creative, and open to revision as new evidence or needs appear. They place the highest value on animal models that match the tissue-level behavior of human osteoporosis, allowing the detailed match of mechanisms to be provided by subsequent molecular and/or enzyme-level characterization. Skeletal research in the ovariectomized (OVX) rat is a critical achievement by bench scientists that validates a highly relevant preclinical model (Wronski et al. 1988, 1989a) that led to efficacious treatments for human osteoporosis. That effort, completed in the late 1980s, was made possible by the dynamic characterization of human bone remodeling of the 1960s (Frost 1969) and the bone mass and metabolic data collected about human postmenopausal osteoporosis during the 1970s (Recker et al. 1978; Nordin et al. 1976; Jensen et al. 1982). The introduction of standardized ex vivo tests of trabecular and cortical bone strength as surrogates for both osteoporotic fracture and safety endpoints in the early 1990s further refined these animal models (Li et al. 1991; Turner and Burr 1993; Toolan et al. 1992).

Imaging techniques may further reduce the length of animal experiments by providing extremely precise measurements that focus on bone regions showing the greatest post-OVX change (Bouxsein et al. 2010; Kinney et al. 1995; Clark and Badea 2014; Breen et al. 1998). Relevant physiologic conditions studied completely should always be the hallmark of animal bone studies. For example, when an *adult* disease process is under study, using a growing animal for convenience makes little sense. Conversely, when attainment of peak bone mass is under study, choosing a growing animal is best. Contributions from growth cartilage activity never exist in the adult human skeleton. Nerve resection experiments (Wakley et al. 1988) best model the motionless, *denervated* limb, not the motionless limb. Combining extreme calcium deprivation and/or glucocorticoids with estrogen deficiency just to create or accelerate bone loss produces a situation with no clinical relevance for any of the conditions (Geusens et al. 1991; Castañeda et al. 2006; Augat et al. 2003; Li et al. 1993). Though in vivo measurements of bone mass and microarchitecture in the animal skeleton are now routine (Amman et al. 1992; Black et al. 1989), reports are usually considered incomplete without quantitative histologic characterization of the accompanying tissue-level events.

4.3.1 Summary

Using relevant physiologic conditions in an adult animal to produce a consistent, though incomplete, set of symptoms (e.g., no osteoprotic fractures) is the correct choice during preclinical testing. Animal models that work well for both estrogen-deficiency and glucocorticoid osteopenia now exist. Additional experimental time, new endpoints, and more precise analysis tools are preferred during model refinement to introducing irrelevant physiology or treatments (i.e., combining estrogen deficiency, glucocorticoids, disuse, etc.). Models that give sporadic results across laboratories or require convoluted manipulations enjoy limited acceptance. Given the existence of validated osteoporosis animal models today, using an irrelevant animal model is inexcusable and wasteful.

4.3.2 Postmenopausal Osteoporosis

In vivo animal models of the skeleton have been addressed often (Jerome 1998; Bellino 2000; Bonucci and Ballanti 2014; Bonjour et al. 1999; Hartke 1999; Turner 2001; Rissanen and Halleen 2010; Syed and Melim 2011; Barlet et al. 1994; Ibrahim et al. 2013; Sophocleous and Idris 2014; Jee and Yao 2001; Kimmel 2002; Lelovas et al. 2008; Kimmel 1996; Miller et al. 1995; Simpson and Murray 2015; Cesnjaj et al. 1991). This chapter not only affirms many of their important points but also encompasses additional animal models and secondary osteoporoses.

4.3.2.1 Growth and Adult Phases

An in vivo animal model of osteoporosis should exhibit both growing and adult skeletal phases of meaningful duration. Low peak bone mass is an established risk factor for osteoporotic fracture (Lofman et al. 2000; Drake et al. 2015). A single BMD measurement at menopause is the best quantitative predictor of future fracture risk (Hui et al. 1989; Ross et al. 1988). That value approximates peak bone mass, because premenopausal bone mass changes are minor (Recker et al. 1992; Rodin et al. 1990).

Growth processes, principally bone modeling influenced by nutrition, physical activity, and heredity (Johnston et al. 1992; Jones et al. 1977; Liel et al. 1988; Smith et al. 1973; Mitchell and Streeten 2013; Ralston 2007; Richards et al. 2012; Karasik and Cohen-Zinder 2012), determine peak bone mass. Adult skeletal processes, predominantly bone remodeling, but some modeling to change bone shape in response to altered physical activity patterns, determine bone quantity during young adulthood and beyond (Frost 1983). Accurate animal models of osteoporosis should have both growth and adult phases of sufficient duration to permit useful studies.

4.3.2.2 Role of Estrogen

4.3.2.2.1 Menstrual/Estrus Cyclicity

Humans not only have a menarche and regular, frequent ovulatory cycles but also experience bone loss at cessation of ovarian function. The positive linkage of regular menses to greater bone mass is reenforced by data from amenorrheic individuals (Drinkwater et al. 1984; Marcus et al. 1985; Klibanski and Greenspan 1986) and the BMD rise that occurs upon resumption of normal menses (Klibanski and Greenspan 1986; Drinkwater et al. 1986). Mammals with regular, frequent ovulatory cycles routinely experience estrogen-deficiency bone loss. Regularly cycling female mammals may integrate an estrogen-related component of bone into the skeleton at puberty (Garn 1970; Gilsanz et al. 1988) that disappears at menopause (Frost 1987). Thus, animals with more frequent cycles tend to display more rapid bone loss in response to estrogen deficiency.

4.3.2.2.2 Natural Menopause

Women undergo a natural menopause of 2–7 years duration (Treloar 1981). Though aside from rate of bone loss, no differences in bone behavior between surgical and natural menopause are known (Cann et al. 1980; Hartwell et al. 1990); naturally postmenopausal women retain widely variant levels of circulating 17-β-estradiol long after their last period (Cummings et al. 1998; Longcope et al. 1984; Cauley et al. 1999), which affect both their fracture and breast cancer risk.

4.3.2.2.3 Bone Loss and Rise in Turnover Rate After Estrogen Deficiency

Following onset of estrogen deficiency, bone loss transiently accelerates in multiple sites (Lindquist et al. 1983; Hedlund and Gallagher 1989; Wasnich et al. 1983) and then decelerates into a plateau phase (Nordin et al. 1976). The bone loss is most pronounced in cancellous and endocortical regions (Nilas et al. 1984; Keshawarz and Recker 1984). Estrogen status is more important in determining bone quantity in the adult woman than age (Richelson et al. 1984; Nilas and Christiansen 1987). Estrogen deficiency enlarges Haversian canals and thins the cortex (Brockstedt et al. 1993). The cancellous and endocortical bone loss is accompanied by increased bone turnover (Recker et al. 1978, 2004; Stepan et al. 1987) and a marked, transient negative calcium balance (Heaney et al. 1978). In an accurate animal model of postmenopausal osteoporosis, these behaviors should be easily monitored by similar methods and endpoints used in humans.

4.3.2.2.4 Skeletal Response to Estrogen Replacement

Oophorectomized or menopausal women given prompt estrogen replacement experience a smaller rise in turnover (Recker et al. 1978; Steiniche et al. 1989), less bone loss (Jensen et al. 1982; Wasnich et al. 1983; Christiansen and Lindsay 1990; Christiansen et al. 1980; Ettinger et al. 1985), and fewer fractures (Jensen et al. 1982; Ettinger et al. 1985; Lindsay et al. 1976) than those who receive no estrogen replacement (Christiansen and Lindsay 1990). This response is demonstrated well by histomorphometric techniques in humans and animals (Wronski et al. 1988; Recker et al. 2004; Steiniche et al. 1989; Turner et al. 1987) and is so certain in adult women that an accurate OVX animal model must have an identical response to estrogen replacement.

4.3.2.3 Skeletal Response to Osteoporosis Treatment Agents

The tissue-level effects of agents proven to reduce osteoporotic fracture risk, including alendronate (Liberman et al. 1995), risedronate (Wallach et al. 2000), teriparatide (Dobnig 2004), zoledronic acid (Black et al. 2007), raloxifene (Ettinger et al. 1999), and denosumab (Cummings et al. 2009), have been well studied. Accurate animal models should yield data predictive of these agents' effects in humans.

4.3.2.4 Development of Osteoporotic Fractures and Steady-State Osteopenia

While the most frequent consequence of osteoporosis is low trauma spine fracture, the most serious is hip fracture (Kanis and McCloskey 1992; Cummings et al. 1985). An animal model that develops estrogen deficiency-related fragility fractures could facilitate preclinical anti-fracture efficacy trials. Though animal models are excellent for detecting estrogen-deficiency bone loss and elevated turnover rate, none exhibit fragility fracture. The lack of such fractures in preclinical models emphasizes the importance of low peak bone mass (Lofman et al. 2000), falls (Tinetti et al. 1994), vitamin D levels (Korkmaz et al. 2014), and extended observation periods (>5,000 patient-years), as factors that contribute to the continued need for osteoporotic fracture as an endpoint in human trials. One animal model, the SAM Mouse, with low peak bone mass displays fragility fractures without estrogen deficiency (Matsushita et al. 1986). Falls, relevant degrees of vitamin D deficiency, and long observation periods are difficult to implement in animal models.

4.3.2.5 Bone Loss and Decreased Formation After Decreased Mechanical Usage

Older humans experience both loss of cancellous and cortical bone and reduced bone formation rate unrelated to estrogen deficiency (Schaadt and Bohr 1988; Riggs and Melton 1983). These changes come during a life phase when generalized declines in physical activity also occur. While extreme physical inactivity during bedrest causes marked bone loss (Minaire et al. 1974; Krolner and Toft 1983), the impact of long-term, mild inactivity on the skeleton is neither well understood nor easily assessed. Disuse osteopenia should be demonstrable in an accurate animal model of osteoporosis but is not crucial to its success.

4.3.2.6 Remodeling

Remodeling is the principal bone cellular process that influences bone quantity and strength in the adult skeleton (Frost 1983; Parfitt 1983). Osteoclasts and osteoblasts function sequentially to first remove pre-existing bone tissue and soon after replace it in situ with an approximately equal amount of new bone tissue. Remodeling rate is quantitated in humans by biomarkers of resorption and formation (Bonjour et al. 2014; Szulc 2012; Burch et al. 2014; Devogelaer et al. 2011) and dynamic histomorphometry of iliac crest and rib biopsy specimens obtained after in vivo labeling with bone-seeking fluorochromes (Frost 1969; Parfitt 1983). Remodeling activity in humans has practical importance, as high remodeling rate tends to be a risk factor for osteoporotic fracture (Johansson et al. 2014). Most agents that decrease osteoporotic fracture incidence also decrease remodeling rate (Recker et al. 1978, 1996; Nordin et al. 1976; Jensen et al. 1982; Steiniche et al. 1989; Liberman et al. 1995; Wallach et al. 2000; Black et al. 2007; Ettinger et al. 1999; Cummings et al. 2009).

Adult humans have both cancellous and Haversian bone remodeling (Frost 1983). Cancellous and endocortical surfaces display much higher remodeling rates than cortical surfaces, accounting for most remodeling rate variation in the skeleton. While cancellous remodeling plays the most important role in renewing bone of the spine, wrist, and greater trochanter, Haversian remodeling is probably more important in the femoral neck. Remodeling rates among cancellous sites vary by an order of magnitude (Kimmel and Jee 1982). Though the extent to which estrogen deficiency elevates Haversian remodeling in humans has never been defined in quantitative histologic studies, increased cortical porosity appears to be a risk factor for postmenopausal osteoporosis and play a role in cortical bone strength with teriparatide treatment (Ahmed et al. 2015; Hansen et al. 2013; Bala et al. 2015).

In theory, an accurate animal model should display considerable steady-state Haversian remodeling, because of its importance for maintaining cortical bone strength. It would be ideal if changes in Haversian remodeling and cortical bone caused by osteoporosis treatments were first revealed preclinically. However, as a practical matter, Phase IV follow-up in humans remains the optimal way to assure long-term cortical bone safety of osteoporosis treatments. Thus, preclinical studies are mainly relied upon to assess cancellous bone remodeling and bone strength.

4.3.2.7 Handling

Convenience for animal models is denominated as timeframe compression, purchase cost, availability, housing requirements, handling difficulties, and the necessity for designing/implementing/validating new analysis procedures.

In menopausal women, the accelerated estrogen-deficiency bone loss phase lasts 5–8 years. The time from attainment of peak bone mass until the development of fragility fractures in humans is 25 or more years. An effective animal model that displays peak bone mass followed by estrogen-deficiency bone loss should compress both times by an order of magnitude or more.

Using a fully accurate animal model can be more difficult than doing a human study. For example, if an intervention requires active subject cooperation, animals may be unable to comply. On the other hand, clinical research problems like recruitment, lost sampling units, and compliance are generally nonissues in animal experiments. Animals convenient for some researchers with specialized facilities and expertise are a poor choice for others who lack those tools.

4.3.2.8 In Vivo Measures and Histomorphometry

Most bone measurements that are staples of human studies can be applied in some form in preclinical studies. In some cases, when crucial bone sites in humans cannot be readily sampled, animal experimentation can partially fill that void.

4.3.3 Glucocorticoid Osteoporosis (GIOP)

The clinical features of human GIOP are cancellous osteopenia with elevated risk of vertebral and rib fractures (Frenkel et al. 2015; Briot et al. 2014; Seibel et al. 2013). It occurs in 30–50% of humans who take 10 mg or more prednisone daily (Seibel et al. 2013; Dempster 1989). Bone fragility in GIOP increases rapidly during the first 3–6 months of treatment and is significantly greater than indicated by bone mass measurement, highlighting a different pathogenesis from postmenopausal osteoporosis. Even maintenance doses of prednisone (e.g., 2.5 mg/day) leave bone formation suppressed and fracture risk elevated. GIOP is characterized at the bone tissue level by normal numbers of unusually thin trabeculae and a very low bone formation rate. Though secondary hyperparathyroidism due to decreased intestinal calcium absorption has been observed, low bone formation rate is the primary bone metabolic abnormality (Dempster 1989; Weinstein 2011; Bressot et al. 1979).

4.4 Animal Model Evaluation

4.4.1 Postmenopausal Osteoporosis

Animals that should receive initial consideration are mouse, rat, rabbit, dog, pig, sheep, and nonhuman primate species. Because the above criteria are directed at tissue-level outcomes in humans, the animal models will be evaluated on how they duplicate those outcomes. Deviation from the human condition in molecular mechanisms of development of one or more conditions may well exist. The fit of each animal model to the above criteria is summarized (Table 4.1).

4.4.1.1 Mouse

The two principal uses today for mice in bone research are to characterize (1) skeletal phenotypes in genetically altered (knockout and transgenic) mice and (2) effects of interventions surrounding osteopenic states.

Strain-specific post-OVX bone loss in mice is known (Iwaniec et al. 2006; Bouxsein et al. 2005; Klinck and Boyd 2008). The mouse is an excellent model of osteopetrosis (Walker 1972; Felix et al. 1990), osteoclast and stromal cell ontogeny (Lennon and Micklem 1986), and human cytokines and bone marrow (Kyoizumi et al. 1992; Passeri et al. 1993; Masuzawa et al. 1994; Takahashi et al. 1994; Jilka et al. 1992; Miyaura et al. 1997; Simonet et al. 1997). The mouse is very popular in skeletal research for the ease with which its genome can be systematically manipulated (Seidlova-Wuttke et al. 2012; Bouxsein et al. 2009; Goswami et al. 2009; Martin-Millan et al. 2010; Craft et al. 2012; Inose et al. 2011; Wang et al. 2013a; Zaiss et al. 2010; Poli et al. 1994), enabling direct examination of how genes control the activities of bone cells functioning within their intermediary organizations of modeling and remodeling. Mice with, respectively, low and high peak bone mass (Beamer et al. 1996) lend themselves to genetic investigations using F2 cross and F1 backcross studies of extremes of a continuous phenotype (Lander and Botstein 1989).

The mouse is inherently attractive as an in vivo model for tissue-level osteoporosis pharmacology research, because its small body weight may enable using only ~10% as much compound during similar experiments as in rats. Much as in the rat, cancellous (Poli et al. 1994; Broulik 1991; Onoe et al. 2000; Yamamoto et al. 1998), but not cortical (Edwards et al. 1992), bone loss occurs soon after OVX in strains like C57BL, C57Bl/6 J/129, and Swiss-Webster. Estrogen-deficiency bone loss is prevented by low doses of 17-β-estradiol (Jilka et al. 1992; Fujioka et al. 2004; Mödder et al. 2004; Zhao et al. 2013). Cancellous and cortical bone loss also occurs in the mouse vertebrae (Bar-Shira-Maymon et al. 1989a; Bar-Shira-Maymon et al. 1989b) and femur (Weiss et al. 1991) during ages 12–24 months.

Table 4.1 Relevance assessment of in vivo animal models with respect to recognized skeletal characteristics of postmenopausal osteoporotic humans

		AA	AAA	A	X	X	A	AA
Attribute	Human	Mouse	Rat	Rabbit	Dog	Pig	Sheep	Primate
Growth/adult phases	Yes	OK	OK	Yes	Yes	Yes	Yes	Yes
Role of estrogen								
Cyclicity	28 days	Inducible	4–5 days	21 days	205 days	21 days	21 days	21–28 days
Natural menopause	Yes	Yes[c]	Yes[c]	NK	No	NK	NK	Yes
Estrogen-deficiency turnover⇑ and bone loss	Yes	Yes	Yes	Yes	No	Poor	Yes	Yes
Estrogen response	Turnover⇓	[d]Turnover⇓	Turnover⇓	Turnover⇓	No	NK	NK	Turnover⇓
Response to proven anti-osteoporotic agents	Yes	Some	Yes	Some	Some	NK	Some	Yes
Development of osteoporotic fractures	Yes	No[e]	No[e]	No[e]	No[e]	No[e]	No[e]	No[e]
Remodeling								
Cancellous	Yes	Yes	Yes	Yes	Yes	Yes	Yes	Yes
Haversian	Yes[a]	No	Low levels; inducible	Yes	Yes	Yes	Yes	Yes
Timeframe compression	No	Yes	Yes	Yes	N/A	Some	Some	Some
Convenience (see postmenopausal osteoporosis [section G])	OK[b]	Yes	Yes	[b]	[b]	[b]	[b]	[b]

In vivo measures								
DXA	Yes	Yes	Yes	Yes	Yes	Yes	Yes	Yes
Biochemical markers	Yes	Yes	Yes	~Yes	NK	NK	NK	Yes
Histomorphometry								
Iliac crest	Yes	No	No	NK	Yes	NK	Yes	Yes
Vertebral body	No	Yes	Yes	Yes[a]	Yes	Yes	Yes	Yes
Proximal femur	No	NK	Yes	Yes	Yes	NK	NK	Yes

AAA highest relevance/utility, *AA* high relevance/utility, *A* relevant when needed, *X* not recommended, *NK* not known

[a]Anatomic study site difficult

[b]Special housing/handling required

[c]Perpetual diestrus

[d]At low doses; formation rises with high doses

[e]*Ex vivo* biomechanical tests as surrogate

Estrogen administration increases bone formation with woven bone deposition through an estrogen receptor-mediated mechanism (Edwards et al. 1992; Urist et al. 1950; Samuels et al. 1999; Bain et al. 1993). However, estrogen-stimulated bone formation in mice occurs at 17-β-estradiol doses above 0.05 mg/kg/day subcutaneously (SC). When 17-β-estradiol doses of ~0.01 mg/kg/day 2–3×/week SC are used in the OVX mouse (Jilka et al. 1992; Fujioka et al. 2004; Mödder et al. 2004; Zhao et al. 2013), estrogen replacement reduces turnover and stops bone loss as in the OVX rat and perimenopausal human.

Many mouse skeletal experiments now involve studies of transgenic and knockout animals, providing valuable data about how genes affect bone cells as they function in their usual intermediary organization. Some mouse strains, including C57BL and C57BL/6J/129 that have been used to make most knockout and transgenic mice, develop estrogen-deficiency bone loss that is prevented by appropriate estrogen replacement, making them suitable for some types of pharmacologic intervention. The use of the CreLox promoter to make conditional knockouts that limit the activity of altered genes to specific periods during the lifespan (e.g., adulthood) makes possible experiments that may identify specific genes that affect peak bone mass and control bone behavior during adulthood (Plück 1996; Ray et al. 2000). The similar lifespan of mice to rats, the similar cost for experimentation, and the lower requirements for compounds suggest that mouse bone studies can and should continue to increase in number.

4.4.1.1.1 Cautionary Notes

The bone formation response to estrogen at endocortical surfaces is a high-dose phenomenon. However, its existence makes studying bone formation responses in the mouse risky, since this formation pathway does not exist in larger animals. Experiments with transgenic and knockout mice are the most likely method to find the molecular pathways by which estrogen influences tissue-level behavior of the human skeleton.

The small size of mouse bones is a problem. The best bone sampling sites are the distal femur and lumbar vertebral bodies for cancellous bone and the mid-femur for cortical bone. Mouse bones contain small amounts of mineral that require specialized (slower speed, smaller collimator) pDXA, rather than conventional human DXA, to obtain adequate precision (Dickson et al. 2004; Brommage 2003; Fuchs et al. 2013). Peripheral quantitative computed tomography (pQCT) is useful (Beamer et al. 1996; Ferretti et al. 1994) but provides a less direct link to human densitometry measurements than pDXA. MicroCT (Bouxsein et al. 2010; Castañeda et al. 2006), with its three-dimensional images, 8–12 μm pixel resolution, and analytical software, is the method of choice for evaluating microarchitectural properties. Despite the existence of specific bone mass and microarchitecture evaluation techniques for mouse bone, the typical sampling sites contain relatively little bone tissue for histomorphometric analysis of bone formation and resorption rates. For example, intact C57BL/6J/129 mice are osteopenic in the distal femoral and proximal tibial metaphyses with a bone volume (BV/TV) of 5–8 %, about one-third that in rats, making intergroup differences in bone mass and turnover inherently difficult to study. Moreover, the mouse distal femur or proximal tibia contains only ~5 to 7 mm^3 of total cancellous bone volume, while the vertebral body contains ~15 mm^3. For comparison, the rat distal femur or proximal tibia contains ~20 to 25 mm^3 of cancellous bone volume, while the vertebral body contains ~35 mm^3 of evaluable bone; and a human transilial biopsy specimen contains ~225 mm^3 of cancellous bone. Because of these factors, extreme care is required during mouse bone analysis to avoid Type II errors, due to the inherent inability to examine the same minimum amounts of bone mass, volume, and surface in mouse bone, as in larger animals and humans. To address this problem, the use of single genders, 12–13 mice/group, and sampling of both left and right long bones and multiple vertebral bodies is recommended, particularly during the phenotyping of knockout and transgenic mice. The problems created by low cancellous bone mass in the long bone metaphyses of C57BL/6J mice may be completely overcome by housing growing mice at 32 °C throughout their lifespan (Iwaniec et al. 2015, 2016).

4.4.1.2 Rat

The rat has long yielded data relevant to the adult human skeleton, including the first evidence that osteoclasts ingest bone (Arnold and Jee 1957) and early evidence of the hematogenous origin of osteoclasts (Gothlin and Ericsson 1973; Andersen and Matthiessen 1966). The rat was once held unsuitable as an adult human skeletal model because many epiphyseal growth cartilages in *male* rats remain open past age 30 months causing it to exhibit "continuous growth and lack of remodeling" (Dawson 1925). However, bone elongation ceases and epiphyseal closure ensues at important sampling sites in *female* rats by age 6–9 months (Acheson et al. 1959; Joss et al. 1963; Turner et al. 1989; Kimmel 1992a). Periosteal expansion in the female rat continues until it reaches peak bone mass at age 10 months (Schapira et al. 1992; Li et al. 1992). The healthy lifespan for intact female rats is 21–24 months, giving it an appreciable lifespan both before and after peak bone mass attainment. It exhibits cancellous remodeling (Wronski et al. 1999). The rat is an accurate model of adult human skeletal disease, providing female rats of at least age 6 months are used and no information on Haversian remodeling is required (Frost and Jee 1992).

From age 9 weeks, female rats have a regular estrus cycle during which estrogen levels spike for 18 h every 4 days (Butcher et al. 1974). During ages 12–24 months, the fraction of rats in constant diestrus rises gradually (Lu et al. 1979) as cancellous bone loss gradually occurs (Schapira et al. 1992). While this is not true menopause, peaks in estrogen cease as bone loss occurs, associating rat "menopause" with cancellous bone loss.

Following OVX, loss of cancellous bone mass and strength begins immediately and then decelerates in a site-specific fashion after 3–4 months, to enter a plateau phase (Shane et al. 2014; Wronski et al. 1989a, 1989b; Kalu 1991; Wronski 1992;

Kimmel and Wronski 1990). The rapid bone loss phase is marked by increased bone turnover (Wronski et al. 1986, 1989a; Black et al. 1989). These metabolic events mimic well bone changes that accompany estrogen deficiency in humans. Only some cancellous bone sites in the rat experience such bone loss (Ito et al. 1993; Li et al. 1996a), further tightening the parallel of the rat and human, since osteoporotic fragility fractures and osteopenia in humans are limited to a few sites (Wasnich et al. 1983; Cummings 1991). DXA, the most widely applied bone mass measurement method in humans, works well in rats (Amman et al. 1992; Kimmel and Wronski 1990).

OVX rats given prompt estrogen replacement experience neither a rise in turnover (Wronski et al. 1988; Turner et al. 1987; Kalu et al. 1991) nor bone loss (Wronski et al. 1988; Turner et al. 1987; Kalu et al. 1991; Garner et al. 1992). Antiresorptive agents like bisphosphonates (Seedor et al. 1991; Bauss et al. 2002; Gasser et al. 2008), SERMs (Evans et al. 1994, 1996), and calcitonin (Mazzuoli et al. 1990) also block the rise in turnover and bone loss in both rats and humans (Liberman et al. 1995; Johnston et al. 2000; Eastell et al. 2010; Szucs et al. 1992; Reginster et al. 2006). Rats never exhibit a bone formation response to 17-β-estradiol.

The rat, like all other animal models of osteoporosis, has no estrogen deficiency fragility fractures. This shortcoming can be partially overcome by ex vivo biomechanical compression testing of the vertebral body (Li et al. 1991; Turner and Burr 1993; Toolan et al. 1992), three-point bending testing of the mid-femoral shaft, and shearing testing of the femoral neck (Ferretti et al. 1994). Rat vertebral body strength appears to be a reasonable surrogate for human vertebral body fracture risk. Such preclinical biomechanical testing may have avoided the problems first revealed during the clinical investigation of sodium fluoride as a treatment for osteoporosis (Kleerekoper et al. 1991; Riggs et al. 1990).

Adult rats lose bone following immobilization. Multiple methods of permanent and temporary immobilization exist (Wakley et al. 1988; Svesatikoglou and Larsson 1976; Thomaidis and Lindholm 1976; Thompson and Rodan 1988; Li et al. 1990; Lindgren and Mattson 1977). Acute (Thompson and Rodan 1988), chronic (Li et al. 1990), and recovery-related (Lindgren and Mattson 1977; Maeda et al. 1993; Lane et al. 1994) bone changes related to disuse are easily studied.

Adult female rats have sufficient cancellous bone remodeling to permit useful experiments (Wronski et al. 1999; Frost and Jee 1992; Vignery and Baron 1980; Erben 1996). The remodeling/modeling ratio rises with age. Reversal lines at the base of cancellous osteons are absent at age 4 months (Erben 1996), suggesting a predominance of modeling. However, studies of older rats confirm that bone activity undergoes a transition to stable bone remodeling activity with a neutral balance of resorption and formation during adulthood (Wronski et al. 1999; Erben 1996), suggesting the presence of remodeling.

Haversian remodeling levels in rats approximate zero at the tibiofibula junction or mid-femur (Baylink et al. 1970; Morey and Baylink 1978). However, processes resembling Haversian remodeling are induced by some anabolic agents (Jee et al. 1990) and stressful metabolic conditions (Ruth 1953; Miller et al. 1986a; deWinter and Steendijk 1975). It is not known whether these agents/conditions accelerate

Haversian remodeling in humans. The rat's low levels of Haversian remodeling preclude its use for analysis of Haversian remodeling, especially of agents that suppress such remodeling.

In 3-month-old OVX rats, the phase of accelerated estrogen-deficiency bone loss lasts 4–5 months in the proximal tibial metaphysis (Wronski et al. 1989a), a 20-fold timeframe compression when compared with estrogen-deplete women. Experiments with 6–9-month-old OVX rats create extremely reliable estrogen-deficiency bone loss within 6 weeks. The rat reaches peak bone mass by age 10 months, a 30-fold timeframe compression when compared to the adult human. The rat is also among the most convenient of experimental animals to handle and house.

During OVX rat experiments, it is wise to restrict food to limit OVX-induced weight gain. This may be either pair feeding to a sham group (conservative) or body weight restriction to a sham group (aggressive). Either type of food restriction speeds cancellous bone loss (Wronski et al. 1987; Roudebush et al. 1993) and may create faster, more reliable cortical bone loss. Investigators should never choose "retired breeders." The variable skeletal status of retired breeders that often become osteopenic during breeding is likely to influence negatively experimental outcomes. Virgin female rats that are skeletally mature and not osteopenic are much more acceptable for prevention studies than retired breeders, because they have considerably more cancellous bone whose loss can be prevented after OVX (Binkley and Kimmel 1994). The ideal experimental group size is 10–11 to allow for some attrition and loss of samples. Treatment during prevention studies should commence within 2 days of OVX, because estrogen-deficiency bone loss begins immediately (Wronski et al. 1988; Campbell et al. 2008). Treatment during restoration studies should commence 14–16 weeks after OVX to assure that bone loss has stabilized in long bone metaphyses and bone changes are due to stimulated formation, rather than inhibition of further ongoing bone loss (Wronski et al. 1988). Waiting until 9–12 months post-OVX will result in long bone metaphyses with such low initial bone mass that too little pre-existing bone surface exists for bone formation-stimulating agents to express efficacy.

The OVX rat is an excellent model that replicates the most important clinical features of the estrogen-deficient adult human skeleton. Its site-specific development of estrogen-preventable cancellous osteopenia in 6–9-month-old female rats is the most widely replicated, easily detected, and relevant physiologic response in in vivo skeletal research. Ample time exists for experimental designs that either prevent estrogen-deficiency bone loss or restore bone lost due to estrogen deficiency. The rat's low levels of Haversian remodeling present little problem when testing agents for their ability to prevent loss or rebuild lost cancellous bone. The lack of fragility fractures can be partially compensated by biomechanical testing. Rats are convenient for most investigators; virgin females aged 6–7 months at the start of an experiment are the optimal choice. All compounds should be tested first in prevention mode, because compounds that only have anti-resorptive efficacy will appear inactive when tested in treatment mode. Six weeks is the optimal time post-OVX for necropsy sampling in prevention mode studies. Sixteen weeks post-OVX is an appropriate time for beginning treatment mode experiments in the osteopenic

OVX rat (Li et al. 1999), but responses in extremely osteopenic bone sites have been studied at up to 15 months post-OVX (Qi et al. 1995). Existing laboratory measurement tools of biochemistry, densitometry, histomorphometry, and mechanical testing are readily applicable. Experimental situations may exist in which the extreme estrogen deficiency of the OVX rat may not exactly match the variable estrogen levels in naturally postmenopausal women (Cummings et al. 1998; Longcope et al. 1984; Cauley et al. 1999). The adult female rat is superior to the adult female mouse for pharmacology studies because its bones provide ~4× more tissue to sample than mouse bones.

4.4.1.3 Guinea Pig, Ferret, and Cat

Though occasional reports using guinea pigs, ferrets, and cats in osteoporosis research have appeared (Ferguson and Hartles 1970; Vanderschueren et al. 1992; Mackey et al. 1995), few formal studies of estrogen-deficiency bone loss exist. Seven-month-old guinea pigs do not lose bone by 4 months post-OVX (Vanderschueren et al. 1992). Estrogen deficiency alone does not cause bone loss in cats (Ferguson and Hartles 1970). The ferret, weighing less than 1 kg, has Haversian remodeling. Its normal skeletal physiology, including the accumulation of estrogen-dependent bone that seems to accompany normal cyclicity in other mammals, is dependent on a regular light cycle. It also exhibits expected changes in bone remodeling rate and bone volume during treatment with PTH (1–34) (Mackey et al. 1995). However, estrogen deficiency by itself does not cause bone loss in ferrets.

4.4.1.4 Rabbit

Adult rabbits have abundant Haversian remodeling and an appropriate parathyroid hormone response (Hirano et al. 2000), but their estrus cycle is unlike that of humans. One 3 month study shows that rabbits exhibit estrogen-deficiency bone loss in the lumbar vertebral body (Pennypacker et al. 2011). While one additional 4 month study that assayed the femur was promising (Sevil and Kara 2010), three others with durations of 1.5–4 months that used the femur and vertebral bodies report no OVX-related bone loss (Castañeda et al. 2008; Baofeng et al. 2010; Liu et al. 2012).

Existing studies of estrogen-deficiency bone loss suggest that significant osteopenia may develop in lumbar vertebral bodies at 3 months post-OVX. Long bone metaphyses are extremely osteopenic in intact rabbits, indicating that the absolute amount of bone available to be lost from such sites after OVX is likely to be small and thus difficult to detect. However, though rabbit lumbar vertebral bodies are also small with only about 50 mm^3 of bone volume, they appear to respond appropriately to OVX. The adult OVX rabbit yields inconsistent data up to 4 months post-OVX but is potentially worth testing at longer time periods.

4.4.1.5 Dog

The adult dog is generally a reliable model of the adult human skeleton. ^{239}Pu-injected beagle studies not only were the first to describe adult cancellous bone remodeling (Arnold and Jee 1957) but also pointed to the hematogenous origin of osteoclasts (Jee and Nolan 1963). The ratio of cortical to cancellous bone is similar to humans (Gong et al. 1964; Johnson 1964). Its age of peak bone mass is about 3 years. Haversian and cancellous osteons remodel with similar morphology, though faster in dogs (Frost 1969; Kimmel and Jee 1982). Skeletal responsiveness parallels the adult human for glucocorticoids (Dempster 1989; Weinstein 2011; Bressot et al. 1979; Jett et al. 1970), uremia (Ritz et al. 1973; Malluche et al. 1987), bisphosphonates (Allen et al. 2010; Helm et al. 2010), disuse (Minaire et al. 1974; Krolner and Toft 1983; Uhthoff et al. 1985; Jaworski and Uhthoff 1986), and parathyroid hormone excess (Parfitt 1976; Podbesek et al. 1983).

However, the oophorectomized beagle yields inconsistent, usually negative, results across multiple laboratories. Most experiments lack significant findings (Kimmel 1992b; Shaw et al. 1994; Boyce et al. 1990; Wilson et al. 1998; Frost 2000; Malluche et al. 1986, 1988; Nakamura et al. 1992; Dannucci et al. 1987; Shen et al. 1992; Karambolova et al. 1985, 1987; Snow and Anderson 1986; Snow et al. 1984; Koyama et al. 1984). The most complete experiment suggests that estrogen-deficiency bone loss at a rate of ~6–9 % occurs annually. Though the bone turnover pattern in two suggests a transient rise after OVX (Boyce et al. 1990; Wilson et al. 1998), one is confounded by a regional acceleratory phenomenon due to the use of serial rib biopsies (Wilson et al. 1998; Frost 2000). In contrast, publications from one group (Malluche et al. 1986, 1988) suggest that bone formation falls rapidly to 50% of sham with no transient increase (Recker et al. 1978, 2004; Stepan et al. 1987). The data also indicate that estrogen suppresses turnover with uncertain effects on bone mass (Karambolova et al. 1987; Snow and Anderson 1986).

Estrogen exposure and menstrual/estrus cyclicity is much lower than in other animals. 17-β-estradiol levels are usually very low, rising only *twice yearly* for several weeks (Concannon et al. 1975; Bell et al. 1971). 17-β-estradiol spikes for 18 h every 4 days in rats (Butcher et al. 1974) and 1–2 days monthly in women and nonhuman primates (Baird and Guevara 1969; Reed et al. 1986; Longcope et al. 1989). Cumulative estrogen exposure in dogs is only one-fourth that in humans with much less frequent peaks. It is similar to primates without peaks. This difference could contribute to the dog's developing a negligible estrogen-dependent compartment of cancellous bone.

The adult beagle, an excellent adult human skeletal model except for estrogen deficiency, has Haversian remodeling. Though osteopenia has occasionally been reported in the OVX beagle, poor interlaboratory reproducibility makes it an unreliable model. Since most experiments were conducted before the wide availability of DXA and MicroCT, most studies apply relatively weak bone mass measurement methods (Shaw et al. 1994; Boyce et al. 1990; Wilson et al. 1998; Frost 2000; Malluche et al. 1986, 1988; Nakamura et al. 1992; Dannucci et al. 1987; Shen et al. 1992; Karambolova et al. 1985, 1987; Snow and Anderson 1986; Snow et al. 1984;

Koyama et al. 1984). The inherent inability to determine the dose of an agent that halts estrogen-deficiency bone loss in the dog markedly increases the difficulty of placing most bone pharmacology experiments that use dogs into proper perspective. Beagles are less estrogen replete than women and may have a smaller estrogen-dependent compartment of bone in their skeleton. Because of its uncertain development of estrogen-deficiency osteopenia and the existence of other reliable large animal models for this characteristic, the use of the dog in postmenopausal osteoporosis research is not recommended.

4.4.1.6 Pig

The pig has both growing and adult skeletal phases and a regular estrus cycle that is somewhat shorter than humans. The age of peak bone mass is older than 3 years. In one 6 month study of combined 0.75 % calcium diet and OVX, BV/TV was 15 % lower, and spine BMD was 6 % lower than in sham pigs fed a 0.9 % calcium diet (Li et al. 1993). In two other OVX only studies, no significant bone loss was seen at 10–20 months post-OVX (Scholz-Ahrens et al. 1996; Sipos et al. 2011). Though pigs have also been used successfully to study fluoride and exercise effects on the skeleton (Kragstrup et al. 1989; Raab et al. 1991), existing data indicate that OVX alone does not cause consistent bone loss in the pig.

4.4.1.7 Sheep

The ewe has both growing and adult skeletal phases with an age of peak bone mass of 5–7 years. Though adult ewes have a regular estrus cycle during the short days of fall and winter, they experience anestrus in spring and summer (Malpaux and Karsch 1990). Fluoride skeletal findings in the adult ewe parallel histologic changes in humans with frequent signs of increased formation, poor mineralization, and osteoblast toxicity (Eriksen et al. 1985; Chavassieux et al. 1991a, 1991b). Glucocorticoid data seem consistent with findings in other large animals and humans (Bressot et al. 1979; Chavassieux et al. 1993). Sheep can be housed readily at most vivariums and pose few handling or ethical concerns.

Data indicate that significant bone loss occurs after ovariectomy in adult ewes in relevant bone sites such as the vertebral body, proximal femur, distal radius, rib, and ilium. One study finds significant bone loss at 3 months post-OVX that does not progress over the next 15 months (Sigrist et al. 2007). Four report significant bone loss at 6 months post-OVX (Jiang et al. 2005; Johnson et al. 2002; Macleay et al. 2004; Chavassieux et al. 2001), and four observe significant bone loss only by 1 year post-OVX (Zhang et al. 2014; Wu et al. 2008; Kreipke et al. 2014; Newton et al. 2004) that does not progress by 2 years after surgery (Kreipke et al. 2014). Two experiments show increased cortical porosity without changes in BMD in typical bone sites at 12 months post-OVX (Healy et al. 2010; Kennedy et al. 2008,

2009). These experiments indicate that bone loss is accompanied by an early phase of elevated turnover (Chavassieux et al. 2001; Newton et al. 2004).

The inconsistences noted in Healy et al. (2010), Kennedy et al. (2008, 2009), and two other early studies (Hornby et al. 1995; Geusens et al. 1996), plus problems related to the photoperiod, cause one review article to be cautious (Oheim et al. 2012) and another to be quite positive (Turner 2002). Considering today's data, the sheep appears to be a worthwhile large animal model of postmenopausal osteoporosis with a humanlike skeletal response to estrogen deficiency in both cancellous and cortical bone. Most experiments report data from animals housed in outdoor pastures with normal seasonal variations in length of day that are known to affect bone outcomes (Healy et al. 2010; Kennedy et al. 2008, 2009). It is likely that housing sheep in laboratory conditions with 16:8 light/dark conditions before and during experiments will produce more consistent data. The potential existence of another useful large animal model of the adult skeleton is particularly important when recognizing the practical problem of persistent ethical concerns over the use of nonhuman primates.

4.4.1.8 Nonhuman Primate

Most of the above models focus on traditional studies that involve a few in vivo measurements and primarily postmortem evaluations. With the nonhuman primate, both terminal and nonterminal evaluations relevant to the treatment and prevention of osteoporosis are possible.

4.4.1.8.1 Terminal Studies

The nonhuman primate has both growing and adult skeletal phases. Peak bone mass occurs at age 9–12 years in cynomolgus and rhesus monkeys (Pope et al. 1989). Like adult women, most nonhuman primates have a regular menstrual cycle of ~28 days duration. Nonhuman primates experience natural menopause shortly after age 20 years (Bellino and Wise 2003; Colman et al. 1999; Packer et al. 1998; Martin et al. 2003).

Nonhuman primates show lower bone mass and strength with increased turnover after OVX (Binkley et al. 1998; Jerome et al. 1986; Miller et al. 1986b; Balena et al. 1993; Itoh et al. 2002; Williams et al. 2013; Ominsky et al. 2011; Lees et al. 2007; Smith et al. 2013; Lundon et al. 1994; Mas et al. 2007), premature menopause (Shively et al. 1991), or GnRH agonist treatment (Mann et al. 1990). Much like in humans, treatment with estrogen (Jayo et al. 1998), bisphosphonates (Binkley et al. 1998; Balena et al. 1993; Itoh et al. 2002), cathepsin K inhibitor (Williams et al. 2013), anti-RANK ligand antibody (Ominsky et al. 2011), SERMs (Lees et al. 2007), or vitamin D analogs (Smith et al. 2013) positively impacts OVX-induced bone loss. Treatment of OVX nonhuman primates with PTH (1–34) also yields similar data to those seen in osteoporotic women (Dobnig 2004; Fox et al. 2008).

Nonhuman primates also experience bone loss with age (Turner 2002; Pope et al. 1989; Grynpas et al. 1993), natural menopause (Colman et al. 1999), and long duration immobilization (Young et al. 1983). Histomorphometric data from iliac bone of nonhuman primates and humans also yield similar values (Pope et al. 1989; Jerome et al. 1986; Schnitzler et al. 1993; Recker et al. 1988). Direct histomorphometric evaluation of the principal sites of osteoporotic fracture in humans, the vertebral body and proximal femur, is extremely difficult in human studies. When such studies are needed, the best alternative is the same anatomic site in the nonhuman primate (Jerome et al. 2012; Stroup et al. 2009; Cusick et al. 2012).

4.4.1.8.2 Nonterminal Studies

Biochemical markers of bone resorption, like urinary deoxypyridinoline, show both age and gender effects in rhesus but only age effects in cynomolgus monkeys (Cahoon et al. 1996; Stroup et al. 2001; Samadfam and Smith 2014). Nonhuman primates can be used in short, nonterminal studies to test the effects of agents on bone biochemical markers of resorption and formation. As in estrogen-deficient humans, cathepsin K antagonists reduce urinary c-telopeptide concentrations in GnRH-treated rhesus monkeys after 5 days. The GnRH-treated nonhuman primate is a model of transient estrogen deficiency (Stroup et al. 2001).

4.4.1.8.3 Summary

The adult nonhuman primate is an excellent model for all aspects of the adult human skeleton (Smith et al. 2009; Pope et al. 1989; Bellino and Wise 2003; Colman et al. 1999; Packer et al. 1998; Martin et al. 2003; Binkley et al. 1998; Jerome et al. 1986, 2012; Miller et al. 1986b; Balena et al. 1993; Itoh et al. 2002; Williams et al. 2013; Ominsky et al. 2011; Lees et al. 2007; Smith et al. 2013; Lundon et al. 1994; Mas et al. 2007; Shively et al. 1991; Mann et al. 1990; Jayo et al. 1998; Fox et al. 2008; Grynpas et al. 1993; Young et al. 1983; Schnitzler et al. 1993; Recker et al. 1988; Stroup et al. 2001; Jerome and Peterson 2001; Cusick et al. 2012; Cahoon et al. 1996). The age of peak bone mass is 9–12 years. Its regular estrus cycles are an excellent analog of menstrual cycles in humans. High turnover with cancellous osteopenia develops within 3–6 months post-OVX. Most currently approved agents have been studied for 16–24 months in nonhuman primates (Binkley et al. 1998; Balena et al. 1993; Itoh et al. 2002; Williams et al. 2013; Ominsky et al. 2011; Lees et al. 2007; Smith et al. 2013). Though housing and care requirements of nonhuman primates limit their use to a small number of facilities, when handled by experienced staff in an appropriate environment, they present few care problems. Studies of agents in nonhuman primates in nonterminal studies using biochemical markers of bone turnover yield relevant screening data for new agents. In testing new anti-osteoporosis treatment agents that will eventually require Phase III trials of human anti-fracture efficacy for successful marketing, because of ethical concerns in some venues, the need for terminal nonhuman primate studies must be carefully considered.

4.4.1.9 Schenk Model (Screen for Anti-resorptive Activity)

Once it was firmly established that the tissue-level action of estrogen replacement therapy for bone was inhibition of resorption (Recker et al. 1978; Heaney et al. 1978), it was apparent that safe, patentable anti-resorptive agents would have excellent potential as pharmaceutical agents for treatment of osteoporosis. The need for a convenient, reliable in vivo preclinical assay for anti-resorptive activity was born.

The long bone metaphysis of the growing rodent was a logical choice. Bone histologists had long been aware of the presence of osteoclasts in three microanatomical regions of the long bone metaphysis: (a) the periosteum of the funnel region, (b) the ends of trabeculae opposite the nearby growth cartilage, and (c) the primary spongiosa just on the metaphyseal side of the growth cartilage metaphyseal junction. Natural in vivo proof of concept of anti-resorptive efficacy came from histologic studies of metaphyseal regions of osteopetrotic mice that showed club-shaped bones with abnormally high quantities of trabecular bone in the metaphysis and overly dense primary spongiosa regions (Walker 1975a, 1975b). From the convenience aspect, no surgery is required, and laboratory rodents aged 4–5 weeks are always readily available.

Any agent that inhibits bone resorption without affecting the bone elongation process could be studied by looking at the overall bone status of a whole metaphyseal region. One of the earliest experiments reported the use of first-generation bisphosphonates in a 10 days experiment, to observe club-shaped bones with high metaphyseal bone volume and a very dense primary spongiosa. This assay showed the simultaneous ability to detect mineralization defects by observing the nearby growth cartilage for lengthening comparable to rickets and sites of bone formation on metaphyseal trabeculae (Schenk et al. 1973). This report was the first to show that experiments using agents that affect the bone elongation process are more difficult to interpret than those using only simple anti-resorptives.

The "Schenk Assay," with its 7–10 days in vivo duration, starting in 4–6-week-old male rats was used to test all three generations of bisphosphonates, including all compounds that eventually became approved osteoporosis or anticancer drugs (Miller and Jee 1975; Sietsema et al. 1989; Schenk et al. 1986; Rowe 1985; Murakami et al. 1994; Mühlbauer et al. 1991; Pataki et al. 1997). The original Schenk Assay relied upon time-consuming histologic analysis. However, once bone densitometry became available, later versions used densitometric (Amman et al. 1992; Kimmel and Wronski 1990), pQCT (McHugh et al. 2003), and MicroCT endpoints (Bouxsein et al. 2010) that yielded data about anti-resorptive activity within a few days of necropsy.

The Schenk Assay was less successful in testing estrogen receptor (ER) ligands for anti-resorptive activity, because ER ligands inhibit bone elongation (Josimovich et al. 1967). Thus, the in vivo bone efficacy of estrogen receptor ligands is best tested first in newly OVXd rodents. A variant of the Schenk Assay exists in genetically altered mice, in which the metaphysis is assayed for high bone volume. This was used to confirm in vivo anti-resorptive activity for anti-RANK ligand antibodies (Kostenuik et al. 2009). Safety assessment pathologists often use the primary spongiosa that is readily visible in standard site sections (e.g., ribs and sternum) as an assay for anti-resorptive activity in general screening of test compounds. The Schenk Assay was adapted as a 10 days assay in growing 4–5-week-old rabbits to test anti-resorptive activity of human cathepsin K inhibitors that are inactive in rats and mice, because of the lack of homology of the human and rodent cathepsin K enzymes (Pennypacker et al. 2013).

4.4.2 *Glucocorticoid Osteoporosis (GIOP)*

Human GIOP is characterized clinically by increased risk of fracture that exceeds what is predicted from BMD values in postmenopausal osteoporotic patients, suggesting poor intrinsic strength of bone tissue. The increased fracture risk is accompanied by very low bone formation rate (Frenkel et al. 2015; Briot et al. 2014; Seibel et al. 2013; Dempster 1989; Weinstein 2011; Bressot et al. 1979). Traditionally, only large animals with Haversian remodeling, such as rabbits, dogs, and sheep, displayed both decreased bone formation and bone loss in response to glucocorticoids (GCs). For several decades, this lack of a reliable small animal model for human GIOP slowed the search for treatments. Rats and mice usually showed reduced bone formation but rarely showed convincing bone loss in response to GCs. Though the preclinical rodent models for adult human GIOP are less established than those for estrogen-deficiency osteopenia, over the past two decades, the adult mouse is now an acceptable small animal model of GIOP.

4.4.2.1 Small Animals

Many mouse and rat GIOP experiments have used growing animals (Geusens et al. 1990; Grahnemo et al. 2015; Lin et al. 2014; Turner et al. 1995; Li et al. 1996b; Smink et al. 2003; Ortoft and Occlude 1988; Aerssens et al. 1994; Ferretti et al. 1993; Zhang et al. 2015; Postnov et al. 2009; Ahmed et al. 2012; Altman et al. 1992; King et al. 1996). They paradoxically show *increased* metaphyseal cancellous bone mass accompanied by reduced rates of bone formation and resorption. GCs decrease both the disappearance rate of mineralized metaphyseal tissue and the bone elongation rate, thereby increasing cancellous mineralized tissue quantity in a smaller bone. GC experiments in small growing animals also show cortical osteopenia and low bone strength (Ortoft and Occlude 1988; Aerssens et al. 1994; Ferretti et al. 1993), due to reduced bone formation at periosteal surfaces. The overall picture differs from adult human GIOP, because of the presence of active bone elongation centers. Furthermore, cortical osteopenia and low bone strength are due to reduced periosteal expansion that has limited importance in the adult human skeleton. All growing animals treated with GCs are poor models for adult human GIOP.

Multiple publications across independent laboratories indicate that adult mice are a useful model for adult human GIOP (Thiele et al., 2012, 2014; Plotkin et al. 1999; Hofbauer et al. 2009; McLaughlin et al. 2002; O'Brien et al. 2004; Takahata et al. 2012; Balooch et al. 2007; Maher et al. 2011; Yao et al. 2015; Conradie et al. 2011; Weinstein et al. 1998). The most repeatable experiments use 60d slow-release prednisolone pellets in 6-month-old C57BL/6/129SvJ, FVbN (3.5 mg), Swiss-Webster (5 mg), or C57Bl/6J (7.5 mg) mice over a 6–8 weeks period (Thiele et al. 2014). These experiments find low bone formation rate and mild-moderate cancellous bone loss in the long bone metaphyses and vertebral bodies. These features can be difficult to detect, because of the small size of mouse bones and their inherent osteopenia, as noted in the discussion of estrogen-deficient mice above. Some publications also report low bone strength (Takahata et al. 2012; Balooch et al. 2007; Maher et al. 2011). The available publications (Thiele et al. 2012, 2014; Plotkin et al. 1999; Hofbauer et al. 2009; McLaughlin et al. 2002; O'Brien et al. 2004; Takahata et al. 2012; Balooch et al. 2007; Maher et al. 2011; Yao et al. 2015; Conradie et al. 2011; Weinstein et al. 1998) combine to suggest that if the adult mouse model is executed as above with group sizes of 12–15 in 6–8 weeks experiments, and precise analytical methods such as MicroCT and bone strength measurement are applied, the adult mouse model of human GIOP can be successful.

There are few encouraging publications about GC skeletal effects in the adult rat. Experiments in older rats generally show only a trend toward, if any decrease in metaphyseal cancellous bone mass and cortical area, with a more uniform and significant finding of decreased bone formation (Nakamura et al. 1996; Sjoden et al. 1984; Goulding and Gold 1988; Wimalawansa et al. 1997; Lindgren and DeLuca 1983; Wang et al. 2013b; Folwarczna et al. 2011; Bitto et al. 2009; Wimalawansa and Simmons 1998). The prednisolone dose is not standardized, and loss of body weight tends to be a problem. GC bone data in older rats have not been uniformly reproduced in multiple laboratories. The adult rat is not as good a model for study of GC skeletal effects as the adult mouse.

Glucocorticoids in adult large animals cause short term increases in remodeling activity and longer term bone loss (Castañeda et al. 2008; Chavassieux et al. 1993). Consistent with the small animal, the formation phase that follows remodeling's resorption phase proceeds slowly (Castañeda et al. 2008; Jett et al. 1970). Rabbit studies suggest that bones are weaker and BMD is lower with GC treatment (Lindgren et al. 1984; Luppen et al. 2002; Waters et al. 2000; Bostrom et al. 2000; Carvas et al. 2010; Handal et al. 2012; Zhang et al. 2012). In dogs, 13 % loss of lumbar spine BMD occurs over 12 months (Jett et al. 1970; Lyles et al. 1993). Glucocorticoids cause both bone loss and reduced bone formation in sheep (Chavassieux et al. 1993, 1997; Ding et al. 2010, 2012; O'Connell et al. 1993). These large animal results are consistent with data from both humans and aged small animal models of GC osteopenia.

4.4.2.2 Summary

The clinical picture of GIOP in humans is one of low bone mass and bone formation rate, and bone fragility that is worse than indicated by bone mass measurements. Growing small animals are a poor model for human GIOP. Adult C57BL/6/129SvJ, FVbN, Swiss-Webster, and C57Bl/6J mice are gaining acceptance as a model of adult human GIOP. Considering that it has historically been easier to find an adult small animal that demonstrates GC-induced low bone strength and bone formation than GC-induced low bone mass, adult rodent GC bone models could potentially provide insight into the non-bone mass aspects of GIOP bone fragility. The rabbit is the large animal model of choice that provides a full spectrum of findings that parallel what occurs in adult human GIOP.

4.5 Summary

The 6–9-month-old OVX female rat is the small animal model of choice to mimic tissue-level findings in estrogen-deficient humans. It is near peak bone mass and can be manipulated to simulate accurately most clinical findings of osteoporosis in the adult female skeleton. Serum biochemistry, histomorphometry, and densitometry, routinely used in humans, can be used. Like all animal models, the rat develops no fragility fractures. However, mechanical testing of rat bones is a reasonably accurate predictor of human bone fragility.

The OVX mouse is an accurate model for tissue-level bone changes when low doses of 17-β-estradiol (<0.01 mg/kg subcutaneously) are used. The adult mouse is gaining acceptance as a model for adult human GIOP. Adult transgenic and knockout mice are the most appropriate method to study molecular mechanisms that control bone mass and bone cell activity that influence modeling and remodeling of cancellous bone.

The rat's low levels of Haversian remodeling do not permit accurate evaluation of cortical bone behavior, making additional studies of cortical bone, either in large animals or humans, a necessity. Though the adult dog is an accurate model of the adult human skeleton, its hormonal dissimilarity causes it to yield inconsistent results in the acutely estrogen-deficient state. Estrogen-deficient nonhuman primates are currently the large animal of choice for quantitative histologic evaluation of Haversian remodeling.

References

Acheson RM, MacIntyre MN, Oldham E. Techniques in longitudinal studies of the skeletal development of the rat. Br J Nutr. 1959;13:283–96.

Aerssens J, Van Audekercke R, Talalaj M, Van Vlasselaer P, Bramm E, Geusens P, Dequeker J. Effect of 1-α-vitamin D_3 on bone strength and composition in growing rats with and without corticosteroid treatment. Calcif Tissue Int. 1994;55:443–50.

Ahmed HH, Morcos NY, Eskander EF, Seoudi DM, Shalby AB. Potential role of leptin against glucocorticoid-induced secondary osteoporosis in adult female rats. Eur Rev Med Pharmacol Sci. 2012;16:1446–52.

Ahmed LA, Shigdel R, Joakimsen RM, Eldevik OP, Eriksen EF, Ghasem-Zadeh A, Bala Y, Zebaze R, Seeman E, Bjørnerem Å. Measurement of cortical porosity of the proximal femur improves identification of women with nonvertebral fragility fractures. Osteoporos Int. 2015;26:2137–46. doi:10.1007/s00198-015-3118-x.

Allen MR, Kubek DJ, Burr DB. Cancer treatment dosing regimens of zoledronic acid result in near-complete suppression of mandible intracortical bone remodeling in beagle dogs. J Bone Miner Res. 2010;25:98–105. doi:10.1359/jbmr.090713.

Altman A, Hochberg Z, Silbermann M. Interactions between growth hormone and dexamethasone in skeletal growth and bone structure of the young mouse. Calcif Tissue Int. 1992;51:298–304.

Amman P, Rizzoli R, Slosman D, Bonjour JP. Sequential and precise in vivo measurement of bone mineral density in rats using dual energy x-ray absorptiometry. J Bone Miner Res. 1992;7:311–7.

Andersen H, Matthiessen ME. The histiocyte in human fetal tissues: its morphology, cytochemistry, origin, function, and fate. Z Zellforsch. 1966;72:193–211.

Arnold JS, Jee WSS. Bone growth and osteoclastic activity as indicated by radioautographic distribution of ^{239}Pu. Am J Anat. 1957;101:367–417.

Augat P, Schorlemmer S, Gohl C, Iwabu S, Ignatius A, Claes L. Glucocorticoid-treated sheep as a model for osteopenic trabecular bone in biomaterials research. J Biomed Mater Res. 2003;66A:457–62.

Bain SD, Jensen E, Celino DL, Bailey MC, Lantry MM, Edwards MW. High-dose gestagens modulate bone resorption and formation and enhance estrogen-induced endosteal bone formation in the ovariectomized mouse. J Bone Miner Res. 1993;8:219–30.

Baird DT, Guevara A. Concentration of unconjugated estrone and estradiol in peripheral plasma in nonpregnant women throughout the menstrual cycle, castrate and postmenopausal women and in men. J Clin Endocrinol. 1969;29:149–56.

Bala Y, Zebaze R, Seeman E. Role of cortical bone in bone fragility. Curr Opin Rheumatol. 2015;27:406–13. doi:10.1097/BOR.0000000000000183.

Balena R, Toolan BC, Shea M, Markatos A, Myers ER, Lee SC, Opas EE, Seedor JG, Klein H, Frankenfield D, Rodan GA. The effects of 2 year treatment with the aminobisphosphonate alendronate on bone metabolism, bone histomorphometry, and bone strength in ovariectomized nonhuman primates. J Clin Invest. 1993;92:2577–86.

Balooch G, Yao W, Ager JW, Balooch M, Nalla RK, Porter AE, Ritchie RO, Lane NE. The aminobisphosphonate risedronate preserves localized mineral and material properties of bone in the presence of glucocorticoids. Arthritis Rheum. 2007;56:3726–37.

Baofeng L, Zhi Y, Bei C, Guolin M, Qingshui Y, Jian L. Characterization of a rabbit osteoporosis model induced by ovariectomy and glucocorticoid. Acta Orthop. 2010;81:396–401. doi:10.3109/17453674.2010.483986.

Bar-Shira-Maymon B, Coleman R, Cohen A, Steinhagen-Thiessen E, Silbermann M. Age-related bone loss in lumbar vertebrae of CW-1 female mice: a histomorphometric study. Calcif Tissue Int. 1989a;44:36–45.

Bar-Shira-Maymon B, Coleman R, Steinhagen-Thiessen E, Silbermann M. Correlation between alkaline and acid phosphatase activities and age-related osteopenia in murine vertebrae. Calcif Tissue Int. 1989b;44:99–207.

Barlet JP, Coxam V, Davicco MJ, Gaumet N. Animal models for postmenopausal osteoporosis. Repro Nutri Dev. 1994;34:221–36.

Bauss F, Lalla S, Endele R, Hothorn LA. Effects of treatment with ibandronate on bone mass, architecture, biomechanical properties, and bone concentration of ibandronate in ovariectomized aged rats. J Rheumatol. 2002;29:2200–8.

Baylink D, Stauffer M, Wergedal J, Rich C. Formation, mineralization, and resorption of bone in vitamin D-deficient rats. J Clin Invest. 1970;49:1122–34.

Beamer WG, Donahue LR, Rosen CJ, Baylink DJ. Genetic variability in adult bone density among inbred strains of mice. Bone. 1996;18:397–403.
Bell ET, Christie DW, Younglai EV. Plasma oestrogen levels during the canine oestrous cycle. J Endocrinol. 1971;51:225–6.
Bellino FL. Nonprimate animal models of menopause: workshop report. Menopause-Jour N Amer Meno Soc. 2000;7:14–24.
Bellino FL, Wise PM. Nonhuman primate models of menopause workshop. Biol Reprod. 2003;68:10–8.
Binkley N, Kimmel D. Effect of age and parity on skeletal response to ovariectomy in rats. J Bone Min Res. 1994;9(Supp 1):S197. (#A270)
Binkley N, Kimmel D, Bruner J, Haffa A, Davidowitz B, Meng C, Schaffer V, Green J. Zoledronate prevents the development of absolute osteopenia following ovariectomy in adult rhesus monkeys. J Bone Miner Res. 1998;13:1775–82.
Bitto A, Burnett BP, Polito F, Levy RM, Marini H, Di Stefano V, Irrera N, Armbruster MA, Minutoli L, Altavilla D, Squadrito F. Genistein aglycone reverses glucocorticoid-induced osteoporosis and increases bone breaking strength in rats: a comparative study with alendronate. Br J Pharmacol. 2009;156:1287–95. doi:10.1111/j.1476-5381.2008.00100.x.
Black D, Farquharson C, Robins SP. Excretion of pyridinium cross-links of collagen in ovariectomized rats as urinary markers for increased bone resorption. Calcif Tissue Int. 1989;44:343–7.
Black DM, Delmas PD, Eastell R, Reid IR, Boonen S, Cauley JA, Cosman F, Lakatos P, Leung PC, Man Z, Mautalen C, Mesenbrink P, Hu H, Caminis J, Tong K, Rosario-Jansen T, Krasnow J, Hue TF, Sellmeyer D, Eriksen EF, Cummings SR, HORIZON Pivotal Fracture Trial. Once-yearly zoledronic acid for treatment of postmenopausal osteoporosis. N Engl J Med. 2007;356:1809–22.
Boggild MK, Gajic-Veljanoski O, McDonald-Blumer H, Ridout R, Tile L, Josse R, Cheung AM. Odanacatib for the treatment of osteoporosis. Expert Opin Pharmacother. 2015;16:1717–26. doi:10.1517/14656566.2015.1064897.
Bonjour JP, Ammann P, Rizzoli R. Importance of preclinical studies in the development of drugs for treatment of osteoporosis: a review related to the 1998 WHO guidelines. Osteoporos Int. 1999;9:379–93.
Bonjour JP, Kohrt W, Levasseur R, Warren M, Whiting S, Kraenzlin M. Biochemical markers for assessment of calcium economy and bone metabolism: application in clinical trials from pharmaceutical agents to nutritional products. Nutr Res Rev. 2014;27:252–74. doi:10.1017/S0954422414000183.
Bonucci E, Ballanti P. Osteoporosis-bone remodeling and animal models. Toxicol Pathol. 2014;42:957–69. doi:10.1177/0192623313512428.
Boonen S, Adachi JD, Man Z, Cummings SR, Lippuner K, Törring O, Gallagher JC, Farrerons J, Wang A, Franchimont N, San Martin J, Grauer A, McClung M. Treatment with denosumab reduces the incidence of new vertebral and hip fractures in postmenopausal women at high risk. J Clin Endocrinol Metab. 2011;96:1727–36. doi:10.1210/jc.2010-2784.
Bostrom MP, Gamradt SC, Asnis P, Vickery BH, Hill E, Avnur Z, Waters RV. Parathyroid hormone-related protein analog RS-66271 is an effective therapy for impaired bone healing in rabbits on corticosteroid therapy. Bone. 2000;26:437–42.
Bouxsein ML, Myers KS, Shultz KL, Donahue LR, Rosen CJ, Beamer WG. Ovariectomy-induced bone loss varies among inbred strains of mice. J Bone Miner Res. 2005;20:1085–92.
Bouxsein ML, Devlin MJ, Glatt V, Dhillon H, Pierroz DD, Ferrari SL. Mice lacking beta-adrenergic receptors have increased bone mass but are not protected from deleterious skeletal effects of ovariectomy. Endocrinology. 2009;150:144–52. doi:10.1210/en.2008-0843.
Bouxsein ML, Boyd SK, Christiansen BA, Guldberg RE, Jepsen KJ, Müller R. Guidelines for assessment of bone microstructure in rodents using micro-computed tomography. J Bone Miner Res. 2010;25:1468–86. doi:10.1002/jbmr.141.
Boyce RW, Franks AF, Jankowsky ML, Orcutt CM, Piacquadio AM, White JM, Bevan JM. Sequential histomorphometric changes in cancellous bone from ovariohysterectomized dogs. J Bone Miner Res. 1990;5:947–53.

Breen SA, Loveday BE, Millest AJ, Waterton JC. Stimulation and inhibition of bone formation: use of pQCT in the mouse in vivo. Lab Anim Sci. 1998;32:467–76.

Bressot C, Meunier PJ, Chapuy MC, Lejeune E, Edouard C, Darby AJ. Histomorphometric profile, pathophysiology and reversibility of corticosteroid-induced osteoporosis. Met Bone Dis Rel Res. 1979;1:303–11.

Briot K, Cortet B, Roux C, Fardet L, Abitbol V, Bacchetta J, Buchon D, Debiais F, Guggenbuhl P, Laroche M, Legrand E, Lespessailles E, Marcelli C, Weryha G, Thomas T. Update of recommendations on the prevention and treatment of glucocorticoid-induced osteoporosis. Joint Bone Spine. 2014;81:493–501.

Brockstedt H, Kassem M, Eriksen EF, Mosekilde L, Melsen F. Age- and sex-related changes in iliac cortical bone mass and remodeling. Bone. 1993;14:681–91.

Brommage R. Validation and calibration of DEXA body composition in mice. Am J Physiol Endocrinol Metab. 2003;285:E454–9.

Broulik PD. Tamoxifen prevents bone loss in OVX mice. Endocr Regul. 1991;25(217):219.

Burch J, Rice S, Yang H, Neilson A, Stirk L, Francis R, Holloway P, Selby P, Craig D. Systematic review of the use of bone turnover markers for monitoring the response to osteoporosis treatment: the secondary prevention of fractures, and primary prevention of fractures in high-risk groups. Health Technol Assess. 2014;18:1–180. doi:10.3310/hta18110.

Butcher RL, Collins WE, Fugo NW. Plasma concentration of LH, FSH, prolactin, progesterone, and estradiol-17-β throughout the four day cycle of the rat. J Endocrinol. 1974;94:1704–8.

Cahoon S, Boden SD, Gould KG, Vailas AC. Noninvasive markers of bone metabolism in the rhesus monkey: normal effects of age and gender. J Med Primatol. 1996;25:333–8.

Campbell GM, Buie HR, Boyd SK. Signs of irreversible architectural changes occur early in the development of experimental osteoporosis as assessed by in vivo micro-CT. Osteoporos Int. 2008;19:1409–19. doi:10.1007/s00198-008-0581-7.

Cann CE, Genant HK, Ettinger B, Gordan GS. Spinal mineral loss in oophorectomized women. J Am Med Assoc. 1980;244:2056–9.

Carvas JS, Pereira RM, Caparbo VF, Fuller P, Silveira CA, Lima LAP, Bonfa E, Mello SBV. A single dose of zoledronic acid reverses the deleterious effects of glucocorticoids on titanium implant osseointegration. Osteoporos Int. 2010;21:1723–9.

Castañeda S, Largo R, Calvo E, Rodríguez-Salvanés F, Marcos ME, Díaz-Curiel M, Herrero-Beaumont G. Bone mineral measurements of subchondral and trabecular bone in healthy and osteoporotic rabbits. Skelet Radiol. 2006;35:34–41.

Castañeda S, Calvo E, Largo R, González-González R, de la Piedra C, Díaz-Curiel M, Herrero-Beaumont G. Characterization of a new experimental model of osteoporosis in rabbits. J Bone Miner Metab. 2008;26:53–9.

Cauley JA, Lucas FL, Kuller LH, Stone K, Browner W, Cummings SR. Elevated serum estradiol and testosterone concentrations are associated with a high risk for breast cancer. Ann Intern Med. 1999;130(4 part 1):270–7. I45

Cesnjaj M, Stavljenic A, Vukicevic S. In vivo models in the study of osteopenias. Eur J Clin Chem Clin Biochem. 1991;29:211–9.

Chapuy MC, Meunier PJ. Prevention and treatment of osteoporosis [see comments]. Aging (Milano). 1995;7:164–73.

Chavassieux P, Pastoureau P, Boivin G, Chapuy MC, Delmas PD, Meunier PJ. Dose effects on ewe bone remodeling of short-term sodium fluoride administration – a histomorphometric and biochemical study. Bone. 1991a;12:421–7.

Chavassieux P, Pastoureau P, Boivin G, Chapuy MC, Delmas PD, Milhaud G, Meunier PJ. Fluoride-induced bone changes in lambs during and after exposure to sodium fluoride. Osteoporos Int. 1991b;2:26–33.

Chavassieux P, Pastoureau P, Chapuy MC, Delmas PD, Meunier PJ. Glucocorticoid-induced inhibition of osteoblastic bone formation in ewes: a biochemical and histomorphometric study. Osteoporos Int. 1993;3:97–102.

Chavassieux P, Buffet A, Vergnaud P, Garnero P, Meunier PJ. Short-term effects of corticosteroids on trabecular bone remodeling in old ewes. Bone. 1997;20:451–5.

Chavassieux P, Garnero P, duBoeuf F, Vergnaud P, Brunner-Ferber F, Delmas PD, Meunier PJ. Effects of a new selective estrogen receptor modulator (MDL-103323) on cancellous and cortical bone in ovariectomized ewes: a biochemical, histomorphometric, and densitometric study. J Bone Miner Res. 2001;16:89–96.

Christiansen C, Christensen MS, McNair P, Hagen C, Stocklund KE, Transbol I. Prevention of early postmenopausal bone loss: controlled 2-year study in 315 normal females. Eur J Clin Investig. 1980;10:273–9.

Christiansen C, Lindsay R. Estrogens, bone loss and preservation. Osteoporos Int. 1990;1:7–13.

Clark DP, Badea CT. Micro-CT of rodents: state-of-the-art and future perspectives. Phys Med. 2014;30:619–34. doi:10.1016/j.ejmp.2014.05.011.

Colman RJ, Kemnitz JW, Lane MA, Abbott DH, Binkley N. Skeletal effects of aging and menopausal status in female rhesus macaques. J Clin Endocrinol Metab. 1999;84:4144–8.

Concannon PW, Hansel W, Visek WJ. The ovarian cycle of the bitch: plasma estrogen, LH and progesterone. Biol Reprod. 1975;13:112–21.

Conradie MM, Cato AC, Ferris WF, de Wet H, Horsch K, Hough S. MKP-1 knockout does not prevent glucocorticoid-induced bone disease in mice. Calcif Tissue Int. 2011;89:221–7. doi:10.1007/s00223-011-9509-x.

Craft CS, Broekelmann TJ, Zou W, Chappel JC, Teitelbaum SL, Mecham RP. Oophorectomy-induced bone loss is attenuated in MAGP1-deficient mice. J Cell Biochem. 2012;113:93–9. doi:10.1002/jcb.23331.

Cummings SR, Kelsey JL, Nevitt MC, O'Dowd KJ. Epidemiology of osteoporosis and osteoporotic fractures. Epidemiol Rev. 1985;7:178–208.

Cummings SR. Epidemiologic studies of osteoporotic fractures: methodologic issues. Calc Tissue Int. 1991;49(Supp):S15–20.

Cummings SR, Nevitt MC, Browner WS, Stone K, Fox KM, Ensrud KE, Cauley J, Black D, Vogt TM. Risk factors for hip fracture in white women. New Eng J Med. 1995;332:767–73.

Cummings SR, Browner WS, Bauer D, Stone K, Ensrud K, Jamal S, Ettinger B. Endogenous hormones and the risk of hip and vertebral fractures among older women. N Engl J Med. 1998;339:733–8.

Cummings SR, San Martin J, McClung MR, Siris ES, Eastell R, Reid IR, Delmas P, Zoog HB, Austin M, Wang A, Kutilek S, Adami S, Zanchetta J, Libanati C, Siddhanti S, Christiansen C, FREEDOM Trial. Denosumab for prevention of fractures in postmenopausal women with osteoporosis. N Engl J Med. 2009;361:756–65. doi:10.1056/NEJMoa0809493.

Cusick T, Chen CM, Pennypacker BL, Pickarski M, Kimmel DB, Scott KR, Duong LT. Odanacatib treatment increases hip bone mass and cortical thickness by preserving endocortical bone formation and stimulating periosteal bone formation in the ovariectomized adult rhesus monkey. J Bone Miner Res. 2012;27:524–37.

Dannucci GA, Martin RB, Patterson-Buckendahl P. Ovariectomy and trabecular bone remodeling in the dog. Calcif Tissue Int. 1987;40:194–9.

Dawson AB. The age order of epiphyseal union in the long bones of the albino rat. Anat Rec. 1925;31:1–17.

Dempster DW. Bone histomorphometry in glucocorticoid-induced osteoporosis. J Bone Miner Res. 1989;4:137–41.

deWinter FR, Steendijk R. The effect of a low-calcium diet in lactating rats; observations on the rapid development and repair of osteoporosis. Calcif Tissue Res. 1975;17:303–16.

Devogelaer JP, Boutsen Y, Gruson D, Manicourt D. Is there a place for bone turnover markers in the assessment of osteoporosis and its treatment? Rheum Dis Clin N Am. 2011;37:365–386., v–vi. doi:10.1016/j.rdc.2011.07.002.

Dickson GR, Luczak M, Wlodarski KH. The limitation of DEXA analysis for bone mass determination in mice. Folia Biol (Krakow). 2004;52:125–9.

Ding M, Cheng L, Bollen P, Schwarz P, Overgaard S. Glucocorticoid induced osteopenia in cancellous bone of sheep: validation of large animal model for spine fusion and biomaterial research. Spine (Phila Pa 1976). 2010;35:363–70. doi:10.1097/BRS.0b013e3181b8e0ff.

Ding M, Danielsen CC, Overgaard S. The effects of glucocorticoid on microarchitecture, collagen, mineral and mechanical properties of sheep femur cortical bone. J Tissue Eng Regen Med. 2012;6:443–50. doi:10.1002/term.448.
Dobnig H. A review of teriparatide and its clinical efficacy in the treatment of osteoporosis. Expert Opin Pharmacother. 2004;5:1153–62.
Drake MT, Clarke BL, Lewiecki EM. The pathophysiology and treatment of osteoporosis. Clin Ther. 2015;37:1837–50. doi:10.1016/j.clinthera.2015.06.006.
Drinkwater BL, Nilson K, Chesnut CH, Brenner WJ, Shainholtz S, Southworth MB. Bone mineral content of amenorrheic and eumenorrheic athletes. N Engl J Med. 1984;311:277–81.
Drinkwater BL, Nilson K, Ott S. Bone mineral density after resumption of menses in amenorrheic athletes. J Am Med Assn. 1986;256:380–2.
Eastell R, Lang T, Boonen S, Cummings S, Delmas PD, Cauley JA, Horowitz Z, Kerzberg E, Bianchi G, Kendler D, Leung P, Man Z, Mesenbrink P, Eriksen EF, Black DM. Effect of once-yearly zoledronic acid on the spine and hip as measured by quantitative computed tomography. Osteoporos Int. 2010;21:1277–85. doi:10.1007/s00198-009-1077-9.
Edwards MW, Bain SD, Bailey MC, Lantry MM, Howard GA. 17-β-estradiol stimulation of end-osteal bone formation in the ovariectomized mouse: an animal model for the evaluation of bone-targeted estrogens. Bone. 1992;13:29–34.
Erben RG. Trabecular and endocortical bone surfaces in the rat: modeling or remodeling? Anat Rec. 1996;246:39–46.
Eriksen EF, Mosekilde L, Melsen F. Effect of sodium fluoride, calcium, phosphate, and vitamin D_2 on trabecular bone balance and remodeling in osteoporotics. Bone. 1985;6:381–9.
Ettinger B, Genant HK, Cann CE. Long-term estrogen replacement therapy prevents bone loss and fractures. Ann Intern Med. 1985;102:319–24.
Ettinger B, Black DM, Mitlak BH, Knickerbocker RK, Nickelsen T, Genant HK, Christiansen C, Delmas PD, Zanchetta JR, Stakkestad J, Gluer CC, Krueger K, Cohen FJ, Eckert S, Ensrud KE, Avioli LV, Lips P, Cummings SR. Reduction of vertebral fracture risk in postmenopausal women with osteoporosis treated with raloxifene. Results from a 3 year randomized clinical trial. J Am Med Assoc. 1999;282:637–45.
Evans G, Bryant HU, Magee D, Sato M, Turner RT. The effects of raloxifene on tibial histomorphometry in ovariectomized rats. Endocrinology. 1994;134:2283–8.
Evans GL, Bryant HU, Magee DE, Turner RT. Raloxifene inhibits bone turnover and prevents further cancellous bone loss in adult OVX rats with established osteopenia. Endocrinology. 1996;137:4139–44.
Felix R, Cecchini MG, Fleisch H. Macrophage colony stimulating factor restores in vivo bone resorption in the op/op osteopetrotic mouse. Endocrinology. 1990;127:2592–4.
Ferguson HW, Hartles RL. The combined effects of calcium deficiency and ovariectomy on the bones of young adult cats. Calcif Tissue Res. 1970;4(Suppl 1):140–1.
Ferretti JL, Delgado CJ, Capozza RF, Cointry G, Monturi E, Roldan E, Perez-Lloret A, Zanchetta JR. Protective effects of disodium etidronate and pamidronate against the biomechanical repercussion of betamethasone-induced osteopenia in growing rat femurs. Bone and Min. 1993;20:265–76.
Ferretti JL, Vazquez SO, Delgado CJ, Capozza R, Cointry G. Biphasic dose-response curves of cortisol effects on properties of rat femur diaphyses as described by peripheral quantitative computerized tomography (pQCT) and bending tests. Bone. 1994;16:119–24.
Folwarczna J, Pytlik M, Sliwiński L, Cegieła U, Nowińska B, Rajda M. Effects of propranolol on the development of glucocorticoid-induced osteoporosis in male rats. Pharmacol Rep. 2011;63:1040–9.
Fox J, Newman MK, Turner CH, Guldberg RE, Varela A, Smith SY. Effects of treatment with PTH (1–84) on quantity and biomechanical properties of thoracic vertebral trabecular bone in ovariectomized rhesus monkeys. Calcif Tissue Int. 2008;82:212–20. doi:10.1007/s00223-008-9108-7.
Frenkel B, White W, Tuckermann J. Glucocorticoid-induced osteoporosis. Adv Exp Med Biol. 2015;872:179–215. doi:10.1007/978-1-4939-2895-8_8.

Frost HM. Tetracycline-based histological analysis of bone remodeling. Calcif Tissue Res. 1969;3:211–39.
Frost HM. The skeletal intermediary organization. Metab Bone Dis Rel Res. 1983;4:281–90.
Frost HM. The mechanostat: a proposed pathogenic mechanism of osteoporoses and the bone mass effects of mechanical and nonmechanical agents. Bone and Min. 1987;2:73–85.
Frost HM. The Utah paradigm of skeletal physiology: an overview of its insights for bone, cartilage and collagenous tissue organs. J Bone Miner Metab. 2000;18:305–16.
Frost HM, Jee WSS. On the rat model of human osteopenias and osteoporoses. Bone Miner. 1992;18:227–36.
Fuchs H, Gau C, Hans W, Gailus-Durner V, Hrabe de Angelis M. Long-term experiment to study the development, interaction, and influencing factors of DEXA parameters. Mamm Genome. 2013;24:376–88. doi:10.1007/s00335-013-9477-9478.
Fujioka M, Uehara M, Wu J, Adlercreutz H, Suzuki K, Kanazawa K, Takeda K, Yamada K, Ishimi Y. Equol, a metabolite of daidzein, inhibits bone loss in ovariectomized mice. J Nutr. 2004;134:2623–7.
Garn SM. The early gain and later loss of cortical bone. Springfield: CC Thomas; 1970.
Garner SC, Anderson JJB, Mar MH, Parikh I. Estrogens reduce bone loss in the ovariectomized, lactating rat model. Bone and Min. 1992;15:19–31.
Garnero P, Hausherr E, Chapuy MC, Marcelli C, Grandjean H, Muller C, Cormier C, Breart G, Meunier PJ, Delmas PD. Markers of bone resorption predict hip fracture in elderly women: the EPIDOS prospective study. J Bone Miner Res. 1996;11:1531–8.
Gasser JA, Ingold P, Venturiere A, Shen V, Green JR. Long-term protective effects of zoledronic acid on cancellous and cortical bone in the ovariectomized rat. J Bone Miner Res. 2008;23:544–51.
Genant HK, Engelke K, Prevrhal S. Advanced CT bone imaging in osteoporosis. Rheumatology (Oxford). 2008;47(Suppl 4):iv9–16. doi:10.1093/rheumatology/ken180.
Geusens P, Dequeker J, Nijs J, Bramm E. Effect of ovariectomy and prednisolone on bone mineral content in rats: evaluation by single photon absorptiometry and radiogrammetry. Calcif Tissue Int. 1990;47:243–50.
Geusens P, Schot LPC, Nijs J, Dequeker J. Calcium-deficient diet in ovariectomized dogs limits the effects of 17-ß-estradiol and nandrolone decanoate on bone. J Bone Miner Res. 1991;6:791–8.
Geusens P, Boonen S, Nijs J, Jiang Y, Lowet G, Van Auderkercke R, Huyghe C, Caulin F, Very JM, Dequeker J, Van der Perre G. Effect of salmon calcitonin on femoral bone quality in adult ovariectomized ewes. Calcif Tissue Int. 1996;59:315–20.
Gilsanz V, Gibbens DT, Roe TF, Carlson M, Senac MO, Boechart MI, Huang HK, Schulz EE, Libanati CR, Cann CC. Vertebral bone density in children: effect of puberty. Radiology. 1988;166:847–50.
Gong JK, Arnold JS, Cohn SH. Composition of trabecular and cortical bone. Anat Rec. 1964;149:325–31.
Goswami J, Hernández-Santos N, Zuniga LA, Gaffen SL. A bone-protective role for IL-17 receptor signaling in ovariectomy-induced bone loss. Eur J Immunol. 2009;39:2831–9. doi:10.1002/eji.200939670.
Gothlin G, Ericsson JLE. On the histogenesis of the cells in fracture callus. Virch Arch Abt B Zellpath. 1973;12:318–29.
Goulding A, Gold E. Effects of chronic prednisolone treatment on bone resorption and bone composition in intact and OVX rats and in OVX rats receiving 17-β-estradiol. Endocrinology. 1988;122:482–7.
Grahnemo L, Jochems C, Andersson A, Engdahl C, Ohlsson C, Islander U, Carlsten H. Possible role of lymphocytes in glucocorticoid-induced increase in trabecular bone mineral density. J Endocrinol. 2015;224:97–108. doi:10.1530/JOE-14-0508.
Grynpas MD, Huckell CB, Reichs KJ, Derousseau CJ, Greenwood C, Kessler MJ. Effect of age and osteoarthritis on bone mineral in rhesus monkey vertebrae. J Bone Miner Res. 1993;8:909–17.
Handal JA, John TK, Goldstein DT, Khurana JS, Saing M, Braitman LE, Samuel SP. Effect of atorvastatin on the cortical bones of corticosteroid treated rabbits. J Orthop Res. 2012;30:872–6. doi:10.1002/jor.22030.

Hansen S, Hauge EM, Beck Jensen JE, Brixen K. Differing effects of PTH 1-34, PTH 1-84, and zoledronic acid on bone microarchitecture and estimated strength in postmenopausal women with osteoporosis: an 18-month open-labeled observational study using HR-pQCT. J Bone Miner Res. 2013;28:736–45. doi:10.1002/jbmr.1784.
Hartke JR. Preclinical development of agents for the treatment of osteoporosis. Toxicol Pathol. 1999;27:143–7.
Hartwell D, Riis BJ, Christiansen C. Changes in vitamin D metabolism during natural and medical menopause. J Clin Endocrinol Metab. 1990;71:127–32.
Healy C, Kennedy OD, Brennan O, Rackard SM, O'Brien FJ, Lee TC. Structural adaptation and intracortical bone turnover in an ovine model of osteoporosis. J Orthop Res. 2010;28:248–51. doi:10.1002/jor.20961.
Heaney RP, Recker RR, Saville PD. Menopausal changes in calcium balance performance. J Lab Clin Med. 1978;92:953–63.
Hedlund LR, Gallagher JC. The effect of age and menopause on bone mineral density of the proximal femur. J Bone Miner Res. 1989;4:639–42.
Helm NB, Padala S, Beck FM, D'Atri AM, Huja SS. Short-term zoledronic acid reduces trabecular bone remodeling in dogs. Eur J Oral Sci. 2010;118:460–5. doi:10.1111/j.1600-0722.2010.00762.x.
Hirano T, Burr DB, Cain RL, Hock JM. Changes in geometry and cortical porosity in adult, ovary intact rabbits after 5 months treatment with LY333334 (hPTH 1-34). Calcif Tissue Int. 2000;66:456–60.
Hofbauer LC, Zeitz U, Schoppet M, Skalicky M, Schüler C, Stolina M, Kostenuik PJ, Erben RG. Prevention of glucocorticoid-induced bone loss in mice by inhibition of RANKL. Arthritis Rheum. 2009;60:1427–37. doi:10.1002/art.24445.
Hornby SB, Ford SL, Mase CA, Evans GP. Skeletal changes in the ovariectomised ewe and subsequent response to treatment with 17 beta estradiol. Bone. 1995;17(4 Suppl):389S–94S.
Hui SL, Slemenda CW, Johnston CC Jr. Baseline measurement of bone mass predicts fracture in white women. Ann Intern Med. 1989;111:355–61.
Ibrahim N, Mohamad S, Mohamed N, Shuid AN. Experimental fracture protocols in assessments of potential agents for osteoporotic fracture healing using rodent models. Curr Drug Targets. 2013;14:1642–50.
Inose H, Zhou B, Yadav VK, Guo XE, Karsenty G, Ducy P. Efficacy of serotonin inhibition in mouse models of bone loss. J Bone Miner Res 2011;26:2002–11. doi:10.1002/jbmr.439.
Ito H, Ke HZ, Jee WSS, Sakou T. Anabolic responses of an adult cancellous bone site to prostaglandin estrogen in the rat. Bone and Min. 1993;21:219–36.
Itoh F, Kojima M, Furihata-Komatsu H, Aoyagi S, Kusama H, Komatsu H, Nakamura T. Reductions in bone mass, structure, and strength in axial and appendicular skeletons associated with increased turnover after ovariectomy in mature cynomolgus monkeys and preventive effects of clodronate. J Bone Miner Res. 2002;17:534–43.
Iwaniec UT, Yuan D, Power RA, Wronski TJ. Strain-dependent variations in the response of cancellous bone to ovariectomy in mice. J Bone Miner Res. 2006;21:1068–74.
Iwaniec UT, Philbrick K, Wong C, Olson D, Kahler-Quesada A, Branscum A, Turner RT. Thermoneutral housing prevents premature age-related cancellous bone loss in mice. J Bone Min Res. 2015;30:S364. abstract MO0102
Iwaniec UT, Philbrick KA, Wong CP, Gordon JL, Kahler-Quesada AM, Olson DA, Branscum AJ, Sargent JL, DeMambro VE, Rosen CJ, Turner RT. Room temperature housing results in premature cancellous bone loss in growing female mice: implications for the mouse as a preclinical model for age-related bone loss. Osteoporosis Intl. 2016;27:3091–101. EPub Jul-2016
Jaworski ZFG, Uhthoff HK. Reversibility of nontraumatic disuse osteoporosis during its active phase. Bone. 1986;7:431–9.
Jayo MJ, Register TC, Carlson CS. Effects on bone of oral hormone replacement therapy initiated 2 years after ovariectomy in young adult monkeys. Bone. 1998;23:361–6.
Jee WSS, Nolan PD. Origin of osteoclasts from the fusion of phagocytes. Nature. 1963;200:225–6.
Jee WSS, Mori S, Li XJ, Chan S. Prostaglandin estrogen enhances cortical bone mass and activates intracortical bone remodeling in intact and ovariectomized female rats. Bone. 1990;11:253–66.

Jee WS, Yao W. Overview: animal models of osteopenia and osteoporosis. J Musculoskelet Neuronal Interact. 2001;1:193–207.
Jensen GF, Christiansen C, Transbol I. Fracture frequency and bone preservation in postmenopausal women treated with estrogen. Obstet Gynecol. 1982;69:493–6.
Jerome CP. Primate models of osteoporosis. Lab Anim Sci. 1998;48:618–22.
Jerome CP, Peterson PE. Nonhuman primate models in skeletal research. Bone. 2001;29:1–6.
Jerome C, Kimmel DB, McAlister JA, Weaver DS. Effects of ovariectomy on iliac trabecular bone in baboons (Papio Anubis). Calcif Tissue Int. 1986;39:206–8.
Jerome C, Missbach M, Gamse R. Balicatib, a cathepsin K inhibitor, stimulates periosteal bone formation in monkeys. Osteoporos Int. 2012;23(1):339–49. doi:10.1007/s00198-011-1593-2.
Jett S, Wu K, Duncan H, Frost HM. Adrenalcorticosteroid and salicylate actions on human and canine haversian bone formation and resorption. Clin Orthop Relat Res. 1970;68:310–5.
Jiang Y, Zhao J, Geusens P, Liao EY, Adriaensens P, Gelan J, Azria M, Boonen S, Caulin F, Lynch JA, Ouyang X, Genant HK. Femoral neck trabecular microstructure in ovariectomized ewes treated with calcitonin: MRI microscopic evaluation. J Bone Miner Res. 2005;20:125–30.
Jilka RL, Hangoc G, Girasole G, Passeri G, Williams DC, Abrams JS, Boyce B, Broxmeyer H, Manolagas SC. Increased osteoclast development after estrogen loss: mediation by interleukin-6. Science. 1992;257:88–91.
Johansson H, Odén A, Kanis JA, EV MC, Morris HA, Cooper C, Vasikaran S, IFCC-IOF Joint Working Group on Standardisation of Biochemical Markers of Bone Turnover. A meta-analysis of reference markers of bone turnover for prediction of fracture. Calcif Tissue Int. 2014;94:560–7. doi:10.1007/s00223-014-9842-y.
Johnson LC. Composition of trabecular and cortical bone in humans. In: Frost HM, editor. Bone biodynamics. Boston: Little-Brown; 1964. p. 543–654.
Johnston CC, Miller JZ, Slemenda CW, Reister TK, Hui S, Christian JC, Peacock M. Calcium supplementation increases in bone mineral density in children. New Eng J Med. 1992;327:82–7.
Johnston CC Jr, Bjarnason NH, Cohen FJ, Shah A, Lindsay R, Mitlak BH, Huster W, Draper MW, Harper KD, Heath H 3rd, Gennari C, Christiansen C, Arnaud CD, Delmas PD. Long-term effects of raloxifene on bone mineral density, bone turnover, and serum lipid levels in early postmenopausal women: three-year data from 2 double-blind, randomized, placebo-controlled trials. Arch Intern Med. 2000;160:3444–50.
Johnson RB, Gilbert JA, Cooper RC, Parsell DE, Stewart BA, Dai X, Nick TG, Streckfus CF, Butler RA, Boring JG. Effect of estrogen deficiency on skeletal and alveolar bone density in sheep. J Periodontol. 2002;73:383–91.
Jones HH, Priest JD, Hayes WC, Tichenor CC, Nagel DA. Humeral hypertrophy in response to exercise. J Bone Jt Surg. 1977;59A:204–8.
Josimovich JB, Mintz DH, Finster JL. Estrogenic inhibition of growth hormone-induced tibial epiphyseal growth in hypophysectomized rats. Endocrinology. 1967;81:1428–30.
Joss EE, Sobel EH, Zuppinger KA. Skeletal maturation in rats with special reference to order and time of epiphyseal closure. Endocrinology. 1963;72:117–22.
Kalu DN. The ovariectomized rat as a model of postmenopausal osteopenia. Bone and Min. 1991;15:175–91.
Kalu DN, Liu CC, Salerno E, Hollis BW, Echon R, Ray M. Skeletal response of ovariectomized rats to low and high doses of 17-β-estradiol. Bone and Min. 1991;14:175–87.
Kanis JA, McCloskey EV. Epidemiology of vertebral osteoporosis. Bone. 1992;13:S1–S10.
Karambolova KK, Snow GR, Anderson C. Differences in periosteal and corticoendosteal bone envelope activities in spayed and intact beagles: a histomorphometric study. Calcif Tissue Int. 1985;37:665–8.
Karambolova KK, Snow GR, Anderson C. Effects of continuous 17-β-estradiol administration on the periosteal and corticoendosteal envelope activity in spayed beagles. Calcif Tissue Int. 1987;40:12–5.
Karasik D, Cohen-Zinder M. Osteoporosis genetics: year 2011 in review. Bonekey Rep. 2012;1:114. doi:10.1038/bonekey.2012.114.

Kennedy OD, Brennan O, Mahony NJ, Rackard SM, O'Brien FJ, Taylor D, Lee CT. Effects of high bone turnover on the biomechanical properties of the L3 vertebra in an ovine model of early stage osteoporosis. Spine (Phila Pa 1976). 2008;33:2518–23. doi:10.1097/BRS.0b013e318186b292.

Kennedy OD, Brennan O, Rackard SM, Staines A, O'Brien FJ, Taylor D, Lee TC. Effects of ovariectomy on bone turnover, porosity, and biomechanical properties in ovine compact bone 12 months postsurgery. J Orthop Res. 2009;27:303–9. doi:10.1002/jor.20750.

Keshawarz NM, Recker RR. Expansion of the medullary cavity at the expense of cortex in post-menopausal osteoporosis. Metab Bone Dis Rel Res. 1984;5:223–8.

Khan AA, Morrison A, Hanley DA, Felsenberg D, McCauley LK, O'Ryan F, Reid IR, Ruggiero SL, Taguchi A, Tetradis S, Watts NB, Brandi ML, Peters E, Guise T, Eastell R, Cheung AM, Morin SN, Masri B, Cooper C, Morgan SL, Obermayer-Pietsch B, Langdahl BL, Al Dabagh R, Davison KS, Kendler DL, Sándor GK, Josse RG, Bhandari M, El Rabbany M, Pierroz DD, Sulimani R, Saunders DP, Brown JP, Compston J, International Task Force on Osteonecrosis of the Jaw. Diagnosis and management of osteonecrosis of the jaw: a systematic review and international consensus. J Bone Miner Res. 2015;30:3–23. doi:10.1002/jbmr.2405.

Kimmel DB, Jee WSS. A quantitative histologic study of bone turnover in young adult beagles. Anat Rec. 1982;203:31–45.

Kimmel DB, Wronski TJ. Non-destructive measurement of bone mineral in femurs from ovariectomized rats. Calcif Tissue Int. 1990;46:101–10.

Kimmel DB. Quantitative histologic changes in the proximal tibial epiphyseal growth cartilage of aged female rats. Cells Mater. 1992a;1(Supp):11–8.

Kimmel DB. The oophorectomized beagle as an experimental model for estrogen-deficiency bone loss in the adult human. Cells and Materials (Supplement). 1992b;1:75–84.

Kimmel DB. A current assessment of in vivo animal models of osteoporosis. In: Marcus R, Feldman D, Kelsey J, editors. Osteoporosis. San Diego: Academic; 1996. p. 671–90.

Kimmel DB. Animal models in osteoporosis research. In: Bilezikian JP, Raisz LG, Rodan GA, editors. Principles of bone biology. 2nd ed. New York: Elsevier; 2002. p. 1635–55.

King CS, Weir EC, Gundberg CW, Fox J, Insogna KL. Effects of continuous glucocorticoid infusion on bone metabolism in the rat. Calcif Tissue Int. 1996;59:184–91.

Kinney JH, Lane NE, Haupt D. Three dimensional in vivo microscopy of sequential changes in trabecular architecture following estrogen deficiency in female rats. J Bone Miner Res. 1995;10:264–70.

Kleerekoper M, Peterson EL, Nelson DA, Phillips E, Schork MA, Tilley BC, Parfitt AM. A randomized trial of sodium fluoride as a treatment for postmenopausal osteoporosis. Calcif Tissue Int. 1991;49:155–61.

Klibanski A, Greenspan SL. Increase in bone mass after treatment of hyperprolactinemic amenorrhea. N Engl Jour Med. 1986;315:542–6.

Klinck J, Boyd SK. The magnitude and rate of bone loss in ovariectomized mice differs among inbred strains as determined by longitudinal in vivo micro-computed tomography. Calcif Tissue Int. 2008;83:70–9. doi:10.1007/s00223-008-9150-5.

Korkmaz N, Tutoglu A, Korkmaz I, Boyaci A. The relationships among vitamin D level, balance, muscle strength, and quality of life in postmenopausal patients with osteoporosis. J Phys Ther Sci. 2014;26:1521–6. doi:10.1589/jpts.26.

Kostenuik PJ, Nguyen HQ, McCabe J, Warmington KS, Kurahara C, Sun N, Chen C, Li L, Cattley RC, Van G, Scully S, Elliott R, Grisanti M, Morony S, Tan HL, Asuncion F, Li X, Ominsky MS, Stolina M, Dwyer D, Dougall WC, Hawkins N, Boyle WJ, Simonet WS, Sullivan JK. Denosumab, a fully human monoclonal antibody to RANKL, inhibits bone resorption and increases BMD in knock-in mice that express chimeric (murine/human) RANKL. J Bone Miner Res. 2009;24:182–95. doi:10.1359/jbmr.081112.

Koyama T, Nagaami A, Makita T, Konno T, Akazawa H, Takahashi H. Effect of 1-α,24(R)-dihydroxyvitamin D3 on trabecular bone remodeling in adult beagle dogs. Jap Jour Bone Min. 1984;2:249–57.

Kragstrup J, Richards A, Fejerskov O. Effects of fluoride on cortical bone remodeling in the growing domestic pig. Bone. 1989;10:421–4.

Kreipke TC, Rivera NC, Garrison JG, Easley JT, Turner AS, Niebur GL. Alterations in trabecular bone microarchitecture in the ovine spine and distal femur following ovariectomy. J Biomech. 2014;47:1918–21. doi:10.1016/j.jbiomech.2014.03.025.

Krolner B, Toft B. Vertebral bone loss: an unheeded side effect of therapeutic bed rest. Clin Sci. 1983;64:537–40.

Kyoizumi S, Baum CM, Kaneshima H, McCune JM, Yee EJ, Namikawa. Implantation and maintenance of functional human bone marrow in SCI-hu mice. Blood. 1992;79:1704–11.

Lander ES, Botstein D. Mapping Mendelian factors underlying quantitative traits using RFLP linkage maps. Genetics. 1989;121:185–99.

Lane N, Maeda H, Cullen DM, Kimmel DB. Cancellous bone behavior in hindlimb immobilized rats during and after naproxen treatment. Bone Miner. 1994;26:43–59.

Lees C, Shen V, Brommage R. Effects of lasofoxifene on bone in surgically postmenopausal cynomolgus monkeys. Menopause. 2007;14:97–105.

Lelovas PP, Xanthos TT, Thoma SE, Lyritis GP, Dontas IA. The laboratory rat as an animal model for osteoporosis research. Comp Med. 2008;58:424–30.

Lennon JE, Micklem HS. Stromal cells in long-term murine bone marrow culture: FACS studies and origin of stromal cells in radiation chimeras. Exp Hematol. 1986;14:287–92.

Li XJ, Jee WSS, Chow SY, Woodbury DM. Adaptation of cancellous bone to aging and immobilization in the adult rat: a single photon absorptiometry and histomorphometry study. Anat Rec. 1990;227:12–24.

Li XJ, Jee WSS, Ke HZ, Mori S, Akamine T. Age related changes of cancellous and cortical bone histomorphometry in female Sprague-Dawley rats. Cells Mater (Suppl). 1992;1:25–37.

Li M, Shen Y, Qi H, Wronski TJ. Comparative study of skeletal response to estrogen depletion at red and yellow marrow sites in rats. Anat Rec. 1996a;245:472–80.

Li M, Shen Y, Halloran BP, Baumann BD, Miller K, Wronski TJ. Skeletal response to corticosteroid deficiency and excess in growing male rats. Bone. 1996b;19:81–8.

Li M, Liang H, Shen Y, Wronski TJ. Parathyroid hormone stimulates cancellous bone formation at skeletal sites regardless of marrow composition in ovariectomized rats. Bone. 1999;24:95–100.

Liberman UA, Weiss SR, Broll J, Minne HW, Quan H, Bell NH, Rodriguez-Portales JR, Downs RW, Dequeker J, Favus M, Seeman E, Recker RR, Capizzi T, Santora AC, Lombardi A, Shah RV, Hirsch LJ, Karpf DB. Effect of oral alendronate on bone mineral density and the incidence of fractures in postmenopausal osteoporosis. New Eng J Med. 1995;333:1437–43.

Liel Y, Edwards J, Shary J, Spicer KM, Gordon L, Bell NH. The effects of race and body habitus on bone mineral density of the radius, hip, and spine in premenopausal women. J Clin Endocrinol Metab. 1988;66:1247–50.

Lin S, Huang J, Zheng L, Liu Y, Liu G, Li N, Wang K, Zou L, Wu T, Qin L, Cui L, Li G. Glucocorticoid-induced osteoporosis in growing rats. Calcif Tissue Int. 2014;95:362–73. doi:10.1007/s00223-014-9899-7.

Lindgren JU, DeLuca HF. Oral 1,25$(OH)_2D_3$: an effective prophylactic treatment for glucocorticoid osteopenia in rats. Calcif Tissue Int. 1983;35:107–10.

Lindgren JU, Mattson S. The reversibility of disuse osteoporosis. Calcif Tissue Res. 1977;23:179–84.

Lindgren JU, DeLuca HF, Mazess RB. Effects of 1,25$(OH)_2D_3$ on bone tissue in the rabbit: studies on fracture healing, disuse osteoporosis, and prednisone osteoporosis. Calcif Tissue Int. 1984;36:591–5.

Lindquist O, Bengtsson C, Hansson T, Jonsson R. Changes in bone mineral content of the axial skeleton in relation to aging and the menopause. Scand J Clin Lab Invest. 1983;43:333–8.

Lindsay R, Hart DM, Aitken JM. Long-term prevention of osteoporosis by estrogen. Lancet. 1976;1:1036–941.

Liu X, Lei W, Wu Z, Cui Y, Han B, Fu S, Jiang C. Effects of glucocorticoid on BMD, microarchitecture and biomechanics of cancellous and cortical bone mass in OVX rabbits. Med Eng Phys. 2012;34:2–8. doi:10.1016/j.medengphy.2011.06.010.

Lofman O, Larsson L, Toss G. Bone mineral density in diagnosis of osteoporosis – reference population, definition of peak bone mass, and measured site determine prevalence. J Clin Dens. 2000;3:177–86.

Longcope C, Baker RS, Hui SL, Johnston CC. Androgen and estrogen dynamics in women with vertebral crush fractures. Maturitas. 1984;6:309–18.

Longcope C, Hoberg L, Steuterman S, Baran D. The effect of ovariectomy on spine bone mineral density in rhesus monkeys. Bone. 1989;10:341–4.

Lu KH, Hopper BR, Vargo TM, Yen SSC. Chronological changes in sex steroid, gonadotropin, and prolactin secretion in aging female rats displaying different reproductive states. Biol Reprod. 1979;21:193–203.

Lundon K, Dumitriu M, Grynpas M. The long-term effect of ovariectomy on the quality and quantity of cancellous bone in young macaques. Bone and Min. 1994;24:135–49.

Luppen CA, Blake CA, Ammirati KM, Stevens ML, Seeherman HJ, Wozney JM, Bouxsein ML. Recombinant human bone morphogenetic protein-2 enhances osteotomy healing in glucocorticoid-treated rabbits. J Bone Miner Res. 2002;17:301–10.

Lyles KW, Jackson TW, Nesbitt T, Quarles LD. Salmon calcitonin reduces vertebral bone loss in glucocorticoid-treated beagles. Am J Phys. 1993;264:E938–42.

Mackey MS, Stevens ML, Ebert DC, Tressler DL, Combs KS, Lowry CK, Smith PN, McOsker JE. The ferret as a small animal model with BMU-based remodeling for skeletal research. Bone. 1995;17(4 Suppl):191S–6S.

Macleay JM, Olson JD, Turner AS. Effect of dietary-induced metabolic acidosis and ovariectomy on bone mineral density and markers of bone turnover. J Bone Miner Metab. 2004;22:561–8.

Maeda H, Kimmel DB, Lane N, Raab D. The musculoskeletal response to immobilization and recovery. Bone. 1993;14:153–9.

Maher JR, Takahata M, Awad HA, Berger AJ. Raman spectroscopy detects deterioration in biomechanical properties of bone in a glucocorticoid-treated mouse model of rheumatoid arthritis. J Biomed Opt. 2011;16:087012. doi:10.1117/1.3613933.

Malluche HH, Faugere MC, Rush M, Friedler RM. Osteoblastic insufficiency is responsible for maintenance of osteopenia after loss of ovarian function in experimental beagle dogs. Endocrinology. 1986;119:2649 55.

Malluche HH, Faugere MC, Friedler RM, Matthews C, Fanti P. Calcitriol, parathyroid hormone, and accumulation of aluminum in dogs with renal failure. J Clin Invest. 1987;79:754–61.

Malluche HH, Faugere MC, Friedler RM, Fanti P. 1,25 dihydroxyvitamin D_3 corrects bone loss but suppresses bone remodeling in ovariohysterectomized beagle dogs. Endocrinology. 1988;122:1998–2006.

Malpaux B, Karsch FJ. A role for short days in sustaining seasonal reproductive activity in the ewe. J Reprod Fertil. 1990;90:555–62.

Mann DR, Gould KG, Collins DC. A potential primate model for bone loss resulting from medical oophorectomy or menopause. J Clin Endocrinol Metab. 1990;71:105–10.

Marcus R, Cann C, Madvig P, Minkoff J, Goddard M, Bayer M, Martin M, Gaudiani L, Haskell W, Genant H. Menstrual function and bone mass in elite women distance runners. Ann Intern Med. 1985;102:158–63.

Martin LJ, Carey KD, Comuzzie AG. Variation in menstrual cycle length and cessation of menstruation in captive raised baboons. Mech Ageing Dev. 2003;124:865–71.

Martin-Millan M, Almeida M, Ambrogini E, Han L, Zhao H, Weinstein RS, Jilka RL, O'Brien CA, Manolagas SC. The estrogen receptor-alpha in osteoclasts mediates the protective effects of estrogens on cancellous but not cortical bone. Mol Endocrinol. 2010;24:323–34. doi:10.1210/me.2009-0354.

Mas ID, Biscardi A, Schnitzler CM, Ripamonti U. Bone loss in the ovariectomized baboon Papio Ursinus: densitometry, histomorphometry and biochemistry. J Cell Mol Med. 2007;11:852–67.

Masuzawa T, Miyaura C, Onoe Y, Kusano K, Ohta H, Nozawa S, Suda T. Estrogen deficiency stimulates B lymphopoiesis in mouse bone marrow. J Clin Invest. 1994;94:1090–7.

Matsushita M, Tsyboyama T, Kasai R, Okumura H, Yamamuro T, Higuchi K, Higuchi K, Coin A, Yonezu T, Utani A, Umezawa M, Takeda T. Age-related changes in bone mass in the senescence-accelerated mouse (SAM). Am J Pathol. 1986;125:276–83.

Mazzuoli GF, Tabolli S, Bigi F, Valtorta C, Minisola S, Diacinti D, Scarnecchia L, Bianchi G, Piolini M, Dell'Acqua S. Effects of salmon calcitonin on the bone loss induced by ovariectomy. Calcif Tissue Int. 1990;47:209–14.

McHugh NA, Vercesi HM, Egan RW, Hey JA. In vivo rat assay: bone remodeling and steroid effects on juvenile bone by pQCT quantification in 7 days. Am J Physiol Endocrinol Metab. 2003;284:E70–5.

McLaughlin F, Mackintosh J, Hayes BP, McLaren A, Uings IJ, Salmon P, Humphreys J, Meldrum E, Farrow SN. Glucocorticoid-induced osteopenia in the mouse as assessed by histomorphometry, microcomputed tomography, and biochemical markers. Bone. 2002;30:924–30.

Miller SC, Jee WSS. Ethane-1-hydroxy-1,1-diphosphonate (EHDP) effects on growth and modeling of the rat tibia. Calcif Tissue Res. 1975;18:215–31.

Miller SC, Shupe JG, Redd EH, Miller MA, Omura TH. Changes in bone mineral and bone formation rates during pregnancy and lactation in rats. Bone. 1986a;7:283–7.

Miller LC, Weaver DS, McAlister JA, Koritnik DR. Effects of ovariectomy on vertebral trabecular bone in the cynomolgus monkey. Calcif Tissue Int. 1986b;38:62–5.

Miller SC, Bowman BM, Jee WSS. Available animal models of osteopenia – small and large. Bone. 1995;17(Supp 4):117S–23S.

Minaire P, Meunier PJ, Edouard C, Bernard J, Courpron P, Bourret J. Quantitative histologic data on disuse osteoporosis. Calcif Tissue Res. 1974;17:57–73.

Mitchell BD, Streeten EA. Clinical impact of recent genetic discoveries in osteoporosis. Appl Clin Genet. 2013;6:75–85. doi:10.2147/TACG.S52047.

Miyaura C, Onoe Y, Inada M, Maki K, Ikuta K, Ito M, Suda T. Increased B lymphopoiesis by interleukin 7 induces bone loss in mice with intact ovarian function: similarity to estrogen deficiency. Proc Nat Acad Sci (USA). 1997;94:9360–5.

Mödder U, Sanyal A, Kearns AE, Sibonga JD, Nishihara E, Xu J, O'Malley BW, Ritman EL, Riggs BL, Spelsberg TC, Khosla S. Effects of loss of steroid receptor coactivator-1 on the skeletal response to estrogen in mice. Endocrinology. 2004;145:913–21.

Morey ER, Baylink DJ. Inhibition of bone formation during spaceflight. Science. 1978;201:1138–41.

Mosekilde L, Sogaard, Danielsen CC, Torring O, MHL N. The anabolic effects of human parathyroid hormone (hPTH) on rat vertebral body mass are also reflected in the quality of bone, assessed by biomechanical testing: a comparison study between hPTH-(1-34) and hPTH-(1-84). Endocrinology. 1991;129:421–8.

Mosekilde L, Weisbrode SE, Safron JA, Stills HF, Jankowsky ML, Ebert DC, Danielsen CC, Sogaard CH, Franks AF, Stevens ML, Paddock CL, Boyce RW. Calcium-restricted ovariectomized Sinclair S-1 minipigs: an animal model of osteopenia and trabecular plate perforation. Bone. 1993;14:379–82.

Mühlbauer RC, Bauss F, Schenk R, Janner M, Bosies E, Strein K, Fleisch H. BM 21.0955, a potent new bisphosphonate to inhibit bone resorption. J Bone Miner Res. 1991;6:1003–11.

Murakami H, Nakamura T, Tsurukami H, Abe M, Barbier A, Suzuki K. Effects of tiludronate on bone mass, structure, and turnover at the epiphyseal, primary, and secondary spongiosa in the proximal tibia of growing rats after sciatic neurectomy. J Bone Miner Res. 1994;9:1355–64.

Nakamura T, Nagai Y, Yamato H, Suzuki K, Orimo H. Regulation of bone turnover and prevention of bone atrophy in ovariectomized beagle dogs by the administration of 24R,25$(OH)_2D_3$. Calc Tissue Int. 1992;50:221–7.

Nakamura H, Nitta T, Hoshino T, Koida S. Glucocorticoid-induced osteopenia in rats: histomorphometric and microarchitectural characterization and calcitonin effect. Biol Pharm Bull. 1996;19:217–9.

Newton BI, Cooper RC, Gilbert JA, Johnson RB, Zardiackas LD. The ovariectomized sheep as a model for human bone loss. J Comp Pathol. 2004;130:323–6.

Nilas L, Borg J, Christiansen C. Different rates of loss of trabecular and cortical bone after the menopause. In: Christiansen C, Arnaud CD, Nordin BEC, Parfitt AM, Peck WA, Riggs BL, editors. Osteoporosis I. Copenhagen: Glostrup Hospital; 1984. p. 161–3.

Nilas L, Christiansen C. Bone mass and its relationship to age and the menopause. J Clin Endocrinol Metab. 1987;65:697–702.

Nordin BEC, Horsman A, Brook R, Williams DA. The relationship between oestrogen status and bone loss in post-menopausal women. Clin Endocrinol. 1976;5:353S–61S.
O'Brien CA, Jia D, Plotkin LI, Bellido T, Powers CC, Stewart SA, Manolagas SC, Weinstein RS. Glucocorticoids act directly on osteoblasts and osteocytes to induce their apoptosis and reduce bone formation and strength. Endocrinology. 2004;145:1835–41.
O'Connell SL, Tresham J, Fortune CL, Farrugai W, McDougall JG, Scoggin BA, Wark JD. Effects of prednisolone and deflazacort on osteocalcin metabolism in sheep. Calcif Tissue Int. 1993;53:117–21.
Oheim R, Amling M, Ignatius A, Pogoda P. Large animal model for osteoporosis in humans: the ewe. Eur Cell Mater. 2012;24:372–85.
Ominsky MS, Stouch B, Schroeder J, Pyrah I, Stolina M, Smith SY, Kostenuik PJ. Denosumab, a fully human RANKL antibody, reduced bone turnover markers and increased trabecular and cortical bone mass, density, and strength in ovariectomized cynomolgus monkeys. Bone. 2011;49:162–73. doi:10.1016/j.bone.2011.04.001.
Onoe Y, Miyaura C, Ito M, Ohta H, Nozawa S, Suda T. Comparative effects of estrogen and RLX on B lymphopoiesis and bone loss induced by sex steroid deficiency in mice. J Bone Miner Res. 2000;15:541–9.
Ortoft G, Occlude H. Reduced strength of rat cortical bone after glucocorticoid treatment. Calcif Tissue Int. 1988;43:376–82.
Packer C, Tatar M, Collins A. Reproductive cessation in female mammals. Nature. 1998;392:807–11.
Parfitt AM. The actions of PTH on bone – relation to bone remodeling and turnover, calcium homeostasis, and metabolic bone disease, PTH and osteoblast relationships between bone turnover and bone loss, and the state of bones in primary hyperparathyroidism. Metabolism. 1976;25:1033–69.
Parfitt AM. The physiologic and clinical significance of bone histomorphometric data. In: Recker RR, editor. Bone histomorphometry: techniques and interpretation. Boca Raton: CRC Press; 1983. p. 143–224.
Passeri G, Girasole G, Jilka RL, Manolagas SC. Increased interleukin-6 production by murine bone marrow and bone cells after estrogen withdrawal. Endocrinology. 1993;133:822–8.
Pataki A, Müller K, Green JR, Ma YF, Li QN, Jee WSS. Effects of short-term treatment with the bisphosphonates zoledronate and pamidronate on rat bone: a comparative histomorphometric study on the cancellous bone formed before, during, and after treatment. Anat Rec. 1997;249:458–68.
Pennypacker BL, Duong LT, Cusick TE, Masarachia PJ, Gentile MA, Gauthier JY, Black WC, Scott BB, Samadfam R, Smith SY, Kimmel DB. Cathepsin K inhibitors prevent bone loss in estrogen-deficient rabbits. J Bone Miner Res. 2011;26:252–62. doi:10.1002/jbmr.223.
Pennypacker BL, Oballa RM, Levesque S, Kimmel DB, Duong LT. Cathepsin K inhibitors increase distal femoral bone mineral density in rapidly growing rabbits. BMC Musculoskelet Disord. 2013;14:344. doi:10.1186/1471-2474-14-344.
Plotkin LI, Weinstein RS, Parfitt AM, Roberson PK, Manolagas SC, Bellido T. Prevention of osteocyte and osteoblast apoptosis by bisphosphonates and calcitonin. J Clin Invest. 1999;104:1363–74.
Plück A. Conditional mutagenesis in mice: the Cre/loxP recombination system. Int J Exp Pathol. 1996;77:269–78.
Podbesek RD, Edouard C, Meunier PJ, Parsons JA, Reeve J, Stevenson RW, Zanelli JM. Effect of two treatment regimes with synthetic human parathyroid hormone fragment on bone formation and the tissue balance of trabecular bone in greyhounds. Endocrinology. 1983;112:1000–6.
Poli V, Balena R, Fattori E, Markatos A, Yamamoto M, Tanaka H, Ciliberto G, Rodan GA, Costantini F. Interleukin-6 deficient mice are protected from bone loss caused by estrogen deficiency. EMBO Jour. 1994;13:1189–96.
Pope NS, Gould KG, Anderson DC, Mann DR. Effects of age and sex on bone density in the rhesus monkey. Bone. 1989;10:109–12.

Postnov A, De Schutter T, Sijbers J, Karperien M, De Clerck N. Glucocorticoid-induced osteoporosis in growing mice is not prevented by simultaneous intermittent PTH treatment. Calcif Tissue Int. 2009;85:530–7. doi:10.1007/s00223-009-9301-3.

Qi H, Li M, Wronski TJ. A comparison of the anabolic effects of parathyroid hormone at skeletal sites with moderate and severe osteopenia in aged ovariectomized rats. J Bone Miner Res. 1995;10:948–55.

Raab DM, Crenshaw T, Kimmel DB, Smith EL. Cortical bone response in adult swine after exercise. J Bone Miner Res. 1991;6:741–9.

Ralston SH. Genetics of osteoporosis. Proc Nutr Soc. 2007;66:158–65.

Ray MK, Fagan SP, Brunicardi FC. The Cre-loxP system: a versatile tool for targeting genes in a cell- and stage-specific manner. Cell Transplant. 2000;9:805–15.

Recker RR, Heaney RP, Saville PD. Menopausal changes in remodeling. J Lab Clin Med. 1978;92:964–71.

Recker RR, Kimmel DB, Parfitt AM, Davies KM, Keshawarz N, Hinders S. Static and tetracycline-based bone histomorphometric data from 34 normal post-menopausal females. J Bone Miner Res. 1988;3:133–44.

Recker RR, Lappe JM, Davies KM, Kimmel DB. Change in bone mass immediately before menopause. J Bone Miner Res. 1992;7:857–62.

Recker RR, Hinders S, Davies KM, Heaney RP, Stegman MR, Lappe JM, Kimmel DB. Correcting calcium nutritional deficiency prevents spine fractures in elderly women. J Bone Miner Res. 1996;11:1961–6.

Recker R, Lappe J, Davies KM, Heaney R. Bone remodeling increases substantially in the years after menopause and remains increased in older osteoporosis patients. J Bone Miner Res. 2004;19:1628–33.

Reed MJ, Beranek PA, James VHT. Oestrogen production and metabolism in perimenopausal women. Maturitas. 1986;8:29–34.

Reginster JY, Adami S, Lakatos P, Greenwald M, Stepan JJ, Silverman SL, Christiansen C, Rowell L, Mairon N, Bonvoisin B, Drezner MK, Emkey R, Felsenberg D, Cooper C, Delmas PD, Miller PD. Efficacy and tolerability of once-monthly oral ibandronate in postmenopausal osteoporosis: 2 year results from the MOBILE study. Ann Rheum Dis. 2006;65:654–61.

Richards JB, Zheng HF, Spector TD. Genetics of osteoporosis from genome-wide association studies: advances and challenges. Nat Rev Genet. 2012;13:576–88. doi:10.1038/nrg3228.

Richelson LS, Wahner HW, Melton LJ, Riggs BL. Relative contributions of aging and estrogen deficiency to postmenopausal bone loss. N Engl J Med. 1984;311:1273–5.

Riggs BL, Melton LJ III. Heterogeneity of involutional osteoporosis: evidence for two osteoporosis syndromes. Am J Med. 1983;75:899–901.

Riggs BL, Hodgson SF, O'Fallon WM, Chao EY, Wahner HW, Muhs JM, Cedel SL, Melton LJ. Effect of fluoride treatment on the fracture rate in postmenopausal women with osteoporosis. New Engl J Med. 1990;322:802–9.

Rissanen JP, Halleen JM. Models and screening assays for drug discovery in osteoporosis. Expert Opin Drug Discovery. 2010;5:1163–74. doi:10.1517/17460441.2010.532484.

Ritz E, Krempien B, Mehls O, Malluche HH. Skeletal abnormalities in chronic renal insufficiency before and during maintenance hemodialysis. Kidney Int. 1973;4:116–27.

Rodin A, Murby B, Smith MA, Caleffi M, Fentimen I, Chapman MG, Fogelman I. Premenopausal bone loss in the lumbar spine and femoral neck: a study of 225 Caucasian women. Bone. 1990;11:1–5.

Ross PD, Wasnich RD, Vogel JM. Detection of prefracture spinal osteoporosis using bone mineral absorptiometry. J Bone Miner Res. 1988;3:1–11.

Ross PD, Genant HK, Davis JW, Miller PD, Wasnich RD. Predicting vertebral fracture incidence from prevalent fractures and bone density among non-black, osteoporotic women. Osteoporos Int. 1993;3:120–6.

Roudebush RE, Magee DE, Benslay DN, Bendele AM, Bryant HU. Effect of weight manipulation on bone loss due to ovariectomy and the protective effects of estrogen in the rat. Calcif Tissue Int. 1993;53:61–4.

Rowe DJ. Effects of a fluorinated bisphosphonate on bone remodeling in vivo. Bone. 1985;6:433–7.
Ruth E. An experimental study of the Haversian-type vascular channels. Anat Rec. 1953;112:429–55.
Samadfam R, Smith SY. Effects of aging on bone turnover markers and bone density regulating hormones in cynomolgus monkeys. J Bone Miner Res. 2014;27:S206. Abstract SA0460.
Samuels A, Perry MJ, Tobias JH. High-dose estrogen induces de novo medullary bone formation in female mice. J Bone Miner Res. 1999;14:178–86.
Schaadt O, Bohr H. Different trends of age-related diminution of bone mineral content in the lumbar spine, femoral neck, and femoral shaft in women. Calcif Tissue Int. 1988;42:71–6.
Schapira D, Lotan-Miller R, Barzilai D, Silbermann M. The rat as a model for studies of the aging skeleton. Cells Mater (Suppl). 1992;1:181–8.
Schenk R, Merz WA, Mühlbauer R, Russell RG, Fleisch H. Effect of ethane-1-hydroxy-1,1-diphosphonate (EHDP) and dichloromethylene diphosphonate (cl 2 MDP) on the calcification and resorption of cartilage and bone in the tibial epiphysis and metaphysis of rats. Calcif Tissue Res. 1973;11:196–214.
Schenk R, Eggli P, Fleisch H, Rosini S. Quantitative morphometric evaluation of the inhibitory activity of new aminobisphosphonates on bone resorption in the rat. Calcif Tissue Int. 1986;38:342–9.
Scholz-Ahrens KE, Delling G, Jungblut PW, Kallweit E, Barth CA. Effect of ovariectomy on bone histology and plasma parameters of bone metabolism in nulliparous and multiparous sows. Z Ernahrungswiss. 1996;35:13–21.
Schnitzler CM, Ripamonti U, Mesquita JM. Histomorphometry of iliac crest trabecular bone in adult male baboons in captivity. Calcif Tissue Int. 1993;52:447–54.
Seedor JG, Quartuccio HA, Thompson DD. The bisphosphonate alendronate (MK-217) inhibits bone loss due to ovariectomy in rats. J Bone Miner Res. 1991;6:339–46.
Seibel MJ, Cooper MS, Zhou H. Glucocorticoid-induced osteoporosis: mechanisms, management, and future perspectives. Lancet Diabetes Endocrinol. 2013;1:59–70. doi:10.1016/S2213-8587(13)70045-7.
Seidlova-Wuttke D, Nguyen BT, Wuttke W. Long-term effects of ovariectomy on osteoporosis and obesity in estrogen-receptor-alpha-deleted mice. Comp Med. 2012;62:8–13.
Sevil F, Kara ME. The effects of ovariectomy on bone mineral density, geometrical, and biomechanical characteristics in the rabbit femur. Vet Comp Orthop Traumatol. 2010;23:31–6. doi:10.3415/VCOT-08-09-0085.
Shane E, Burr D, Abrahamsen B, Adler RA, Brown TD, Cheung AM, Cosman F, Curtis JR, Dell R, Dempster DW, Ebeling PR, Einhorn TA, Genant HK, Geusens P, Klaushofer K, Lane JM, McKiernan F, McKinney R, Ng A, Nieves J, O'Keefe R, Papapoulos S, Howe TS, van der Meulen MC, Weinstein RS, Whyte MP. Atypical subtrochanteric and diaphyseal femoral fractures: second report of a task force of the American Society for Bone and Mineral Research. J Bone Miner Res. 2014;29:1–23. doi:10.1002/jbmr.1998.
Shaw JA, Wilson SC, Bruno A, Paul EM. Comparison of primate and canine models for bone ingrowth experimentation, with reference to the effect of ovarian function on bone ingrowth potential. J Orthop Res. 1994;12:268–73.
Shen V, Dempster DW, Birchman R, Mellish RWE, Church E, Kohn D, Lindsay R. Lack of changes in histomorphometric, bone mass, and biochemical parameters in ovariohysterectomized dogs. Bone. 1992;13:311–6.
Shively CA, Jayo MJ, Weaver DS, Kaplan JR. Reduced vertebral bone mineral density in socially subordinate female cynomolgus macaques. Am J Primatol. 1991;24:135–8.
Sietsema WK, Ebetino FH, Salvagno AM, Bevan JA. Antiresorptive dose-response relationships across three generations of bisphosphonates. Drugs Exp Clin Res. 1989;15:389–96.
Sigrist IM, Gerhardt C, Alini M, Schneider E, Egermann M. The long-term effects of ovariectomy on bone metabolism in sheep. J Bone Miner Metab. 2007;25:28–35.
Simonet WS, Lacey DL, Dunstan CR, Kelley M, Chang MS, Luthy R, Nguyen HQ, Wooden S, Bennett L, Boone T, Shimamoto G, DeRose M, Elliott R, Colombero A, Tan HL, Trail G, Sullivan J, Davy E, Bucay N, Renshaw Gegg L, Hughes TM, Hill D, Pattison W, Campbell P,

Boyle WJ. Osteoprotegerin: a novel secreted protein involved in the regulation of BMD. Cell. 1997;89:309–19.
Simpson AH, Murray IR. Osteoporotic fracture models. Curr Osteoporos Rep. 2015;13:9–15. doi:10.1007/s11914-014-0246-8.
Sipos W, Kralicek E, Rauner M, Duvigneau CJ, Worliczek HL, Schamall D, Hartl RT, Sommerfeld-Stur I, Dall'Ara E, Varga P, Resch H, Schwendenwein I, Zysset P, Pietschmann P. Bone and cellular immune system of multiparous sows are insensitive to ovariectomy and nutritive calcium shortage. Horm Metab Res. 2011;43:404–9. doi:10.1055/s-0031-1277154.
Sjoden G, Johnell O, DeLuca HF, Lindgren JU. Effects of 1-α-OHD_2 and 1-α-OHD_3 in rats treated with prednisolone. Acta Endocrinol. 1984;106:564–8.
Smink JJ, Gresnigt MG, Hamers N, Koedam JA, Berger R, Van Buul-Offers SC. Short-term glucocorticoid treatment of prepubertal mice decreases growth and IGF-I expression in the growth plate. J Endocrinol. 2003;177:381–8.
Smith DM, Nance WE, Ke WK, Christian JC, Johnston CC. Genetic factors in determining bone mass. J Clin Invest. 1973;52:2800–8.
Smith SY, Jolette J, Turner CH. Skeletal health: primate model of postmenopausal osteoporosis. Am J Primatol. 2009;71:752–65. doi:10.1002/ajp.20715.
Smith SY, Doyle N, Boyer M, Chouinard L, Saito H. Eldecalcitol, a vitamin D analog, reduces bone turnover and increases trabecular and cortical bone mass, density, and strength in ovariectomized cynomolgus monkeys. Bone. 2013;57:116–22. doi:10.1016/j.bone.2013.06.005.
Snow GR, Cook MA, Anderson C. Oophorectomy and cortical bone remodeling in the beagle. Calcif Tissue Int. 1984;36:586–90.
Snow GR, Anderson C. The effects of 17-β-estradiol and progestagen on trabecular bone remodeling in oophorectomized dogs. Calcif Tissue Int. 1986;39:198–205.
Sophocleous A, Idris AI. Rodent models of osteoporosis. Bonekey Rep. 2014;3:614. doi:10.1038/bonekey.2014.109.
Steiniche T, Hasling C, Charles P, Eriksen EF, Mosekilde L, Melsen F. A randomized study on the effects of estrogen/gestagen or high dose oral calcium on trabecular bone remodeling in postmenopausal osteoporosis. Bone. 1989;10:313–20.
Stepan JJ, Pospichal J, Presl J, Pacovsky V. Bone loss and biochemical indices of bone remodeling in surgically-induced postmenopausal women. Bone. 1987;8:279–84.
Stroup GB, Lark MW, Veber DF, Bhattacharyya A, Blake S, Dare LC, Erhard KF, Hoffman SJ, James IE, Marquis RW, Ru Y, Vasko-Moser JA, Smith BR, Tomaszek T, Gowen M. Potent and selective inhibition of human cathepsin K leads to inhibition of bone resorption in vivo in a nonhuman primate. J Bone Miner Res. 2001;16:1739–46.
Stroup GB, Kumar S, Jerome CP. Treatment with a potent cathepsin K inhibitor preserves cortical and trabecular bone mass in ovariectomized monkeys. Calcif Tissue Int. 2009;85(4):344–55. doi:10.1007/s00223-009-9279-x.
Svesatikoglou JA, Larsson SE. Changes in composition and metabolic activity of the skeletal parts of the extremity of the adult rat following below-knee amputation. Acta Chir Scand. 1976;467:9–15.
Syed FA, Melim T. Rodent models of aging bone: an update. Curr Osteoporos Rep. 2011;9:219–28. doi:10.1007/s11914-011-0074-z.
Szucs J, Horvath C, Kollin E, Szathmari M, Hollo I. Three-year calcitonin combination therapy for postmenopausal osteoporosis with crush fractures of the spine. Calcif Tissue Int. 1992;50:7–10.
Szulc P. The role of bone turnover markers in monitoring treatment in postmenopausal osteoporosis. Clin Biochem. 2012;45:907–19. doi:10.1016/j.clinbiochem.2012.01.022.
Takahashi K, Tsuboyama T, Matsushita M, Kasai R, Okumura H, Yamamuro T, Okamoto Y, Toriyama K, Kitagawa K, Takeda T. Modification of strain-specific femoral bone density by bone marrow-derived factors administered neonatally: a study on the spontaneously osteoporotic mouse, SAMP6. Bone and Min. 1994;24:245–55.
Takahata M, Maher JR, Juneja SC, Inzana J, Xing L, Schwarz EM, Berger AJ, Awad HA. Mechanisms of bone fragility in a mouse model of glucocorticoid-treated rheumatoid

arthritis: implications for insufficiency fracture risk. Arthritis Rheum. 2012;64:3649–59. doi:10.1002/art.34639.
Thiele S, Ziegler N, Tsourdi E, De Bosscher K, Tuckermann JP, Hofbauer LC, Rauner M. Selective glucocorticoid receptor modulation maintains bone mineral density in mice. J Bone Miner Res. 2012;27:2242–50. doi:10.1002/jbmr.1688.
Thiele S, Baschant U, Rauch A, Rauner M. Instructions for producing a mouse model of glucocorticoid-induced osteoporosis. Bonekey Rep. 2014;3:552. doi:10.1038/bonekey.2014.47.
Thomaidis VT, Lindholm TS. The effect of remobilization on the extremity of the adult rat after short-term immobilization in a plaster cast. Acta Chir Scand. 1976;467:36–9.
Thompson DD, Rodan GA. Indomethacin inhibition of tenotomy-induced bone resorption in rats. J Bone Miner Res. 1988;3:409–14.
Tinetti ME, Baker DI, McAvay G, Claus EB, Garrett P, Gottschalk M, Koch ML, Trainor K, Horwitz RI. A multifactorial intervention to reduce the risk of falling among elderly people living in the community. N Engl J Med. 1994;331:821–7.
Toolan BC, Shea M, Myers ER, Borchers RE, Seedor JG, Quartuccio H, Rodan G, Hayes WC. Effects of 4-amino-1-hydroxybutylidene bisphosphonate on bone biomechanics in rats. J Bone Miner Res. 1992;7:1399–406.
Treloar AE. Menstrual cyclicity and the pre-menopause. Maturitas. 1981;3:249–64.
Turner AS. Animal models of osteoporosis – necessity and limitations. Eur Cell Mater. 2001;22:66–81.
Turner AS. The sheep as a model for osteoporosis in humans. Vet J. 2002;163:232–9.
Turner CH, Burr DB. Basic biomechanical measurements of bone: a tutorial. Bone. 1993;14:595–608.
Turner RT, Vandersteenhoven JJ, Bell NH. The effects of ovariectomy and 17-β-estradiol on cortical bone histomorphometry in growing rats. J Bone Miner Res. 1987;2:115–22.
Turner RT, Hannon KS, Demers LM, Buchanan J, Bell NH. Differential effects of gonadal function on bone histomorphometry in male and female rats. J Bone Miner Res. 1989;4:557–63.
Turner RT, Hannon KS, Greene VS, Bell NH. Prednisone inhibits formation of cortical bone in sham-operated and OVX rats. Calcif Tissue Int. 1995;56:311–5.
Uhthoff HK, Sekaly G, Jaworski ZFG. Effect of long-term nontraumatic immobilization on metaphyseal spongiosa in young adult and old beagle dogs. Clin Orthop Relat Res. 1985;192:278–83.
Urist MR, Budy AM, McLean FC. Endosteal bone formation in estrogen-treated mice. J Bone Jt Surg. 1950;32A:143–62.
Vanderschueren D, Van Herck E, Suiker AM, Allewaert K, Visser WJ, Geusens P, Bouillon R. Bone and mineral metabolism in the adult guinea pig: long-term effects of estrogen and androgen deficiency. J Bone Miner Res. 1992;7:1407–15.
Vignery A, Baron R. Dynamic histomorphometry of alveolar bone remodeling in the adult rat. Anat Rec. 1980;196:191–200.
Wakley GK, Baum BL, Hannon KS, Turner RT. The effects of tamoxifen on the osteopenia induced by sciatic neurotomy in the rat: a histomorphometric study. Calcif Tissue Int. 1988;43:383–8.
Walker DG. Congenital osteopetrosis in mice cured by parabiotic union with normal siblings. Endocrinology. 1972;91:916–20.
Walker DG. Control of bone resorption by hematopoietic tissue. The induction and reversal of congenital osteopetrosis in mice through use of bone marrow and splenic transplants. J Exp Med. 1975a;142:651–63.
Walker DG. Bone resorption restored in osteopetrotic mice by transplants of normal bone marrow and spleen cells. Science. 1975b;190:784–5.
Wallach S, Cohen S, Reid DM, Hughes RA, Hosking DJ, Laan RF, Doherty SM, Maricic M, Rosen C, Brown J, Barton I, Chines AA. Effects of risedronate treatment on bone density and vertebral fracture in patients on corticosteroid therapy. Calcif Tissue Int. 2000;67:277–85.
Wang F, Wang PX, Wu XL, Dang SY, Chen Y, Ni YY, Gao LH, Lu SY, Kuang Y, Huang L, Fei J, Wang ZG, Pang XF. Deficiency of adiponectin protects against ovariectomy-induced osteoporosis in mice. PLoS One. 2013a;8:e68497. doi:10.1371/journal.pone.0068497.

Wang FS, Chuang PC, Lin CL, Chen MW, Ke HJ, Chang YH, Chen YS, Wu SL, Ko JY. MicroRNA-29a protects against glucocorticoid-induced bone loss and fragility in rats by orchestrating bone acquisition and resorption. Arthritis Rheum. 2013b;65:1530–40. doi:10.1002/art.37948.

Wasnich R, Yano K, Vogel J. Postmenopausal bone loss at multiple skeletal sites: relationship to estrogen use. J Chronic Dis. 1983;36:781–90.

Waters RV, Gamradt SC, Asnis P, Vickery BH, Avnur Z, Hill E, Bostrom M. Systemic corticosteroids inhibit bone healing in a rabbit ulnar osteotomy model. Acta Orthop Scand. 2000;71:316–21.

Weinstein RS. Glucocorticoid-induced bone disease. N Engl J Med. 2011;365:62–70.

Weinstein RS, Jilka RL, Parfitt AM, Manolagas SC. Inhibition of osteoblastogenesis and promotion of apoptosis of osteoblasts and osteocytes by glucocorticoids. Potential mechanisms of their deleterious effects on bone. J Clin Inv. 1998;102:274–82.

Weiss A, Arbell I, Steinhagen-Thiessen E, Silbermann M. Structural changes in aging bone: osteopenia in the proximal femurs of female mice. Bone. 1991;12:165–72.

Williams DS, McCracken PJ, Purcell M, Pickarski M, Mathers PD, Savitz AT, Szumiloski J, Jayakar RY, Somayajula S, Krause S, Brown K, Winkelmann CT, Scott BB, Cook L, Motzel SL, Hargreaves R, Evelhoch JL, Cabal A, Dardzinski BJ, Hangartner TN, Duong LT. Effect of odanacatib on bone turnover markers, bone density and geometry of the spine and hip of ovariectomized monkeys: a head-to-head comparison with alendronate. Bone. 2013;56:489–96. doi:10.1016/j.bone.2013.06.008.

Wilson AK, Bhattacharyya MH, Miller S, Mani A, Sacco-Gibson N. Ovariectomy-induced changes in aged beagles: histomorphometry of rib cortical bone. Calcif Tissue Int. 1998;62:237–43.

Wimalawansa SJ, Chapa MT, Yallampalli C, Zhang R, Simmons DJ. Prevention of corticosteroid-induced bone loss with nitric oxide donor nitroglycerin in male rats. Bone. 1997;21:275–80.

Wimalawansa SJ, Simmons DJ. Prevention of corticosteroid-induced bone loss with alendronate. Proc Exp Biol Med. 1998;217:162–7.

Wronski TJ. The ovariectomized rat as an animal model for postmenopausal bone loss. Cell Mater. 1992;1(Supp 1):69–74.

Wronski TJ, Walsh CC, Ignaszewski LA. Histologic evidence for osteopenia and increased bone turnover in ovariectomized rats. Bone. 1986;7:119–24.

Wronski TJ, Schenck PA, Cintron M, Walsh CC. Effect of body weight on osteopenia in ovariectomized rats. Calcif Tissue Int. 1987;40:155–9.

Wronski TJ, Cintron M, Doherty AL, Dann LM. Estrogen treatment prevents osteopenia and depresses bone turnover in ovariectomized rats. Endocrinology. 1988;123:681–6.

Wronski TJ, Dann LM, Scott KS, Cintron M. Long-term effects of ovariectomy and aging on the rat skeleton. Calcif Tissue Int. 1989a;45:360–6.

Wronski TJ, Dann LM, Horner SL. Time course of vertebral osteopenia in ovariectomized rats. Bone. 1989b;10:295–301.

Wronski TJ, Pun S, Liang H. Effects of age, estrogen depletion, and parathyroid hormone treatment on vertebral cancellous wall width in female rats. Bone. 1999;25:465–8.

Wu ZX, Lei W, Hu YY, Wang HQ, Wan SY, Ma ZS, Sang HX, Fu SC, Han YS. Effect of ovariectomy on BMD, micro-architecture and biomechanics of cortical and cancellous bones in a sheep model. Med Eng Phys. 2008;30:1112–8. doi:10.1016/j.medengphy.2008.01.007.

Yamamoto M, Fisher JE, Gentile M, Seedor JG, Leu CT, Rodan SB, Rodan GA. The integrin ligand echistatin prevents bone loss in OVX mice and rats. Endocrinology. 1998;139:1411–9.

Yao W, Dai W, Jiang L, Lay EY, Zhong Z, Ritchie RO, Li X, Ke H, Lane NE. Sclerostin-antibody treatment of glucocorticoid-induced osteoporosis maintained bone mass and strength. Osteoporos Int. 2016;27:283–94.

Young DR, Niklowitz WJ, Steele CR. Tibial changes in experimental disuse osteoporosis in the monkey. Calcif Tissue Int. 1983;35:304–8.

Zaiss MM, Sarter K, Hess A, Engelke K, Böhm C, Nimmerjahn F, Voll R, Schett G, David JP. Increased bone density and resistance to ovariectomy-induced bone loss in FoxP3-transgenic mice based on impaired osteoclast differentiation. Arthritis Rheum. 2010;62:2328–38. doi:10.1002/art.27535.

Zhang KJ, Zhang J, Kang ZK, Xue XM, Kang JF, Li YW, Dong HN, Liu DG. Ibandronate for prevention and treatment of glucocorticoid-induced osteoporosis in rabbits. Rheumatol Int. 2012;32:3405–11. doi:10.1007/s00296-011-2074-9.

Zhang Y, Li Y, Gao Q, Shao B, Xiao J, Zhou H, Niu Q, Shen M, Liu B, Hu K, Kong L. The variation of cancellous bones at lumbar vertebra, femoral neck, mandibular angle and rib in ovariectomized sheep. Arch Oral Biol. 2014;59:663–9. doi:10.1016/j.archoralbio.2014.03.013.

Zhang J, Song J, Shao J. Icariin attenuates glucocorticoid-induced bone deteriorations, hypocalcemia and hypercalciuria in mice. Int J Clin Exp Med. 2015;8:7306–14.

Zhao Q, Liu X, Zhang L, Shen X, Qi J, Wang J, Qian N, Deng L. Bone selective protective effect of a novel bone-seeking estrogen on trabecular bone in ovariectomized rats. Calcif Tissue Int. 2013;93:172–83. doi:10.1007/s00223-013-9739-1.

Part II
Methods in Bone Research

Chapter 5
Biochemical Markers of Bone Turnover

Susan Y. Smith and Rana Samadfam

Abstract The use of clinically relevant biomarkers provides an important noninvasive, sensitive, rapid, and real-time tool to monitor bone activity at the whole skeleton level and to investigate potential mechanisms. The goal of this chapter is to describe the biochemical markers of bone turnover, consisting of markers of bone formation and bone resorption, considered important when conducting safety assessments in a preclinical setting. The bone formation markers used are osteocalcin, bone alkaline phosphatase, and procollagen type I propeptides, and the bone resorption markers are tartrate-resistant acid phosphatase, deoxypyridinoline, N-telopeptides, and C-telopeptides. Markers can be included in any study type and are available for most animal species routinely used for safety assessments. Changes in bone markers signal perturbations in bone and as such are useful as a screening tool in early drug development or to monitor temporal changes in a chronic setting. When combined with additional (in vivo) end-points such as calcium and phosphorus measures, analysis of the calciotropic hormones, fibroblast growth factor 23, insulin-like growth factor-1, and bone densitometry, they provide important mechanistic information to further characterize the effects of a compound on the skeleton.

Keywords Bone formation • Osteocalcin (OC) • Bone alkaline phosphatase (BAP) • Procollagen type I propeptides (PINP, PICP) • Bone resorption • TRACP5b (or TRAP5b) • Deoxypyridinoline (DPD) • N-telopeptides (NTx) • C-telopeptides (CTx)

S.Y. Smith (✉)
SY Smith Consulting, Montreal, QC, Canada
e-mail: sysmithconsulting@gmail.com

R. Samadfam
Charles River Preclinical Services, Senneville, QC, Canada

S.Y. Smith et al. (eds.), *Bone Toxicology*, Molecular and Integrative Toxicology,
DOI 10.1007/978-3-319-56192-9_5

5.1 Introduction

By definition a biomarker is "a characteristic that is objectively measured and evaluated as an indicator of normal biological processes, pathogenic processes or pharmacological responses to a therapeutic intervention." A wide range of biomarkers are used today; every biological system has its own specific biomarkers and bone is no exception. Many of these biomarkers are relatively easy to measure and can be performed as part of routine preclinical and clinical evaluations. The skeleton is continually undergoing remodeling and modeling during growth, adaptation, and maintenance. These processes include bone formation and bone resorption either as a coupled, cyclical process (remodeling), or independently on different surfaces (modeling). During bone remodeling, bone resorption and formation occur within organized cellular units called the basic multicellular unit (BMU); thus, levels of bone turnover markers reflect resorption and formation within a BMU and the number of BMUs. While biochemical markers are used to assess bone formation and bone resorption, they cannot be used to differentiate between modeling/remodeling nor between trabecular or cortical bone turnover; however, they do provide an effective real-time measure of the function of the bone machinery: the osteoblasts and the osteoclasts. An understanding of the lineage of the bone cells (osteoclasts, osteoblasts) is also important to appreciate the potential mechanism underlying any changes observed in the markers (refer to Chap. 2: Bone Physiology and Biology).

Biochemical markers reflecting bone turnover can be measured in blood and urine and are key parameters used to assess bone disease or to monitor drug treatment, providing a real-time, rapid, and sensitive assessment of bone turnover at the whole skeleton level. Generally, changes in biochemical markers of bone turnover are larger than seen using histomorphometric parameters (see Chap. 8), with the advantage that they can be easily repeated. Bone turnover markers can be added as an end-point to any study type (short- and long-term), with any species, and using any route of test article administration. These markers have been successfully employed in preclinical studies varying from proof-of-concept (Ominsky et al. 2010) to carcinogenicity testing (Jolette et al. 2006). They can be used to assess the therapeutic equivalence of various dose levels or regimens of the same drug, or for comparison with a biosimilar, and can provide information regarding the temporal pattern of a response to treatment. These data provide a profile reflecting the mechanism of action of the compound such as the "anabolic window" induced by anabolic drugs or "uncoupling" as observed using cathepsin K inhibitors. Understanding short-term vs long-term effects allows appropriate sampling times to be incorporated into future studies (preclinical and clinical). Bone markers can be used as a pharmacodynamic (PD) end-point and when combined with pharmacokinetic (PK) evaluations provide a PK/PD profile (e.g., Ominsky et al. 2010) that can facilitate the design of future longer-term preclinical studies and clinical trials. Markers can also be monitored during an off-treatment or recovery phase.

The balance between bone resorption and formation impacts bone mass and ultimately bone strength; therefore, biochemical markers are especially of value when combined with other in vivo bone biomarkers such as bone densitometry. In appropriately designed mechanistic studies, they can be used to explore the effect of a drug on the relationship between bone mass and strength and establish whether a new drug has the potential to improve "bone quality" or not (e.g., see Kostenuik et al. 2015). Importantly, most bone turnover markers used in preclinical studies can also be used in a clinical setting and are, therefore, an excellent translational tool. Clinically, however, bone mineral density (BMD) remains the main criterion for diagnosis of bone disease and to monitor treatment, although bone turnover markers do provide important real-time information regarding disease status, drug efficacy, and compliance.

Except for compounds intended to affect bone, drug classes used in many therapeutic areas are known to influence bone turnover and increase fracture risk, for example, the thiazolidinediones (TZDs) used in the treatment of diabetes (Kumar et al. 2013; Samadfam et al. 2012; Smith et al. 2015). As more is understood regarding the role of the skeleton and its cross talk with other organ systems, use of markers as a tool to identify a signal that may indicate bone is a target tissue has proven an important end-point in hazard identification and risk assessment. Bone markers are widely employed in safety studies early in drug development for many different drug classes from diverse therapeutic areas. While biochemical markers can be used as the primary end-point to discriminate any perturbation in bone, additional biochemical parameters are often utilized to provide a more comprehensive assessment of the mechanism underlying any changes. These additional evaluations may include measurement of the calciotropic hormones, sex steroid hormones, and clinical chemistry parameters including total alkaline phosphatase and calcium/phosphorus levels. Pharmacokinetic parameters measured coincident with biochemical markers may also provide evidence to support use of a specific bone marker as a PD end-point. This chapter will discuss the principal markers of bone turnover and associated parameters used routinely in preclinical safety assessments where bone is under investigation as a potential target tissue.

5.2 Biochemical Markers of Bone Turnover

A panel of biochemical markers, consisting minimally of two formation and two resorption markers, is normally employed to assess bone turnover since a single marker, unless used as an established PD marker, may not be sufficient to capture a treatment effect (depending on the mechanism), and does not provide information on the net balance of bone formation vs bone resorption. Bone markers typically used in a preclinical setting include osteocalcin (OC), procollagen type I N-terminal propeptide (PINP) and bone-specific alkaline phosphatase (BAP) as bone formation markers and N-telopeptides (NTx), C-telopeptides (CTx), deoxypyridinoline (DPD), and tartrate resistance acid phosphatase 5b (TRAP5b) as bone resorption markers (Table 5.1). The markers therefore consist of two groups: enzymes which reflect the metabolic activity of the bone cells (BAP and TRAP5b) and the bone matrix components (OC, PINP, NTx, CTx, DPD) (Fig. 5.1). Parameters can be measured in blood and/or urine, and for some markers either matrix can be used, depending on availability (Table 5.1). These markers are further described in the excellent overview by Szulc and Bauer (2013). Much of the information below is based on human markers, but is considered relevant since human kits are cross-validated to analyze samples from most laboratory animal species, except for rodents for which specific kits are available.

5.2.1 Bone Formation Markers

5.2.1.1 Osteocalcin (OC)

Osteocalcin (gamma-carboxyglutamic acid-containing protein (BGLAP)) is mainly secreted by osteoblasts, osteocytes, and odontoblasts and is specific for bone and dentin (Szulc and Bauer 2013). OC makes up to 15% of the noncollagenous protein in bone, although the level is variable depending on the animal species, age, and bone type (Boskey and Robey 2013). OC is thought to regulate the growth of apatite crystals and may act as a specific signal from the bone matrix stimulating migration and adhesion of osteoclasts (Chenu et al. 1994). Although its exact function in bone remains uncertain, its role in the mineralization and remodeling of bone may be attributed to its binding to one or more surfaces of the growing crystal thereby blocking growth in one or more directions or limiting crystal size. This protein contains 47–50 amino acid residues, including 3 glutamic acid residues (molecular weight 5,200–5,900), depending on the species. It undergoes γ-carboxylation which yields a protein (Gla-OC) with a high binding affinity to hydroxyapatite crystals in mineralizing matrix. Carboxylation is vitamin K dependent and in most species occurs at all 3 glutamic acid sites. The undercarboxylated osteocalcin (Glu-OC), with a low binding affinity to hydroxyapatite, circulates in blood and has recently been recognized as a novel hormone regulating energy metabolism, vascular calcification, and the male reproductive system, at least in the rodent (Karsenty 2006; Lee et al. 2007). In addition to its potential role in energy metabolism, OC has also been suggested to function as a neuropeptide.

Specific immunoassays have been developed to measure Gla- and Glu-OC. Nonspecific assays measure total OC, which includes both the Gla- and Glu-OC. In circulation, OC exists primarily as the carboxylated form and to a lesser degree in the undercarboxylated form. During bone resorption, OC in the bone matrix is released into the circulation as fragments and intact molecules and is further metabolized in the circulation and peripheral organs such as the kidney, liver, and lungs

Table 5.1 Biochemical markers of bone turnover routinely used in preclinical studies

Activity	Marker	Abbreviation	Matrix[a]	Species	Assay type	Storage	Comment
Bone formation	Osteocalcin	OC	Serum	Rat	ELISA	−20 °C	Recommended as part of a bone marker panel
				Mouse	Luminex		
				NHP	EIA		Assay not feasible in dog serum
				Rabbit	EIA		
	Bone-specific alkaline phosphatase	BAP	Serum	NHP	ELISA	−20 °C	Recommended as part of a bone marker panel
				Dog			Total ALP used for rodents
	Procollagen type I N-terminal propeptide	PINP	Serum	Rat	EIA	−20 °C	Recommended as part of a bone marker panel
				Mouse	EIA		
				NHP	RIA		
Bone resorption	Tartrate-resistant acid phosphatase 5b	TRAP5b	Serum	Rat	ELISA	−20 °C	Marker of osteoclast activity and number
				Mouse			
				NHP			
	N-telopeptides	NTx	Urine	NHP	ELISA	−20 °C	Urine assay most robust
				Dog			No assay available for rodents
				Rabbit			Recommended as part of a bone marker panel
				Pig			
			Serum	NHP		−20 °C	
	C-telopeptides	CTx	Urine	Rat	ELISA	−20 °C	Recommended as part of a bone marker panel
				NHP			
				Dog			
			Serum	Rat		−20 °C	
				Mouse			
				NHP			
	Free deoxypyridinoline	DPD	Urine	Rat	ELISA	−20 °C	Recommended as part of a bone marker panel
				NHP			
				Dog			

NHP nonhuman primates, *EIA* enzyme immunoassay, *ELISA* enzyme-linked immunosorbent assay, *RIA* radioimmunoassay, *MMPs* matrix metalloproteinases

[a]Urinary creatinine analyzed on the same sample and used to correct for urine concentration

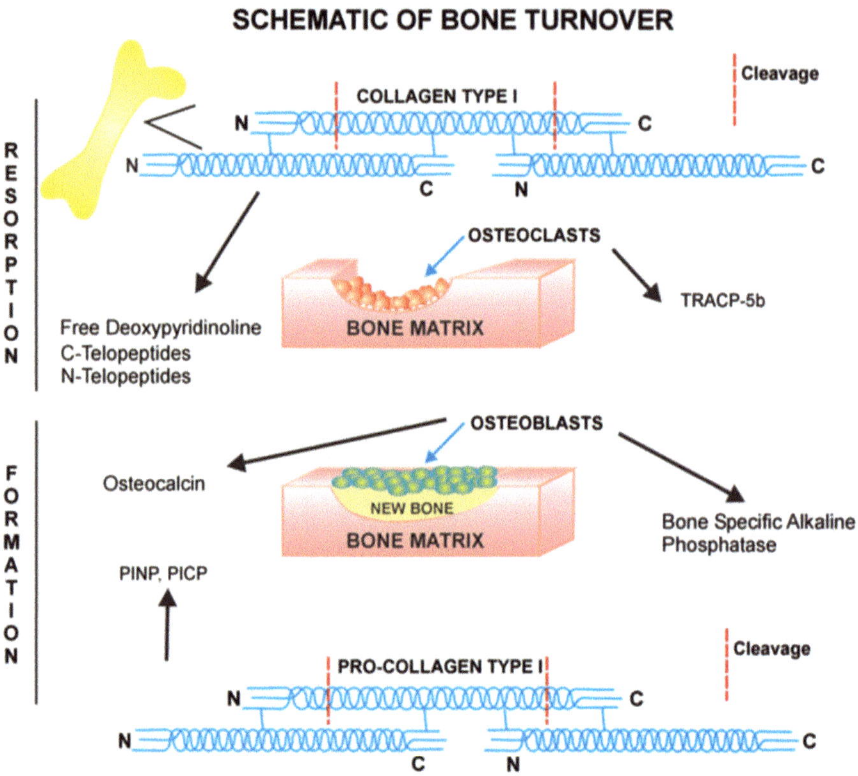

Fig. 5.1 Schematic illustrating origin of biochemical markers of bone turnover (From Gunson et al. 2013 in Haschek and Rousseaux's Handbook of Toxicologic Pathology, with permission)

(Price et al. 1981; Farrugia and Melick 1986; Aliberti et al. 2000). Strong correlations between OC and CTx levels, a marker of bone resorption, in rat and human cell cultures, suggest that OC can also be used as an index of bone turnover instead of a specific marker of bone formation (Ivaska et al. 2004). These authors showed that during bone resorption, it appeared that the acidification step and not the cathepsin-mediated degradation was the major source of intact OC bound to calcium via the Gla residues. After demineralization, proteolytic enzymes, especially cathepsins, cleave collagen and other matrix proteins, including OC, into smaller fragments. It is worth noting here that rodents have sufficient vitamin K for all tissue functions due to the ability to synthesize vitamin K in the gut, whereas humans rely on a diet rich in green vegetables and plant oils to maintain adequate vitamin K status (Booth and Rajabi 2008). In humans, the carboxylation status correlates with vitamin K intake, and the level of undercarboxylated OC (uOC) is used to determine vitamin K deficiency.

The fraction of newly synthesized OC that is released into the circulation can be measured by immunoassay. Serum levels of OC correlate well with bone formation and osteoblast number as determined by histomorphometry and bone mineral densitometry measurements (Brown et al. 1987, 1984; Eastell et al. 1998). Intact OC represents approximately one-third of circulating OC in human adult blood, with one-third made up of a large N-terminal-mid molecule fragment and one-third several small fragments (Garnero et al. 1994). At room temperature, a significant fraction of intact OC is converted into the N-mid fragment reflecting the labile nature of the intact molecule. Similar heterogeneity can be expected in blood from the various animal species and highlights the importance of standardizing collection procedures to ensure appropriate comparisons can be made between groups and temporally. Human data is reported to be more robust when the N-mid fragment assay is used (measures intact and N-mid fragments); however, assessments in rodents showed intact and N-mid assays gave comparable results with similar precision (unpublished data), possibly reflecting the more rigid conditions under which samples are collected in preclinical studies. In blood, OC has a short half-life and is rapidly cleared by the kidney (Price et al. 1981) and to a lesser extent by the liver (Farrugia and Melick 1986); therefore, changes in kidney and liver function may impact circulating levels of OC and its fragments, an important aspect to consider in the context of a toxicology study. Kits intended for human sample analysis are cross-validated for use with most species, except rats, which require rat-specific kits for sample analysis.

5.2.1.2 Bone Alkaline Phosphatase

Alkaline phosphatases (ALPs) are plasma membrane-bound glycoproteins that catalyze the hydrolysis of phosphate monoesters to release inorganic phosphate (Sharma et al. 2014). These enzymes are usually confined to the cell surface and are widely distributed in nature. Human ALPs can be classified into at least four tissue-specific isoforms or isozymes according to the tissue to be expressed, which include placental ALP (only found in humans and great apes (Fishman 1990)), intestinal ALP (all mammals), germ cell ALP, and a combined group from liver/bone/kidney ALP collectively termed tissue non-specific alkaline phosphatase: TNSALP (Whyte et al. 1995). Bone and liver ALP constitute about 95% of the total ALP activity in human serum, with BAP contributing approximately half (Sharma et al. 2014). The isoforms of TNSALP differ only by posttranslational modifications (N-linked glycosylation) (Szulc and Bauer 2013). Little is known about the physiological function of ALPs in most tissues except that the bone isoenzyme has long been associated with mineralization (Wuthier and Register 1995) and is a hallmark marker signaling bone formation. It is interesting to note that for animals with a skeleton, the structure of bone-specific alkaline phosphatase (BAP) contains one calcium atom in the metal-binding site (in addition to two zinc atoms and one magnesium), but where there is no skeleton to mineralize a calcium site does not exist (Mornet et al. 2001). BAP hydrolyzes inorganic pyrophosphate which is a natural inhibitor of bone mineralization. BAP also provides inorganic phosphate (from pyrophosphate and organic phosphomonoesters) for the synthesis of hydroxyapatite (Hessle et al. 2002; Wennberg et al. 2000). ALP is the major enzyme-hydrolyzing pyrophosphate, although phosphate transport is regulated by a number of other enzymes. High total ALP levels are associated with bone, liver, and other diseases; therefore, total ALP levels should be interpreted in the context of other tissue-specific enzymes before attributing changes to the bone isoform. For example, high total ALP levels could be due to bile duct obstruction or could reflect active bone formation or disease that affects blood calcium levels (such as hyperparathyroidism) (Sharma et al. 2014). Low levels or absence of total ALP could reflect bone disease also, such as hypophosphatasia, or other diseases such as hypothyroidism, achondrodysplasia, or anemia, for example (Sharma et al. 2014). Mice with null mutations for TNSALP show increased osteoid and defective growth plate development underscoring the importance of BAP in mineralization (Fedde et al. 1999).

Several isoforms of BAP exist in humans, as differentiated by HPLC: B/I (70% of bone ALP and 30% of intestinal ALP), B1, B2, and B1x (Linder et al. 2009). The B1x isoform is present in normal bone tissue but is detectable in serum only in patients with severe renal insufficiency (Magnusson et al. 1999, 2002b). The bone isoforms have similar enzymatic activity and only differ by the carbohydrate and sialic acid content (Magnusson and Farley 2002). B1 and B2 ALP activity is higher in trabecular bone than cortical; B1 activity is approximately twofold higher than B2 in cortical bone, while B2 activity is approximately twofold higher activity than B1 in trabecular bone (Magnusson et al. 1999, 2002b). These isoforms are also found in various animal species, including mice (Linder et al. 2013), which have identical BAP isoforms as reported for humans. The variability in catalytic activity among these isoforms, therefore, seems to extend across mammalian species, which is an important observation when interpreting and translating animal data to humans. Worthy of note is that although the main sources of ALP in humans are bone and liver, in mice liver ALP could not be detected (intestinal isoforms were detected) (Linder et al. 2013); therefore, measuring ALP in mice is not useful when studying hepatocytes and liver function.

Despite the existence of several ALP isoforms, total ALP remains a useful and sensitive surrogate marker of bone formation when other enzymes reflecting liver function are normal. Where bone-specific ALP assays are available then these should certainly be used to confirm data, but currently no reliable assay exists for rodents; therefore, total ALP, as part of a panel with other bone biomarkers, is often used to assess bone formation activity. Human bone-specific immunoassay kits based on monoclonal antibodies are commercially available for sample analysis. However, a low level of cross-reactivity does occur between the bone and liver isozymes since they have identical protein cores and differ only by posttranslational glycosylation (Magnusson et al. 2002a, report). Currently, available monoclonal/immunoassays for human bone ALP have no affinity for rodent bone. Other assays have been used to test rodent samples, for example, the wheat-germ lectin precipitation assay (Rosalki 1994), but this assay showed spurious results under certain conditions (unpublished data) and was therefore considered unreliable for use in regulated studies. Human immunoassays for BAP have been successfully cross-validated for use in other laboratory animal species and are highly recommended as part of a panel of bone marker assays. Separation of isozymes can also be performed using standard electrophoresis techniques.

5.2.1.3 Procollagen Type I N- Propeptides and C-Terminal Propeptides

Collagen type I makes up 90% of the bone protein matrix and is synthesized by osteoblasts as an immature procollagen molecule with propeptide extensions which are subsequently cleaved during maturation, before assembly into fibrils. Both N-terminal (the short signal peptide) (PINP) and C-terminal (terminal extension peptide) (PICP) peptides of procollagen are found in the circulation and used as markers of bone formation; however, the N-terminal is the most widely used. In bone, type I collagen acts as a scaffold and binds to other proteins that initiate hydroxyapatite deposition. Type I collagen is the predominant collagen of the skin, tendon, and bone. As the metabolism of these tissues is normally slower than that of bone, most of the circulating serum PICP and PINP is considered to originate from bone (Szulc and Bauer 2013). Conditions where there is active fibrosis of the liver, lungs, or heart, however, may contribute significantly to the circulating levels of PICP and PINP (Maceira et al. 2002; Querejeta et al. 2004). Assays directed at the cleaved procollagen molecules in serum therefore normally provide a specific measurement of collagen maturation and bone formation. Clinically, PINP is considered the most sensitive marker of bone formation and has proven useful to monitor effects of compounds directly targeting bone turnover. Both PINP and PICP are degraded by the endothelial cells of the liver but are taken up by different receptors. PICP is taken up by the mannose receptor and PINP by the scavenger receptor (Smedsrød et al. 1990; Melkko et al. 1994). These differences in the metabolic regulation of their catabolism may be related to their clinical utility. While PINP is a sensitive marker of increases in bone turnover attributed to menopause, PICP is more sensitive to longitudinal growth in children (Tapanainen et al. 1997; Szulc et al. 2000). In either case, impaired liver function may impact on the level of circulating propeptides.

Structural differences are noted for PICP and PINP. Following cleavage, PICP is stabilized by disulfide bonds between the chains and thus retains its trimeric structure, whereas the trimeric PINP is unstable at 37 °C and is quickly transformed to stable monomers (low molecular weight peptides of α-1 and 2 chains) (Melkko et al. 1996). The intact trimeric form is measured by a specific assay and shows increases to a similar extent as ALP and OC in response to natural or surgical menopause (Peris et al. 1999). Serum PINP levels increase rapidly in response to bone formation-stimulation and is the marker of choice to monitor anabolic therapies (Glover et al. 2009; Tsujimoto et al. 2011). Early increases in PINP significantly correlated with subsequent increases in bone mineral density (Chen et al. 2005). PICP failed to show increases similar to other bone formation markers after the menopause and levels did not correlate with subsequent bone loss (Hassager et al. 1993), limiting the utility of this marker to clinically monitor osteoporosis bone disease and therapy. PICP has, however, been shown to significantly correlate with longitudinal growth in children and adolescents and is reported to be useful for monitoring growth hormone (GH) in children and in GH deficiency states or with osteogenesis imperfecta (Tapanainen et al. 1997; Szulc et al. 2000; Marini et al. 2003). In addition to endocrine factors such as GH, IGF-1, sex steroids, and leptin, chondrocyte proliferation is regulated by a multitude of paracrine factors including bone morphogenetic protein gradient or C-type natriuretic factor. PICP may prove useful as a marker of bone growth in conditions where these paracrine factors may also be affected.

PINP assays have been successfully validated and are in use for most species routinely used for preclinical bone safety assessments, except for mini-pigs (Table 5.1). PINP is a valuable marker to add to a panel of markers during skeletal safety assessments based on rapid changes observed in response to various classes of compounds affecting bone turnover and bone formation specifically, important assay considerations with respect to stability, and sample volume requirements and has shown no diurnal variation (Luchavova et al. 2011). Selection of PICP as a marker of bone metabolism should be done in the context of what is known regarding the mechanism of action of the compound under test and the age of the test system.

5.2.2 Bone Resorption Markers

5.2.2.1 Tartrate-Resistant Acid Phosphatase

Tartrate-resistant acid phosphatase (TRACP) is a family of enzymes synthesized in bone, spleen, and lungs; the isoform 5b (TRAP5b) is specifically produced and released into circulation by the osteoclast (Hayman et al. 2000; Oddie et al. 2000). It is the most well-characterized enzymatic marker only produced by osteoclasts (Hayman et al. 2000; Halleen et al. 2001) and is considered to reflect osteoclast number and metabolic activity. Selection of this marker as part of a panel of bone markers may be relevant for specific drug classes which are thought to inhibit collagen type I breakdown, while osteoclasts remain viable, as with cathepsin K (CatK) inhibitors such as odanacatib. The selective inhibition of CatK leaves the osteoclasts intact but unable to degrade type I collagen; therefore, type I collagen fragments decline but TRAP5b levels remain stable (Eisman et al. 2011). However, the changes observed with this marker following menopause, osteoporosis therapy, or in bone diseases characterized by high turnover are not always consistent with changes observed with other bone resorption markers measuring collagen type I degradation, suggesting it is less sensitive as a bone turnover marker than other resorption markers (Hannon et al. 1998; Nenonen et al. 2005; Tähtelä et al. 2005; Eisman et al. 2011). Notably, TRAP5b activity measured by immunoassay is not influenced by liver or kidney function. TRAP5b is inactivated rapidly during circulation and degraded into fragments, before clearance from the circulation through the liver.

Therefore, renal dysfunction does not affect enzymatically active intact TRAP5b levels. Even in liver failure, inactive fragments accumulate in blood, while enzymatically active intact TRAP5b molecules do not accumulate (from TRAP5b Technical Bulletin n.d., Human (Quidel®)).

TRAP5b has several procedural issues that may limit its use. At room temperature, the enzyme loses more than 20% of its activity per hour and is unstable in frozen samples, and in blood it may be bound to alpha 2-macroglobulin which influences its enzymatic activity and deteriorates its binding with assay antibodies (Brehme et al. 1999). Since platelets release acid phosphatase during clotting, serum levels are normally greater than plasma levels. Nonetheless, use of this marker under rigid conditions in a preclinical regulated setting has yielded important data; therefore, it is often used in a panel of bone markers or as a specific PD marker for molecules such as CatK inhibitors.

5.2.2.2 Collagen Pyridinium Crosslinks, N-Telopeptides, and C-Telopeptides

Pyridinoline (PYD) and deoxypyridinoline (DPD) are two nonreducible pyridinium crosslinks formed during the extracellular maturation of collagen, creating interchain bonds stabilizing the molecule in the extracellular matrix (Szulc and Bauer 2013). The highest PYD content is found in articular cartilage, while DPD content is minimal in this tissue. Both are found in the tendon, dentin, and aorta but are absent from skin. The formation of PYD and DPD is time-dependent; therefore, their content increases with age and is higher in old interstitial bone than in younger osteonal bone (Nyman et al. 2006). PYD and DPD are released during the breakdown of mature crosslinked collagen and excreted in urine as free (estimated to be approximately 40–45%) and peptide-bound forms. The proportion of free crosslinks is twofold lower in serum than urine and renal clearance is fourfold higher for free than for peptide-bound crosslinks (Colwell and Eastell 1996). This suggests that crosslink-containing peptides are taken up by tubular cells where they are cleaved to free forms and then re-excreted into the tubular lumen and excreted in urine, a process that appears to be rate-limiting in high turnover states (Garnero et al. 1995; Colwell and Eastell 1996). Once released from bone by osteoclastic resorption, DPD is not reused and is not further metabolized. Levels in bone, blood, or urine are not influenced by dietary sources nor by degradation of newly formed collagen during synthesis (Szulc and Bauer 2013). They do, however, display circadian rhythms (Greenspan et al. 1997). Various factors modulate the circadian rhythm including fasting and food intake, among others (Clowes et al. 2002; Qvist et al. 2002). In humans, bone resorption markers rise overnight and show declines early to midmorning; the amplitude of this decline is most notable following food intake. The effect on bone formation is weak. The postprandial decrease in bone resorption is likely mediated by glucagon-like peptide 2 which is stimulated by food intake (Henriksen et al. 2003). Similar rhythms in bone markers have been reported for various animal species but these are not as extensively documented. To minimize variability and ensure data are consistently obtained under experimental conditions, sample collection should be rigidly controlled with animal's food deprived overnight or several hours prior to sample collection; samples should be obtained during a specific time interval for all groups on all occasions. Measurement of total PYR and DPD initially involved high-performance liquid chromatography (HPLC) following sample hydrolysis and extraction (Fraser 1998); however, these HPLC assays have largely been replaced by immunoassays which measure free crosslinks. Currently, immunoassays for free DPD in urine and related peptide fragments N-telopeptides (NTx) and C-telopeptides (CTx, ICTP) measured in blood or urine are the most widely used markers of bone resorption. When measured in urine, these analytes (free DPD, NTx, and CTx) are corrected for urinary creatinine content using the same sample. They are relatively stable in urine and do not require any specific collection requirements other than clean procedures to avoid fecal contamination. These analytes are highly specific for bone since other DPD-containing tissues represent a relatively minor source compared to bone and are not remodeled at rates as high as bone.

NTx in serum and urine is measured by an immunoassay based on the antibody raised against an epitope on the α-2 chain of type I collagen (Hanson et al. 1992). Digests of skin collagen show similar reactivity in this assay. This analyte is stable at room temperature and during freeze-thaw cycles. Notably this assay cannot be used in rodents; mice and rats show no consistent cross-reactivity in either serum or urine. Other animal species show excellent sensitivity and specificity when NTx is measured in urine, although serum assays are often less robust and data generally has less discriminatory power (unpublished data).

Matrix metalloproteinases (MMPs) produce carboxyterminal telopeptide of type I collagen (ICTP), which is further degraded by CatK to release CTx (Garnero et al. 2003; Sassi et al. 2000). In parallel, CatK releases NTx at the N-terminal end. Since ICTP is promptly degraded to CTx, CTx and/or NTx are normally the resorption markers of choice unless CatK activity is affected, either genetically or with CatK inhibitors such as odanacatib. CatK inhibitors decrease levels of CTx and NTx, but not ICTP (Masarachia et al. 2012). ICTP has also been shown to be a useful marker of bone resorption in metastatic bone cancer (prostate or lung cancer, multiple myeloma with osteolytic lesions) where MMP levels are increased (Kong et al. 2007; Lein et al. 2009). MMPs facilitate spread by stimulating cytokines involved in the metastatic process and degradation of basement membranes of blood vessels (Lynch 2011).

CTx is measured by an immunoassay based on the antibody raised against an epitope of the α-1 chain of type I collagen. The amino acid sequence of this epitope contains lysine (K) involved in a trivalent crosslink and the aspartate-glycine sequence. This sequence may undergo β-isomerization and racemization generating four isomers: the native form and three age-related forms. These processes are related to the maturation and aging of collagen. Immunoassays were therefore developed to recognize newly synthesized bone and progressively older bone (Cloos and Fledelius 2000). Used together, the ratio of α/β-CTx in bone can provide information useful in the diagnosis of a defect in the mineralization process or to follow treatment. Immunoassays use antibodies against molecules containing two α-CTx chains (αCrossLaps) typical of native new collagen or two β-CTx chains typical of aged collagen. Only the β-isomer of CTx is measured in the serum CrossLaps assay, while α- and β-isomers of CTx are measured in the urine CrossLaps assay; β-isomer is measured in the urine β-CrossLaps assay. While the reader should be aware of the sophisticated markers currently available, these new assays may be useful clinically but they have yet to find a place in preclinical use for routine bone safety testing.

Clinically, CTx can be measured in both serum and urine with similar outcome. Often the choice of matrix in a preclinical setting may be directed by sample availability and the robustness of the assay. Needless to say, blood sample volume in small lab animals may be limiting and urine a good option. On the other hand, it is often useful to have multiple markers measured using the same sample. This has become more feasible with the use of Luminex multiplex-based assay systems where a panel of markers can be obtained from a single sample. Clinically, resorption markers are normally measured in serum primarily due to the ease of sampling. In preclinical studies, it can also be useful to mimic what is intended to be done clinically. Where distinct differences in outcome and robustness between serum and urine assays used preclinically have been established, based on discrimination or the working range, then the most sensitive assay is recommended. Serum and urinary CTx assays seem to perform equally well in most lab animal species.

One other bone resorption marker to mention which has been used clinically is the alpha1 helicoidal peptide 620–633 of type I collagen which correlates well with CTx (Garnero and Delmas 2003). Although it shows similar activity and specificity as CTx, there appears to be no advantage to adding this marker routinely when established markers such as CTx are available and used widespread both clinically and preclinically.

5.3 Guidelines for the Use of Bone Markers in Preclinical Studies

A panel of bone markers, consisting ideally of two formation and two resorption markers, provides the most robust evaluations, allowing the "body of evidence" to direct interpretation; relying on data from a single marker can be limiting. In short-term or proof-of-concept studies, markers can be used as the primary output measure to provide information on mechanism. Whenever possible though, bone marker assays should be combined with bone densitometry measurements to fully interpret both datasets.

Marker data is most reliable from a well-powered study with data generated ideally from at least eight samples per analyte. The validation of the assay and well-controlled procedures for sample collection are likely the most critical steps to a successful outcome. Use of quality control standards and generation of reference values by species/strain, sex, and age, where appropriate, are valuable additional tools. Standard criteria for assay validation apply to establish sensitivity, specificity, stability (room temperature and freeze-thaw cycles), linearity, and precision, both inter- and intra-assay. In addition, a robust assay must also be validated to assess the appropriate biological response. Assays that have passed all the standard validation criteria have been observed to fail to discriminate the appropriate biological response. The reasons for this are normally not known but likely relate to factors within the matrix interfering with the assay. An example of this was seen recently with the use of an estradiol assay (Doyle et al. 2016). Although estradiol is not considered a bone marker, the lesson is learned nonetheless. Two assays, both meeting all requirements for standard validation, were used to compare results using the same samples for analysis. The traditional radiometric estradiol assay showed a clear separation between ovarian intact female nonhuman primates and samples from ovariectomized females, as well as gonad intact males. Estrogen peaks in ovulating animals could easily be seen. However, an immunoassay under test to move away from radiometric methods, while passing all routine assay criteria, failed to discriminate differences between these populations.

Under laboratory conditions it is far easier to maintain animals on the same diet and under the same housing conditions without influence from other drug treatments, etc. than it is for humans. Preclinical safety studies are not prone to the same compliance issues as clinical studies where treatment may be sporadic. Samples from experimental animals can be obtained under the same fasted conditions and at the same time of day, minimizing experimental error. Samples analyzed under these rigid conditions can be reliable and provide definitive data, especially if the study is well powered. Marker data from rat studies are normally relatively homogeneous allowing elimination of pre-study sampling and facilitating cross-sectional interpretation of data. Longitudinal assessments are typically required for larger animal species. Large animal safety studies usually have smaller group sizes, with only three to five animals/sex/group. There may be sex differences, so data should be analyzed by gender separately and combined as considered appropriate. Data from non-rodent species are normally more heterogeneous than rat data and require robust pre-study sampling to establish baseline levels to monitor changes over time for individual animals. Multiple sampling occasions are often needed to provide definitive data. Bone marker levels also decline with age, even in the context of relatively short-term studies, depending on species and the age of the test system at the start of treatment. Blood samples need to be processed promptly and consistently; most are used to harvest serum which can be frozen in aliquots for subsequent analysis, depending on stability. This allows samples to be obtained for multiple analyses that can be decided upon based on initial outcome measures. Normally urine samples from rats can be collected overnight by jar (at ambient temperature), and morning samples can be obtained similarly from non-rodent species, or directly from the bladder by cystocentesis, if the study setting permits. Blood and urine samples are normally collected under fasted conditions. Where studies include injection of fluorochrome labels, samples should be collected before labels are given, or at least 2 days afterward to allow adequate clearance since the fluorochrome may interfere with the assay performance, depending on the assay.

Knowledge of the PK profile of the drug under test can aid in the study design to determine the most appropriate time for blood/urine sampling for bone markers, depending on the drug class. These data may identify a "window of opportunity" to optimize sample collection to capture effects on markers that may be associated with drug blood levels, especially where small molecules are metabolized relatively quickly. Markers analyzed using the same PK samples can be used to develop a PK/PD profile which is helpful to determine the timing for sample collection. It may also provide information on the most sensitive marker which can be applied in future studies, without the need to repeatedly analyze a full panel of markers in all studies.

Bone markers can be useful in dose range finding studies to set dose levels for subsequent definitive studies or to determine optimal dose levels to monitor a pharmacological effect. They have successfully been employed to assess the efficacy of biosimilar molecules, for example, PTH analogues. Bone turnover markers can also be used to assess recovery from treatment effects, for example, where turnover rates are elevated or may be reduced or undetectable; the return of marker levels to within the expected range can signal recovery of bone turnover. This strategy has proven effective, for example, in monitoring recovery of bone from cytotoxic agents under test for cancer treatment. Similarly, they are extensively used to monitor for the development of anti-drug antibody (ADA), where a specific bone PD marker may reflect loss of pharmacological activity due to antibody production.

5.4 Use of Additional End-Points

5.4.1 Clinical Pathology Parameters

There are several additional analyses that can be included in a study when an effect on bone has been determined or is suspected. Routine clinical pathology end-points are included in most drug development safety studies; therefore, basic information on calcium and phosphorus in blood and urine are normally obtained and provide useful information for the interpretation of the bone markers. Additional parameters may be included to measure fractional excretion of these and other electrolytes, as appropriate. Magnesium should be included in the list of clinical chemistry parameters as well as fluoride or iron, depending upon what is known of the mechanism of action of the compound, or its structure.

5.4.2 Calciotropic Hormones and Fibroblast Growth Factor 23

Once changes in biochemical markers of bone turnover are established, understanding the mechanism driving these changes can be facilitated by the analysis of the calciotropic hormones: parathyroid hormone (PTH), 1,25 dihydroxyvitamin D3 and calcitonin, and the phosphate-metabolizing hormone: fibroblast growth factor 23

Table 5.2 Additional assays used in preclinical studies

Activity	Assay	Abbreviation	Matrix	Species	Assay type	Storage	Comment
Calciotropic Hormones	Parathyroid hormone	PTH(1–84)	Serum	Rat	Luminex	−80 °C	Calcium and phosphate regulation
				Mouse	ELISA		
				NHP	ELISA		
				Rabbit	ELISA		
	1,25-dihydroxyvitamin D_3	1,25 vitamin D	Serum	Rat	RIA	−20 °C	Active form of vitamin D
				NHP			Also called calcitriol
				Dog			Acts in concert with PTH for calcium regulation
							Decreased by FGF23
	25-hydroxyvitamin D_2	25-OH vitamin D	Serum	Rat	RIA	−20 °C (mouse: fresh)	Substrate for 1-alpha-hydroxylase in the kidney, converted to active vitamin D_3
				Mouse			
				NHP			
				Dog			
	Calcitonin	Calcitonin	Serum	Rat	ELISA	−20 °C	Reduces serum calcium
				NHP			Opposes effects of PTH
Other	Fibroblast growth factor 23	FGF23	Serum	Rat	ELISA	−20 °C	Bone-derived hormone
				Mouse			Phosphate regulation
				NHP			
	Insulin-like growth factor-1	IGF-1	Serum	Rat	ELISA	−20 °C	Reflects growth
				Mouse			Binds to several binding proteins that affect activity
				NHP			

NHP nonhuman primates, *EIA* enzyme immunoassay, *ELISA* enzyme-linked immunosorbent assay, *RIA* radioimmunoassay

(FGF23) (Table 5.2). Cross talk between the kidneys, intestines, bone, and parathyroid gland is key to maintaining mineral homeostasis. Calcium and phosphate are tightly regulated by these hormones, and these hormones also regulate each other (Rowe 2015; Penido and Alon 2012). The reader is referred to the Kidney-Bone chapter for a detailed explanation of calcium/phosphate metabolism and the action of these hormones. Under normal physiological conditions, the net effect of these bone- and kidney-derived regulators, as well as PTH and calcitonin, will lead to stable serum calcium and phosphate levels, while also providing calcium and phosphate pools for bone mineralization.

5.4.3 *Insulin-Like Growth Factor-1*

The safety assessment of potent new molecules from therapeutic areas including diabetes and obesity that affect energy metabolism has led to new challenges in drug development. These include, for example, the fibroblast growth factor 21 (FGF21) mimetics and sodium-glucose transporter (SGLT1/2) inhibitors. Use of these molecules in a toxicology setting where supra-pharmacological doses are used can lead to complex findings attributable in large part to exaggerated pharmacology. Maintaining animals under study in an appropriate nutritional state can become dose-limiting due to the effects of starvation and malnutrition. The skeleton is a very metabolically active tissue and consequently a target tissue. Additional end-points in studies using these drug classes benefit from the addition of markers that not only reflect metabolism in general (such as the lipid profile, insulin, and glucose levels), but it can also be important to include markers that are mediators of skeletal growth such as insulin-like growth factor-1 (IGF-1) and its binding proteins (Table 5.2). IGF-1 is a nutrient sensor which is modulated by available energy; in scenarios such as starvation, circulating and skeletal IGF-1 are reduced and bone formation slows. Circulating IGF-1 is primarily generated by the liver in response to growth hormone, although virtually all tissues have been demonstrated to express IGF-1. IGFs are bound in the skeletal matrix and are one of the most abundant noncollagen proteins. As such, matrix IGF-1 is released during bone resorption and may play a role in the activation and proliferation of mesenchymal stem cells. IGF-1 plays a role in skeletal development and mineral acquisition, in mitochondrial function during chondrocyte development and function, as well as osteoblast proliferation and differentiation, and may act as a coupling factor in remodeling and be required for normal osteoclast differentiation (Guntur and Rosen 2013). Global loss of IGF-1 affects growth and skeletal gains at all ages, resulting in short stature and slender and weaker bones. Longitudinal studies have demonstrated the importance of serum IGF-1 in transverse bone growth and periosteal bone apposition (Govoni et al. 2007), as well as in bone adaptation to increases in body weight. In mice, IGF-1 levels have been associated with specific bone phenotypes: low levels in C57B6 mice are associated with lower bone density than C3H mice with higher circulating levels of IGF-1. Liver-specific deletion of IGF-1 resulting in reduced serum IGF-1 levels has shown that the decrease had minimal impact on trabecular bone but decreased cortical bone and bone strength (Guntur and Rosen 2013). Differential effects of IGF-1 have been also reported with respect to age. Mouse studies suggest serum IGF-1 is essential for bone accrual during the postnatal growth phase but depletion after peak bone acquisition in the older adult mouse (age 16 weeks) is compartment specific and does not have a detrimental effect on cortical bone mass (Courtland et al. 2011). These findings may have important implications for the interpretation of data generated from safety studies in animals of varying ages and in different species.

5.4.4 Bone Densitometry

Bone densitometry techniques allow a direct in vivo measurement of bone mass. Measurement of bone mass (bone mineral content (BMC) and bone mineral density (BMD)) are excellent end-points to assess skeletal effects of treatment although they provide no information regarding mechanism. Measurement of biochemical markers of bone turnover provides some mechanistic insight as to how changes in bone mass occur. Bone densitometry measurements reflect the net effect of changes in bone markers over time. While biochemical markers reflect real-time bone turnover, changes in bone densitometry develop over time, and the length of time depends on the test system species, age, the duration of treatment, and the site measured. In vivo bone densitometry measures are generally only useful in studies of 28-day duration, or longer, and are performed under anesthesia. Bone densitometry can also be performed ex vivo on specimens retained at necropsy. Provision can be made in the protocol to perform this activity depending on the primary outcome measures (markers and clinical pathology).

Bone densitometry is typically performed using dual energy X-ray absorptiometry (DXA) and/or peripheral quantitative computed tomography (pQCT). For further information on these and other imaging techniques, the reader is referred to the following chapter.

5.4.5 Markers of Bone Quality

In addition to markers of bone turnover, future bone safety assessments may need to look to markers of bone quality. This may be relevant during the late stages of drug development or as a post-marketing activity. Bone turnover markers can speak to bone safety in a preclinical setting, while markers of bone quality that contribute to bone strength might add a new dimension. Such markers currently available include analysis of the posttranslational modifications of type I collagen, such as the degree of crosslinking (e.g., see Saito et al. 2015), β-isomerization, and accumulation of advanced glycation end-points (AGEs) (Boskey and Robey 2013), the tissue mineral density (assessed by micro-CT; e.g., see Ominsky et al. 2017), the ratio of urinary α/β CTx (Hoshino et al. 1999), and serum levels of undercarboxylated OC (Knapen et al. 2000). The importance of other aspects of bone quality may include measurement of sclerostin levels and other measures of osteocyte function. Given our understanding that the osteocyte, the most prolific bone cell accounting for 95% of bone cells, is responsible for the mechanotransduction of signals and the orchestration of osteoblast lineage cell and osteoclast function, this could be an important area of new growth.

5.5 Interpretation of Bone Turnover Markers and Bone Mass

In a controlled and well-powered study, bone turnover markers can be used to "predict" the effects of the compound on bone mass or bone quality (where measures of bone densitometry are not included initially). Importantly, it is the net effect of a compound on the balance of bone resorption and bone formation that determines the outcome. Tables 5.3A and 5.3B provide a simplified guide for the interpretation of changes in bone markers and the potential impact on BMC and/or BMD in the juvenile and mature skeleton, respectively. Decreases in bone formation markers with no effect on bone resorption (or increases in bone resorption markers) would be consistent with bone loss and/or decreases in gain in bone mass (in growing animals). On the other hand, increases in bone formation markers with no effect on bone resorption markers (or decreases in resorption markers) would be consistent with an increase in bone mass. When bone densitometry data is available, the changes in bone markers provide some mechanistic insight. In rapidly growing animals, increases in bone mass (bone mineral content, BMC) could be attributed to an increase in growth and size of the skeleton (with changes consistent with body weight effects). Under normal physiological conditions in bone remodeling, bone resorption and formation are coupled, meaning a decrease in one will be followed by a decrease in the other (or an increase in one will follow with an increase in the other), highlighting the importance of the selected time points for bone marker analyses and also the net effect on bone turnover. In a high bone turnover state, where both bone resorption and bone formation markers are increased, if the increases in bone formation occur first and/or the magnitude of the increases in bone formation is greater than the increases in bone resorption, the net effect on bone mass will likely be positive. If the increases in bone resorption occur first and/or the magnitude of the increases in bone resorption is greater than the bone formation, the net effect on bone mass will likely be negative. PTH is a classic example which results in a high bone turnover state with two different outcomes on bone mass and bone quality depending on the frequency of dosing. Daily injections of PTH (intermittent dosing) result in increases in bone mass (BMC and BMD), whereas continuous administration by infusion results in bone loss and decreases in bone quality. On the other hand, in a low bone turnover state, where bone resorption and bone formation

Table 5.3A Interpreting changes in biochemical markers in preclinical studies

	Change		Anticipated effect on bone mass					
			Relative to baseline		Relative to controls			
Mature skeleton	BF	BR	BMC	BMD	BMC	BMD	Fragility	Reference
High bone T/O/ anabolics	↑	–	↑	↑	↑	↑	↓	Abaloparatide: Varela et al. (2017)
	–	↑	↓	↓	↓	↓	↑	No reference
	↑	↑	–	↓/–	–	↓/–	↑/–	No reference
	↑	↑	↑	↑	↑	↑	↓	PTH(1–84):Fox et al. (2006)
	↑	↑	↓	↓	↓	↓	↑	PTH(1–34) Infusion: Samadfam et al. (2008)
Low bone T/O/ antiresorptives	↓	–	↓	↓	↓	↓	↑	No reference
	–	↓	↑	↑	↑	↑	↓	Eldecalcitol: Takeda et al. (2015)
	↓	↓	↑	↑	↑	↑	↓	Denosumab: Ominsky et al. (2011); SGLT1/2 inhibitors: Samadfam et al. (2013)
	↓	↓	↓/–	↓/–	↓	↓	↑	No reference
	↓	↓	↑	↑	↑	↑	↓	Bisphosphonates: Smith et al. (2003)
Uncoupled	↑	↓	↑	↑	↑	↑	↓	Sclerostin antibody: Schett and Bozec (2014)
	↓	↑	↓	↓	↓	↓	↑	Rosiglitazone: Kumar et al. (2013);
								FGF21: Wei et al. (2012)
								Glucocorticoids

markers are decreased, or decreases in bone resorption occur first, the net effect on bone mass will likely be positive. Clinically available antiresorptive compounds are good examples of compounds that decrease bone turnover with a positive outcome on bone mass (Samadfam et al. 2007). However, for a compound with an unknown effect on bone, as frequently occurs for compounds tested in toxicology studies, changes in bone turnover markers may not always be predictive of changes in bone mass. The changes in serum (in particular calcium and phosphorus) and urine biochemistry parameters, body weight (growth), and hormones (if available) are useful to facilitate interpretation of these data. In these scenarios, the relationship and

Table 5.3B Interpreting changes in biochemical markers in preclinical studies[a] (cont'd)

Juvenile skeleton	Change		Anticipated effect on bone mass						Fragility
			Relative to baseline			Relative to controls			
	BF	BR	Size	BMC	BMD	Size	BMC	BMD	
High bone T/O/anabolics	↑	–	↑	↑	↑	↑	↑	↑	↓
	–	↑	↑	↑	↓/–	–	↓	↓	↑
	↑	↑	↑	↑	↑	↑/–	↑/–	↓/–	↓/–
	↑	↑	↑	↑	↑	↑	↑	↑	↓
	↑	**↑**	↑/–	↑/–	↓/–	↑/–	↓/–	↓	↑/–
Low bone T/O/ antiresorptives	↓	–	–	↓	↓	↓	↓	↓	↑
	–	↓	↑	↑	↑	↑/–	↑	↑	↓
	↓	↓	↑/–	↑/–	↑/–	↓	↓	↑/–	↓/–
	↓	↓	↑/–	↑/–	↑/–	↓	↓	↑	↓
	↓	**↓**	↑/–	↑	↑	↓	↑	↑	↓
Uncoupled	↑	↓	↑	↑	↑	↑	↑	↑	↓
	↓	↑	–	↓	↓	↓	↓	↓	↑

BF bone formation, *BR* bone resorption, *BMC* bone mineral content, *BMD* bone mineral density, *Bone T/O* bone turnover

↑, increase; **↑**, greater increase; ↓, decrease; **↓**, greater decrease; – normal level, i.e., comparable to baseline levels or concurrent age-matched control

[a]Relationships are explored in this table to indicate the potential outcome on bone mass in scenarios where bone markers are affected. The reader should be aware that the outcomes depicted will be affected by imbalances in calcium and phosphorus levels and in the calciotropic hormone levels (PTH, vitamin D3, and FGF23). This table is not intended to cover all scenarios. The outcome will also depend on the species, site selected, and the type of bone affected (trabecular vs cortical or both).

correlations between changes in bone turnover and bone mass become even more complex if the compound has an effect on kidney or liver function that may influence the clearance of biomarkers. The magnitude of the changes in bone turnover markers depends on the mechanism of action, degree of inhibition of bone resorption and/or degree of activation of bone formation, the duration, and route of compound administration.

5.6 Conclusions

Biochemical markers of bone turnover provide a noninvasive, real-time, rapid, and sensitive clinically translatable tool that can be used to discriminate effects of a compound on the skeleton. Blood and/or urine samples can be collected for analysis for all preclinical safety study types, irrespective of species, duration, and route of compound administration. Markers provide mechanistic insight during drug development and are often used as a screening tool to determine whether more in-depth analyses of bone are required. In a chronic setting, analysis of samples can be performed on multiple occasions, providing important information on the time-course of any effects. Used as a pharmacodynamic marker, they can also signal issues arising from the development of anti-drug antibody or can be used to monitor recovery or progression of changes during an off-treatment period.

References

Aliberti G, Pulignano I, Proietta M, Tritapepe L, Cigognetti L, Menichetti A, Russo A, de Michele LV, Corvisieri P, Minisola S. Osteocalcin metabolism in the pulmonary circulation. Clin Physiol. 2000;20(2):122–5.

Booth SL, Rajabi A. Determinants of vitamin K status in humans. Vitam Horm. 2008;78:1–22.

Boskey AL, Robey PG. The regulatory role of matrix proteins in mineralization of bone. In: Marcus R, Feldman D, Dempster DW, Luckey M, Cauley JA, editors. Osteoporosis. New York: Elsevier; 2013. p. 235–58.

Brehme CS, Roman S, Shaffer J, Wolfert R. Tartrate-resistant acid phosphatase forms complexes with α2-macroglobulin in serum. J Bone Miner Res. 1999;14(2):311–8.

Brown JP, Delmas PD, Malaval L, Edouard C, Chapuy MC, Meunier PJ. Serum bone Gla-protein: a specific marker for bone formation in postmenopausal osteoporosis. Lancet. 1984;1:1091–3.

Brown JP, Delmas PD, Arlot M, Meunier PJ. Active bone turnover of the cortico-endosteal envelope in postmenopausal osteoporosis. J Clin Endocrinol Metab. 1987;64:954–9.

Chen P, Satterwhite JH, Licata AA, Lewiecki EM, Sipos AA, Misurski DM, Wagman RB. Early changes in biochemical markers of bone formation predict BMD response to teriparatide in postmenopausal women with osteoporosis. J Bone Miner Res. 2005;20(6):962–70.

Chenu C, Colucci S, Grano M, Zigrino P, Barattolo R, Zambonin G, Baldini N, Vergnaud P, Delmas PD, Zallone AZ. Osteocalcin induces chemotaxis, secretion of matrix proteins, and calcium-mediated intracellular signaling in human osteoclast-like cells. J Cell Biol. 1994;127(4):1149–58.

Cloos PAC, Fledelius C. Collagen fragments in urine derived from bone resorption are highly racemized and isomerized: a biological clock of protein aging with clinical potential. Biochem J. 2000;345(3):473–80.

Clowes JA, Hannon RA, Yap TS, Hoyle NR, Blumsohn A, Eastell R. Effect of feeding on bone turnover markers and its impact on biological variability of measurements. Bone. 2002;30(6):886–90.

Colwell A, Eastell R. The renal clearance of free and conjugated pyridinium cross-links of collagen. J Bone Miner Res. 1996;11(12):1976–80.

Courtland H-W, Elis S, Wu Y, Sun H, Rosen CJ, Jepsen KJ, Yakar S. Serum IGF-1 affects skeletal acquisition in a temporal and compartment-specific manner. PLoS One. 2011;6(3):e14762.

Doyle, N, S. Cotton, E. Lesage, A. Varela, S. Lavallee and S. Y. Smith. Pitfalls in biomarker analyses: lessons learnt using an ELISA assay to measure estradiol levels in ovariectomized female cynomolgus monkeys. Endocrine Society annual meeting (ENDO2016). Poster session: bench to bedside – female reproductive endocrinology and female reproductive tract. Endocr Rev. 2016;37(2):Abstract FRI 191.

Eastell R, Delmas PD, Hodgson SF, Eriksen EF, Mann KG, Riggs BL. Bone formation rate in older normal women: concurrent assessment with bone histomorphometry, calcium kinetics, and biochemical markers. J Clin Endocrinol Metab. 1998;67:741–8.

Eisman JA, Bone HG, Hosking DJ, McClung MR, Reid IR, Rizzoli R, Resch H, Verbruggen N, Hustad CM, DaSilva C, Petrovic R, Santora AC, Ince BA, Lombardi A. Odanacatib in the treatment of postmenopausal women with low bone mineral density: three-year continued therapy and resolution of effect. J Bone Miner Res. 2011;26(2):242–51.

Farrugia W, Melick RA. Metabolism of osteocalcin. Calcif Tissue Int. 1986;39(4):234–8.
Fedde KN, Blair L, Silverstein J, Coburn SP, Ryan LM, Weinstein RS, Waymire K, Narisawa S, Millán JL, Macgregor GR, Whyte MP. Alkaline phosphatase knock-out mice recapitulate the metabolic and skeletal defects of infantile hypophosphatasia. J Bone Miner Res. 1999;14(12):2015–26.
Fishman WH. Alkaline phosphatase isozymes: recent progress. Clin Biochem. 1990;23:99–104.
Fox J, Miller MA, Newman MK, Metcalfe AF, Turner CH, Recker RR, Smith SY. Daily treatment of aged ovariectomized rats with human parathyroid hormone (1–84) for 12 months reverses bone loss and enhances trabecular and cortical bone strength. Calcif Tissue Int. 2006;79(4):262–72.
Fraser WD. The collagen crosslinks pyridinoline and deoxypyridinoline: a review of their biochemistry, physiology, measurement, and clinical applications. J Clin Ligand Assay. 1998;21(2):102–10.
Garnero P, Delmas PD. An immunoassay for type I collagen α1 helicoidal peptide 620–633, a new marker of bone resorption in osteoporosis. Bone. 2003;32(1):20–6.
Garnero P, Grimaux M, Seguin P, Delmas PD. Characterization of immunoreactive forms of human osteocalcin generated in vivo and in vitro. J Bone Miner Res. 1994;9:255–64.
Garnero P, Gineyts E, Arbault P, Christiansen C, Delmas PD. Different effects of bisphosphonate and estrogen therapy on free and peptide-bound bone cross-links excretion. J Bone Miner Res. 1995;10(4):641–9.
Garnero P, Ferreras M, Karsdal MA, Nicamhlaoibh R, Risteli J, Borel O, Qvist P, Delmas PD, Foged NT, Delaissé JM. The type I collagen fragments ICTP and CTX reveal distinct enzymatic pathways of bone collagen degradation. J Bone Miner Res. 2003;18(5):859–67.
Glover SJ, Eastell R, McCloskey EV, Rogers A, Garnero P, Lowery J, Belleli R, Wright TM, John MR. Rapid and robust response of biochemical markers of bone formation to teriparatide therapy. Bone. 2009;45(6):1053–8.
Govoni KE, Lee SK, Chung Y-S, Behringer RR, Wergedal JE, Baylink DJ, Mohan S. Disruption of insulin-like growth factor-I expression in type IIalphaI collagen-expressing cells reduces bone length and width in mice. Physiol Genomics. 2007;30:354–62.
Greenspan SL, Dresner-Pollak R, Parker RA, London D, Ferguson L. Diurnal variation of bone mineral turnover in elderly men and women. Calcif Tissue Int. 1997;60(5):419–23.
Gunson D, Gropp KE, Varela A. Bone and joints. In: Haschek WM, Rousseaux CG, Wallig MA, Bolon B, Ochoa R, editors. Haschek and Rousseaux's handbook of toxicologic pathology, Systems Toxicologic Pathology, vol. III. 3rd ed. Cambridge: Elsevier; 2013. p. 2761–858.
Guntur AR, Rosen CJ. Review: IGF-1 regulation of key signaling pathways. BoneKEy Rep. 2013;2:437.
Halleen JM, Alatalo SL, Janckila AJ, Woitge HW, Seibel MJ, Väänänen HK. Serum tartrate-resistant acid phosphatase 5b is a specific and sensitive marker of bone resorption. Clin Chem. 2001;47(3):597–600.
Hannon R, Blumsohn A, Naylor K, Eastell R. Response of biochemical markers of bone turnover to hormone replacement therapy: impact of biological variability. J Bone Miner Res. 1998;13(7):1124–33.
Hanson DA, Weis MAE, Bollen AM, Maslan SL, Singer FR, Eyre DR. A specific immunoassay for monitoring human bone resorption: quantitation of type I collagen cross-linked N-telopeptides in urine. J Bone Miner Res. 1992;7(11):1251–8.
Hassager C, Fabbri-Mabelli G, Christiansen C. The effect of the menopause and hormone replacement therapy on serum carboxyterminal propeptide of type I collagen. Osteoporos Int. 1993;3(1):50–2.
Hayman AR, Bune AJ, Bradley JR, Rashbass J, Cox TM. Osteoclastic tartrate-resistant acid phosphatase (Acp 5): its localization to dendritic cells and diverse murine tissues. J Histochem Cytochem. 2000;48(2):219–27.
Henriksen DB, Alexandersen P, Bjarnason NH, Vilsbøll T, Hartmann B, Henriksen EEG, Byrjalsen I, Krarup T, Holst JJ, Christiansen C. Role of gastrointestinal hormones in postprandial reduction of bone resorption. J Bone Miner Res. 2003;18(12):2180–9.

Hessle L, Johnson AK, Anderson HC, Narisawa S, Sali A, Goding JW, Terkeltaub R, Millán JL. Tissue-nonspecific alkaline phosphatase and plasma cell membrane glycoprotein-1 are central antagonistic regulators of bone mineralization. Proc Natl Acad Sci. 2002;99(14):9445–9.

Hoshino H, Takahashi M, Kushida K, Ohishi T, Inoue T. The relationships between the degree of β-isomerization of type I collagen degradation products in the urine and aging, menopause and osteoporosis with fractures. Osteoporos Int. 1999;9(5):405–9.

Ivaska KK, Hentunen TA, Vääräniemi J, Ylipahkala H, Pettersson K, Väänänen HK. Release of intact and fragmented osteocalcin molecules from bone matrix during bone resorption in vitro. J Biol Chem. 2004;279:18361–9.

Jolette J, Wilker CE, Smith SY, Doyle N, Hardisty JF, Metcalfe AJ, Marriott TB, Fox J, Wells DS. Defining a noncarcinogenic dose of recombinant human parathyroid hormone 1-84 in a 2-year study in Fischer 344 rats. Toxicol Pathol. 2006;34:929–40.

Karsenty G. Convergence between bone and energy homeostases: review leptin regulation of bone mass. Cell Metab. 2006;4(5):341–8.

Knapen MHJ, Hellemons-Boode BSP, Langenberg-Ledeboer M, Knottnerus JA, Hamulyák K, Price PA, Vermeer C. Effect of oral anticoagulant treatment on markers for calcium and bone metabolism. Haemostasis. 2000;30(6):290–7.

Kong QQ, Sun TW, Dou QY, Li F, Tang Q, Pei FX, Tu CQ, Chen ZQ. Beta-CTX and ICTP act as indicators of skeletal metastasis status in male patients with non-small cell lung cancer. Int J Biol Markers. 2007;22(3):214.

Kostenuik PJ, Smith SY, Samadfam R, Jolette J, Zhou L, Ominsky MS. Effects of denosumab, alendronate, or denosumab following alendronate on bone turnover, calcium homeostasis, bone mass and bone strength in ovariectomized cynomolgus monkeys. J Bone Miner Res. 2015;30(4):657–69.

Kumar S, Hoffman SJ, Samadfam R, Mansell P, Jolette J, Smith SY, Guldberg RE, Fitzpatrick LA. The effect of rosiglitazone on bone mass and fragility is reversible and can be attenuated with Alendronate. J Bone Miner Res. 2013;28(7):1653–65.

Lee NK, Sowa H, Hinoi E, Ferron M, Ahn JD, Confavreux C, Dacquin R, Mee PJ, McKee MD, Jung DY, Zhang Z, Kim JK, Mauvais-Jarvis F, Ducy P, Karsenty G. Endocrine regulation of energy metabolism by the skeleton. Cell. 2007;130(3):456–69.

Lein M, Miller K, Wirth M, Weißbach L, May C, Schmidt K, Haus U, Schrader M, Jung K. Bone turnover markers as predictive tools for skeletal complications in men with metastatic prostate cancer treated with zoledronic acid. Prostate. 2009;69(6):624–32.

Linder CH, Narisawa S, Millán JL, Magnusson P. Glycosylation differences contribute to distinct catalytic properties among bone alkaline phosphatase isoforms. Bone. 2009;45:987–93.

Linder CH, Englund UH, Narisawa S, Millán JL, Magnusson P. Isozyme profile and tissue-origin of alkaline phosphatases in mouse serum. Bone. 2013;53(2):399–408.

Luchavova M, Zikan V, Michalska D, Raska I, Kubena AA, Stepan JJ. The effect of timing of teriparatide treatment on the circadian rhythm of bone turnover in postmenopausal osteoporosis. Eur J Endocrinol. 2011;164(4):643–8.

Lynch CC. Matrix metalloproteinases as master regulators of the vicious cycle of bone metastasis. Bone. 2011;48(1):44–53.

Maceira AM, Barba J, Varo N, Beloqui O, Díez J. Ultrasonic backscatter and serum marker of cardiac fibrosis in hypertensives. Hypertension. 2002;39(4):923–8.

Magnusson P, Farley JR. Differences in sialic acid residues among bone alkaline phosphatase isoforms: a physical, biochemical, and immunological characterization. Calcif Tissue Int. 2002;71(6):508–18.

Magnusson P, Larsson L, Magnusson M, Davie MWJ, Sharp CA. Isoforms of bone alkaline phosphatase: characterization and origin in human trabecular and cortical bone. J Bone Miner Res. 1999;14(11):1926–33.

Magnusson P, Ärlestig L, Paus E, Di Mauro S, Testa MP, Stigbrand T, et al. "Monoclonal antibodies against tissue-nonspecific alkaline phosphatase." Report of the ISOBM TD9 workshop. Tumour Biol. 2002a;23:228–48.

Magnusson P, Sharp CA, Farley JR. Different distributions of human bone alkaline phosphatase isoforms in serum and bone tissue extracts. Clin Chim Acta. 2002b;325:59–70.

Marini JC, Hopkins E, Glorieux FH, Chrousos GP, Reynolds JC, Gundberg CM, Reing CM. Positive linear growth and bone responses to growth hormone treatment in children with types III and IV osteogenesis imperfecta: high predictive value of the carboxyterminal propeptide of type I procollagen. J Bone Miner Res. 2003;18(2):237–43.

Masarachia PJ, Pennypacker BL, Pickarski M, Scott KR, Wesolowski GA, Smith SY, Samadfam R, Goetzmann JE, Scott BB, Kimmel DB, Duong LT. Odanacatib reduces bone turnover and increases bone mass in the lumbar spine of skeletally mature ovariectomized rhesus monkeys. J Bone Miner Res. 2012;27(3):509–23.

Melkko J, Hellevik T, Risteli L, Risteli J, Smedsrod B. Clearance of NH2-terminal propeptides of types I and III procollagen is a physiological function of the scavenger receptor in liver endothelial cells. J Exp Med. 1994;179(2):405–12.

Melkko J, Kauppila S, Niemi S, Risteli L, Haukipuro K, Jukkola A, Risteli J. Immunoassay for intact amino-terminal propeptide of human type I procollagen. Clin Chem. 1996;42(6 Suppl):947–54.

Mornet E, Stura E, Lia-Baldin AS, Stigbrand T, Menez A, Le DuH MH. Structural evidence for a functional role of human tissue non specific alkaline phosphatase in bone mineralization. J Biol Chem. 2001;276:31171–8.

Nenonen A, Cheng S, Ivaska KK, Alatalo SL, Lehtimäki T, Schmidt-Gayk H, Uusi-Rasi K, Heinonen A, Kannus P, Sievänen H, Vuori I, Väänänen HK, Halleen JM. Serum TRACP 5b is a useful marker for monitoring Alendronate treatment: comparison with other markers of bone turnover. J Bone Miner Res. 2005;20(10):1804–12.

Nyman JS, Roy A, Acuna RL, Gayle HJ, Reyes MJ, Tyler JH, Dean DD, Wang X. Age-related effect on the concentration of collagen crosslinks in human osteonal and interstitial bone tissue. Bone. 2006;39(6):1210–7.

Oddie GW, Schenk G, Angel NZ, Walsh N, Guddat LW, de Jersey J, Cassady AI, Hamilton SE, Hume DA. Structure, function, and regulation of tartrate-resistant acid phosphatase. Bone. 2000;27(5):575–84.

Ominsky MS, Vlasseros F, Jolette J, Smith SY, Stouch B, Doellgast G, Gong J, Gao Y, Cao J, Graham K, Tipton B, Cai J, Deshpande R, Zhou L, Hale MD, Lightwood DJ, Henry AJ, Popplewell AG, Moore AR, Robinson MK, Lacey DL, Simonet WS, Paszty C. Two doses of sclerostin antibody in cynomolgus monkeys increases bone formation, bone mineral density, and bone strength. J Bone Miner Res. 2010;25(5):948–59.

Ominsky MS, Stouch B, Schroeder J, Pyrah I, Stolina M, Smith SY, Kostenuik PJ. Denosumab, a fully human RANKL antibody, reduced bone turnover markers and increased trabecular and cortical bone mass, density, and strength in ovariectomized cynomolgus monkeys. Bone. 2011;49(2):162–73.

Ominsky MS, Boyd SK, Varela A, Jolette J, Felx M, Doyle N, Mellal N, Smith SY, Locher K, Buntich S, Pyrah I, Boyce RW. Romosozumab improves bone mass and strength while maintaining bone quality in ovariectomized cynomolgus monkeys. J Bone Miner Res. 2017;32:1–14.

Penido MG, Alon US. Phosphate homeostasis and its role in bone health. Pediatr Nephrol. 2012;27:2039–48.

Peris P, Alvarez L, Monegal A, GuaÑabens N, DurÁn M, Pons F, Martínez de Osaba MJ, Echevarría M, Ballesta AM, Muñoz-Gómez J. Biochemical markers of bone turnover after surgical menopause and hormone replacement therapy. Bone. 1999;25(3):349–53.

Price PA, Williamson MK, Lothringer JW. Origin of a vitamin K-dependent bone protein found in plasma and its clearance by kidney and bone. J Biol Chem. 1981;256:12760–6.

Querejeta R, López B, González A, Sánchez E, Larman M, Martínez Ubago JL, Díez J. Increased collagen type I synthesis in patients with heart failure of hypertensive origin. Relat Myocardial Fibros. 2004;110(10):1263–8.

Qvist P, Christgau S, Pedersen BJ, Schlemmer A, Christiansen C. Circadian variation in the serum concentration of C-terminal telopeptide of type I collagen (serum CTx): effects of gender, age, menopausal status, posture, daylight, serum cortisol, and fasting. Bone. 2002;31(1):57–61.

Rosalki SB. Bone-origin alkaline phosphatase in plasma by wheat-germ lectin methods in bone disease. Clin Chim Acta. 1994;226(2):143–50.

Rowe PS. A unified model for bone-renal mineral and energy metabolism. Curr Opin Pharmacol. 2015;22:64–71.

Saito M, Grynpas MD, Burr DB, Allen MR, Smith SY, Doyle N, Amizuka N, Hasegawa T, Kida Y, Marumo K, Saito H. Treatment with eldecalcitol positively affects mineralization, microdamage, and collagen crosslinks in primate bone. Bone. 2015;73:8–15.

Samadfam R, Xia Q, Goltzman D. Pretreatment with anticatabolic agents blunts but does not eliminate the skeletal anabolic response to parathyroid hormone in oophorectomized mice. Endocrinology. 2007;148(6):2778–87.

Samadfam R, Xia Q, Miao D, Hendy GN, Goltzman D. Exogenous PTH and endogenous 1,25-Dihydroxyvitamin D are complementary in inducing an anabolic effect on bone. J Bone Miner Res. 2008;23:1257–66.

Samadfam R, Awori M, Bénardeau A, Bauss F, Sebokova E, Wright M, Smith SY. Combination treatment with pioglitazone and fenofibrate attenuates pioglitazone-mediated acceleration of bone loss in ovariectomized rats. J Endocrinol. 2012;212(2):179–86.

Samadfam R, Doyle N, Kissner T, Krupp E, Smith SY. Anti-diabetes drug class of SGLT1 inhibitors increases bone mass in young and adult female Sprague-Dawley rats by decreasing bone turnover. Can J Diabetes. 2013;37:S6–S11.

Sassi ML, Eriksen H, Risteli L, Niemi S, Mansell J, Gowen M, Risteli J. Immunochemical characterization of assay for carboxyterminal telopeptide of human type I collagen: loss of antigenicity by treatment with cathepsin K. Bone. 2000;26(4):367–73.

Schett G, Bozec A. Removing the bone brake. Cell Metab. 2014;20:394–5.

Sharma U, Pal D, Prasad R. Alkaline phosphatase: an overview. Indian J Clin Biochem. 2014;29(3):269–78.

Smedsrød B, Melkko J, Risteli L, Risteli J. Circulating C-terminal propeptide of type I procollagen is cleared mainly via the mannose receptor in liver endothelial cells. Biochem J. 1990;271(2):345–50.

Smith SY, Recker RR, Hannan M, Muller R, Bauss F. Intermittent intravenous administration of the bisphosphonate ibandronate prevents bone loss and maintains bone strength and quality in ovariectomized cynomolgus monkeys. Bone. 2003;32:45–55.

Smith SY, Samadfam R, Chouinard L, Awori M, Bénardeau A, Bauss F, Guldberg RE, Sebokova E. Effects of pioglitazone and fenofibrate co-administration on bone biomechanics and histomorphometry in ovariectomized rat. J Bone Miner Metab. 2015;33(6):625–41.

Szulc P, Bauer DC. Biochemical markers of bone turnover. In: Marcus R, Feldman D, Dempster DW, Luckey M, Cauley JA, editors. Osteoporosis. New York: Elsevier; 2013. p. 1573–610.

Szulc P, Seeman E, Delmas PD. Biochemical measurements of bone turnover in children and adolescents. Osteoporos Int. 2000;11(4):281–94.

Tähtelä R, Seppänen J, Laitinen K, Katajamäki A, Risteli J, Välimäki MJ. Serum tartrate-resistant acid phosphatase 5b in monitoring bisphosphonate treatment with clodronate: a comparison with urinary N-terminal telopeptide of type I collagen and serum type I procollagen aminoterminal propeptide. Osteoporos Int. 2005;16(9):1109–16.

Takeda S, Smith SY, Tamura T, Saito H, Takahashi F, Samadfam R, Haile S, Doyle N, Endo K. Long-term treatment with Eldecalcitol (1a, 25-Dihydroxy-2b- (3-hydroxypropyloxy) vitamin D3) suppresses bone turnover and leads to prevention of bone loss and bone fragility in ovariectomized rats. Calcif Tissue Int. 2015;96(1):45–55.

Tapanainen P, Knip M, Risteli L, Kemppainen L, Kaar M, Risteli J. Collagen metabolites in the prediction of response to GH therapy in short children. Eur J Endocrinol. 1997;137(6):621–5.

TRAP5b Technical Bulletin. http://www.tecomedical.com/en/laboratory-ivd-kits-reagents/bone-and-cartilage-parameters/bone-metabolism/TRAP5b-Human-Quidel. (n.d.).

Tsujimoto M, Chen P, Miyauchi A, Sowa H, Krege JH. PINP as an aid for monitoring patients treated with teriparatide. Bone. 2011;48(4):798–803.

Varela A, Chouinard L, Lesage E, Smith SY, Hattersley G. One year of abaloparatide, a selective activator of the PTH1 receptor, increased bone formation and bone mass in osteopenic ovariectomized rats without increasing bone resorption. J Bone Miner Res. 2017;32(1):24–33.

Wei W, Dutchak PA, Wang X, Wang X, Ding X, Wang X, Bookout AL, Goetz R, Mohammadi M, Gerard RD, Dechow PC, Mangelsdorf DJ, Kliewer SA, Wan Y. Fibroblast growth factor 21 promotes bone loss by potentiating the effects of peroxisome proliferator-activated receptor γ. Proc Natl Acad Sci U S A. 2012;109(8):3143–8.

Wennberg C, Hessle L, Lundberg P, Mauro S, Narisawa S, Lerner UH, Millán JL. Functional characterization of osteoblasts and osteoclasts from alkaline phosphatase knockout mice. J Bone Miner Res. 2000;15(10):1879–88.

Whyte MP, Landt M, Ryan LM, Mulivor RA, Henthorn PS, Fedde KN, Mahuren JD, Coburn SP. Alkaline phosphatase: placental and tissue-nonspecific isoenzymes hydrolyze phosphoethanolamine, inorganic pyrophosphate, and pyridoxal 50-phosphate substrate accumulation in carriers of hypophosphatasia corrects during pregnancy. J Clin Investig. 1995;95:1440–5.

Wuthier RE, Register T. Role of alkaline phosphatase, a polyfunctional enzyme, in mineralizing tissues. In: Butler WT, editor. Chemistry and biology of mineralized tissues. Birmingham: EBSCO Media; 1995. p. 113–24.

Chapter 6
Skeletal Imaging

Aurore Varela

Abstract During the past two decades, the use and refinements of imaging modalities have markedly increased taking now a significant place in drug development. The practical use of preclinical imaging represents an important step in the refinement of animal studies allowing longitudinal assessments with powerful, non-invasive, and clinically translatable tools to monitor drug effects. This chapter reviews the different imaging modalities used in the preclinical studies to evaluate the skeleton and bone tissue: X-rays, dual-energy X-ray absorptiometry (DXA), peripheral quantitative computed tomography (pQCT), micro-CT, high-resolution pQCT (HR-pQCT), magnetic resonance imaging (MRI), positron emission tomography (PET), and single photon emission computed tomography (SPECT), to understand the technologies, their applications and limitations, and important considerations for including these different imaging technologies in toxicology rodent and non-rodent studies, good laboratory practice (GLP) validation, to actually integrate imaging approaches into safety assessment in drug development. Ongoing research and development focusing on integrating different imaging modalities and increasing resolution will certainly facilitate the translational applications from anatomical to molecular and functional evaluations to improve the efficacy in toxicology preclinical studies and accelerate drug development.

Keywords Radiography • DXA • pQCT • Micro-CT • HR-pQCT • MRI • PET • SPECT • Growth • BMC • BMD • FEA • Architecture • Trabecular • Cortical

6.1 Introduction

In vivo imaging techniques such as radiology, osteodensitometry, tomography, and magnetic resonance and nuclear imaging are largely used in skeletal clinical medicine for diagnosis and guided therapy and play an increasing role in clinical trials, being an integral part of the drug development process. Imaging sciences have

A. Varela, DVM, MSc, DABT (✉)
Charles River Preclinical Services, Senneville, QC, Canada
e-mail: aurore.varela@crl.com

S.Y. Smith et al. (eds.), *Bone Toxicology*, Molecular and Integrative Toxicology,
DOI 10.1007/978-3-319-56192-9_6

known an incredible development during the last decades from two-dimensional to three-dimensional techniques and from anatomical to quantitative and molecular applications, and many techniques, such as magnetic resonance imaging (MRI), positron emission tomography (PET), single photon emission computed tomography (SPECT), and computed tomography (CT), have become indispensable. The applications of in vivo translational imaging are now extending further into drug discovery and development and have the potential to considerably accelerate the process, reduce the cost, significantly affect the drug development process, and comply with the 3R as noninvasive techniques. It is important to understand the technologies, their applications and limitations. Imaging technology includes a range of modalities from radiology, osteodensitometry, CT, MRI, PET, to SPECT. Bone mineral density (BMD) can be evaluated in vivo or ex vivo using dual-energy X-ray absorptiometry (DXA) and/or peripheral quantitative computed tomography (pQCT) or by micro-computed tomography (micro-CT). Finite element analysis (FEA) of CT even allows to predict bone strength, reaching another level in noninvasive bone assessment. These noninvasive and quantitative techniques provide not only anatomical evaluation but also functional and molecular information that can access the mechanisms of drug action or its toxicity. This overview provides a review of the main skeletal imaging techniques, reviews important considerations for applications of imaging in toxicology rodent and non-rodent studies, and the challenges of GLP validation, to actually integrate imaging approaches into safety assessment in drug development.

6.2 Radiology

Radiology remains an essential tool of skeleton evaluation, and provides a simple and commonly available technology, although it requires a skilled and experienced veterinarian or veterinary radiologist. It is a straightforward means for examining the size, shape, skeletal abnormalities, and overall density of the entire bony skeleton or large subdivisions thereof, in a stereotypical orientation. Longitudinal evaluations on skeletal growth and development and growth plate assessment are also key in juvenile toxicology studies.

Radiological diagnosis can effectively support histological diagnosis and often is helpful in deciding which bone locations should be examined histologically to best characterize the disease process or treatment effect. Furthermore, radiographs can help to define whether or not a lesion causing clinical signs of local lameness is actually part of a systemic disease process. A radiograph of a bone tumor, particularly when interpreting the results from tumor biopsies, is as essential as the histologic section in rendering a final diagnosis and prognosis. The radiograph can give information concerning the extent of the lesion, its distribution within the bone, areas of bone formation and destruction, the quantity of periosteal reactive bone, the location of the tissue specimen, and initial evidence regarding its spread either locally or systemically. Factors to be considered in viewing the radiograph are alterations in bone density and geometry, lesion location and distribution, periosteal reactions, and soft tissue changes.

Radiography is the method of choice to scan the entire skeleton in vivo of small animals over the course of the study and before scheduled termination to detect any lesions at skeletal sites that are not routinely harvested and that would not have been detected otherwise. Radiographic data are of particular importance for carcinogenicity studies (Jolette et al. 2006) (Figs. 6.1 and 6.2) but also are warranted in subchronic and chronic nonclinical studies when direct or indirect effects

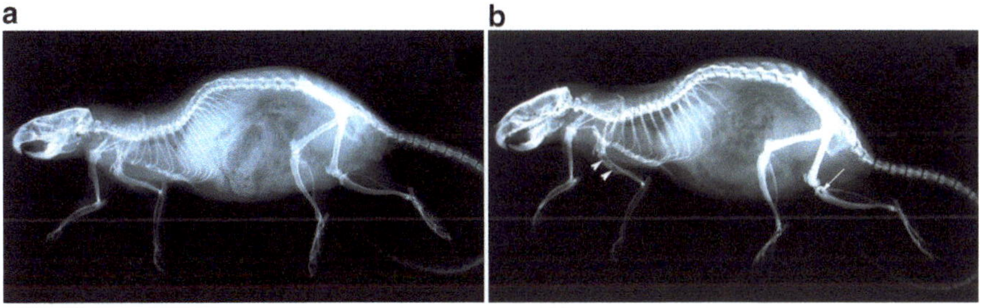

Fig. 6.1 Radiological assessment for noninvasive identification of additional skeletal lesions that might require histopathological evaluation, especially "occult" lesions from sites not routinely sampled at necropsy or that would otherwise have gone undetected. Whole-body radiographs in lateral recumbency of (**a**) a control female, and (**b**) a female treated for 13.5 months with recombinant human parathyroid hormone (PTH) 1–84 at 150 mg/kg/day shows a generalized increase in skeletal radiodensity, which is most evident in the femurs as a loss of the medullary cavity shadow, as well as localized bone loss in the humerus (*arrowheads*) and tibia (*arrow*) as indicated by multifocal sites of reduced radiodensity. These lucent areas subsequently were diagnosed as multicentric osteosarcoma (Figure reproduced from Jolette et al. (2006), with permission)

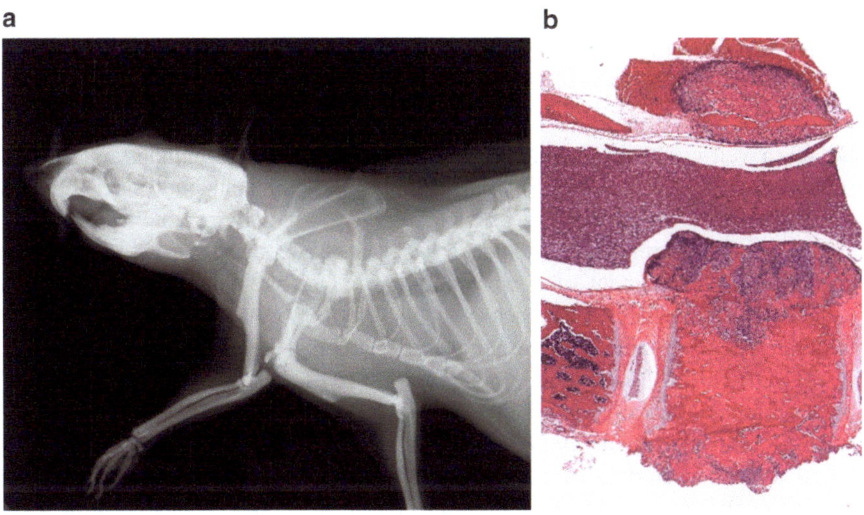

Fig. 6.2 Occult osteosarcomas in the axial skeleton that were detected radiologically in a rat treated for an extended period with recombinant human PTH. (**a**, **b**) Comparable radiograph and corresponding microscopic findings of an occult osteosarcoma affecting a vertebral body. (**b**) H&E stain (Reproduced from Gunson et al. 2013b with permission)

on bone tissue are suspected. In large animals, pretreatment radiographs will show any preexisting bone lesions. Radiographs also can be performed ex vivo on skeletal segments. High-resolution cabinet radiography systems constitute an additional tool for imaging mice in vivo or excised small segments or carcasses at necropsy. Such benchtop units allow magnification of up to fivefold for evaluation of specific skeletal regions in rodents.

In juvenile toxicity studies, serial radiographs obtained at intervals in growing animals provide an accurate record from which to measure bone growth in rodent (Fig. 6.3) (Pouliot et al. 2013) and non-rodents (Fig. 6.4) (Boyce et al. 2014) and assess epiphyseal closure. Calibrated measuring devices are included in the radiograph digital system to facilitate measurements. Measurements of long bones (femur and/or tibia, length and diameter) and the spine (lumbar vertebrae) provide a good estimation of the skeletal growth. Other sites can be included if specific targets are known. Precise anatomical landmarks need to be defined for measurements to obtain reproducible results (reproducibility with coefficient of variation below 2 %). Physeal closure occurs at different ages depending on the species, sex, on the bone and on the proximal or distal physis due to different patterns of development and growth (Table 6.1 and Fig. 6.5). The sequence of growth plate closure is generally similar in mammals (Geiger et al. 2014); tibial crest and distal ulna physes are generally the latest to close. In rodents, the physis does not close until old age at certain site like the distal femur, proximal tibia, distal radius, and ulna; however, growth stop and physis become inactive after 6–7 months (Walker and Kember 1972; Roach et al. 2003; Martin et al. 2003). Bone formation and bridges forming in the physis prevent any further growth even if the physis is present. The rapid phase of growth occurs before 3 months of age.

6.3 Dual Energy X-Ray Absorptiometry (DXA)

DXA is a well-established technology for osteodensitometry in humans and animals (FDA 1994). It is widely used in clinical trials and preclinical studies because it is the approved diagnostic tool for osteoporosis, the radiation dose is low, the precision is good, it is relatively inexpensive and available, and scan times are short. Despite the development of new technologies, DXA remains one of the main osteodensitometry techniques because of its role in the diagnosis of osteoporosis, fracture risk assessment, and monitoring treatment response in humans (ISCD 2015; NOF 2014; Blake et al. 2013). DXA allows a two-dimensional (2D) measurement of apparent areal bone mineral density (aBMD in g/cm^2). The attenuation of two different photon energy X-ray beams is measured. Machines are calibrated based on two materials of defined composition, one mineral (hydroxyapatite, $Ca_5(PO_4)_3$ OH) and a homogeneous soft tissue of a set area adjacent to the bone area. The bone mapping is performed by edge detection within the region of interest (ROI), defining the bone area cm^2 (except for the whole body for which the total body area is measured). Each pixel BMD is average to calculate the aBMD; then the

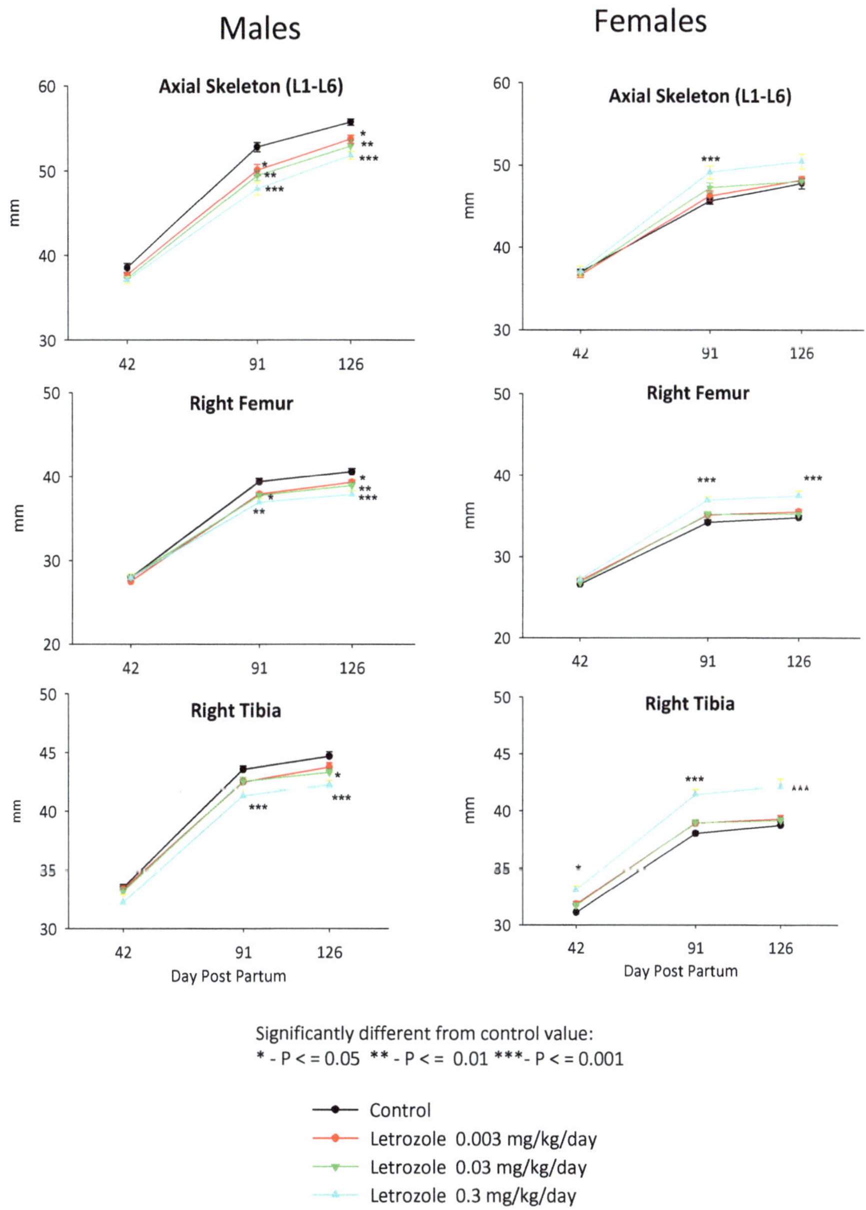

Fig. 6.3 Skeletal growth in juvenile rats: lumbar spine, femur, and tibia lengths measured on radiographs (Modified from Pouliot et al. 2013)

bone mineral content (BMC) is calculated, measuring the total hydroxyapatite, $Ca_5(PO_4)_3$ OH mass in the bone ROI. It provides an assessment of bone mass and density at the entire skeleton level (whole body) and large areas of the skeleton at clinical relevant sites (i.e., lumbar vertebrae, femur, proximal femur distal femur,

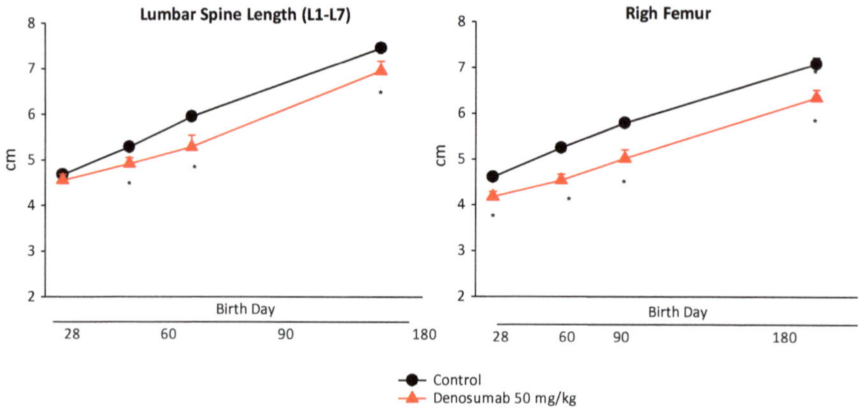

Fig. 6.4 Skeletal growth radiological assessment (Modified from Boyce et al. 2014)

distal radius, proximal tibia), Fig. 6.6, but DXA does not distinguish between cortical and trabecular bone. Whole-body and regional scans are also used to study body composition (whole body and regional body) by measuring lean and fat mass (Schoeller et al. 2005).

In preclinical studies, in vivo DXA is a good translational tool for prediction of clinical trial outcome (Smith et al. 2011; Fox et al. 2007a, b; Ominsky et al. 2011; Kumar et al. 2013). Longitudinal assessment in anesthetized animals from baseline allows a powerful evaluation of individual changes over time, especially in large animals. In preclinical and especially in GLP studies, standardized procedures and good precision of machines and operators are very important. Precision errors (typically 1–2 %, percentage coefficient of variation) are very small in laboratory animals and comparable to human studies (Shepherd et al. 2006; ISCD 2015). Accuracy errors that may impact individual scans in humans (Blake and Fogelman 2008) have less relevance in controlled longitudinal preclinical studies unless changes in body weight and/or body composition occur over time or vs. controls. However, limitations and disadvantages of DXA as an investigational tool are well known: two-dimensional (2D) projection of a three-dimensional (3D) structure (effect of bone size and shape), inability to make geometrical evaluations, cannot distinguish effects on trabecular and cortical bone density and dimensions, the effects of body size and composition on the BMD measurement, and unable to differentiate low bone mass due to osteomalacia or osteoporosis (Bolotin 2007). The effect of size and shape is particularly a problem for interpreting results in juvenile populations (animals and humans, Binkovitz and Henwood 2007). The conversion of X-ray attenuation to BMD values depends on quantity and composition of bone marrow and surrounding soft tissues (Blake and Fogelman 2008; Yu et al. 2012). Interpretation of data should take into account the normal variability in bone measures especially in non-rodents, the gender, age, species and strain, and body composition.

Table 6.1 Age at physeal closure in usual laboratory animals

Species/sites	Age at physeal closure	
Cynomolgus monkey[a]		
	Male	Female
Humerus proximal – distal	6 years–3 years 5 months	4 years 9 months–2 years 3 months
Radius proximal – distal	5 years 3 months–5 years 3 months	3 years 9 months–5 years 9 months
Ulna proximal – distal	5 years–6 years 6 months	4 years 6 months–5 years 9 months
Femur proximal – distal	6 years–5 years 3 months	4 years 9 months–4 years 9 months
Tibia proximal – distal	5 years–5 years 3 months	5 years–4 years 9 months
Fibula proximal – distal	6 years–5 years 3 months	4years 9months–4 years 9 months
Dog[b]		
Humerus proximal – distal	10/12 months–6/8 months	
Radius proximal – distal	9/10 months–10/12 months	
Femur proximal – distal	6/13 months–6/11 months	
Tibia proximal – distal	6/11 months–5/8 months	
Rat[b]		
Humerus proximal – distal	52 weeks[e]–6 weeks	
Radius proximal – distal	8/14 weeks–104 weeks[e]	
Femur proximal – distal	104 weeks[e]–15/17 weeks	
Tibia proximal – distal	104 weeks[e]–>16 weeks	
Rabbit[c]		
Femur proximal – distal	19/24 weeks	
Tibia proximal – distal	22/32 weeks	
Fibula proximal – distal	23/32 weeks	
Gottingen minipigs[d]		
Femur and lumbar vertebrae start to close at 25 and 21 months of age, respectively; the complete closure occurs around 3.5 years old		

[a]Fukuda et al. (1978)
[b]Zoetis and Hurtt (2003)
[c]Kaweblum et al. (1994)
[d]Tsutsumi et al. (2004)
[e]In rats, rapid growth occurs between 1 and 5 weeks of age but declines by skeletal maturity at 11.5–13 weeks. Until 26 weeks of age, longitudinal growth still continues but virtually ceases thereafter

In preclinical toxicology studies, DXA is an easy and effective technique to monitor the response to treatment, either positive or negative effects on bone mass, for compounds such as corticosteroids, vitamin A and retinoic acids (Rohde and DeLuca 2003), peroxisome proliferator-activated receptor-γ (PPAR-γ) agonist (Lazarenko et al. 2007; Kumar et al. 2013), Fig. 6.7, and Fibroblast Growth Factor 21 (FGF21). In animal studies, the proton pump inhibitors (PPIs), omeprazole, reduced bone density by DXA (Cui et al. 2001) and in humans (Bahtiri et al. 2015).

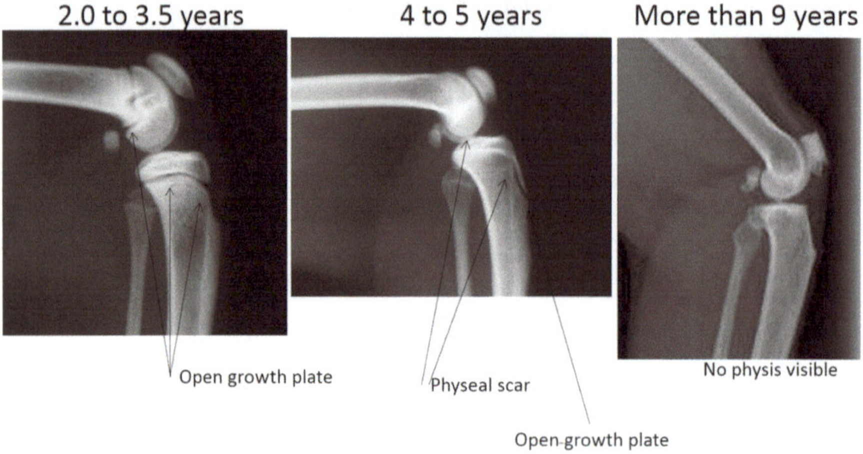

Fig. 6.5 Skeletal growth radiological assessment. Evaluation of skeletal maturity based on the growth plate closure (proximal tibia and distal femur) can be performed in vivo. Physeal closures can be scored semiquantatively (From widely open physis to physeal scar and physis closed) and qualitatively for abnormalities

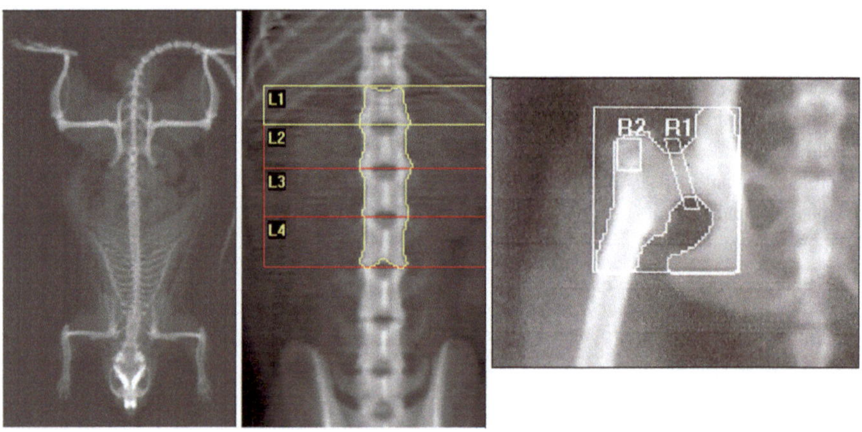

Fig. 6.6 DXA scan images of the rat whole body, lumbar spine (L1–L4) and proximal femur/hip in nonhuman primates. BMD is obtained for the entire skeleton or specific regions. Positioning of the animals, placement and size of region of interest (ROI), and analysis mode are key elements for precise and standardized procedures

Fluoxetine decreases BMD in juvenile mice and is known to increase fracture risk in humans (Prozac package insert). In juvenile toxicology studies, DXA allows to evaluate bone size in two dimensions and bone mass. Areal BMD increases with height and weight during skeletal growth in young animals. DXA is not measuring the true volumetric bone density, but a part of the bone size and mass. Bone size is important to assess skeletal growth and development in juvenile toxicology studies

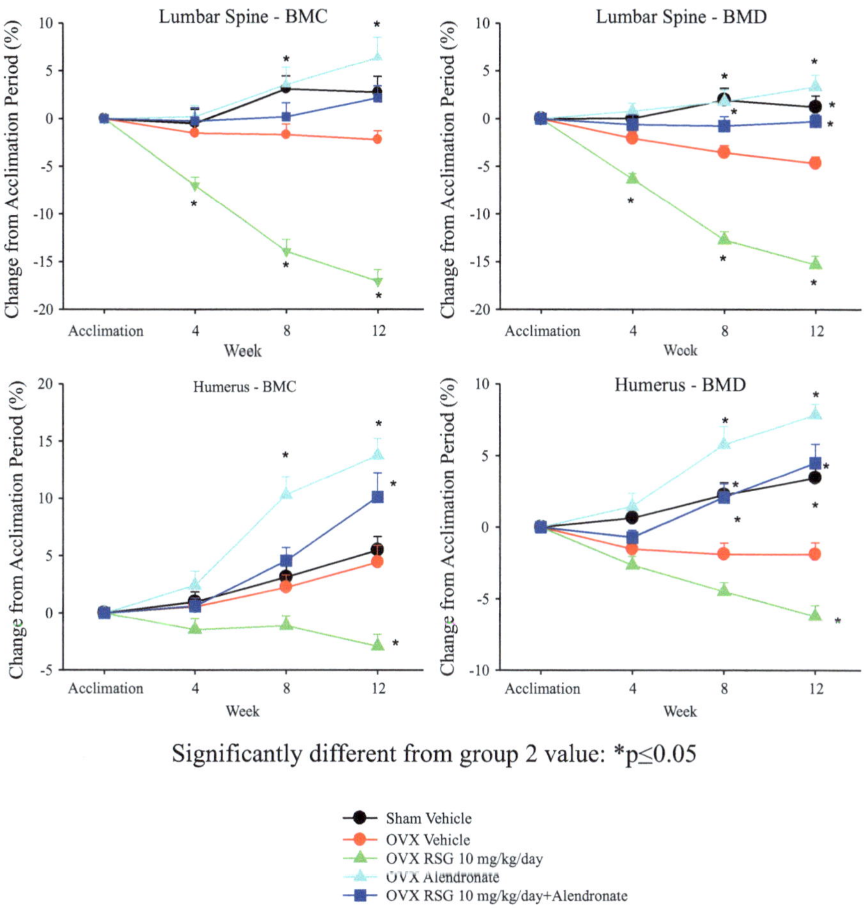

Fig. 6.7 Changes over time (as percent change from baseline) in bone mineral content (BMC, *left*) and density (BMD, *right*) for lumbar spine and humerus as measured by DXA during the treatment phase with different doses of rosiglitazone, a PPAR gamma agonist. Changes in bone parameters associated with RSG administration are reversible on treatment cessation or preventable by coadministration with an antiresorptive agent, alendronate. *OVX* ovariectomized, *RSG* rosiglitazone, *ALN* alendronate (Modified from Kumar et al. 2013)

(de Mello et al. 2012; Pouliot et al. 2013). Adjustment for weight, height, and bone area and size can be done to further discriminate changes in body size and overall growth vs. a direct effect on bone.

Despite the known limitations of DXA, correlation with bone strength is very good and it remains an important densitometry tool to measure the entire skeleton or large segment of it. More specific questions on bone size and geometry, or to distinguish cortical and trabecular bone can be addressed by 3D CT techniques as pQCT and/or micro-CT.

6.4 Quantitative Computed Tomography (QCT)

Quantitative computed tomography is a CT techniques that provides a three-dimensional (3D) volumetric bone density (vBMD) and that does not have the DXA limitations. Therefore, pQCT is frequently used in preclinical studies and clinical trials to complement DXA measurements. DXA scanning permits evaluations of the entire body and large areas of the skeleton, whereas peripheral QCT (pQCT) scans provide information from a single (or multiple) slice(s) of bone at specific skeletal areas, normally the proximal tibia and distal radius. It allows the separation of the trabecular and cortical regions for compartment-specific measurements and as well the analysis of geometric parameters (typically at sites in the diaphysis). As a 3D CT technique, pQCT scans measure a true volumetric BMD. Geometric measurements include periosteal circumference, endocortical circumference, and cortical thickness. The ability to analyze the trabecular bone compartment and quantify bone geometry increases the sensitivity of in vivo monitoring, allowing differentiation of treatment-induced changes in the cortical and trabecular compartments, treatment effects on cortical thickness, porosity, and density. QCT scans are performed on a single (or multiple) slice(s) of bone. Analysis routinely is undertaken at the proximal tibia and distal radius similar to humans. Reference phantoms are scanned as quality controls. They have a known composition, which allows conversion of the CT Hounsfield units (HU) into calcium hydroxyapatite (Ca-HAP) equivalent BMD. Typical single-slice pQCT X-ray tubes have 0.2–0.8 or 1-mm slice thickness in plane pixel sizes of 100–300 μm. BMD and other parameters are measured at the metaphysis (at a specific % of the total bone length, mostly trabecular) and at a more diaphyseal location (cortical site). As with any other CT technology, the partial volume effect is important to consider (Zemel et al. 2008) to avoid any under-estimation of density measurement. Specific scan settings, placement of the slice, the volume of interest VOIs and parameters of analysis are crucial for sound evaluation. QCT is less affected by size, but it is sensitive to bone marrow changes, and the amount of fat in the marrow cavity (Glüer and Genant 1989). QCT may not provide the complete answer, because it is not possible to distinguish low bone organ density from low tissue mineralization (e.g., in rickets).

Peripheral QCT is largely used in preclinical studies to evaluate the effect of bone agents on bone mass and geometry (Smith et al. 2011; Fox et al. 2007a, b; Ominsky et al. 2011). The techniques have also been used to study a variety of disorders with detrimental effects on bone, corticosteroids (Ferretti et al. 1995), PPAR γ (Kumar et al. 2013), and exacerbated pharmacology (Boyce et al. 2014), Fig. 6.8. In juvenile toxicology studies, pQCT provides an important asset to assess skeletal growth and development in measuring bone size and geometry, and bone mass accrual (Pouliot et al. 2013; Boyce et al. 2014), Fig. 6.9.

6.5 Micro-CT and HR-pQCT

Skeletal imaging has passed to the twenty-first century with the development of high-resolution micro-CT. Compared to pQCT, micro-CT reaches another scale of a few microns (nominal resolution or pixel size) because of a microfocus X-ray tube

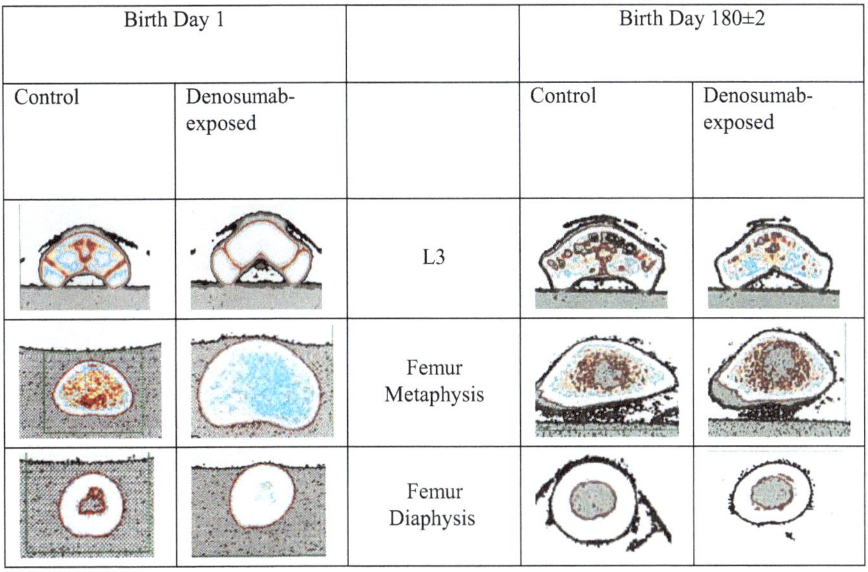

Fig. 6.8 Representative pQCT scans of the femur distal metaphysis and diaphysis and lumbar vertebra, in control (first and third column) and denosumab-exposed in utero (second and last column) infants at birth day (BD) 1 and denosumab-exposed infants at BD180. At BD1, bone mass was increased at the distal femur metaphysis and lumbar vertebra in the denosumab-exposed group (note the denser aspect of the trabecular compartment, white and/or blue, the increased femoral metaphysis cross-sectional area, reflecting the metaphyseal flaring that occurred with denosumab exposure). At the femur diaphysis, the thickness of the cortex was increased in association with a marked decrease in endosteal circumference, but the cortical density was decreased (due to the presence of primary spongiosa in the marrow canal). Six month after birth, bone mass and bone geometry were comparable to controls, except the cortical thickness that decreased due to endocortical expansion (Modified from Boyce et al. 2014)

and therefore can provide volumetric measurements of bone microstructure (trabecular architecture: relative bone volume, trabecular number, trabecular surface area, trabecular thickness, and trabecular separation), cortical porosity and geometry, and the volumetric/apparent and tissue BMD for both cortical and trabecular bone compartments. Some of these static parameters have been historically measured in 2D with histomorphometry, but micro-CT gives a unique opportunity to evaluate an entire volume with additional architectural parameters (trabeculae shape, platelike or rodlike, connectivity and the degree of anisotropy of the bone meshwork). For BMD assessment, calibration using an appropriate phantom of known density is required to convert pixel brightness values on the CT image to a measure of mineral density (typically mg/cm^3 of calcium hydroxyapatite). Images are binarized. Variables are directly determined from 3D distance transformation techniques. BMD parameters assume a constant tissue mineral density (1,200 mg Ca-HAP/ml usually), and micro-CT can provide both material (tissue BMD) and compartment density (trabecular or cortical).

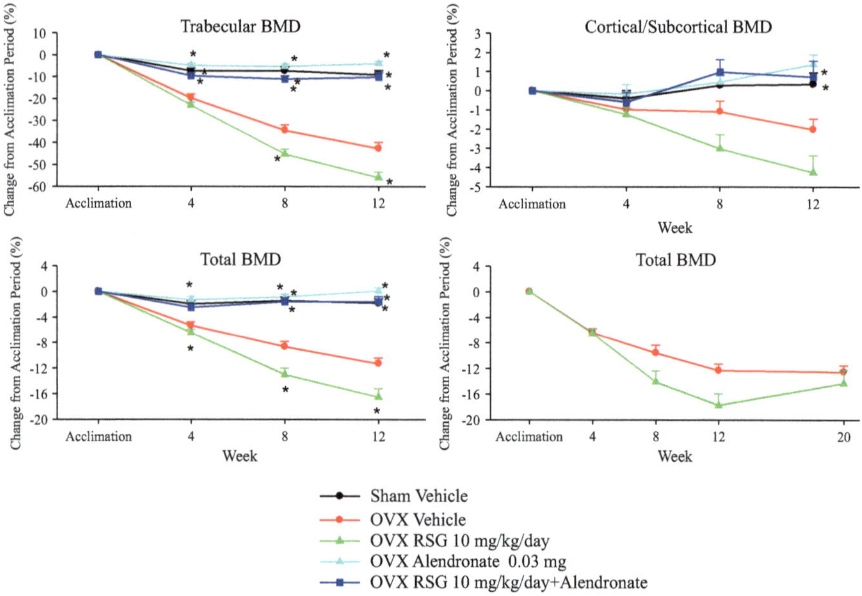

Fig. 6.9 Changes over time (as percent change from baseline) in trabecular, cortical/subcortical, and total bone mineral density (BMD) at tibial metaphysis over time as measured by pQCT during the treatment phase with different doses of rosiglitazone, a PPAR gamma agonist. Changes in bone parameters associated with RSG administration are reversible on treatment cessation or preventable by coadministration with an antiresorptive agent, alendronate. *OVX* ovariectomized, *RSG* rosiglitazone, *ALN* alendronate (Modified from Kumar et al. 2013)

6.5.1 Micro-CT

Thus, micro-CT is a powerful new tool to assess changes in bone microarchitecture in 3D over relatively large regions and provides a greater sensitivity than is possible with conventional 2D static histomorphometry, Fig. 6.10. In vivo, micro-CT allows precise measurements of bone mass and architecture (Boyd et al. 2006). Ex vivo fresh, frozen, or fixed specimens can also be scanned by micro-CT without destruction, thus allowing the same specimen to be used for additional evaluations. Scanning is rapid, allowing high throughput of specimens. In proof-of-concept studies, trabecular architecture or cortical porosity parameters can be reported quickly for more timely strategic decision-making in early drug development and also provide key translational information for bone quality assessment in later-stage development (Fox et al. 2008), Fig. 6.11. Additionally, evaluation of bone tissue mineral density TMD can provide important information on the mineralization processes and bone biomechanics (Kazakia et al. 2008).

6.5.2 HR-pQCT

Recently, a multislice high-resolution pQCT (HR-pQCT or Xtreme CT) device has been marketed for clinical research. This in vivo micro-CT scanner allows fast scanning time while reaching an isotropic voxel size and slice thickness of 30 μm (82 μm

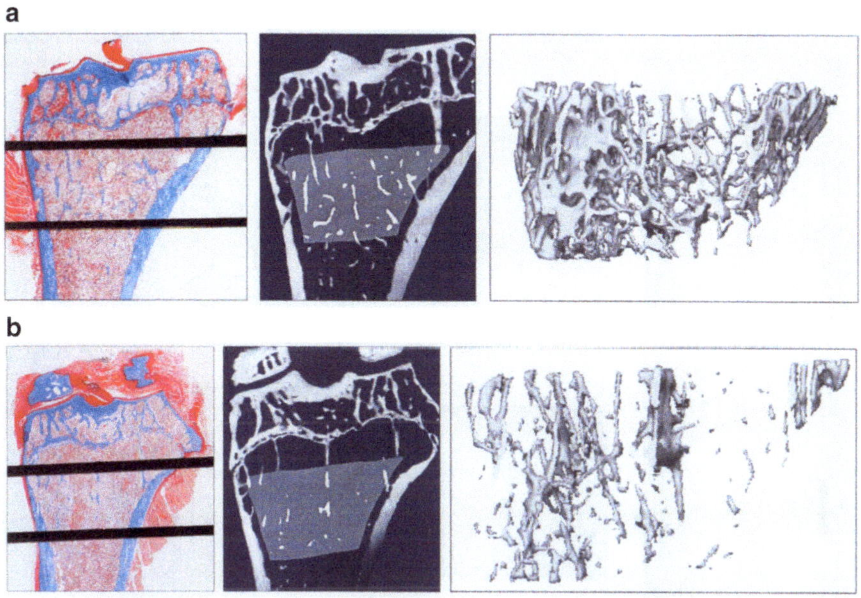

Fig. 6.10 Comparison of static histomorphometry (*left column*, modified Goldner's trichrome stain) with high-resolution microcomputed tomography (micro-CT) in two (2D, *middle* column) or three (3D, *right* column) dimensions as tools for assessing bone microarchitecture. Relative to an age-matched sham control animal (**a**), an osteopenic rat (**b**) has fewer trabeculae in both 2D and 3D representations. Note the similarities of the trabecular representations for histomorphometry and 2D micro-CT of the proximal tibia (coronal orientation) (Figure reproduced from Haschek and Rousseaux's Handbook of Toxicologic Pathology, 3rd Ed. Haschek WM, Rousseaux CG, Wallig MA, Bolon B, Ochoa R, Mahler MW, eds. Elsevier, Amsterdam (2013), Fig. 63.15, p. 2795, with permission)

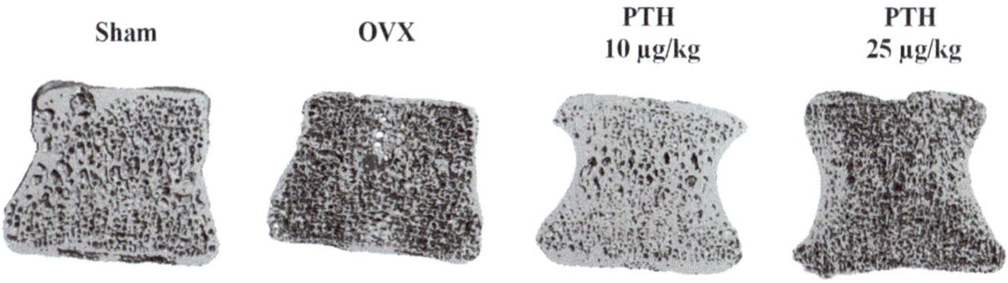

Fig. 6.11 Representative micro-CT scans from the middle of the T10 vertebral body from sham or OVX rhesus monkeys treated daily with vehicle or PTH(1–84) (5, 10, or 25 μg/kg) for 16 months. Trabecular bone content is lower in the OVX control monkey and increased in PTH(1–84)-treated animals. Trabecular number and connectivity density increased, particularly in monkeys given the 25 μg/kg dose of PTH(1–84). BV/TV increases were similar to histomorphometry results at the lumbar spine validating extrapolation of results from lumbar to thoracic spine (Modified from Fox et al. 2008)

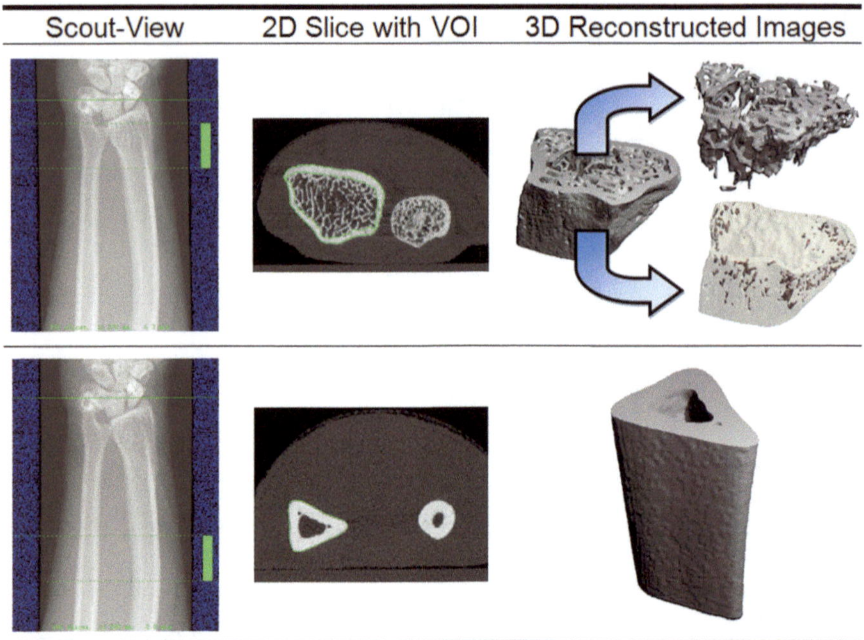

Fig. 6.12 HR-pQCT of the distal radius metaphysis (*top*) and diaphysis (*bottom*) allows visualization and the quantification of BMD, tissue mineral density (TMD), trabecular architecture, and cortical porosity

used in humans), sufficient to depict microarchitecture assessment over time in large animals. Measurements are obtained at the distal radius or the distal tibia within a large region of 10 mm length approximately depending on the scan setting, Fig. 6.12. The precision of densitometric variables is high (MacNeil and Boyd 2008) except for cortical porosity due to the minimal voxel size reached. Cortical porosity and tissue mineral density (TMD) can be biased by partial volume effects. HR-pQCT provides assessments of drug efficacy and also safety. Decreases in bone mass and related bone strength were reported with exemetase, an oral steroid aromatase inhibitor (Cheung et al. 2012).

6.5.3 *Finite Element Analysis*

CT and micro-CT allow to reach another level in bone noninvasive assessment with finite element analysis (FEA) used to predict biomechanical strength of the bone, Fig. 6.13. As in clinical studies, FEA developed from computed tomography (CT) images of bones are useful in preclinical rodent studies assessing treatment effects on bone strength in rodents (Nyman et al. 2015) and in nonhuman primates (NHP)

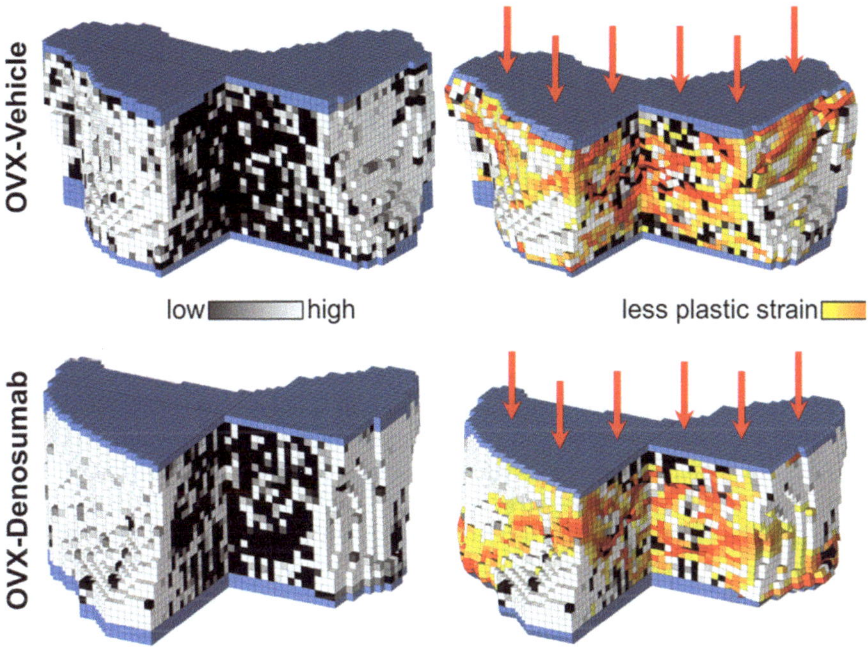

Fig. 6.13 Finite element modeling of the vertebral bodies under a compression load. The areas of high plastic strain (*red*) are likely those where fractures occur first; grayscale values indicate bone mineral density (Modified from Lee et al. 2016b)

(Jayakar et al. 2012). Even at a lower resolution, FEA correlated well with actual biomechanical strength measurement in preclinical studies, allowing to discriminate drug effect in preclinical studies (Lee et al. 2016) which correlated well with results reported in humans (Keaveny et al. 2014).

6.5.4 *Fetal Examination*

Micro-CT imaging found a special place in skeletal preclinical imaging for examining fetal skeletons in regulatory developmental toxicology studies. Assessments of the fetal skeleton are routinely conducted in preclinical safety studies in drug development. Examination of fetal skeletons after staining with alizarin red is currently the standard procedure for detection of developmental abnormalities. However, micro-CT, with 3D high resolution, fast scanning time, and carrousel loading for high throughput, is a practical and reliable imaging alternative for ex vivo fetal evaluations (Winkelmann and Wise 2009). Recently, a guidance paper for micro-CT fetal examinations was proposed to advance the acceptability of this technology to provide data to regulatory agencies (Solomon et al. 2016), highlighting the importance of the scanner calibration and maintenance, scan settings, image processing, procedure precision, along with GLP validation of technologies, software, and procedures, definition of the raw data, and standard operating procedures (SOPs).

6.6 MRI

Magnetic resonance imaging (MRI) has not been largely used for skeletal imaging both in clinical and preclinical studies, but it can be used in addition to X-rays and computed tomography to detect small or stress fractures and metastatic fractures. MRI found its principal application in soft tissue imaging because of the large range of soft tissue contrast possible, but it is also applicable to differentiate different types of fractures. New developments led to the assessment of trabecular and cortical bone (Patsch et al. 2011). However, imaging bone microstructure requires high-resolution MRI and a compromise on signal-to noise ratio, spatial resolution, and imaging time. For trabecular microarchitecture assessment, 3T MRI correlates well with micro-CT for trabecular structural parameters and trabecular microarchitecture (Phan et al. 2006), but binarization of the MRI images to obtain these measurement is difficult (Majumdar et al. 1996). Measurements of cortical bone and cortical porosity can also be performed (Gomberg et al. 2005). So application of MRI to preclinical bone imaging is limited, but current active research for clinical application could lead to a larger interest in the future for this technology which lacks ionizing radiation.

6.7 PET and SPECT

These platforms are very promising tools for investigating functional and molecular processes in vivo with new tracers becoming available as biomarkers, especially for bone tumor studies in small animal models (Doré-Savard et al. 2014). As instrument availability grows and prices fall, these methods may see increasing use in specialized studies of skeletal toxicity as well. In bone scintigraphic nuclear imaging, radiobisphosphonate bone scans following the administration of 99m-technetium-labeled ethylene diphosphonate may be advantageous as this agent is preferentially incorporated into the skeleton at metabolically active sites associated with bone formation. Bone scintigraphy reveals enhanced radiobisphosphonate deposition due to changes in blood flow or increased osteoblastic activity. The nuclear bone scan competes with functional imaging using positron emission tomography (PET), single photon emission computed tomography (SPECT), and optical imaging in the ability to visualize abnormal metabolism in bones (Tremoleda et al. 2011). A key advantage of nuclear bone scans in laboratory animals is that they are considerably less expensive. Optical imaging (fluorescence and bioluminescence), PET, and SPECT usually are used to visualize specific molecular processes in association

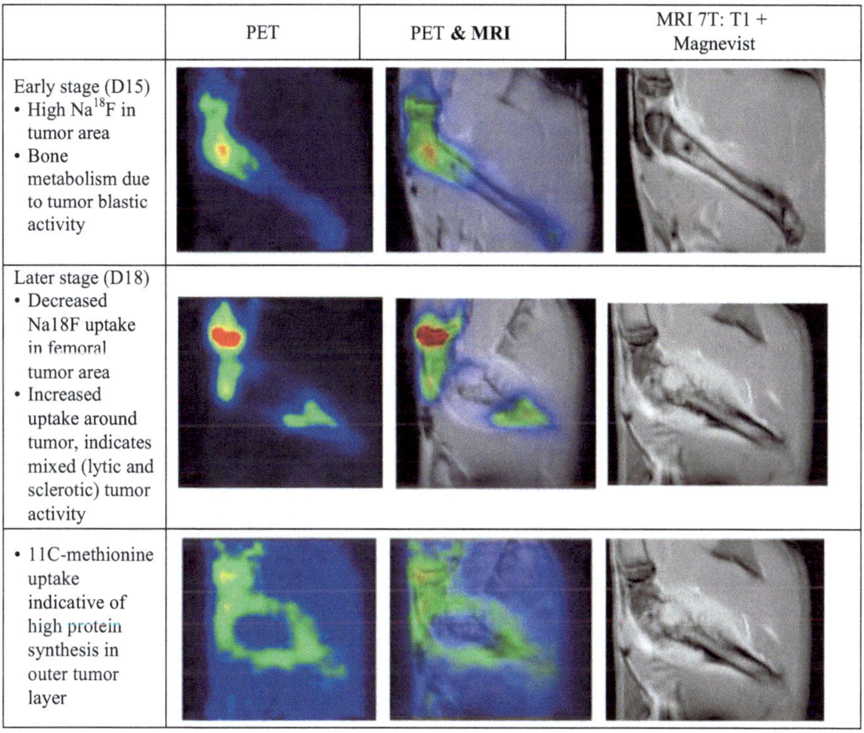

Fig. 6.14 MRI and PET for bone cancer (Modified from Doré-Savard et al. 2014)

with their anatomic localization rather than to view bone structures per se. A number of bone-specific and tumor-specific tracers for positron emission tomography/computed tomography (PET/CT) are now available. At present, many manufacturers provide multimodal systems combining the advantages of anatomical modalities such as computed tomography (CT) and magnetic resonance imaging (MRI) with the functional applications of PET and SPECT. As in the clinical market, common combinations for small animal imaging are SPECT/CT, PET/CT, and PET/MRI (Fig. 6.14).

6.8 Integrating Skeletal Imaging in Preclinical Studies

Implementing these technologies into a good laboratory practice environment requires rigorous testing, validation, and documentation to ensure the reproducibility of data. Standardized reproducible procedures and instrument precision are key for rigorous studies.

Standard safety evaluation usually involves X-rays and densitometry assessments (DXA and/or pQCT), whereas more mechanistic studies could require the use the micro-CT or HR-pQCT or other imaging modalities such as MRI or PET/SPECT.

6.8.1 Calibration, Phantom, and Quality Control

Scanners should have excellent long-term precision because changes in bone can occur over a long period of time. Maintenance and verification of the calibration needs to be performed on a regular basis. Calibration has to be very stable. There are instrument quality control procedures provided by manufacturers to detect any long-term drifts should they occur. They have to be effective, monitored over time, and performed even when scanners are not in use. Cross-calibration between multiple densitometry machines can be performed to allow quantitative comparisons and use of more than one scanner per study. Imaging facilities need to establish and enforce corrective action thresholds that trigger a call for service, maintain service logs, and comply with government inspections, radiation surveys, and regulatory requirements (ISCD 2015).

6.8.2 Positioning/Landmark and Precision Assessment

Given the importance of such data, these techniques must be both precise and reproducible. Good precision depends on scanners being operated by skilled and appropriately trained staff and that rigorous quality assurance protocols or standard operating procedures are in place. Precision errors measure the reproducibility of densitometry measures, and structural results can be shown to be reproducible by performing repeated scans on a representative group of subjects or phantoms. Precision is usually expressed in terms of the percentage coefficient of variation (CV). Standardization of positioning for each specific site to scan and use of clear anatomical landmarks are very important. To reach a statistically significant conclusion when monitoring change over time in vivo, precision error should be as small as possible.

6.8.3 Scanning Parameters and Image Analysis

Defining scan parameters is critical to properly assess bone structure depending on the site and the species. Scan parameters and image analysis processing need to be clearly defined for each type of scan (Bouxsein et al. 2010). Another important challenge for preclinical imaging is the small size of animals or specimens that require high resolution and sensitivity in smaller fields of view while maintaining a reasonable acquisition time. Radiation exposure is important to consider for in vivo scanning, especially the cumulative dose for the longitudinal studies. The level of radiation exposure depends on the resolution, the scanned site (proximity to radiation-sensitive organs), and the size of the scanned region. The radiation exposure for DXA, pQCT, and micro-CT measurements is small (about 3 μ Sv effective dose for radial HR-pQCT, Cabal et al. 2013).

6.8.4 GLP Environment and Validation of Imaging Software

Good laboratory practice (GLP) requirements must be considered with imaging technologies used in regulatory toxicity studies (FDA Regulations on Software/Computers/Electronic Records: 21 CFR Part 58: Good Laboratory Practices and 21 CFR Part 11 Electronic Records & Signatures). Digital imaging validation presents a real challenge in a regulated environment. Data integrity is key to ensure accuracy, reliability, consistent intended performance, and the ability to discern invalid or altered records. Software validation of systems must define the intended use of the system (e.g., consultative, illustrative, image analysis), limit to specified functionality, define regulatory issues and mitigations, write good requirements, test for intended use, use and maintenance procedures, and define limitations to the use of the system. Ideally, the system audit trail records must contain the following information: identity of operator entering the transaction, date and local time of the transaction, old (previous) entry, new entry, action taken (i.e., create, modify, delete), and reason for the change. However, procedural controls can be put in place for systems that lack audit trail or for which the audit trial does not fulfill all requirements. Imaging equipment and archived electronic data need to be housed under appropriate environmental conditions (e.g., protected from heat, water, and electromagnetic forces). Vendor assessment is also required to verify the quality management program, documentation to support development, and testing of the system development life cycle, procedures, testing, software versioning control, bug reporting, and problem resolution.

When included in preclinical regulated studies, protocols have to detail the equipment, software systems (e.g., image reconstruction, image evaluation), settings, scan times and resolution, contrast agents if used, and archiving. Study data should be transferred to the archives at the close of the study, including electronic data. Archived computer data are housed under appropriate environmental conditions and there should be controlled procedures for adding or removing material.

6.8.5 Study Design and Practical Aspects

Bone densitometry (DXA or/and pQCT) measurements can be included easily in most toxicity study designs of 28 days' duration or longer in rodents and 3 months' duration or longer for large animals and during the recovery phase. For toxicity studies of 3–6 months' duration that include bone densitometry measurements, skeletally mature monkeys of 6 years or older are recommended in order to minimize the range of age-related changes among the animals on the study. Densitometry measurements can be performed longitudinally in vivo as well as on excised bones ex vivo. For short-term studies, ex vivo assessment may be sufficient to perform an initial skeletal evaluation, especially in rodents. Ex vivo densitometry assessments are specifically important before biomechanical testing of bones to measure bone density and geometry and thereafter calculate material bone strength properties and analyze correlation between bone mass/density and bone strength parameters. Pretreatment data are of importance for interpreting densitometry data, especially for large animal studies. Baseline or pretreatment data are important to calculate individual changes at each time point. The percent change from pretreatment baseline calculated for each individual provides more powerful data (vs. absolute values at each occasion) and helps to overcome limitations due to individual animal variability and the small group sizes normally used in toxicology studies for large animal species. Adjustment strategies can be taken to correct for body weight or for a specific bone length, especially if an important effect on body weight or growth is expected.

Typically for monitoring individual animals, an adequate interval of time, 1 month in rodents and 3 months in NHP, is required between measurements (examples of studies using these scan intervals can be found in Varela et al. 2016; Ominsky et al. 2016), unless particularly large changes in BMD are expected (as used in a proof-of-concept study reported by Ominsky et al. 2010). Loss or gain in bone mass, alterations in bone geometry, and ultimately shifts in biomechanical properties are cumulative, so important changes that may be detected later reflect the net adjustments in bone metabolism that occurred over the entire study period. DXA and QCT are generally used together to provide complementary assessments of bone density and geometry for different parts of the skeleton. For juvenile toxicity studies, pQCT is of particular interest for in vivo growth evaluations over time as the evolving geometry of developing bone can be assessed readily with this technique. Animals are imaged under anesthesia, which helps to restrain their gross motion. For some technology, the implementation of gating methods has markedly improved the acquisitions, minimizing any interference effects due to the physiological movement. Anesthesia for densitometry procedures should be questioned if drug-specific clinical signs are expected (e.g., morbidity affecting cardiovascular and respiratory functions, kidney or hepatic failure) because of the inherent risks associated with anesthesia in animals undergoing physiological stress. Moreover, close monitoring of anesthesia is required for the animal welfare. Where possible, both DXA and pQCT scans or different imaging techniques can be performed during the same anesthesia session to limit the number of anesthesia exposures. When testing biologic agents, a lack of pharmacological or toxicological responses for these densitometric endpoints may be related to the development of antidrug antibodies. Different endpoints are often used in combination or applied using a tiered approach. For a tier-based approach to work efficiently, provision is made in the protocol to retain serum and/or urine as well as frozen and fixed samples of specific bones for possible future densitometric and/or biomechanical and/or histomorphometric assessments. In these cases, additional analyses will be conducted depending on initial in vivo results from bone markers alone or from bone markers combined with in vivo DXA and/or pQCT readings.

6.9 Conclusion

In vivo imaging represents a powerful tool to assess toxicity to bone. Imaging stands out as one of the most promising noninvasive and translational techniques that can significantly improve decision-making in the early phase of drug development, to kill compounds that are destined to fail in a later phase; the go/no-go decisions can be made earlier based on pertinent information. Imaging technologies can improve the drug development process, not only with the development of safer and effective drugs but also with reducing timelines. Better prediction of toxicology at an earlier stage will certainly limit the large contribution of drug failure for adverse effect in later-stage development. Specific imaging assessments are essential in preclinical toxicology studies for some class of drugs due to the limitations of usual bone histology. Adverse effects of drugs on the skeleton can result in decreases in bone mass or geometry and ultimately affect bone strength. Imaging can be complementary to other different methods but consist of a powerful in vivo tool. General toxicology studies in which skeletal changes are in question or specialized studies designed for the evaluation of skeleton often incorporate additional endpoints for analysis. Bone mass measurement by densitometry is a primary endpoint of drug efficacy in both clinical trials and nonclinical studies and also is employed as a means to quantify any adverse effects on bone mass.

Although imaging has not yet a major place in safety pharmacology and toxicology studies, several applications exist in skeletal toxicity. Same technologies and the same physiological and pathological parameters can be quantified for both pharmacology and toxicology applications, confirming the role of in vivo imaging as a translational biomarker for both efficacy and safety assessment. The future trends will certainly be in molecular and multimodality imaging (PET and MRI), combining high-sensitivity and molecular techniques with high spatial resolution and morphological techniques.

References

Bagi CM, Berryman E, Moalli MR. Comparative bone anatomy of commonly used laboratory animals: implications for drug discovery. Comp Med. 2011;61(1):76–85.

Bahtiri E, Islami H, Hoxha R, Qorraj-Bytyqi H, Rexhepi S, Kreshnik H, Thaçi K, Thaçi S, Karakulak C. Esomeprazole use is independently associated with significant reduction of BMD: 1-year prospective comparative safety study of four proton pump inhibitors. J Bone Miner Metab. 2015;34(5):571–9.

Binkovitz LA, Henwood MJ. Pediatric DXA: technique and interpretation. Pediatr Radiol. 2007;37(1):21–31. doi:10.1007/s00247-006-0153-y.

Blake GM, Fogelman I. How important are BMD accuracy errors for the clinical interpretation of DXA scans? J Bone Miner Res. 2008;23(4):457–62.

Blake G, Adams JE, Bishop N. Chapter 30. DXA in adults and children 251 primer on the metabolic bone diseases and disorders of mineral metabolism, Eighth Edition. Edited by Clifford J. Rosen. 2013 American Society for Bone and Mineral Research. Published 2013 by John Wiley & Sons, Inc.

Bolotin HH. DXA in vivo BMD methodology: an erroneous and misleading research and clinical gauge of bone mineral status, bone fragility, and bone remodelling. Bone. 2007;41(1):138–54.

Bonnet N, Benhamou CL, Brunet-Imbault B, Arlettaz A, Horcajada MN, Richard O, Vico L, Collomp K, Courteix D. Severe bone alterations under beta2 agonist treatments: bone mass, microarchitecture and strength analyses in female rats. Bone. 2005;37(5):622–33.

Bonnick SL, Johnston CC Jr, Kleerekoper M, Lindsay R, Miller P, Sherwood L, Siris E. Importance of precision in bone density measurements. J Clin Densitom. 2001;4(2):105–10.

Bouxsein ML, Boyd SK, Christiansen BA, Guldberg RE, Jepsen KJ, Müller R. Guidelines for assessment of bone microstructure in rodents using micro–computed tomography. J Bone Miner Res. 2010;25:1468–86. (Free full text available from http://onlinelibrary.wiley.com/doi/10.1002/jbmr.141/full).

Boyce RW, Varela A, Chouinard L, Bussiere JL, Chellman GJ, Ominsky MS, Pyrah IT. Infant cynomolgus monkeys exposed to denosumab in utero exhibit an osteoclast-poor osteopetrotic like skeletal phenotype at birth. Bone. 2014;64:314–25.

Boyd SK, Davison P, Müller R, Gasser JA. Monitoring individual morphological changes over time in ovariectomized rats by in vivo micro-computed tomography. Bone. 2006;39(4):854–62.

Cabal A, Jayakar RY, Sardesai S, Phillips EA, Szumiloski J, Posavec DJ, Mathers PD, Savitz AT, Scott BB, Winkelmann CT, Motzel S, Cook L, Hargreaves R, Evelhoch JL, Dardzinski BJ, Hangartner TN, McCracken PJ, Duong LT, Williams DS. High-resolution peripheral quantitative computed tomography and finite element analysis of bone strength at the distal radius in ovariectomized adult rhesus monkey demonstrate efficacy of odanacatib and differentiation from alendronate. Bone. 2013;56(2):497–505.

Castillo AB, Tarantal AF, Watnik MR, Martin RB. Tenofovir treatment at 30 mg/kg/day can inhibit cortical bone mineralization in growing rhesus monkeys (*Macaca mulatta*). J Orthop Res. 2002;20:1185–9.

Cheung AM, Tile L, Cardew S, Pruthi S, Robbins J, Tomlinson G, Kapral MK, Khosla S, Majumdar S, Erlandson M, Scher J, Hu H, Demaras A, Lickley L, Bordeleau L, Elser C, Ingle J, Richardson H, Goss PE. Bone density and structure in healthy postmenopausal women treated with exemestane for the primary prevention of breast cancer: a nested substudy of the MAP.3 randomised controlled trial. Lancet Oncol. 2012;13(3):275–84.

Cui GL, Syversen U, Zhao CM, et al. Long-term omeprazole treatment suppresses body weight gain and bone mineralization in young male rats. Scand J Gastroenterol. 2001;36:1011–5.

de Mello WG, de Morais SR, Dornelles RC, Kagohara Elias LL, Antunes-Rodrigues J, Bedran de Castro JC. Effects of neonatal castration and androgenization on sexual dimorphism in bone, leptin and corticosterone secretion. Bone. 2012;50(4):893–900.

Doré-Savard, et al. Mammary cancer bone metastasis follow-up using multimodal small-animal MR and PET imaging. J Nucl Med. 2014;54:944–52.

European Medicines Agency CHMP. Guideline on the evaluation of medicinal products in the treatment of primary osteoporosis. London: European Medicines Agency; 2006.

Ferretti JL, Gaffuri O, Capozza R, Cointry G, Bozzini C, Olivera M, Zanchetta JR, Bozzini CE. Dexamethasone effects on mechanical, geometric and densitometric properties of rat femur diaphyses as described by peripheral quantitative computerized tomography and bending tests. Bone. 1995;16(1):119–24.

Food and Drug Administration; Osteoporosis: Nonclinical Evaluation of Drugs Intended for Treatment Guidance for Industry. Draft guidance 2016. http://www.fda.gov/downloads/Drugs/GuidanceComplianceRegulatoryInformation/Guidances/UCM506366.pdf.

Fox J, Miller MA, Newman MK, Turner CH, Recker RR, Smith SY. Treatment of skeletally mature ovariectomized rhesus monkeys with PTH(1-84) for 16 months increases bone formation and density and improves trabecular architecture and biomechanical properties at the lumbar spine. J Bone Miner Res. 2007a;22:260–73.

Fox J, Miller MA, Newman MK, Recker RR, Turner CH, Smith SY. Effects of daily treatment with parathyroid hormone 1-84 for 16 months on density, architecture and biomechanical properties of cortical bone in adult ovariectomized rhesus monkeys. Bone. 2007b;41:321–30.

Fox J, Newman MK, Turner CH, Guldberg RE, Varela A, Smith SY. Effects of treatment with parathyroid hormone 1–84 on quantity and biomechanical properties of thoracic vertebral trabecular bone in Ovariectomized rhesus monkeys. Calcif Tissue Int. 2008;82:212–20.

French JM. Imaging and morphology in reproductive toxicology – progress to date and future directions. Reprod Toxicol. 2014;48:37–40. pii: S0890-6238(14)00052-5. doi:10.1016/j.reprotox.2014.03.008. [Epub ahead of print].

Fukuda S, Cho F, Honjo S. Bone growth and development of secondary ossification centers of extremities in the cynomolgus monkey (*Macaca fascicularis*). Jikken Dobutsu. 1978;27(4):387–97.

Geiger M, Forasiepi AM, Koyabu D, Sánchez-Villagra MR. Heterochrony and post-natal growth in mammals – an examination of growth plates in limbs. J Evol Biol. 2014;27(1):98–115.

Gluer CC. Monitoring skeletal changes by radiological techniques. J Bone Miner Res. 1999;14(11):1952–62.

Glüer CC, Genant HK. Impact of marrow fat on accuracy of quantitative CT. J Comput Assist Tomogr. 1989;13(6):1023–35.

Gomberg BR, Saha PK, Wehrli FW. Method for cortical bone structural analysis from magnetic resonance images. Acad Radiol. 2005;12(10):1320–32.

Gunson D, Gropp KE, Varela A. Bone and joints. In: Haschek WM, Rousseaux CG, Wallig MA, editors. Haschek and Rousseaux's handbook of toxicologic pathology. Elsevier Inc., Academic Press; 2013a. p. 2761–858.

Gunson D, Gropp KE, Varela A. Toxicologic pathology of bones and joints. In: Haschek WM, Rousseaux CG, editors. Handbook of toxicologic pathology. 3rd ed. San Diego: Academic Press; 2013b.

Hiligsmann A, Papaioannou D, Pierroz D, Silverman SL, P. Szulc, the epidemiology and quality of life working group of IOF. Osteoporos Int. 2013;24:2763–4. (http://link.springer.com/article/10.1007/s00198-013-2413-7).

International Society for Clinical Densitometry (ISCD). Official positions adults & pediatric official positions. West Hartford: International Society for Clinical Densitometry; 2015. Available from: http://www.iscd.org.

Jayakar RY, Cabal A, Szumiloski J, Sardesai S, Phillips EA, Laib A, Scott BB, Pickarski M, Duong LT, Winkelmann CT, McCracken PJ, Hargreaves R, Hangartner TN, Williams DS. Evaluation of high-resolution peripheral quantitative computed tomography, finite element analysis and biomechanical testing in a pre-clinical model of osteoporosis: a study with odanacatib treatment in the ovariectomized adult rhesus monkey. Bone. 2012;50(6):1379–88.

JMHW. Guideline concerning the clinical evaluation method for anti-osteoporosis agents, pharmaceutical examination No. 742. In: Management BoH, editor: issued by the Prefectural Bureau Chief, Bureau of Health Management, Section of Examination and Control, Bureau of Drug Safety, Japanese Ministry of Health and Welfare; 1999.

Johansson S, Lind PM, Hakansson H, Oxlund H, Orberg J, Melhus H. Subclinical hypervitaminosis A causes fragile bones in rats. Bone. 2002;31(6):685–9.

Johnson C, Winkelmann CT, Wise LD. Considerations for conducting imaging studies in support of developmental toxicology studies for regulatory submission. Reprod Toxicol. 2014;48:41–3.

Jolette J, Wilker CE, Smith SY, Doyle N, Hardisty JF, Metcalfe AJ, Marriott TB, Fox J, Wells DS. Defining a noncarcinogenic dose of recombinant human parathyroid hormone 1-84 in a 2-year study in Fischer 344 rats. Toxicol Pathol. 2006;34:929–40.

Kaweblum M, Aguilar MC, Blancas E, Kaweblum J, Lehman WB, Grant AD, Strongwater AM. Histological and radiographic determination of the age of physeal closure of the distal femur, proximal tibia, and proximal fibula of the New Zealand white rabbit. J Orthop Res. 1994;12(5):747–9.

Kazakia GJ, Burghardt AJ, Cheung S, Majumdar S. Assessment of bone tissue mineralization by conventional x-ray microcomputed tomography: comparison with synchrotron radiation microcomputed tomography and ash measurements. Med Phys. 2008;35(7):3170–9.

Keaveny TM, McClung MR, Genant HK, Zanchetta JR, Kendler D, Brown JP, Goemaere S, Recknor C, Brandi ML, Eastell R, Kopperdahl DL, Engelke K, Fuerst T, Radcliffe HS, Libanati C. Femoral and vertebral strength improvements in postmenopausal women with osteoporosis treated with denosumab. J Bone Miner Res. 2014;29(1):158–65.

Kumar S, Hoffman SJ, Samadfam R, Mansell P, Jolette J, Smith SY, Guldberg RE, Fitzpatrick LA. The effect of rosiglitazone on bone mass and fragility is reversible and can be attenuated with alendronate. J Bone Miner Res. 2013;28(7):1653–65.

Lazarenko OP, Rzonca SO, Hogue WR, Swain FL, Suva LJ, Lecka-Czernik B. Rosiglitazone induces decreases in bone mass and strength that are reminiscent of aged bone. Endocrinology. 2007;148(6):2669–80. Epub 2007 Mar 1.

Lee DC, Varela A, Kostenuik PJ, Ominsky MS, Keaveny TM. Finite element analysis of Denosumab treatment effects on vertebral strength in ovariectomized cynomolgus monkeys. J Bone Miner Res. 2016a; doi:10.1002/jbmr.2830.

Lee DC, Varela A, Kostenuik PJ, Ominsky MS, Keaveny TM. Finite element analysis of denosumab treatment effects on vertebral strength in ovariectomized cynomolgus monkeys. J Bone Miner Res. 2016b;31(8):1586–95.

Li X, Ominsky MS, Warmington KS, et al. Sclerostin antibody treatment increases bone formation, bone mass, and bone strength in a rat model of postmenopausal osteoporosis. J Bone Miner Res. 2009;24(4):578–88.

MacNeil JA, Boyd SK. Improved reproducibility of high-resolution peripheral quantitative computed tomography for measurement of bone quality. Med Eng Phys. 2008;30(6):792–9. doi:10.1016/j.medengphy.2007.11.003. Epub 2008 Mar 4.

Majumdar S, Newitt D, Mathur A, Osman D, Gies A, Chiu E, Lotz J, Kinney J, Genant H. Magnetic resonance imaging of trabecular bone structure in the distal radius: relationship with X-ray tomographic microscopy and biomechanics. Osteoporos Int. 1996;6(5):376–85.

Martin EA, Ritman EL, Turner RT. Time course of epiphyseal growth plate fusion in rat tibiae. Bone. 2003;32(3):261–7.

McLaughlin F, Mackintosh J, Hayes BP, McLaren A, Uings IJ, Salmon P, Humphreys J, Meldrum E, Farrow SN. Glucocorticoid-induced osteopenia in the mouse as assessed by histomorphometry, microcomputed tomography, and biochemical markers. Bone. 2002;30:924–30.

National Osteoporosis Foundation. Clinician's guide to prevention and treatment of osteoporosis. Washington, DC: National Osteoporosis Foundation; 2014. Available from: https://my.nof.org/bone-soruce/education/clinicians-guide-to-the-prevention-and-treatment-of-osteoporosis.

Nyman JS, Uppuganti S, Makowski AJ, Rowland BJ, Merkel AR, Sterling JA, Bredbenner TL, Perrien DS. Predicting mouse vertebra strength with micro-computed tomography-derived finite element analysis. Bonekey Rep. 2015;4:664. doi:10.1038/bonekey.2015.31. eCollection 2015.

Ominsky MS, Vlasseros F, Jolette J, Smith SY, Stouch B, Doellgast G, Gong J, Gao Y, Cao J, Graham K, Tipton B, Cai J, Deshpande R, Zhou L, Hale MD, Lightwood DJ, Henry AJ, Popplewell AG, Moore AR, Robinson MK, Lacey DL, Simonet WS, Paszty C. Two doses of sclerostin antibody in cynomolgus monkeys increases bone formation, bone mineral density, and bone strength. J Bone Miner Res. 2010;25:948–59.

Ominsky MS, Stouch B, Schroeder J, et al. Denosumab, a fully human RANKL antibody, reduced bone turnover markers and increased trabecular and cortical bone mass, density, and strength in ovariectomized cynomolgus monkeys. Bone. 2011;49:162–73.

Ominsky MS, Boyd SK, Varela A, Jolette J, Felx M, Doyle N, Mellal N, Smith SY, Locher K, Buntich S, Pyrah I, Boyce RW. Romosozumab improves bone mass and strength while maintaining bone quality in ovariectomized cynomolgus monkeys. J Bone Miner Res. 2016; doi:10.1002/jbmr.3036. [Epub ahead of print].

Patsch JM, Burghardt AJ, Kazakia G, Majumdar S. Noninvasive imaging of bone microarchitecture. Ann N Y Acad Sci. 2011;1240:77–87.

Phan CM, Matsuura M, Bauer JS, Dunn TC, Newitt D, Lochmueller EM, Eckstein F, Majumdar S, Link TM. Trabecular bone structure of the calcaneus: comparison of MR imaging at 3.0 and 1.5 T with micro-CT as the standard of reference. Radiology. 2006;239(2):488–96.

Pouliot L, Schneider M, DeCristofaro M, Samadfam R, Smith SY, Beckman DA. Assessment of a nonsteroidal aromatase inhibitor, letrozole, in juvenile rats. Birth Defects Research (Part B). 2013;98(374–390)

Roach HI, Mehta G, Oreffo ROC, Clarke NMP, Cooper C. Temporal analysis of rat growth plates: cessation of growth with age despite presence of a physis. J Histochem Cytochem. 2003;51:373.

Rohde CM, DeLuca H. Bone resorption activity of all-*trans* retinoic acid is independent of vitamin D in rats. J Nutr. 2003;133:777–83.

Schoeller DA, Tylavsky FA, Baer DJ, Chumlea WC, Earthman CP, Fuerst T, Harris TB, Heymsfield SB, Horlick M, Lohman TG, Lukaski HC, Shepherd J, Siervogel RM, Borrud LG. QDR 4500A dual-energy X-ray absorptiometer underestimates fat mass in comparison with criterion methods in adults. Am J Clin Nutr. 2005;81(5):1018–25.

Shepherd JA, Fan B, Lu Y, Lewiecki EM, Miller P, Genant HK. Comparison of BMD precision for prodigy and Delphi spine and femur scans. Osteoporos Int. 2006;17(9):1303–8.

Smith BB, Cosenza ME, Mancini A, Dunstan C, Gregson R, Martin SW, Smith SY, Davis H. A toxicity profile of osteoprotegerin in the cynomolgus monkey. Int J Toxicol. 2003a;22:403–12.

Smith SY, Recker RR, Hannan M, Muller R, Bauss F. Intermittent intravenous administration of the bisphosphonate ibandronate prevents bone loss and maintains bone strength and quality in ovariectomized cynomolgus monkeys. Bone. 2003b;32:45–55.

Smith SY, Jolette J, Turner CH. Skeletal health: primate model of postmenopausal osteoporosis. Am J Primatol. 2009;71:1–14.

Smith SY, Varela A, Jolette J. Nonhuman primate models of osteoporosis. In: Duque G, Watanabe K, editors. Osteoporosis research, animals models. London: Springer; 2011. p. 135–57.

Solomon HM, Makris SL, Alsaid H, Bermudez O, Beyer BK, Chen A, Chen CL, Chen Z, Chmielewski G, AM DL, de Schaepdrijver L, Dogdas B, French J, Harrouk W, Helfgott J, Henkelman RM, Hesterman J, Hew KW, Hoberman A, Lo CW, McDougal A, Minck DR, Scott L, Stewart J, Sutherland V, Tatiparthi AK, Winkelmann CT, Wise LD, Wood SL, Ying X. Micro-CT imaging: developing criteria for examining fetal skeletons in regulatory developmental toxicology studies – A workshop report. Regul Toxicol Pharmacol. 2016;77:100–8.

Tremoleda JL, Khalil M, Gompels LL, Wylezinska-Arridge M, Vincent T, Gsell W. Imaging technologies for preclinical models of bone and joint disorders. EJNMMI Res. 2011;1(1):11.

Tsutsumi H, Katagiri K, Takeda S, Nasu T, Igarashi S, Tanigawa M, Mamba K. Standardized data and relationship between bone growth and bone metabolism in female Göttingen minipigs. Exp Anim. 2004;53(4):331–7.

Varela A, Chouinard L, Lesage E, Guldberg R, Smith SY, Kostenuik PJ, Hattersley G. One year of abaloparatide, a selective peptide activator of the PTH1 receptor, increased bone mass and strength in ovariectomized rats. Bone. 2016;95:143–50.

Walker KV, Kember NF. Cell kinetics of growth cartilage in the rat tibia. II. Measurements during ageing. Cell Tissue Kinet. 1972;5:409–19.

Wei W, Dutchak PA, Wang X, Ding X, Wang X, Bookout AL, Goetz R, Mohammadi M, Gerard RD, Dechow PC, Mangelsdorf DJ, Kliewer SA. Fibroblast growth factor 21 promotes bone loss by potentiating the effects of peroxisome proliferator-activated receptor g. Proc Natl Acad Sci U S A. 2012;109:3143–8.

Winkelmann CT, Wise LD. High-throughput micro-computed tomography imaging as a method to evaluate rat and rabbit fetal skeletal abnormalities for developmental toxicity studies. J Pharmacol Toxicol Methods. 2009;59(3):156–65.
Wise LD, Xue D, Winkelmann CT. Micro-computed tomographic evaluation of fetal skeletal changes induced by all-trans-retinoic acid in rats and rabbits. Birth Defects Research (Part B). 2010;89:408–17.
Ying, X., Barlow, N., Feuston, M., 2011. Micro-CT and volumetric imaging in developmental toxicology. In: Gupta, R.C. (Ed.), Reproductive and Developmental Toxicology. Elsevier, Inc, London, pp. 983–1000.
Yu EW, Thomas BJ, Brown JK, Finkelstein JS. Simulated increases in body fat and errors in bone mineral density measurements by DXA and QCT. J Bone Miner Res. 2012;27(1):119–24.
Zemel B, Bass S, Binkley T, Ducher G, Macdonald H, McKay H, Moyer-Mileur L, Shepherd J, Specker B, Ward K, Hans D. Peripheral quantitative computed tomography in children and adolescents: the 2007 ISCD Pediatric Official Positions. J Clin Densitom. 2008;11(1):59–74.
Zoetis T, Hurtt ME. Species comparison of anatomical and functional renal development. Birth Defects Res B Dev Reprod Toxicol. 2003 Apr;68(2):111–20.

Chapter 7
Biomechanics

Angela S.P. Lin, Gabrielle Boyd, Aurore Varela, and Robert E. Guldberg

Abstract This chapter provides guidelines for biomechanics evaluations related to preclinical bone toxicology studies. The measurement of bone mechanical properties in pharmaceutical testing is critical to understanding the potential effects of that pharmaceutical on bone health and fracture risk. The specific objectives of each study are factors to be considered in deciding which preclinical model and test type(s) to utilize, and it is important to maintain consistency in test approaches throughout each study. Therefore, to assist in standardization and experimental design decision-making, specific mechanical testing procedures for achieving functional outcome measures are outlined in this chapter. Test modes and measurements that will be described are related to compressive, bending, torsion, and shear properties of bone.

Keywords Bone biomechanics • Functional mechanical testing • Preclinical pharmaceutical effects • Fracture risk

7.1 Introduction

The skeletal system provides critical functionality including structural support, body movement, internal organ protection, mineral regulation, and hematopoeisis. Bone consists of two tissue types – cortical (or compact) and cancellous (or trabecular) – that possess different microstructural characteristics in order to carry out different skeletal system functions. Cortical bone forms the outer shell of most bones and consists of densely packed cylindrical units called osteons or haversian systems.

A.S.P. Lin (✉) • R.E. Guldberg
Woodruff School of Mechanical Engineering, Georgia Institute of Technology, Atlanta, GA, USA

Petit Institute for Bioengineering & Bioscience, Georgia Institute of Technology, Atlanta, GA, USA
e-mail: angela.lin@gatech.edu; Robert.guldberg@me.gatech.edu

G. Boyd
Musculoskeletal Research, Preclinical Services, Charles River, Senneville, Quebec, Canada

A. Varela
Charles River Preclinical Services, Senneville, QC, Canada

S.Y. Smith et al. (eds.), *Bone Toxicology*, Molecular and Integrative Toxicology, DOI 10.1007/978-3-319-56192-9_7

Cancellous bone is contained in the interior of the cortical shell and consists of a porous network of trabeculae with the spaces between occupied by the bone marrow and blood vessels.

Mechanically, bones provide protection and stability to the body by being resistant to forces exerted upon them through motion and impact. Cortical bone makes up approximately 80 % of adult human skeletal mass and has high density, stiffness, and strength, thus providing the bulk of the protective qualities of the skeleton. Trabecular rods and plates contribute by distributing stresses and strains and dissipating energy. Additionally, bone modeling and remodeling are affected by the local mechanical environment and by local bone microstructure. Mechanical stimuli experienced by both cortical and cancellous bone signal local cellular responses via mechanotransduction to regulate the adaptation of bone tissue (Duncan and Turner 1995; Frost 1987a, b, 2003).

From a compositional standpoint, bone contains many active components including various cell types, organic matrix, inorganic mineral, and water. Together with more macroscopic structural characteristics, arrangement and organization of these compositional elements contribute to the mechanical integrity of bone tissue. Osteoblasts and osteoclasts are the cell types responsible for bone formation and resorption, respectively, continually changing the microarchitecture of cortical and cancellous bone. These and other cells are surrounded by extracellular matrix material, for which type I collagen is the predominant protein. Its helical structure allows for combination of many collagen molecules into fibrils that provide bone tissue its elastic properties, tensile strength, and toughness. Also embedded in this organic matrix are minerals and salts, largely hydroxyapatite (HA), which contribute to the compressive strength and stiffness of bone.

Preclinical efficacy and toxicology research focuses not only on the effects of pharmaceuticals with an intended target of improving musculoskeletal function but also the possible deleterious responses of bone to therapeutics intended for other applications. Characterizing structural, compositional, and mechanical properties and their interactions can provide valuable information regarding functional responses of bone tissue to stimuli and therapeutics. Commonly studied anatomical sites for assessing skeletal changes include cortical and cancellous bone in vertebrae and long bones such as femora, tibiae, and radii. These load-bearing bones are susceptible to structural weakening and fragility fractures with age and disease, thus making them interesting targets for assessing changes with delivery of pharmaceuticals and therapeutics. Bone microarchitecture and mineral density are usually quantified using X-ray-based techniques such as microcomputed tomography (micro-CT), peripheral quantitative computed tomography (pQCT), and dual-energy X-ray absorptiometry (DXA). Histology and biomarker analyses are commonly used to get a more complete understanding of the bone resorption and formation components affected. Mechanical evaluation allows further assessment of effects on the functional performance of bone in its structural support role.

Pharmaceutical development programs can be limited by assumptions that safety and efficacy data derived from clinical studies in adults or preclinical studies in adult animals are applicable to the pediatric population. However, developmental processes occurring in pediatric patients may contribute to different drug responses compared to adult patients. These developmental processes, including growth and changes in metabolism, body composition, and organ functional capacity, may also be adversely affected by pharmaceuticals exhibiting no detectable effects in mature systems. Because of these potential differences, studies with juvenile animals have shown utility in providing more complete understanding of age-related pharmaceutical toxicity and efficacy during growth and development. Though resulting mechanical properties evaluated for juvenile compared to adult animals may be different, the general procedures and anatomical sites used will be similar.

The purpose of this chapter is to provide guidelines for biomechanics evaluations related to preclinical bone toxicology studies. Specific mechanical testing configurations and procedures to achieve functional outcome measures for various applications will be described.

7.2 General Biomechanical Assessment Concepts

7.2.1 Standardized Testing Methodology

Maintaining standardized testing methodology throughout a study is critical for consistency. Therefore, several basic methodology norms should be followed:

- The testing system used to perform biomechanical property assessment should undergo routine maintenance and calibration.
- The load cell capacity should be selected to appropriately reflect the test magnitude. For example, a load cell capacity of 10 kN would not provide adequate sensitivity for 3-point bending tests on mouse femurs but is appropriate for compression tests on nonhuman primate (NHP) lumbar vertebrae.
- All specimens should be stored and prepared for testing in the same manner.
- Positioning and orientation of the bones within the fixture should be consistent
- Acceptable levels of precision in the test equipment should be verified before and after biological sample testing with objects of known material properties (i.e., stiffness), such as acrylic tubes or rods.
- Strain rate should be carefully controlled and remain constant across the study.

7.2.2 Biomechanical Measures

Standard biomechanical parameters that are dependent on size and shape of the bone, referred to as extrinsic parameters, include stiffness, yield load, peak load, and work to failure (or area under the curve, AUC). As overall bone integrity is a

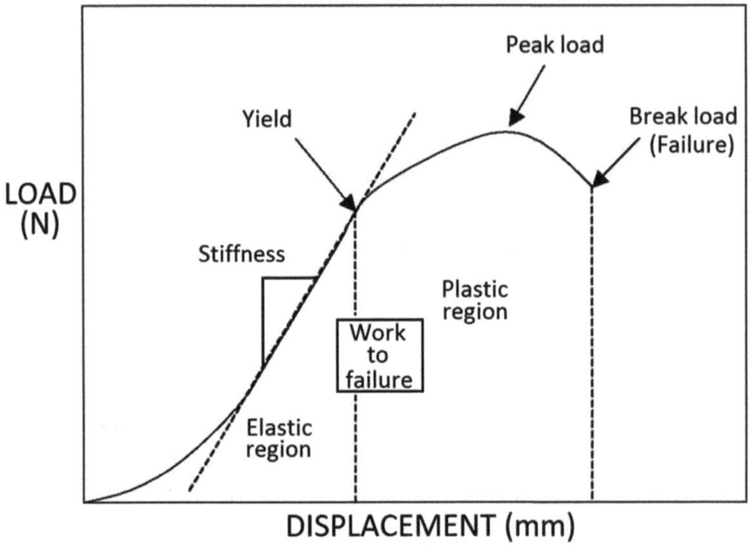

Fig. 7.1 Illustration of load-displacement curve for bone mechanical testing, depicting elastic and plastic regions, stiffness, yield point, peak load, failure point, and work to failure

function of bone mass, mineral density, and its geometric distribution, a complete assessment of bone quality will normally include the determination of geometry and density parameters measured using X-ray-based imaging techniques (Hildebrand et al. 1999; Hernandez and Keaveny 2006; Muller 2009). Geometric measurements are subsequently used to derive intrinsic parameters (those independent of bone size and shape), such as ultimate stress (apparent strength for compression), modulus, and toughness (Turner and Burr 1993; Bone Mechanics Handbook 2001).

Unlike soft tissues, the biomechanical behavior of bone can be approximated by linear elastic assumptions. The elastic region of the load-displacement curve (Fig. 7.1) corresponds to the region where the load increases linearly with the displacement (or deformation). Theoretically, if the load is removed from the specimen during this phase, the bone should recover its initial shape due to its elastic properties. In this phase, stiffness is defined as the slope of the load-displacement curve. Subsequently, after the yield point, the plastic region represents the phase where permanent damage is inflicted to the bone, until complete failure occurs.

The area under the curve (AUC) to failure point represents work to failure. Higher work to failure values reflect tougher bone and suggest greater fracture resistance because the bone can absorb more energy from a traumatic event, like a fall, before it breaks. On the other hand, skeletal fragility and proclivity to fracture are often associated with brittleness of the bone and reduced capacity to absorb energy.

7.3 Appropriate Mechanical Testing and Data Analysis Procedures

After harvest and prior to mechanical testing, it is recommended bone specimens be stored intact at −20 °C in gauze saturated with 0.9 % physiologic saline solution and then thawed completely just before testing. It is recommended that the number of specimen freeze-thaw cycles be minimized and kept constant within a given study. Other methods of storage can be considered if necessary but should be kept constant and may alter biomechanical test results. Due to its water content, bone displays limited viscoelasticity, and mechanical properties will therefore vary slightly with varying strain rate. It has been shown that increasing strain rate by an order of magnitude can increase bone strength measurements by 15 % (Pugh et al. 1973; Turner and Burr 1993; Carter and Hayes 1976, 1977). For simulating physiological conditions, strain rate should be kept in a range of 0.01–0.08 %/s (Turner and Burr 1993; Rubin and Lanyon 1982). Additionally, calculations of intrinsic mechanical properties require specimen-specific geometric information, thus measuring these with X-ray-based imaging procedures (micro-CT, pQCT) or other methods is recommended prior to testing (Goldstein et al. 1993; Hildebrand et al. 1999).

Guidelines for sample preparation, testing, and data analyses for understanding failure properties from various modes of testing are described in this section. These guidelines are meant to provide insight on possible ways to achieve testing results in the described modes as well as provide examples of species-specific test configurations. Clearly, in order to accommodate variations in sample characteristics and equipment, hardware setup and testing parameters may need to be adjusted for individual needs. It is also important to note that bone mechanical properties and structural geometry will vary depending on many factors such as anatomical site, primary loading direction, sex, age, and storage/handling conditions (Morgan and Keaveny 2001; Morgan et al. 2003; Keaveny and Hayes 1993; Keaveny 2001; Hildebrand et al. 1999; Goldstein et al. 1983; Fyhrie and Christiansen 2015).

7.3.1 Compression

Anatomical sites containing substantial trabecular bone are often loaded in compression in vivo; thus, this mode of testing is relevant for bone samples taken from vertebrae (thoracic or lumbar), long bone metaphyses, and trabecular bone cores from various locations (Athanasiou et al. 2000; Turner and Burr 2001). A simple hardware setup example is illustrated in Fig. 7.2a.

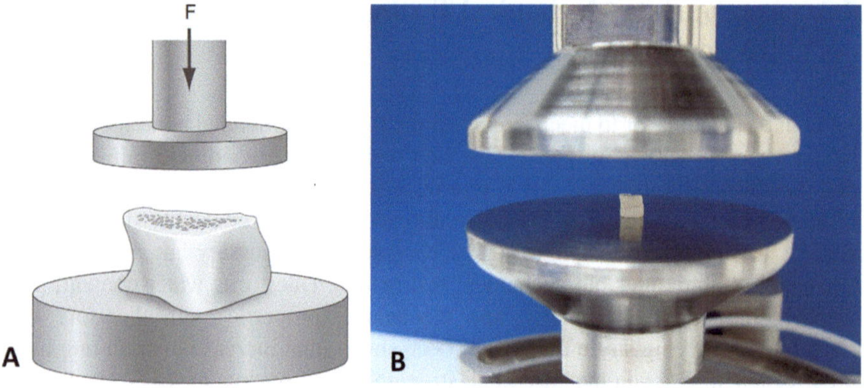

Fig. 7.2 (**a**) General test fixture illustration for compressive testing (vertebrae, vertebral bone cores, long bone distal metaphyses). Load (F) is applied at a constant axial displacement rate. (**b**) Specific test setup example photograph for rat vertebral compression

7.3.1.1 General Testing Procedures

Specimen height and cross-sectional area should be measured prior to testing. The specimen should be placed at the center of the lower compression platen such that the z-axis orientation is replicated for each (e.g., cranial surface always facing upward). With the specimen positioned on the platen, the load channel should be tared (i.e., set to a zero value). The upper compression platen can then be lowered carefully to contact the upper surface of the specimen, and once a contact load of designated constant magnitude is measured, the displacement channel should be tared again. Artifacts due to friction at the platen-sample interfaces, end effects, and edge effects have been shown to produce error in elastic modulus measurements (Keaveny et al. 1993a, 1997; Odgaard and Linde 1991; Un et al. 2006). Two methods of compensating for nonparallel testing surfaces and end/edge effects are to utilize a platen that includes a tilting surface plate (Turner and Burr 1993) or potting the bone/core ends (Kopperdahl and Keaveny 1998; Keaveny et al. 1997). Cyclic preconditioning can also be performed to reduce end effects from specimen processing. These additional methods are often deemed unnecessary assuming surface planing is consistently performed to produce parallel test surfaces. Accuracy of mechanical evaluation can also be improved via direct measurement of strain at the specimen midsection using an extensometer (Keaveny et al. 1993a, 1997). The specimen should be compressed at a constant axial displacement rate until a maximum load is reached and plateau and/or load decrease follows, indicating failure.

7.3.1.2 Vertebral and Bone Core Sample Preparation

Prior to compression testing, whole vertebrae must be prepared to isolate the vertebral body and ensure consistent test specimens. To produce plano-parallel ends, vertebrae are typically fixed in a jig and both endplates removed with a diamond saw under constant saline irrigation. Using a double-bladed saw to cut the two ends simultaneously helps ensure that the resulting cranial and caudal surfaces are parallel to one another. All vertebral processes are also removed carefully using a bone rongeur or diamond saw. After this point, cores can be excised from the vertebral body, if the intent is to investigate trabecular bone mechanical properties only. Trabecular bone cores from vertebrae (and other anatomical sites) are typically taken only in large animal studies due to size constraints. It has been shown that cylindrical trabecular specimens provide more consistent, less variable results compared to samples with rectangular cross-section (Keaveny et al. 1993b); therefore, cylindrical geometry is recommended. A diamond coated hollow drill bit, or trephine, should be used with ample irrigation. When preparing trabecular cores from large animal vertebrae (e.g., NHP, dog), an inner diameter approximately 5 mm for the trephine is suitable for extracting a representative trabecular region.

7.3.1.3 Long Bone Metaphysis Sample Preparation

Long bones should be cleaned of surrounding soft and connective tissues. Then samples should be fixed or clamped such that two parallel cuts can be made to isolate a specific length of the metaphysis with parallel surfaces. Similar double-bladed saw techniques as used with vertebral sample preparation can be used for long bone metaphyses. Another technique that can be used is to create a channel in a metal block and then machine parallel cut lines at various distances. The bone would be stabilized in the channel, and a saw would be used to make cuts using the parallel machined lines as guides.

7.3.1.4 Data Analysis and Outcome Measures

Force and displacement measurements are collected via timed data acquisition with a predetermined appropriate sampling rate. Some investigators utilize externally mounted strain gauges on the specimen to detect displacement changes more locally (instead of using test system crosshead movement measurements). These data are used to generate load-displacement curves and are normalized to stress and strain using cross-sectional area data and original gage length (or specimen height) according to the following Eqs. (7.1 and 7.2):

$$\sigma = \frac{F}{A} \tag{7.1}$$

$$\varepsilon = \frac{\Delta l}{l} \tag{7.2}$$

where

σ = stress (MPa), F = load measured (N), A = cross-sectional area (mm^2), ε = strain, Δl = change in height (mm), and l = original gage length/height (mm) (Athanasiou et al. 2000; Turner and Burr 2001).

These data and curves allow determination of outcome measures such as:

- Peak load (N): Maximum value of force recorded during the test.
- Yield load (N): Force measurement at yield point, where elastic deformation ends and plastic deformation begins. Yield point should be defined using an accepted standard technique, including the 0.2 % offset or 10 % decrease of secant slope methods (Turner and Burr 1993, 2001).
- Stiffness (N/mm): Slope of the linear portion of the load-displacement curve.
- Ultimate strength (MPa): Stress calculated at the point of peak load (maximum stress).
- Yield stress (MPa): Stress calculated at the yield point.
- Modulus (MPa): Slope of the linear portion of the stress-strain curve.
- Work to failure (N·mm): Energy required to reach peak load point, area under the load-displacement curve to the peak load point.
- Toughness (MPa): Area under the stress-strain curve to the ultimate stress point.

In some instances, likely involving vertebral bodies of dense trabecular microarchitecture, the load may not plateau or decrease but continue an upward slope due to trabecular consolidation. In this case, the interpretation of the strength parameters cannot be made with respect to the peak load or its derived parameters (work to failure or toughness). For these specimens, results will include stiffness and yield properties but not those related to ultimate or failure strength.

7.3.2 Bending

Testing of cortical bone regions, such as long bone diaphyses, in bending provides tensile loading on one side and compressive loading on the opposite side of the specimen. This type of loading represents one mode by which intact long bones and fractures undergoing repair may fail in vivo.

7.3.2.1 General Testing Procedures

Long bone diaphyseal bending, or flexural, tests are performed in either the 3-point (Fig. 7.3) or 4-point (Fig. 7.4) configuration (Athanasiou et al. 2000; Turner and Burr 2001; Robertson et al. 2006; Rajaai et al. 2010). All support and loading points

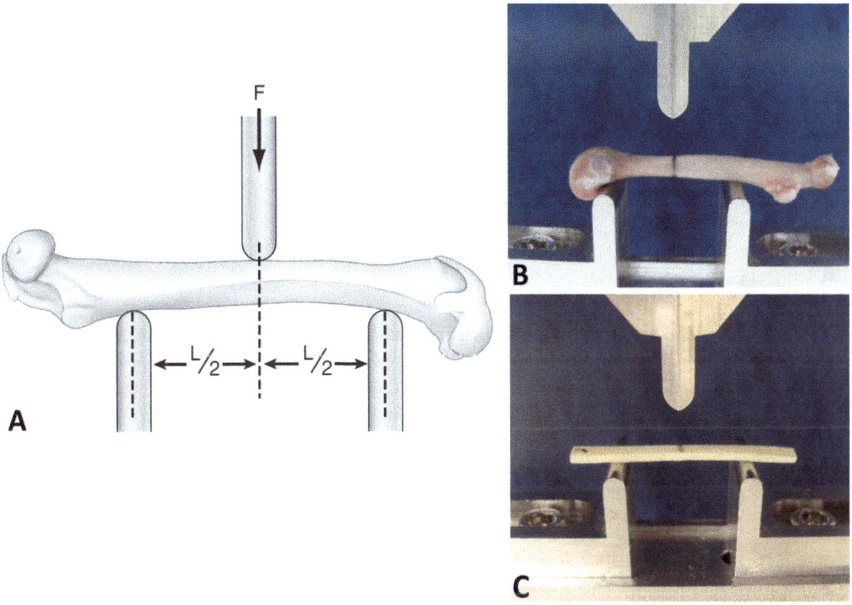

Fig. 7.3 (**a**) General test fixture illustration for long bone 3-point bending. Specimen is positioned such that bone rests in contact with and as level as possible on the lower span supports. Load (F) is applied through the upper testing nose at a constant axial displacement rate. Dimensions labeled: L = lower support span width, L/2 = distance between upper loading nose point of contact and each of the lower span supports. (**b**) Specific fixture and test setup photograph for femoral 3-point bending. Some species-specific span widths (L) are as follows (mm): rat femur female, 15; rat femur male, 20 (for age 1 year and over); mouse femur, 5; rabbit femur juvenile, 30 (New Zealand white rabbits, male or female, approx. 8 weeks old). (**c**) Cortical beams of rectangular cross section can also be machined for 3-point bending tests. In this example photograph, beam was machined from the anterior humerus to dimensions = 35 mm length × 3 mm width × 1 mm depth and positioned periosteal side up with span width $L = 25$ mm

should be rounded to reduce stress concentrations. Materials testing standards (ASTM and ISO) are commonly used resources for providing guidelines on fixture and specimen dimensions (e.g., span width, aspect ratio, radius of curvature for loading noses and supports). However, specifications set out by these guidelines, predominantly created for fabricated metal or polymer materials testing, are sometimes impossible to achieve with whole bone geometries.

The two bending test configurations have unique advantages and disadvantages. The 3-point bending fixture is simpler, with two lower support posts that form the span width (L) and one upper loading nose that applies deflection at the point directly between the lower support posts (Fig. 7.3a). One disadvantage of the 3-point bending configuration is that high localized stresses are produced directly below the loading nose. For 4-point bending, the test fixture includes two lower supports (span width, L_o) and two upper loading supports (loading span width, L_i) instead of one (Fig. 7.4a). This produces constant bending moment and in theory zero transverse shear stresses between the two loading points. However, it requires an even surface

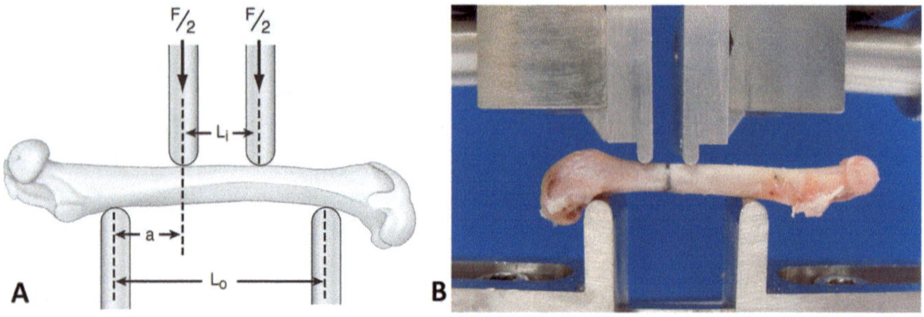

Fig. 7.4 (**a**) General test fixture illustration for long bone 4-point bending. Specimen is positioned such that bone rests in contact with and as level as possible on the lower span supports. Load (F/2 for each) is applied through the upper posts at a constant axial displacement rate. Dimensions labeled: L_o = lower support span width, L_i = upper support span width, a = distance between lower and upper span support. (**b**) Specific fixture and test setup photograph for rodent femoral 4-point bending. Mid-diaphyseal point marked with pencil is positioned halfway between the upper span supports. Rat: L_o = 18 mm, L_i = 6 mm; mouse: L_o = 12 mm, L_i = 4 mm

for each of the two upper loading points to provide equivalent load contributions, which may be challenging for some long bone geometries. Some investigators overcome this challenge by using a fixture that allows the height of the two loading supports to be individually adjusted.

Specimen orientation with respect to the bending fixtures should be consistent across all tests to ensure bending about the same plane. The specimen should be loaded onto the lower supports, load channel tared, and upper loading fixture lowered slowly to make contact. At this point, the displacement channel should be tared. Testing should be performed at constant axial displacement control rate until fracture at the mid-diaphysis occurs. It is useful to record fracture origination and patterns via diagrams or notes for each specimen, including any movement during testing. If fracture occurs away from the mid-diaphyseal region, it should be noted such that variations in resulting data may be explained.

7.3.2.2 Long Bone Sample Preparation

Long bones should be cleaned of as much soft tissue as possible, particularly at the mid-diaphyseal region where loading will be applied. The midpoint of the diaphysis should be marked using pencil or indelible ink to indicate the point at which loading will be applied (for 3-point bending). If performing 4-point bending, this mark will lie halfway between the two upper loading posts. The overall span width contact points can also be marked.

To assess cortical tissue properties specifically, beams of rectangular cross section can be carefully machined with precision diamond saws or milling tools to isolate the cortical tissue from a long bone and then tested in bending (Kuhn et al. 1989; Jepsen et al. 1997). This is typically only performed for cortical bone from large animals.

7.3.2.3 Specific Examples

Figure 7.3b shows an example of rodent femoral 3-point bending test setup. The specimen shown is positioned with anterior side upward, distal end to the left. Diaphyseal position designated for testing is marked in pencil and aligned with the center of the loading nose. Care should be taken to maintain level positioning on the lower span supports. In this example, the femur is shifted distally in part to allow the femoral neck shear testing to be performed on the same specimen.

Figure 7.3c shows a beam with rectangular cross section that was machined from cortical bone (typically from the anterior humerus shaft, 35 mm long × 3 mm wide × 1 mm deep) and set up in the 3-point bending configuration. The central position designated for testing is marked in pencil and aligned with the center of the loading nose. As cortical beams can be machined to be identical in size and dimensions, adjustment of the test fixture dimensions due to geometry is not necessary.

Figure 7.4b shows an example of rodent femoral 4-point bending test setup. The specimen is shown with anterior side upward, distal end to the left. Diaphyseal position designated for testing is marked in pencil and positioned halfway between the upper span supports. Note that the femur is shifted distally to ensure level positioning on the lower span supports and to allow subsequent femoral neck shear testing with the same specimen.

7.3.2.4 Data Analysis and Outcome Measures

Force and displacement measurements are collected via timed data acquisition with a predetermined appropriate sampling rate. Cross-sectional areal moment of inertia and distance from center of mass measurements are typically made using micro-CT imaging and utilized for stress and elastic modulus calculations according to the following equations.

For 3-point bending of whole bone (Eqs. 7.3 and 7.4):

$$\sigma = F\left(\frac{\mathrm{Lc}}{4I}\right) \tag{7.3}$$

$$E = S\left(\frac{L^3}{48I}\right) \tag{7.4}$$

For 3-point bending of machined beams of rectangular cross section (Eqs. 7.5 and 7.6):

$$\sigma = F\left(\frac{3L}{2\mathrm{bd}^2}\right) \tag{7.5}$$

$$E = S\left(\frac{L^3}{4bd^3}\right) \tag{7.6}$$

For 4-point bending (Eqs. 7.7 and 7.8):

$$\sigma = F\left(\frac{ac}{2I}\right) \tag{7.7}$$

$$E = S\left(\frac{a^2}{12I}\right)(3L_o - 4a) \tag{7.8}$$

where

σ = stress (MPa), F = load measured (N), L = span width (3-point bending, Fig. 7.3a, mm), c = distance from cross-sectional center of mass (mm), I = cross-sectional moment of inertia or second moment of area for the mid-diaphysis (mm^4), b = width of test beam (mm), d = depth of test beam (mm), a = horizontal distance between lower span support and upper span support (4-point bending, Fig. 7.4a, mm), E = elastic bending modulus (MPa), S = stiffness (slope of load-deflection curve), and L_o = lower support span width (4-point bending, Fig. 7.4a, mm) (Athanasiou et al. 2000; Turner and Burr 2001).

These data and curves allow determination of outcome measures such as:

- Peak load (N): Maximum value of force recorded during the test.
- Yield load (N): Force measurement at yield point, where elastic deformation ends and plastic deformation begins. Yield point should be defined using an accepted standard technique, such as the 0.2 % offset or 10 % decrease of secant slope methods (Turner and Burr 1993, 2001).
- Displacement to yield (mm): Displacement/deflection required to reach yield point.
- Post-yield deflection (mm): Difference of displacement at break and displacement at yield.
- Stiffness (N/mm): Slope of the linear portion of the load-displacement curve.
- Ultimate strength (MPa): Stress calculated at the point of peak load (maximum stress) (see Eqs. 7.3, 7.5, and 7.7).
- Yield stress (MPa): Stress calculated at the yield point.
- Modulus (MPa): See Eqs. 7.4, 7.6, and 7.8.
- Work to failure (N·mm): Energy required to reach break point, area under the load-displacement curve to the break point.
- Toughness (MPa): Area under the stress-strain curve to the break point.

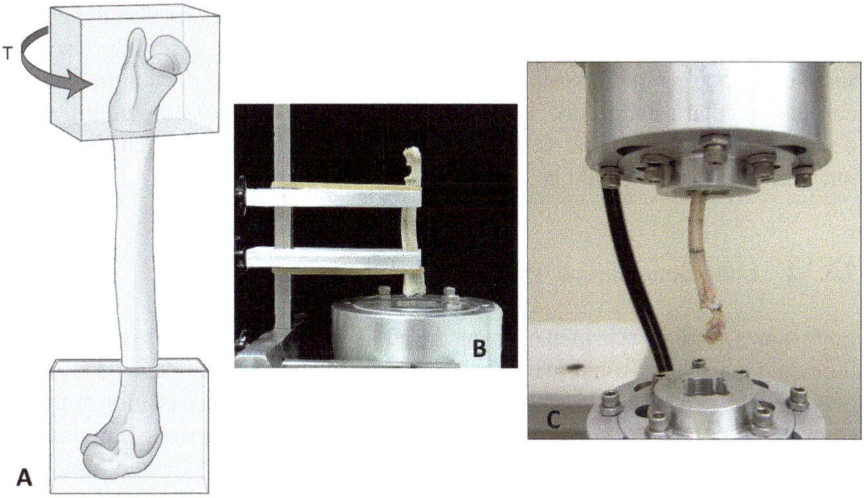

Fig. 7.5 (**a**) For long bone diaphyseal torsion testing, the long bone is potted at each end to secure and prevent any movement of the specimen relative to the fixtures. Torque (T) is applied to one end at a constant angular rotation rate. (**b**) Specific positioning/potting tool for rabbit radius-ulna specimens. Embedding *lines* are marked and at a consistent distance away from the mid-diaphyseal mark (25 mm). Potting tool arms are placed inside the embedding lines, and the specimen is held in place with rubber bands. Tool allows sample to be lowered into the potting block as a stable unit. (**c**) Once one end is potted, the block can be inverted and the remaining end potted in the second block

7.3.3 Torsion

Another relevant mode of failure for cortical regions of whole long bones and long bone fracture or defect healing models is under rotational loads, or torsion, which impart shear stresses across the cross section of the mid-diaphysis of the tested long bone (Athanasiou et al. 2000; Turner and Burr 2001). A generalized test setup is illustrated in Fig. 7.5a.

7.3.3.1 General Testing Procedures

Bone ends should be stabilized in potting blocks as described below in sample preparation such that any loading applied to the block will translate to loading on the bone end (i.e., no movement between block and sample). The torque channel should be tared initially, and then the sample assembly should be carefully loaded into a torsion fixation framework designed to prevent any unprogrammed rotational movement between the blocks and the framework. Retaining anatomical consistency from sample to sample is recommended (e.g., distal end is consistently placed on the lower end).

If there was a fixturing plate or other assembly components used in the sample potting process, those components should be removed prior to testing. Once the sample and potting block assembly is loaded and affixed to the torsion framework, constant angular rotation rate should be applied to the upper fixture until fracture occurs at the weakest point along the diaphysis. If fracture occurs at the potting interface, test results should be excluded.

7.3.3.2 Long Bone Sample Preparation

Long bones should be cleaned of as much soft tissue as possible. This is particularly important for the proximal and distal bone ends, which will be embedded within potting material in metal blocks to ensure that no rotational or translation motion is allowed. Bone ends can also be trimmed to a specific dimension. Lines can be marked with pencil or indelible ink for potting material reference lines. Metal potting blocks, typically with square cross section, should be designed with a recessed hole of large enough dimensions (width, height, diameter, depth) to accommodate the bone end plus sufficient surrounding potting material for secure fixation. Commonly used potting materials include fast curing epoxies and Wood's metal (or Lipowitz's alloy).

The potting material should be prepared (e.g., Wood's metal must be melted within the potting blocks) and bone ends dried just prior to potting. Bone ends should be potted one at a time, and the order should be kept consistent for all specimens. Ensure that the bone retains vertical positioning and minimize the movement it experiences during the curing process. Once the potting material for the first bone end has cured, invert the sample and block carefully and pot the remaining end in the second block. Often, a custom alignment assembly is used to achieve several goals: maintain consistent bone orientation, ensure that weight of the potting blocks does not damage the specimen, and maintain parallel x-y planes and y-z planes for the individual blocks. If heat is dissipated by the potting material, it is recommended that the diaphysis be kept hydrated with saline-saturated strips of gauze or Kimwipes® to prevent desiccation during sample preparation.

7.3.3.3 Specific Example

Figure 7.5b depicts a rabbit radius-ulna specimen held in a vertical position using an alignment tool. The upper and lower alignment supports have channels for the specimen to be inserted, and rubber bands are used to hold the specimen in contact with those channels. The alignment tool allows for a potting mold (bottom) to be moved upward to submerge the lower end of the bone in the contained potting material. Once one bone end is potted, the specimen can be inverted and second bone end potted with the alignment tool or with the load frame, as shown in Fig. 7.5c.

7.3.3.4 Data Analysis and Outcome Measures

Torque and rotational angle measurements are collected via timed data acquisition with a predetermined appropriate sampling rate. Measurement of polar moment of inertia, J, is required for calculations of shear stresses, strains, and modulus and can be acquired from micro-CT imaging or estimated based on assumptions of geometry. When torsional loading is applied, the shear stress varies from zero at the axis of rotation to the maximum value at the periosteal surface. For computing estimated shear stress and elastic shear modulus in whole bone testing, the following equations are used:

$$\tau = \frac{T \times r}{J} \tag{7.9}$$

$$G = \frac{k_t L}{J_{\text{avg}}} \tag{7.10}$$

where

τ = shear stress (MPa), T = torque (N·mm), r = outer radius at measurement of J = polar moment of inertia (pMOI) (mm), G = elastic shear modulus (MPa), k_t = torsional stiffness, L = gage length (mm), and J_{avg} = average pMOI (mm^4) (Athanasiou et al. 2000; Boerckel et al. 2012).

These data and curves allow for determination of outcome measures such as:

- Maximum torque (N·mm): Maximum value of torque recorded during the test
- Torsional stiffness (N·mm/deg): Slope of the torque-rotation curve
- Work to failure (J, N·mm·deg): Area under the torque-rotation curve to fracture point
- Shear strength (MPa): Shear stress at fracture point (Eq. 7.9)
- Elastic shear modulus (GPa): Specimen's response to shear stress (Eq. 7.10)

Extrinsic structural mechanical parameters such as maximum torque, torsional stiffness, and work to failure are typically reported rather than intrinsic material properties such as shear strength and modulus (Athanasiou et al. 2000).

7.3.4 Shear

A test method analogous to cantilever beam shear testing can be utilized to characterize mechanical properties of regions such as the femoral neck. The variations of this test type emulate conditions by which the femoral head experiences load in vivo, and the femoral neck is a relevant site for fracture risk assessment because many instances of age-related fractures occur at that anatomic site (Athanasiou et al. 2000; Turner and Burr 2001).

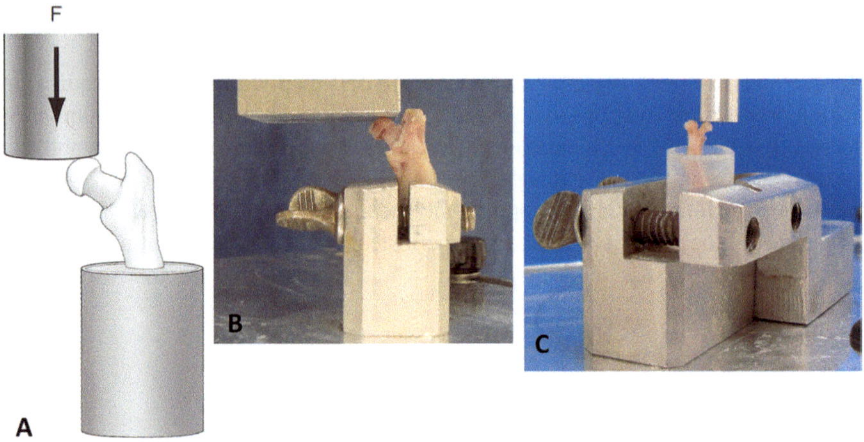

Fig. 7.6 (**a**) General setup illustration for femoral neck shear testing in the axial configuration. Femur is positioned and secured vertically. Load (F) is applied to the femoral head only using a compression platen or block at a constant axial displacement rate. (**b**) Specific test fixture photograph for rat femoral neck shear. Distal end of femur placed vertically in lower fixture and the third trochanter installed in a y-shaped clamp. Upper compression block is controlled at a constant displacement control rate downward until fracture occurs in the femoral neck. (**c**) Specific test fixture and setup photograph for mouse femoral neck shear. Distal end of femur is embedded in acrylic, and then the assembly is positioned and secured into lower fixture clamp. Upper compression block is controlled at a constant displacement control rate downward until fracture occurs in the femoral neck

7.3.4.1 General Testing Procedures

There are two configurations for femoral neck testing – axial (also called standing or weight-bearing) and lateral (or fall) configurations. In animal models, the axial configuration is most commonly used. For human cadaver studies, the lateral configuration is more utilized (Kupke 2010). The axial configuration will be described here, as diagrammed in Fig. 7.6a. The stabilized femur will be positioned such that the femoral head is directly under the loading platen, and the loading platen's downward path should not contact the femoral shaft as a constant axial displacement control rate is applied until fracture in the femoral neck occurs. The test dynamics are analogous to cantilever beam loading on the femoral neck. It is important to note fracture pattern and location, as biological variability (femoral neck length, angle, etc.) will result in some samples that fracture in the femoral head or in the diaphysis. For tests where fracture occurs outside the femoral neck region, data should be excluded.

7.3.4.2 Femoral Neck Sample Preparation

Femora should be cleaned of as much soft tissue as possible, particularly around the femoral head and neck such that visualization of the cartilage and bone at those sites is unobstructed. The distal half of the specimen should be removed using a diamond saw to produce a planar surface. One potting block or clamping device is needed for securing the distal cut end of the femur such that the femoral shaft is positioned vertically with proximal end up. For species larger than rodent, additional restraint at the greater trochanter may be required to prevent bending in the shaft and ensure fracture in the femoral neck.

7.3.4.3 Specific Examples

Figure 7.6b shows a rat proximal femur fixture for femoral neck shear testing. The distal portion of the specimen is placed vertically in the lower fixture clamp and tightened to secure the positioning. Figure 7.6c shows a mouse proximal femur set up for femoral neck shear testing. The distal portion of the specimen is first embedded into acrylic, and then this assembly is positioned into the lower fixture clamp and tightened to secure the acrylic in place.

7.3.4.4 Data Analysis and Outcome Measures

Force and displacement measurements are collected via timed data acquisition with a predetermined appropriate sampling rate. Because loading is imparted in a combination of stress conditions and on both cortical and trabecular bone tissues in this region, intrinsic material property determination is not reliable, and researchers typically report extrinsic structural properties such as:

- Maximum load (N): Maximum value of force recorded during the test
- Fracture load (N): Force value at break point
- Stiffness (N/mm): Slope of the load-displacement curve

7.3.5 Correlations and Further Computational Analyses

As mentioned earlier in this section, imaging techniques are often utilized to assess changes in bone mass, density, and microarchitectural parameters. Measurements of biochemical markers of bone metabolism and remodeling are also used to further understand the mechanisms by which bone changes as an effect of the therapeutic being tested. The relationships between mechanical properties and various measures of bone structure and remodeling offer insight into functionally relevant changes that may be occurring due to disease, degeneration, or pharmaceutical interventions (Keaveny 2001; Cory et al. 2010; Ciarallo et al. 2006). For example, bone quality assessments are standardly represented via correlation analyses between measures such as load vs. BMC or load vs. BMD data (Fig. 7.7). Changes in slope for regression lines between control and treatment groups in correlation plots such as these indicate differences in bone quality.

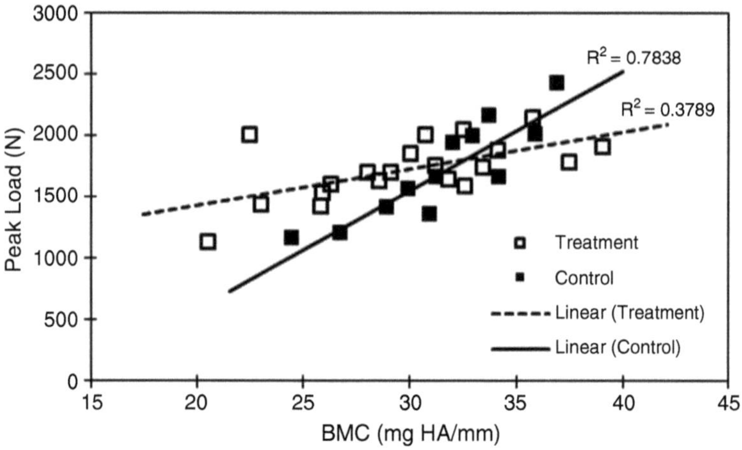

Fig. 7.7 Example of a correlation plot for load vs. BMC for control and treatment groups. Linear regression lines exhibit a significant difference in slope that is indicative of differences in bone quality between control and treatment groups

Additionally, finite element analyses (FEA) can be utilized to model local tissue mechanics based on geometry obtained through imaging techniques, simulated apparent level testing conditions applied to that geometry, and assumed or measured tissue-level material properties (Mizrahi et al. 1993; Niebur et al. 1999, 2000; Keaveny et al. 1994; Prot et al. 2015; McDonald et al. 2010; Van Rietbergen et al. 1996, 2001; Morgan et al. 2004; Yeni et al. 2005; Yeni and Fyhrie 2001; Cohen et al. 2010; Sanyal et al. 2012). When combined with appropriate validation processes, these methodologies have the potential to provide predictions of bone quality and fracture risk in clinical practice where empirical mechanical testing is not feasible yet high-resolution skeletal imaging techniques (e.g., pQCT, micromagnetic resonance imaging (micro-MRI), DXA, or quantitative ultrasound (QUS)) are available (Manhard et al. 2016; (Kopperdahl et al. 2014; Liu et al. 2013; Wang et al. 2015; Cory et al. 2010; Mueller et al. 2011; Weatherholt et al. 2015; Yip et al. 2016; Hosseini et al. 2017; Baroncelli 2008; Klintstrom et al. 2016; Christiansen et al. 2011; Lopez et al. 2012).

7.4 Discussion

7.4.1 Fracture Risk: Structure-Function Relationships and Correlation Analyses

Changes in bone density, bone mass, and microarchitecture due to pharmaceutical interventions are important measures of bone quality, but correlating those properties to structural mechanics allows for a more complete understanding of functional changes. For example, long bone bending test outputs of peak load and ultimate stress can be plotted vs. DXA or pQCT measurements of cortical bone mineral density or mineral content. Regression analyses can be performed to elucidate differences in the relationship between mineralization and bone strength in treatment groups compared to controls. Computing Pearson's correlation coefficient (r) allows for assessment of significant correlative effects. These structure-function relationships are indicators of fracture risk, and it is these alterations in fracture risk that are the most relevant preliminary predictors of safety for translating preclinical animal model evaluations to human clinical trials. Therapies that are shown in preclinical trials to cause no harm to bone quality should also demonstrate no negative effects on fracture risk, while pharmaceuticals with potential adverse effects on bone quality should be shown to reduce fracture risk (Bone 2002).

7.4.2 *Animals Models: Limitations and Decisions*

While animal models provide critical data to inform safety and efficacy prior to moving to clinical trials, there are limitations as well as decisions to be made regarding which animal model(s) will provide the most appropriate information (Osteoporosis Research: Animal Models 2011; Ke and Li 2005; Mosekilde 1995; Turner 2011). Small animal models cost less and are more straightforward to develop. However, these models do not always provide representative mechanistic responses relative to humans. For example, rodents exhibit limited intracortical remodeling and have growth plates that remain open, and these properties may impact certain skeletal morphometric and mechanical results. Large animal models, such as ovine and porcine, present a size scale that is more similar to human but are more expensive and still may not present directly translatable responses. Nonhuman primate models provide the closest genetic background for studying pharmaceutical effects on humans but involve many technical, financial, and ethical challenges. In all models, the study duration should consider the relative rates of bone turnover between animal and human and the intended pharmaceutical target. Decisions about sample size may also depend on the species being utilized and the expected effect size of the pharmaceutical intervention being tested. Additionally, appropriate analogous human-to-animal exposure levels should be determined and tested – generally using three doses (high, intermediate, and low) (Guideline on the Evaluation of Medicinal Products in the Treatment of Primary Osteoporosis 2006).

7.4.3 *Pediatrics: Effects of Growth and Development*

Another consideration is for therapies targeted at pediatric use. Pharmaceuticals may exhibit different safety and efficacy profiles in adults compared to pediatric patients due to metabolic differences and effects of growth and development. The

United States Department of Health and Human Services Food and Drug Administration (FDA) has provided guidance and suggestions for assessing pediatric drug products using juvenile animal studies (Guidance for Industry: Nonclinical Safety Evaluation of Pediatric Drug Products 2006). This documentation outlines issues to consider when determining appropriate juvenile animal models to use as well as human clinical trial translational elements and developmental event analogues between human and various animal species. Some important items to address during experimental design include selecting treatment durations and toxicological endpoints that span at least the significant relevant periods of postnatal development and establishing a dose-response relationship for adverse effects. Additionally, distinguishing between acute and permanent effects of a pharmaceutical may be important depending on the therapeutic target, and this may require inclusion of a recovery group in the study design (Tassinari 2014). The FDA guidance document also suggests that studies include both juvenile male and female animals, as differences in stages of development and other sex-based influences may exist. This recommendation aligns with requirements from the National Institutes of Health (NIH) for incorporating sex as a factor in preclinical study design as part of an effort to develop improved personalized medicine approaches (Clayton and Collins 2014).

7.5 Conclusions

One key element in the process of translating preclinical toxicology studies to human clinical trials is assessing the relationships between bone mechanical properties and microarchitectural or mineral density measures. Preclinical studies that identify changes in the correlations of bone microstructure and mineral density to mechanics measurements due to treatment with a pharmaceutical agent provide valuable insight on the potential of that pharmaceutical to affect bone quality and fracture risk in human clinical applications. This chapter provides guidelines for applying appropriate mechanical testing procedures based on anatomic site or mode of testing, with an overarching goal of maintaining consistency, reproducibility, and repeatability across study groups. Due to all the possible considerations involved in selecting animal models, specific mechanical testing hardware adjustments may be required to accommodate experimental variations. Some specific examples are included within the chapter for demonstration purposes, but it is important to note that there are many factors that need to be considered when setting particular testing parameters (e.g. span widths, test control rates). Examples of factors that could affect these decisions include anatomical differences in shape and size, differences in material properties (e.g. density, stiffness, ductility), sex, age, bone maturity or growth plate closure, and abnormalities in structure at the testing site. These decisions must be made on a study-by-study basis depending on the aims of the work and the particular intrinsic properties of the specimens to be tested.

Acknowledgments The authors would like to thank Jeanne Robertson of Jeanne Robertson Illustration, Raleigh, NC, for providing illustrations for the mechanical testing procedure figures.

References

Athanasiou KA, Zhu C, Lanctot DR, Agrawal CM, Wang X. Fundamentals of biomechanics in tissue engineering of bone. Tissue Eng. 2000;6(4):361–81. doi:10.1089/107632700418083.

Baroncelli GI. Quantitative ultrasound methods to assess bone mineral status in children: technical characteristics, performance, and clinical application. Pediatr Res. 2008;63(3):220–8. doi:10.1203/PDR.0b013e318163a286.

Boerckel JD, Kolambkar YM, Stevens HY, Lin AS, Dupont KM, Guldberg RE. Effects of in vivo mechanical loading on large bone defect regeneration. J Orthop Res: Off Publ Orthop Res Soc. 2012;30(7):1067–75. doi:10.1002/jor.22042.

Bone H. U.S. Guidance for the Development of Drugs for Osteoporosis: rationale, durability and evolution. Endocrinologic & Metabolic Drugs Advisory Committee. Food and Drug Administration, 2002, Silver Spring, MD.

Carter DR, Hayes WC. Bone compressive strength: the influence of density and strain rate. Science. 1976;194(4270):1174–6.

Carter DR, Hayes WC. The compressive behavior of bone as a two-phase porous structure. J Bone Joint Surg Am. 1977;59(7):954–62.

Christiansen BA, Kopperdahl DL, Kiel DP, Keaveny TM, Bouxsein ML. Mechanical contributions of the cortical and trabecular compartments contribute to differences in age-related changes in vertebral body strength in men and women assessed by QCT-based finite element analysis. J Bone Miner Res. 2011;26(5):974–83. doi:10.1002/jbmr.287.

Ciarallo A, Barralet J, Tanzer M, Kremer R. An approach to compare the quality of cancellous bone from the femoral necks of healthy and osteoporotic patients through compression testing and microcomputed tomography imaging. Mcgill J Med. 2006;9(2):102–7.

Clayton JA, Collins FS. Policy: NIH to balance sex in cell and animal studies. Nature. 2014;509(7500):282–3.

Cohen A, Dempster DW, Muller R, Guo XE, Nickolas TL, Liu XS, Zhang XH, Wirth AJ, van Lenthe GH, Kohler T, DJ MM, Zhou H, Rubin MR, Bilezikian JP, Lappe JM, Recker RR, Shane E. Assessment of trabecular and cortical architecture and mechanical competence of bone by high-resolution peripheral computed tomography: comparison with transiliac bone biopsy. Osteoporos Int. 2010;21(2):263–73. doi:10.1007/s00198-009-0945-7.

Cory E, Nazarian A, Entezari V, Vartanians V, Muller R, Snyder BD. Compressive axial mechanical properties of rat bone as functions of bone volume fraction, apparent density and micro-ct based mineral density. J Biomech. 2010;43(5):953–60. doi:10.1016/j.jbiomech.2009.10.047.

Cowin SC, editor. Bone mechanics handbook. 2nd ed. Boca Raton: CRC Press; 2001. hardback ISBN 0-8493-9117-2.

Duncan RL, Turner CH. Mechanotransduction and the functional response of bone to mechanical strain. Calcif Tissue Int. 1995;57(5):344–58.

Frost HM. Bone "mass" and the "mechanostat": a proposal. Anat Rec. 1987a;219(1):1–9. doi:10.1002/ar.1092190104.

Frost HM. The mechanostat: a proposed pathogenic mechanism of osteoporoses and the bone mass effects of mechanical and nonmechanical agents. Bone Miner. 1987b;2(2):73–85.

Frost HM. Bone's mechanostat: a 2003 update. Anat Rec A Discov Mol Cell Evol Biol. 2003;275(2):1081–101. doi:10.1002/ar.a.10119.

Fyhrie DP, Christiansen BA. Bone material properties and skeletal fragility. Calcif Tissue Int. 2015;97(3):213–28. doi:10.1007/s00223-015-9997-1.

Goldstein SA, Wilson DL, Sonstegard DA, Matthews LS. The mechanical properties of human tibial trabecular bone as a function of metaphyseal location. J Biomech. 1983;16(12):965–9.
Goldstein SA, Goulet R, McCubbrey D. Measurement and significance of three-dimensional architecture to the mechanical integrity of trabecular bone. Calcif Tissue Int. 1993;53(Suppl 1):S127–32. discussion S132-123.
Guidance for Industry: Nonclinical Safety Evaluation of Pediatric Drug Products. Rockville: U.S. Department of Health and Human Services; 2006.
Guideline on the Evaluation of Medicinal Products in the Treatment of Primary Osteoporosis. European Medicines Agency Committee for Medicinal Products for Human Use (CHMP). London: 2006.
Hernandez CJ, Keaveny TM. A biomechanical perspective on bone quality. Bone. 2006;39(6):1173–81. doi:10.1016/j.bone.2006.06.001.
Hildebrand T, Laib A, Muller R, Dequeker J, Ruegsegger P. Direct three-dimensional morphometric analysis of human cancellous bone: microstructural data from spine, femur, iliac crest, and calcaneus. J Bone Miner Res. 1999;14(7):1167–74. doi:10.1359/jbmr.1999.14.7.1167.
Hosseini HS, Dunki A, Fabech J, Stauber M, Vilayphiou N, Pahr D, Pretterklieber M, Wandel J, Rietbergen BV, Zysset PK. Fast estimation of Colles' fracture load of the distal section of the radius by homogenized finite element analysis based on HR-pQCT. Bone. 2017;97:65–75. doi:10.1016/j.bone.2017.01.003.
Jepsen KJ, Schaffler MB, Kuhn JL, Goulet RW, Bonadio J, Goldstein SA. Type I collagen mutation alters the strength and fatigue behavior of Mov13 cortical tissue. J Biomech. 1997;30(11–12):1141–7.
Ke HZ, Li XJ. Animal models and study design for osteoporosis research. In: Deng H, Liu Y, editors. Current topics in osteoporosis. Hackensack: World Scientific Publishing Co. Pte. Ltd.; 2005. p. 499–512. doi:10.1142/9789812701220.
Keaveny TM. Strength of trabecular bone. In: Cowin SC, editor. Bone mechanics handbook. 2nd ed. Boca Raton: CRC Press; 2001.
Keaveny TM, Hayes WC. A 20-year perspective on the mechanical properties of trabecular bone. J Biomech Eng. 1993;115(4B):534–42.
Keaveny TM, Borchers RE, Gibson LJ, Hayes WC. Theoretical analysis of the experimental artifact in trabecular bone compressive modulus. J Biomech. 1993a;26(4–5):599–607.
Keaveny TM, Borchers RE, Gibson LJ, Hayes WC. Trabecular bone modulus and strength can depend on specimen geometry. J Biomech. 1993b;26(8):991–1000.
Keaveny TM, Guo XE, Wachtel EF, McMahon TA, Hayes WC. Trabecular bone exhibits fully linear elastic behavior and yields at low strains. J Biomech. 1994;27(9):1127–36.
Keaveny TM, Pinilla TP, Crawford RP, Kopperdahl DL, Lou A. Systematic and random errors in compression testing of trabecular bone. J Orthop Res. 1997;15(1):101–10. doi:10.1002/jor.1100150115.
Klintstrom E, Klintstrom B, Moreno R, Brismar TB, Pahr DH, Smedby O. Predicting trabecular bone stiffness from clinical cone-beam CT and HR-pQCT data; an in vitro study using finite element analysis. PLoS One. 2016;11(8):e0161101. doi:10.1371/journal.pone.0161101.
Kopperdahl DL, Keaveny TM. Yield strain behavior of trabecular bone. J Biomech. 1998;31(7):601–8.
Kopperdahl DL, Aspelund T, Hoffmann PF, Sigurdsson S, Siggeirsdottir K, Harris TB, Gudnason V, Keaveny TM. Assessment of incident spine and hip fractures in women and men using finite element analysis of CT scans. J Bone Miner Res. 2014;29(3):570–80. doi:10.1002/jbmr.2069.
Kuhn JL, Goldstein SA, Choi K, London M, Feldkamp LA, Matthews LS. Comparison of the trabecular and cortical tissue moduli from human iliac crests. J Orthop Res. 1989;7(6):876–84. doi:10.1002/jor.1100070614.
Kupke JS. Characterization of the femoral neck region's response to the rat hindlimb unloading model through tomographic scanning, mechanical testing and estimated strengths. Texas A&M University; 2010, College Station, TX.
Liu XS, Wang J, Zhou B, Stein E, Shi X, Adams M, Shane E, Guo XE. Fast trabecular bone strength predictions of HR-pQCT and individual trabeculae segmentation-based plate and

rod finite element model discriminate postmenopausal vertebral fractures. J Bone Miner Res. 2013;28(7):1666–78. doi:10.1002/jbmr.1919.
Lopez E, Ibarz E, Herrera A, Mateo J, Lobo-Escolar A, Puertolas S, Gracia L. A mechanical model for predicting the probability of osteoporotic hip fractures based in DXA measurements and finite element simulation. Biomed Eng Online. 2012;11:84. doi:10.1186/1475-925X-11-84.
Manhard MK, Nyman JS, Does MD. Advances in imaging approaches to fracture risk evaluation. Transl Res. 2016; doi:10.1016/j.trsl.2016.09.006.
McDonald K, Little J, Pearcy M, Adam C. Development of a multi-scale finite element model of the osteoporotic lumbar vertebral body for the investigation of apparent level vertebra mechanics and micro-level trabecular mechanics. Med Eng Phys. 2010;32(6):653–61. doi:10.1016/j.medengphy.2010.04.006.
Mizrahi J, Silva MJ, Keaveny TM, Edwards WT, Hayes WC. Finite-element stress analysis of the normal and osteoporotic lumbar vertebral body. Spine (Phila Pa 1976). 1993;18(14):2088–96.
Morgan EF, Keaveny TM. Dependence of yield strain of human trabecular bone on anatomic site. J Biomech. 2001;34(5):569–77.
Morgan EF, Bayraktar HH, Keaveny TM. Trabecular bone modulus-density relationships depend on anatomic site. J Biomech. 2003;36(7):897–904.
Morgan EF, Bayraktar HH, Yeh OC, Majumdar S, Burghardt A, Keaveny TM. Contribution of inter-site variations in architecture to trabecular bone apparent yield strains. J Biomech. 2004;37(9):1413–20. doi:10.1016/j.jbiomech.2003.12.037.
Mosekilde L. Assessing bone quality – animal-models in preclinical osteoporosis research. Bone. 1995;17(4):S343–52. doi:10.1016/8756-3282(95)00312-2.
Mueller TL, Christen D, Sandercott S, Boyd SK, van Rietbergen B, Eckstein F, Lochmuller EM, Muller R, van Lenthe GH. Computational finite element bone mechanics accurately predicts mechanical competence in the human radius of an elderly population. Bone. 2011;48(6):1232–8. doi:10.1016/j.bone.2011.02.022.
Muller R. Hierarchical microimaging of bone structure and function. Nat Rev Rheumatol. 2009;5(7):373–81. doi:10.1038/nrrheum.2009.107.
Niebur GL, Yuen JC, Hsia AC, Keaveny TM. Convergence behavior of high-resolution finite element models of trabecular bone. J Biomech Eng. 1999;121(6):629–35.
Niebur GL, Feldstein MJ, Yuen JC, Chen TJ, Keaveny TM. High-resolution finite element models with tissue strength asymmetry accurately predict failure of trabecular bone. J Biomech. 2000;33(12):1575–83.
Odgaard A, Linde F. The underestimation of Young's modulus in compressive testing of cancellous bone specimens. J Biomech. 1991;24(8):691–8.
Osteoporosis Research: Animal Models. Osteoporosis research: animal models. New York: Springer; 2011. doi:10.1007/978-0-85729-293-3.
Prot M, Saletti D, Pattofatto S, Bousson V, Laporte S. Links between mechanical behavior of cancellous bone and its microstructural properties under dynamic loading. J Biomech. 2015;48(3):498–503. doi:10.1016/j.jbiomech.2014.12.002.
Pugh JW, Rose RM, Radin EL. Elastic and viscoelastic properties of trabecular bone: dependence on structure. J Biomech. 1973;6(5):475–85.
Rajaai SM, Saffar KP, JamilPour N. Mechanical properties of long bone shaft in bending. In: Bamidis PD, Pallikarakis N, editors. XII mediterranean conference on medical and biological engineering and computing 2010. IFMBE Proceedings. Berlin: Springer; 2010. vol 29 p. 729–732.
van Rietbergen B. Micro-FE analyses of bone: state of the art. Adv Exp Med Biol. 2001;496:21–30.
Robertson G, Xie C, Chen D, Awad H, Schwarz EM, O'Keefe RJ, Guldberg RE, Zhang X. Alteration of femoral bone morphology and density in COX-2-/- mice. Bone. 2006;39(4):767–72. doi:10.1016/j.bone.2006.04.006.
Rubin CT, Lanyon LE. Limb mechanics as a function of speed and gait: a study of functional strains in the radius and tibia of horse and dog. J Exp Biol. 1982;101:187–211.
Sanyal A, Gupta A, Bayraktar HH, Kwon RY, Keaveny TM. Shear strength behavior of human trabecular bone. J Biomech. 2012;45(15):2513–9. doi:10.1016/j.jbiomech.2012.07.023.

Tassinari M. Signals for bone toxicity in nonclinical safety assessments and regulatory applications for pediatric trials. Paper presented at the Advancing the Development of Pediatric Therapeutics: Pediatric Bone Health, Silver Spring, MD; June 3, 2014.

Turner AS. How to select your animal model for osteoporosis research. Osteoporos Res: Anim Models. 2011:1–12. doi:10.1007/978-0-85729-293-3_1.

Turner CH, Burr DB. Basic biomechanical measurements of bone: a tutorial. Bone. 1993;14(4):595–608.

Turner CH, Burr DB. Experimental techniques for bone mechanics. In: Cowin SC, editor. Bone mechanics handbook, vol. 1. 2nd ed. Boca Raton: CRC Press; 2001.

Un K, Bevill G, Keaveny TM. The effects of side-artifacts on the elastic modulus of trabecular bone. J Biomech. 2006;39(11):1955–63. doi:10.1016/j.jbiomech.2006.05.012.

Van Rietbergen B, Odgaard A, Kabel J, Huiskes R. Direct mechanics assessment of elastic symmetries and properties of trabecular bone architecture. J Biomech. 1996;29(12):1653–7.

Wang J, Zhou B, Liu XS, Fields AJ, Sanyal A, Shi X, Adams M, Keaveny TM, Guo XE. Trabecular plates and rods determine elastic modulus and yield strength of human trabecular bone. Bone. 2015;72:71–80. doi:10.1016/j.bone.2014.11.006.

Weatherholt AM, Avin KG, Hurd AL, Cox JL, Marberry ST, Santoni BG, Warden SJ. Peripheral quantitative computed tomography predicts humeral diaphysis torsional mechanical properties with good short-term precision. J Clin Densitom. 2015;18(4):551–9. doi:10.1016/j.jocd.2014.10.002.

Yeni YN, Fyhrie DP. Finite element calculated uniaxial apparent stiffness is a consistent predictor of uniaxial apparent strength in human vertebral cancellous bone tested with different boundary conditions. J Biomech. 2001;34(12):1649–54.

Yeni YN, Christopherson GT, Dong XN, Kim DG, Fyhrie DP. Effect of microcomputed tomography voxel size on the finite element model accuracy for human cancellous bone. J Biomech Eng. 2005;127(1):1–8.

Yip BH, Yu FW, Wang Z, Hung VW, Lam TP, Ng BK, Zhu F, Cheng JC. Prognostic value of bone mineral density on curve progression: a longitudinal cohort study of 513 girls with adolescent idiopathic scoliosis. Sci Report. 2016;6:39220. doi:10.1038/srep39220.

Chapter 8
Application of Histopathology and Bone Histomorphometry for Understanding Test Article-Related Bone Changes and Assessing Potential Bone Liabilities

Reinhold G. Erben, Jacquelin Jolette, Luc Chouinard, and Rogely Boyce

Abstract Qualitative histopathology remains the gold standard for hazard identification and safety assessment of potential new therapeutics. Although, in most cases, qualitative histopathology has sufficient sensitivity to detect test article-related effects on bone marrow and growth plates in standard toxicity studies, it may lack the sensitivity to detect effects of test articles on key physiological processes in bone tissue such as bone formation, mineralization, and resorption, which often requires chronic dosing to result in structural changes, such as variation in bone mass, that can be appreciated by qualitative assessment. Bone histomorphometry is an important tool that provides sensitive methods that can detect effects of test articles on bone resorption, formation, mineralization, remodeling rates, and growth before structural changes occur. Bone histomorphometry can be used to understand the cellular mechanisms responsible for test article-related structural changes, detected by histopathology or imaging techniques, or can be used prospectively to address a potential target- or class-related theoretical bone liability. In this chapter we review the methods for embedding, sectioning, staining, and analysis of bone sections and provide some general guidance on approaches for the evaluation of bone and use of histomorphometric analyses applied to preclinical safety assessment.

R.G. Erben, MD, DVM (✉)
Department of Biomedical Sciences, University of Veterinary Medicine Vienna, Veterinärplatz 1, 1210 Vienna, Austria
e-mail: Reinhold.Erben@vetmeduni.ac.at

J. Jolette, DVM, MSc, Dip ACVP • L. Chouinard, DVM, DES, DACVP
Department of Pathology, Charles River Montreal ULC, 22022 Transcanadienne, Senneville, QC, Canada, H9X 3R3
e-mail: jacquelin.jolette@crl.com; luc.chouinard@crl.com

R. Boyce, DVM, PhD, DACVP
Comparative Biology and Safety Sciences, Amgen Inc., One Amgen Center Dr, Thousand Oaks, CA 91320, USA
e-mail: rboyce@amgen.com

S.Y. Smith et al. (eds.), *Bone Toxicology*, Molecular and Integrative Toxicology,
DOI 10.1007/978-3-319-56192-9_8

Keywords Bone • Histomorphometry • Bone formation • Bone resorption • Bone remodeling

8.1 Introduction: Preliminary Information from Decalcified Bone Histology – When Is Histomorphometry Needed and What Can Histomorphometry Tell You?

Bone histomorphometry is an important and indispensable tool for assessing the mechanisms by which therapeutic agents act on the skeleton and for assessing the skeletal safety of therapeutic agents. Bone histomorphometry can yield valuable information about bone structure, bone formation, bone resorption, bone mineralization, as well as bone modeling and remodeling activity. While imaging techniques such as dual-energy X-ray absorptiometry (DXA), peripheral quantitative computed tomography (pQCT), and micro-computed tomography (micro-CT) are sensitive methods that can provide excellent information about bone mass and bone structure, these techniques do not explain the tissue-level mechanisms that lead to a bone mass imbalance or altered microstructure. In addition, long-term dosing may be required for structural changes to manifest as a consequence of test article-related effects on bone cell activity. Bone histomorphometry is an essential and sensitive ex vivo tool to assess cellular activity and mineralization in the bone. The strength of bone histomorphometry is that it can provide mechanistic insight into the effects of a test article on the bone to explain structural changes identified by histopathology or imaging. In addition, it can be used as a sensitive tool to prospectively address potential target- or class-related theoretical bone liabilities.

The nomenclature for histomorphometric parameters in this chapter is based on the suggestions of the American Society for Bone and Mineral Research (ASBMR) nomenclature committee (Dempster et al. 2013). For details regarding the calculation of histomorphometric parameters, the reader is referred to other review articles and textbooks (Parfitt et al. 1987; Erben and Glosmann 2012; Dempster et al. 2013).

8.2 Histological Processing of Bone Specimens

Qualitative histopathology remains the gold standard for the assessment of the effects of new therapeutics on tissues and is generally the initial tool for the evaluation of test article-related effects on the bone. Therefore, this chapter includes a brief description of how bone specimens have to be harvested, fixed, embedded, and stained in order to achieve a successful histopathological evaluation of bones in preclinical toxicology studies. The unique characteristics of the bone require additional work, tools, and adapted histology techniques for evaluation compared to soft tissues. For a detailed description of the embedding and staining techniques, the reader is referred to other journal articles and textbooks (Schenk et al. 1984; Erben 1997; Erben and Glosmann 2012).

8.2.1 Tissue Harvesting

Knowledge of the basic anatomy, size, and shape of the bone to be collected is often neglected. A bone-specific diagram can be included in the protocol to facilitate the tissue dissection. Radiography using standard radiography is a very good tool for screening the entire skeleton to identify focal bone changes such as small fractures or occult lesions that are not detected clinically and/or would be missed at necropsy. This may be particularly important in 2-year rodent bioassays to detect potential carcinogenic effects on the bone or in the identification of some test article-related nonsystemic bone effects occurring at non-routinely evaluated sites such as the spine or the humerus. In rodent studies, the whole carcass with the remaining appendicular and axial skeleton can be retained in fixative for potential future X-ray evaluation and/or histopathology. Radiographs are also very useful to target specific sites at necropsy and in the histology laboratory for section orientation.

At necropsy, prior to fixation, the skin, muscle, and soft tissues should be removed taking care not to damage the joint, the periosteum, or other structures. Small bones and growth plates of young animals, especially juvenile rodents, are fragile and can be easily damaged if not handled properly. Inadequate dissection or sampling will result in inappropriate fixation and poor specimen quality. Specimens should be fixed as soon as possible following harvesting. Delays in fixation will cause autolysis and will impair tissue evaluation. The marrow cavity should be exposed to allow better penetration of the fixative solution. A proper tool box that includes an oscillating saw and various bone cutters should be adequate for routine bone collection of rodents. Precision trimming of non-rodent large bones into smaller specimens ensures proper fixation and also accelerates the decalcification process if samples are planned to be decalcified. A rotating circular saw such as the IsoMet® Low Speed Saw (Buehler, Lake Bluff, IL) or a slow band saw like the EXAKT 300 CP system (EXAKT Technologies Inc., Oklahoma City, OK) greatly facilitates trimming of fresh or fixed specimens. Cutting of fresh bones with these saws introduces bone fragments into the nearby bone marrow that will interfere with subsequent evaluation. For that reason, any trimming cut should be done several millimeters away from the planned microtomy surface.

8.2.2 Fixation

Immersion fixation with 10% neutral buffered formalin (NBF) is the method of choice for bone fixation for routine decalcified, paraffin-embedded hematoxylin and eosin (H&E)-stained sections. Fixation time will depend on the specimen size and exposure of the bone marrow. Consequently, trimming of larger specimens at necropsy is highly recommended to allow better penetration of the fixative. A 10:1 formalin-tissue ratio should be used. The penetration of formalin into dense cortical bone is slow (2 mm/24 h) compared to soft tissues (1–4 mm/h). Small specimens (less than 5 mm thick) are completely fixed after 2–4 days; larger specimens are cut into 3–4 mm thick slices with a diamond circular or band saw to complete the fixation. Fixation in 10% NBF is done at room temperature.

For undecalcified histology, long-term preservation of the bone and/or when fluorochrome labels were given to the animals, the bones should be transferred into 70% ethanol once adequately fixed in NBF to avoid leaching of the labels and potential decalcification. Other fixatives such as paraformaldehyde might be needed for special histochemical staining such as tartrate-resistant acid phosphatase (TRAP) staining or immunohistochemistry.

8.2.3 Decalcification

It is an important point to consider that decalcification of a bone specimen will also remove any fluorochrome labels present within the specimen, making dynamic histomorphometric analysis impossible. Fluorochromes are charged fluorescent molecules, such as tetracycline, that are administered in vivo prior to termination and that are deposited at sites of active bone mineralization. Fluorochrome labels are required for the measurement of parameters related to dynamic bone formation. Therefore, decalcified bone specimens are used for routine histopathological evaluation but are only rarely employed for histomorphometric evaluation of the bone. For a complete histomorphometric analysis, undecalcified, plastic-embedded sections are necessary.

Formic acid solution, at a concentration of 5–15%, is the decalcifying solution of choice for routine, paraffin-embedded bone samples. The solution can be prepared in the laboratory or purchased commercially. Formic acid provides relatively rapid, reproducible, and good quality results. Like fixation, specimen thickness and decalcifying solution volume are critical to achieve high-quality results. A 20:1 decalcifying tissue volume ratio is suggested when using the formic acid solution. It is recommended to change the solution on a daily basis until complete decalcification endpoints are achieved. Use of a slow-moving shaker or manual agitation will accelerate the speed of decalcification. Other decalcifying solutions such as ethylenediaminetetraacetic acid (EDTA) or commercial products (Immunocal™) are available and should be used when immunohistochemistry (IHC) or histochemical stains are needed. Testing for completion of the decalcification can be accomplished with three methods: physical testing, chemical testing, and radiography. Physical testing is performed by inserting a needle in the specimen or cutting the specimen with a scalpel blade. This method is not always reliable and may damage the specimen. The method of choice for testing completeness of the decalcification of multiple specimens is the chemical testing. It is accomplished with mixing ammonium solution with the used formic acid solution. If the solution is clear, the decalcification is complete; if the solution is cloudy, the decalcification is incomplete, and the specimen should be placed into a fresh decalcifying solution for another 24 h before repeating the testing. Radiography can be used to determine the completion of the process and can be very useful for large specimens and for teeth. Following decalcification, the samples are rinsed in water to arrest decalcification and to remove the remaining decalcifying solution that could interfere with the processing and staining. Over-decalcification can alter cell morphology especially of osteocytes that could be misinterpreted as necrotic/apoptotic. Incomplete decalcification will result in cutting artifacts such as shattering.

Following decalcification, specimens are dehydrated. Many protocols are used; the most common is a series of increasing strength of alcohol (70–100%) followed by clearing with xylene. The tissues are subsequently infiltrated with paraffin and embedded in molds. Ink or pencils can be used to mark the sample and ensure proper orientation of the specimens.

8.2.4 Processing of Undecalcified Bone Specimens

High-quality undecalcified bone histology is a prerequisite for bone histomorphometric analysis. It is important to consider that high-quality methyl methacrylate-embedded (MMA) sections require thorough dehydration and infiltration of specimens which are both time-consuming processes. Moreover, in contrast to paraffin embedding, polymerization of MMA is a radical-generating, exothermic chemical chain reaction. Therefore, conventional MMA embedding methods almost completely destroy antigenic determinants and enzyme activities by radical-driven, covalent modification of biomolecules in the embedded tissue. For partial preservation of enzyme activities and tissue reactivity to antibodies, it is possible to employ modified MMA embedding methods which use polymerization temperatures at around −20 °C (Erben 1997). These modified MMA embedding procedures are at least partially suitable for histochemistry and immunohistochemistry (Erben 1997), and the histological section quality is comparable to conventional MMA embedding (Erben and Glosmann 2012).

For sectioning of undecalcified bones, both rotary and sledge microtomes can be used. Typically, 5–7-μm-thick sections are used. It is important to use a slide press for drying the sections to achieve good section quality. To increase the reproducibility of the histomorphometric analyses and to reduce the interindividual variance within groups of animals, it is important to standardize the sampling sites.

For toxicologic studies that generally involve an actively growing rat model, it is best to choose the proximal tibia and, possibly, a lumbar vertebra for cancellous bone histomorphometry. The proximal tibia, sectioned along the sagittal or coronal plane, provides a straight epiphyseal plate (in contrast to the distal femur) and abundant active metaphyseal bone, facilitating histomorphometric evaluation (Fig. 8.1a). The body of a cranial lumbar vertebra, generally L1 or L2, sectioned along the sagittal plane, is the preferred site for evaluating the axial skeleton; this leaves the more caudal and larger lumbar vertebrae to be preserved for biomechanical and bone densitometric analyses.

In mice, the distal femoral metaphysis is the preferred site for cancellous bone histomorphometry in the appendicular skeleton, because the femur is less curved

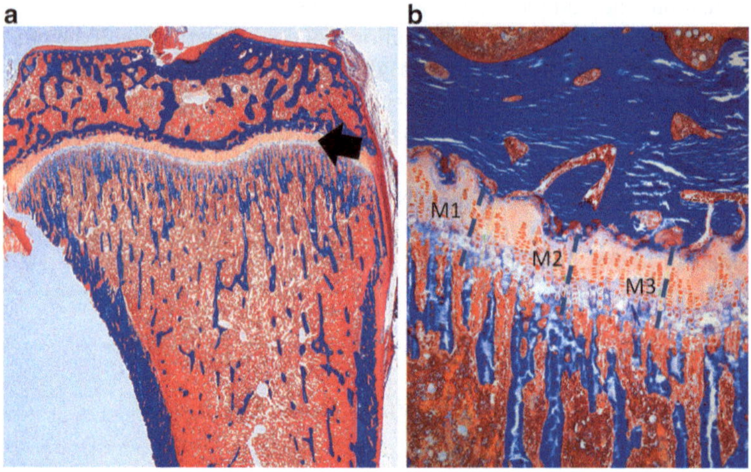

Fig. 8.1 (**a**) Undecalcified coronal section of rat tibia stained with Goldner's trichrome counterstained with aniline blue, illustrating the straight epiphyseal plate (*black arrow*) and abundant metaphyseal cancellous bone (original magnification 10×). (**b**) Undecalcified section of rat tibia stained with Goldner's trichrome counterstained with aniline blue illustrating three linear measurement (M1, M2, M3) for the estimation of growth plate thickness (original magnification 100×)

compared with the murine tibia. Murine vertebrae need to be sectioned in a frontal plane to yield a sufficient amount of cancellous bone available for histomorphometric analysis.

In larger species such as macaques, valuable bone histomorphometry information may be obtained from bone biopsies collected either at the beginning or several months after the start of treatment (Smith et al. 2011). Sampling is obtained under general anesthesia from the ilium (1 × 1 cm full-thickness wedge at the tuber coxa, at 1 cm caudal to the cranio-dorsal iliac spine) and from a rib (targeting the mid-length bony seventh rib), for cancellous and cortical bone analyses, respectively. The contralateral site of these bones may also be biopsied at another time point to study the effects of treatment over time. At each time point of bone procurement, different bone-seeking fluorochromes must be injected in order to differentiate them. Fluorochrome labels persist until resorption of the labeled bone matrix, which may take several years in cortical bone. The iliac biopsies of some monkeys yield scant cancellous bone for evaluation due to closely apposed cortical plates; in such cases, only the fluorochrome label-derived parameters should be used for assessment of an effect. At necropsy, sampling of nonhuman primate bones is generally centered on clinically relevant sites, the lumbar vertebra, proximal femur, and distal radius, particularly when the animal model is used to study osteoporosis. However, in safety assessment of non-rodent species, the proximal tibia and/or a cranial lumbar vertebra is the preferred sampling sites for histomorphometric evaluation.

Slabs of diaphysis carefully sectioned perpendicular to the long axis of bones are used for evaluation of cortical bone. The diaphysis is MMA embedded, cut, and ground to provide unstained 20–80 μm transverse sections of long bones. For rats, the tibial shaft at 2 mm proximal to the insertion of the fibula is the site generally selected for measurements. In mice, the mid-shaft of the femur is a good site. In nonhuman primates, several alternatives may be considered at study end, including the weight-bearing femur or tibia at mid-shaft and/or the non-weight-bearing radius at mid-shaft.

8.2.5 Stains and Evaluation

8.2.5.1 Decalcified Bone Sections

Hematoxylin and eosin (H&E) is the method of choice for staining decalcified bone specimens in standard toxicology studies. It provides excellent cellular and structural details as well as good contrast between the cartilage, bone, and marrow elements of the sections. The H&E stain is reproducible, quick, and the majority of bone abnormalities can be detected and characterized with a good quality 4–5-μm-thick paraffin section. Additionally semiquantitative evaluation is easily performed. Polarized light evaluation should be conducted on bone sections to verify the lamellar organization of the bone matrix and to examine whether woven bone is present.

Several stains can be used to stain the cartilage and evaluate the proteoglycan content. The preferred methods for cartilage staining include the safranin O and toluidine blue stains. Both stains are cationic dyes that stain proteoglycans as well as glycosaminoglycans. It is important to stain all the sections at the same time to avoid variation in staining intensity. Collagen fibers can be visualized in histological sections with picrosirius red staining or Goldner's trichrome staining.

A variant of the H&E stain, the hematoxylin phloxin saffron (HPS) stain, provides better contrast between the bone and collagen and also facilitates identification of osteoclasts on the bone surface. Anti-cathepsin K immunohistochemistry and TRAP staining provide specific osteoclast staining and are valuable tools for osteoclast quantification.

8.2.5.2 Undecalcified Bone Sections

The Goldner's trichrome has been established as a general bone stain for structural evaluation of undecalcified microtome sections for all species. It provides a clear discrimination between the mineralized and unmineralized bone matrix as well as between osteoclasts and osteoblasts (Schenk et al. 1984). The stain may be counterstained with light green, fast green, or aniline blue. As an alternative for this general purpose stain, the von Kossa counterstained with McNeal's tetrachrome is another recommended alternative, especially for rodents. However, the intense black staining of the mineralized matrix will not allow visualizing a disorganized woven bone matrix or the cement lines. In rodents, discrimination of the cement lines, used for wall width measurement, necessitates a toluidine blue stain adapted to demonstrate these linear structures (Schenk et al. 1984).

The Stevenel's blue surface staining method, a good general purpose stain, can be applied to acid-etched plastic-embedded ground sections of diaphyseal bone. It consists of methylene blue stain counterstained with Van Gieson picro-fuchsin (Sterchi and Callis 1993). If the counterstain is omitted, it is possible to observe concurrently most structural elements and the calcein fluorochrome labels. However, it should be noted that Stevenel's blue interferes with the detection of other fluorochrome labels than calcein.

8.3 Species and Age Considerations

A considerable body of histomorphometric data has accumulated for the adult ovariectomized rat and macaque (*Macaca fascicularis*), specifically because they are the preferred animal models of postmenopausal osteoporosis and have been extensively used to meet the regulatory requirements for the safety assessment of bone therapeutics. In the toxicology setting, the most toxicologically and pharmacologically relevant species may not be the rat or monkey for a specific test article. Detailed bone analyses in this setting may be limited only to the rodent (rat or mouse) or may require a non-rodent such as the dog. In addition to the many factors that are considered in determining the most relevant species for toxicological assessments, there are some specific differences between the rodent and non-rodent skeleton that may also need to be considered. Because there is a lack of true Haversian remodeling in rodents (Bagi et al. 2011), questions about the effects of a test article on cortical bone and intracortical remodeling would require a non-rodent species such as monkey or dog. Notably, similar to the adult human, bone remodeling involves the entire cortex in adult dogs, unlike in the adult monkey where primary bone may still occupy variable amounts of the cortex.

Another important consideration is the age of the animals used in standard toxicology studies; typically adolescent animals are used to avoid the confounding issues of age-related findings. These animals are skeletally immature, still growing, and accruing bone mass with bone modeling playing an important role. This is an important consideration when interpreting changes in bone mass in growing animals; reduced bone mass in test article-treated animals may be due to failure to accrue bone rather than true bone loss. Histomorphometric analyses can aid in the interpretation of these bone mass changes. Although the growth plate remains open in the rat for most of its lifespan, longitudinal bone growth and modeling cease by ~1 year of age (Kilborn et al. 2002; Roach et al. 2003). The rapidly growing adolescent rat used in toxicity studies with active longitudinal growth offers the opportunity to examine the effects of test articles on endochondral ossification and longitudinal bone growth, relevant to pediatric indications.

Although rabbits are commonly used in reproduction and juvenile studies, the use of rabbits for assessing potential skeletal effect upon bones is less desirable, due to a scant amount of cancellous bone at clinically relevant sites. For these reasons and using Wessler's definition of an animal model (Wessler 1976), rats should be given first consideration as model to study potential adverse effects on growing bones because of convenience and relevance of that model whenever the appropriate (compatibility with humans) criteria are met.

Incorporation of histomorphometric analyses in standard toxicology studies is highly feasible in rat studies. This may be combined with longitudinal assessments of bone biomarkers and either longitudinal or terminal bone densitometry evaluation. Rat studies have statistical power, because of larger group sizes ($n = 10$) in standard toxicity studies, allowing robust comparison with controls. Due to the high biological variability in non-rodent species for most bone endpoints and the small group size, incorporation of histomorphometric endpoints in non-rodent studies is problematic and often requires dedicated studies. Bone biomarkers are most frequently used in this setting that allow for comparison with pretreatment values to control for high biological variation. However, in subchronic studies of 1–3 months duration, special fluorochrome labeling procedures can be used (quadruple labeling) that allow for comparison of bone formation rates prior to and at the end of dosing (Ominsky et al. 2010), which may be useful to address specific questions. If disturbances in skeletal remodeling are suspected, dedicated non-rodent studies in skeletally mature animals may be required.

8.4 Fluorochrome Labeling

Labeling of the skeleton with fluorochromes allows the quantitative measurement of bone formation rate and of bone remodeling dynamics and, since these are key measurements in bone histomorphometry, should be used for all bone histomorphometric analyses (Frost 1983b; Erben 2003; Dempster et al. 2013). The fluorochromes used for this purpose are fluorescent, calcium-seeking substances that are incorporated into the mineralization front of mineralizing surfaces and that can be visualized in histological sections by excitation with UV or blue light. Standard fluorochromes used for histomorphometry are calcein green, alizarin complexone, tetracyclines, and xylenol orange (Erben 2003). Fluorochromes can be given intravenously, intraperitoneally, intramuscularly, or subcutaneously. The authors generally recommend subcutaneous administration of the substances. However, tetracyclines are recommended to be given intravenously, especially in non-rodents, as the pH of the tetracycline solution has to be acidic to remain in solution and may therefore elicit pain when delivered subcutaneously; other fluorochromes can be given subcutaneously in rodents and non-rodents as their pH can be adjusted to neutrality. Examples of multiple fluorochrome labeling regimens with recommended dosage and injection solution concentration have been published (Paddock et al. 1995; Boyce et al. 1996; Ominsky et al. 2016). Due to their calcium-complexing effects, care must be taken when fluorochromes are injected intravenously. Intravenous dosing of fluorochromes must be conducted slowly, over several

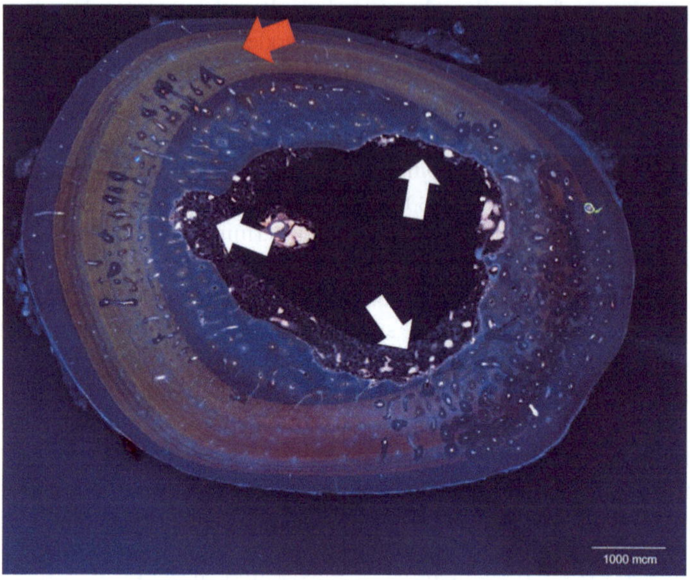

Fig. 8.2 Undecalcified unstained ground cross section of tibia from a cynomolgus monkey viewed in a fluorescent microscope. Diffuse *yellow-brown* staining in the primary cortical bone is incorporated tetracycline administered over a long duration during growth (*red arrow*). There is extensive eroded surface on the endocortical surface (*white arrows*) (scale bar = 1 mm)

minutes, in order to avoid the occurrence of hypocalcemic shock in susceptible animals. Calcein green at 8–20 mg/kg (20 mg/kg in young rodents, 10 mg/kg in aged rats and mice, 8 mg/kg in larger animals) is the authors' preference for animal studies because of its low toxicity and bright fluorescence.

Fluorochromes are bound to mineralized matrix in the bone as complexes with calcium. Therefore, as mentioned above, undecalcified histology is a prerequisite for their later microscopic study. In bone sections, fluorochrome labels appear in the form of bands. The width of the fluorochrome band in a histological section depends on the time the substance circulates in the extracellular fluid, the velocity of mineralization progression, and on the angle at which a three-dimensional label plane has been sectioned. Fluorochromes incorporated into mineralization sites persist for extended periods of time and are not released until osteoclastic resorption of the fluorochrome-containing bone matrix. This long persistence is readily appreciated in nonhuman primates, which are often exposed to tetracyclines or analogs prior to placement on toxicology studies (Fig. 8.2). Some monkeys may have abundant accumulation of this fluorochrome in bones. As this might interfere with recognition of on-study-administered tetracycline, this fluorochrome should not be the preferred one in this species.

Two labels given at different time points are necessary for the calculation of mineral apposition rate (MAR), bone formation rate (BFR), and bone remodeling dynamics. The time interval between the administration of the labels is the interlabel

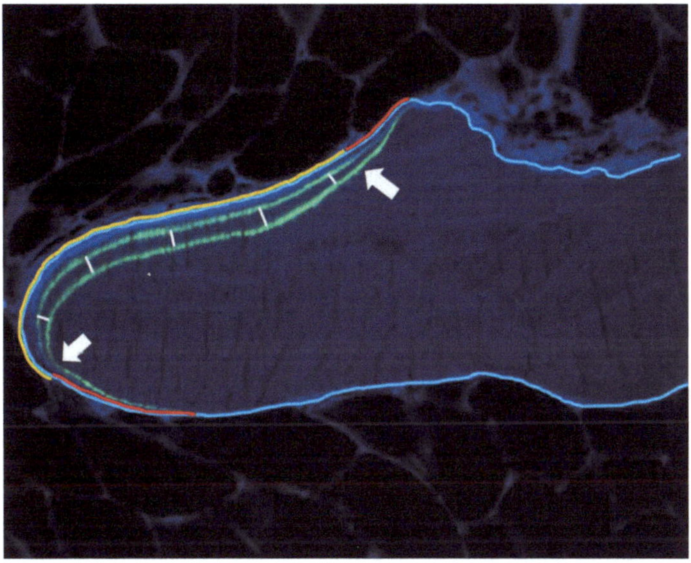

Fig. 8.3 Measurement of interlabel distance between double calcein labels incorporated into an active forming surface in cancellous bone from a cynomolgus monkey. The total bone perimeter is delineated with a turquoise line. The calcein labels are *bright green*. Evenly space linear measurements are made between the labels (*white lines*) which are divided by the interlabel period for calculation of mineral apposition rate. Double-labeled perimeter, or the extent of the bone perimeter that has double labels, is defined by an *orange line* between the two *white arrows*. Single-labeled perimeter is defined by the *red lines* (original magnification 400×)

time (Ir.L.t) which is used for calculation of appositional rates. The mean distance between the labels divided by the Ir.L.t corresponds to the mineral apposition rate (Fig. 8.3).

There are some important considerations when planning the labeling schema in a study: (1) It is important to allow for sufficient time between the last label and the necropsy, so that the label is covered with new bone matrix and will not be lost by elution into solutions used for fixation or processing of the specimen later on. Appropriate minimum time intervals after the last label are 5 days in monkeys, 2 days in rats, and 1 day in mice. (2) To avoid large label escape (failure to incorporate the second label due to an extended interlabel period allowing bone formation to be completed in a specific site) and skewed sampling errors, the marker interval needs to be as short as possible. However, the interlabel period needs to be of sufficient duration to allow for separation of double labels and accurate measurement of MAR. This requires consideration of the formation periods in the animal model being used for histomorphometric analyses. The magnitude of the label escape error is determined by the ratio between the Ir.L.t and the formation period (Ir.L.t/FP ratio). Therefore, the smaller the Ir.L.t /FP ratio is, the greater the proportion of double-labeled remodeling units will be. As a practical rule, the Ir.L.t/FP ratio should be kept below 0.2 to minimize the label escape error (Frost 1983c). Notably, no double labels will be found if the marker interval is longer than the formation period. Another error introduced by an inappropriately long marker interval is the skewed sampling error. It impacts the measurement of the mineral apposition rate (MAR) and of all derived parameters (Frost 1983b). The skewed sampling error results from the fact that organic matrix formation and also MAR are higher during the early phase of the formation period and slow down at the later stages of bone formation. In addition, the bone-forming activity of osteoblasts may temporarily pause during the formation phase. Therefore, a marker interval that is too long will result in erroneously low mean MAR values (Frost 1983b; Erben 2003). We recommend a marker interval of 5 days in rats beyond 3 months of age (Erben 2003). In mice, appropriate marker intervals range from 1 day in 3–4-week-old mice, 2 days in 6–12-week-old mice, and 2 or 3 days in mice greater than 3 months of age (Erben 2003). In adult macaques and dogs, a 10-day interval marker is appropriate.

At doses used for labeling bones, fluorochrome labels are well tolerated, but they should not be considered without any effects in the context of a safety assessment study. Potential transient interferences may be encountered upon the toxicokinetic profile or clinical chemistry parameters; consequently, blood sampling for various analyses should be done prior to and not within a few days after fluorochrome administration. At sites of subcutaneous injection, these substances often induce minimal or mild local irritation. Therefore, different sites of injections must be used for the test/control article and the fluorochrome labels if these substances are administered via the same route.

For a more in-depth description of fluorochrome labeling and of the label escape and skewed sampling errors, the reader is referred to other textbooks (Frost 1983b; Erben 2003).

8.5 Bone Architecture

Although the amount of bone and bone architecture are routinely assessed by bone histomorphometry, it is important to note that 2D assessment of bone structure lacks sensitivity because it is not a 3D global assessment. The current gold standard for measurement of bone mass and architecture are bone imaging techniques. DXA can provide global 2D assessment of bone mass, and pQCT can provide more detailed 3D information on the distribution of mass in cortical and cancellous bone. The advantage of these methods is that they can be performed in vivo for longitudinal assessments of changes in cancellous and cortical bone mass and area. μCT, which is mostly performed ex vivo, provides assumption-free detailed 3D information about cancellous and cortical bone microarchitecture with high precision. This type of information is impossible to obtain by histomorphometry and can only be estimated by more labor-intensive stereological analyses that were employed prior to μCT implementation. However, because X-ray-based μCT analysis detects only mineralized, but not unmineralized, bone matrix, histomorphometric analyses can discriminate between these two contributors to "bone area" and ultimately to the detection of impaired bone mineralization.

8.5.1 Cancellous Bone

Standard sites for cancellous bone histomorphometry in rats and non-rodent species are the proximal tibial metaphysis and the lumbar vertebral bodies. In mice, typical sites are the distal femoral metaphysis and the lumbar vertebral bodies. Cancellous bone histomorphometry is normally performed within the secondary spongiosa, a site more relevant to remodeling-based bone turnover in contrast to the primary spongiosa that is predominated by modeling and dependent on endochondral bone formation. To exclude the primary spongiosa, bone within a certain distance from the growth plate is excluded from the analyses. Depending on the species and age of the animal and the site, 1–0.25 mm distance from the growth plate is generally accepted values, 1 mm for non-rodents species and young rats, 0.5 mm in older rats and young mice, and 0.25 mm in older mice (Erben and Glosmann 2012). To exclude endocortical bone remodeling activity, the authors recommend excluding the cancellous bone within 0.25 mm from the endocortical bone surface from the analyses at all sampling sites.

Analysis of cancellous bone structural parameters is based on only four primary measurements, bone area, bone perimeter, number of individual bone trabeculae, and total area (= the measuring area). From these primary measurements, various other parameters can be calculated. For details, the reader is referred to the specialized literature (Dempster et al. 2013; Erben and Glosmann 2012). In most cases, bone volume (BV/TV), trabecular thickness (Tb.Th), trabecular number (Tb.N), and trabecular separation (Tb.Sp) are used for evaluation. It should be mentioned here that the calculation of trabecular number, trabecular thickness, and trabecular separation from 2D data is indirect indices, calculated from bone area and perimeter, assuming that cancellous bone is structured as a series of parallel plates (Parfitt et al. 1983). This model assumption may not always be entirely correct for the standard sampling sites in rodents and monkeys and under certain experimental manipulations (Boyce et al. 1995a). Therefore, calculation of histomorphometric 3D parameters in cancellous bone of typical species used for toxicological studies is always associated with some systematic error and may deviate from assumption-free parameters measured by μCT.

8.5.2 Cortical Bone

The standard site for cortical bone histomorphometry in rat studies is the tibial shaft at 2 mm proximal to the insertion of the fibula. In mice, the mid-shaft of the femur is usually used. In monkeys, rib biopsies can be used for cortical bone histomorphometry at baseline or during the study, whereas at study end, the mid-shaft of the femur, tibia, or radius is usually employed for cortical bone measurements. Structural parameters can be measured on 20–80-μm-thick micro-ground sections either unstained or stained with toluidine blue or Stevenel's blue. Primary measurements for cortical bone histomorphometry are cross-sectional area (Tt.Ar), cortical area (Ct.Ar), marrow area (Ma.Ar), cortical thickness (Ct.Th), and number and area of intracortical pores (N.Po and Po.Ar). In addition to the absolute values for Ct.Ar, Ma.Ar, and Po.Ar, it is always advisable to calculate the relative values for cortical and marrow areas (Ct.Ar/Tt.Ar and Ma.Ar/Tt.Ar, %) and intracortical pore area (Po. Ar/Ct.Ar, %), because the variance of the relative values is usually less within groups of animals. Relative values are especially helpful when samples of different size need to be compared.

8.6 Cancellous Bone Turnover

8.6.1 Bone Formation

In contrast to analysis of bone structure, histomorphometry is an indispensable tool for the assessment of bone turnover and bone mineralization within a specific site. Fluorochrome-labeled bone specimens can be used to quantitatively assess the rate of osteoblastic matrix synthesis and subsequent mineralization of the newly formed matrix. Fluorochrome-based histomorphometric parameters are also called dynamic histomorphometric parameters. Because dynamic histomorphometry provides a functional readout of osteoblastic bone formation and bone mineralization, these parameters are superior to static morphological parameters for the assessments of osteoblastic activity such as osteoblast perimeter/surface or osteoblast number. As mentioned above, double calcein labeling is routinely used for dynamic histomorphometry.

Primary measurements needed for dynamic histomorphometry are the bone perimeter, the mineralizing bone perimeter (M.Pm), and the mineral apposition rate (MAR). The measurements are made on unstained sections under a fluorescence microscope. The MAR is defined by the mean distance between the labels, divided by the interlabel distance, and expressed as μm/day. There are two ways to measure MAR, either indirectly by tracing along the individual labels or directly by a two-point distance measurement; the latter provides the capability to examine distribution. MAR measurements in sites where broader, "fuzzy" labels are present are best made from midpoint to midpoint, rather than from edge to edge of the labels (Frost 1983a, b). Mineralizing perimeter is the percent surface extent of either the double-labeled perimeter alone (M.Pm = Db.Lb.Pm) or the double-labeled perimeter plus one half of the single-labeled perimeter (M.Pm = Db.Lb.Pm + 5 * S.Lb.Pm) (Fig. 8.3). The latter way to express M.Pm is mathematically more correct than using the double-labeled perimeter alone (Parfitt et al. 1987) and should therefore be used in all non-rodent species. However, to reduce errors based on nonspecific single labels in rodents, many histomorphometric labs use only the double-labeled perimeter to assess mineralizing perimeter in rats and mice, although this strategy underestimates the true M.Pm (Erben and Glosmann 2012). To provide information about the MAR averaged over the entire osteoid surface, which accounts for osteoblast intermittency (the percentage of osteoid associated with an underlying fluorochrome label), the adjusted apposition rate (Aj.Ar) can be calculated in primates and other large animal species as MAR*M.Pm/O.Pm. In rodents, the calculation of the Aj.Ar does not make sense, because the ratio between M.Pm and osteoid perimeter (O.Pm) is usually greater than 1 in rodents, a reflection of the efficiency and rapidity of matrix mineralization in rodents.

The most important parameter for the quantitative assessment of bone formation is the surface-based bone formation rate (BFR), given as BFR/B.Pm = BFR/BS = M.Pm/100 * MAR. It is expressed as $\mu m^2/\mu m$/day or $\mu m^3/\mu m^2$/day and represents the amount of newly formed bone per unit of time and per unit of bone surface. In most cases the surface-based BFR is superior to other ways to express BFR such as per tissue volume or per time, because tissue volume-based BFR depends on the amount of cancellous bone present in the measurement site, and may lead to erroneous conclusions if bone mass is different between the groups of animals. When a test article acts as a strong suppressor of bone turnover, double labels may be absent within the sampling site, although the animal was properly double labeled. In that case the authors recommend to report M.Pm and BFR as zero, whereas MAR should be treated as a missing value.

8.6.2 *Bone Mineralization*

Histomorphometry is clearly the best tool and the gold standard for assessment of impairments in bone mineralization. Histomorphometric parameters used for assessing bone mineralization are based on the measurement of the osteoid width (O.Wi) and osteoid area (O.Ar) (Fig. 8.4). Osteoid is quantified on sections stained with von Kossa and counterstained with McNeal's tetrachrome or with Goldner's trichrome (Schenk et al. 1984), because both stains give a reliable distinction between mineralized bone and osteoid. Osteoid width can either be calculated from osteoid area and osteoid perimeter (the standard way in histomorphometry software packages) or measured directly by a 2-point distance measurement which allows for examination of distribution of O.Wi, often helpful in detecting focal osteomalacia. In non-rodent species, the 3D parameter osteoid thickness (O.Th) can be calculated as O.Th = O.Wi * $\pi/4$ to correct for obliquity of section. Increases in osteoid area/volume (O.Ar/B.Ar, OV/BV) and perimeter/surface (O.Pm/B.Pm, OS/BS) are not necessarily indicative of impaired bone mineralization, because these changes can also occur when bone turnover is enhanced. Rather, impaired bone mineralization is associated with increases in OV/BV and O.Th, as well as with increased osteoid maturation time (Omt = O.Wi/MAR) and mineralization lag time (Mlt = O.Th/Aj.Ar, only used in non-rodent species). Increases in Omt and Mlt alone do not indicate impaired mineralization, as this occurs at the later stages of the formation sites, when osteoblast function is intermittent and is a common finding in postmenopausal osteoporosis. It is the combination of increased Omt and Mlt with increased OV/BV and O.Th that indicates impaired bone mineralization. Therefore, these parameters should be routinely determined when information about bone mineralization is required.

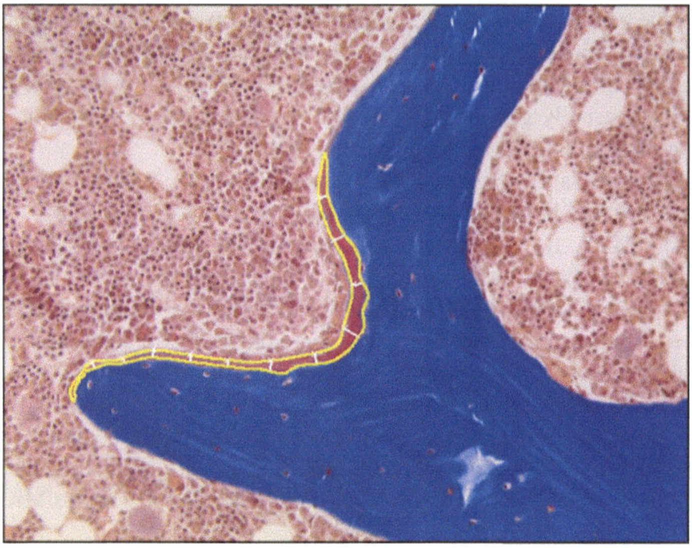

Fig. 8.4 Goldner's trichrome-stained undecalcified section of cancellous bone from a cynomolgus monkey illustrating differential staining of mineralized bone matrix (*blue*) and osteoid (*red*) and measurements for estimation of osteoid volume and osteoid thickness. Osteoid volume is measured by outlining the boundary of the osteoid seam (*yellow line*) for calculation of volume. An osteoid surface measurement would be the length of the *yellow line* at the surface of the bone. Osteoid thickness measurements are illustrated by the evenly spaced perpendicular white lines (original magnification 200×)

8.6.3 Bone Resorption

Osteoclast numbers and the percentage of bone surface covered with osteoclasts (osteoclast perimeter or surface, Oc.Pm/B.Pm, Oc.S/BS) are commonly used to assess bone resorption. However, in contrast to the fluorochrome-based measurement of bone formation rate, histomorphometric parameters of bone resorption are not functional parameters and do not necessarily provide information about the intensity of osteoclastic bone resorption. There are many examples that inhibition of the bone-resorptive capacity of osteoclasts by drugs or genetic deficiencies can result in an upregulation of osteoclast numbers in an attempt to compensate for the deficiency. Hence, it is important to note that histomorphometric parameters of bone resorption are static and do not necessarily reflect the activity of bone-resorbing cells.

In mice, quantification of osteoclasts should be performed on TRAP-stained sections, because osteoclasts are more difficult to recognize in mice compared with other species due to a large proportion of mononuclear osteoclasts and shallow Howship's lacunae. In rats and other commonly used non-rodent laboratory animal species, Goldner's trichrome- or toluidine blue-stained sections are sufficient to identify multinucleated osteoclasts. Since the 3D morphology of osteoclasts varies,

2D osteoclast numbers can be biased by osteoclast size and shape. Osteoclast surface (Oc.S/BS), the percentage of bone surface interfaced with an osteoclast, avoids the bias of cell size and shape. Osteoclast number is generally expressed per mm of bone perimeter (N.Oc/B.Pm). Osteoclasts can only resorb mineralized bone. Therefore, osteoclast number per mm of mineralized bone perimeter (N.Oc/Md.Pm; Md.Pm = B.Pm − O.Pm) or osteoclastic perimeter per mm of mineralized bone perimeter (Oc.Pm/Md.Pm = Oc.S/Md.S) is the best parameters to quantify osteoclast activity, especially when the amount of bone covered with osteoid perimeter differs between different treatment groups. Caution needs to be applied in the interpretation of the parameter osteoclast number per tissue area (N.Oc/T.Ar), because this parameter depends on the amount of cancellous bone present within the specimen.

In non-rodent species, but with some limitations also in rats, the scalloped depressions (Howship's lacunae) lining the trabecular surface can be measured as eroded perimeter (E.Pm) with some reproducibility, whenever the depth of the erosion exceeds at least one bone lamella. E.Pm/B.Pm (=ES/BS) and Oc.Pm/B.Pm (=Oc.S/BS) therefore capture prior and currently active sites of osteoclastic resorption, respectively (Fig. 8.5).

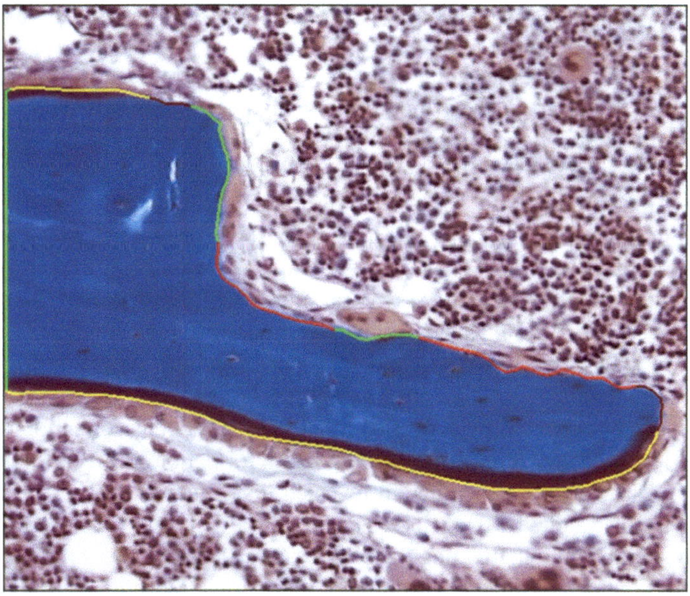

Fig. 8.5 Goldner's trichrome-stained undecalcified section of cancellous bone from a cynomolgus monkey illustrating measurement of resorption parameters. Osteoclast surface, the percentage of bone surface interfaced with osteoclasts, is defined by the *green lines*. Osteoclast-negative eroded surface, the percentage of bone surface that is eroded due to previous osteoclastic resorption and is lined by mononuclear or reversal cells, is defined by the *red lines*. Since eroded surface comprises osteoclast surface, eroded surface is given as the sum of osteoclast surface and osteoclast-negative eroded surface. *Yellow lines* define osteoid and osteoblastic surfaces (original magnification 200×)

8.6.4 Bone Modeling and Remodeling

Standard histomorphometric methods can provide significant data regarding the kinetics of bone remodeling. However, they do not provide an important piece of data in regard to the ultimate consequence of the effects of a test article at the level of the remodeling unit (basic multicellular unit, BMU), that is, what is the net balance between bone resorption and formation at the level of the BMU? Is there a positive or negative BMU bone balance? This is an important concept as metabolic bone disease in the adult skeleton is a consequence largely of alterations in remodeling. For example, it has been demonstrated that the pathogenesis of the accelerated irreversible cancellous bone loss at the menopause is due to increased resorptive cell function resulting in a net loss of bone at each remodeling unit, magnified by the increased rate at which remodeling units are initiated (Eriksen et al. 1999). Potential therapies can correct this imbalance to stabilize bone loss or shift to a positive balance and potentially help restore bone mass. A method known as kinetic reconstruction was developed by Eriksen et al. (1984a, b) and modified by Steiniche et al. (1992) that allows for the determination of the amount of bone removed by resorption (final resorption depth) and the amount of bone deposited by bone formation at the level of the individual remodeling unit formed under the influence of the test article (final wall thickness). These methods can also essentially dissect the kinetics of the remodeling process, providing information as to which phase of the remodeling process and which bone cell populations involved in remodeling are influenced by a test article. This elegant method has provided detailed insights into the effects of various metabolic bone diseases and therapeutic agents on the remodeling site. The reader is referred to several references on the method and its applications (Eriksen 1986; Boyce et al. 1995b, 1996; Storm et al. 1993).

A recent example of the value of kinetic reconstruction was the assessment of the remodeling unit and bone balance in vertebral cancellous bone from cynomolgus monkeys administered with the sclerostin antibody, romosozumab, for 10 or 28 weeks (Boyce et al. 2017). Romosozumab is a bone-forming agent under investigation for the treatment of osteoporosis in postmenopausal women at high risk of fracture. Although it has been demonstrated that romosozumab increases cortical and cancellous bone mass by activating modeling-based bone formation (Ominsky et al. 2014), the contribution of remodeling-based formation to effects on bone mass was unknown. Kinetic reconstruction revealed that romosozumab effected a positive bone balance at the BMU by increasing final wall thickness and decreasing final resorption depth without affecting the kinetics (duration) of bone formation (formation period) but decreasing the duration of resorption (resorption period). Because the effects on modeling-based bone formation were transient and not sustained for the 28-week treatment period, yet bone mass continued to increase in the spine, these data provide evidence that the effects at the BMU level likely contributed to the continued increase in spinal bone mass.

The calculation of remodeling-based parameters requires the measurement of wall width (W.Wi), the mean width of completed remodeling packages. Wall

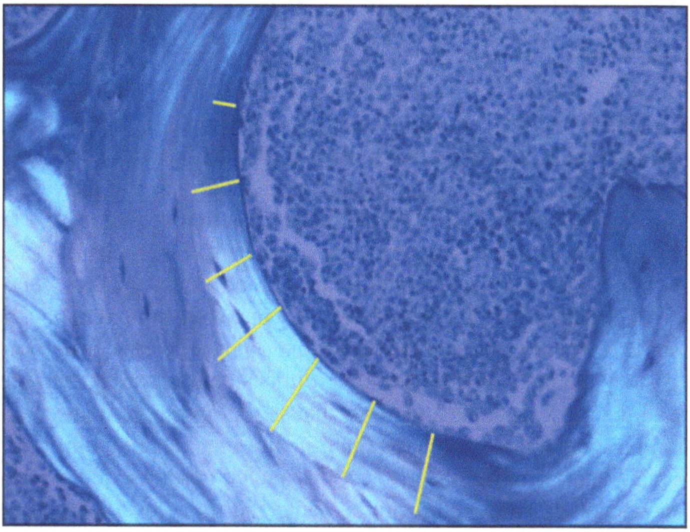

Fig. 8.6 Example of measurement of completed wall thickness in toluidine blue-stained section of cancellous bone from cynomolgus monkey viewed in polarized light. Equidistant perpendicular length measurements (*yellow lines*) from the cement line to the bone surface are made on completed bone packets (original magnification 200×)

thickness (W.Th) represents the same parameter in 3D terminology and is defined by multiplication of W.Wi with π/4 to correct for obliquity of section. The measurement of wall thickness can be performed by a 2-point distance measurement, sampling completed remodeling units, thereby measuring the distance between the cement line and the completed bone surface perpendicular to the bone surface (Fig. 8.6). Alternatively, wall width can be determined indirectly by tracing the bone surface and the cement line in each remodeling unit. To visualize the cement lines, different stains can be employed, usually in combination with polarized light: a cement line stain based on a surface-etching technique, toluidine blue stain, or Goldner's trichrome with fast green as counterstain (Schenk et al. 1984). Typically, at least 15 remodeling units per bone site are measured. The number of measurements required to achieve sufficient precision will ultimately depend on the variance of the measurement and the effect size; generally, collection of enough measurements to achieve a coefficient of error of ~10% is more than adequate. Based on the wall width measurement and on the principle that the surface extent of a certain activity is proportional to the time period occupied by this activity, formation period, resorption period, reversal period, remodeling period, and activation frequency can be calculated. For details of the calculations, the reader is referred to more specialized review articles and other textbooks (Parfitt 1983; Eriksen 1986; Dempster et al. 2013; Bonucci and Ballanti 2014).

8.6.5 Detailed Analysis of Bone Cell Populations: Stereological Analyses

Although standard histomorphometric methods can provide significant data regarding the effects of a test article on the dynamics of bone formation and resorption at the tissue level, there are additional methods available that can provide important mechanistic data and give further insight on the quantitative effects of a test article at the cellular level. This involves applying stereological methods to evaluate bone cell populations. Stereological methods have been widely used in neuroscience, respiratory, and renal research but until recently have had limited application to the bone (Vesterby 1993; Odgaard and Gundersen 1993). The data from stereological analyses convey changes in the true three-dimensional tissue at the whole-organ level. One of the most valuable endpoints that stereological methods can provide is information on total cell number contained in an organ, an endpoint that is highly biased by many factors in routine 2D sections and cell profile counts. Stereological methods have evolved from being cumbersome and time-consuming to being efficient and practical because of combined advances in stereological theory, sampling techniques, software, imaging devices, and automated programmable microtomes. It is beyond the scope of this chapter to discuss the principles of stereological methods, and the reader is referred to several references (Boyce et al. 2010; Cruz-Orive and Weibel 1990; Weibel et al. 2007; Nyengaard 1999; Muhlfeld et al. 2010; Howard and Reed 2005).

An example of the utility and sensitivity of stereological methods to evaluate quantitatively the effects of test articles on specific bone cell populations was recently demonstrated in a study comparing the effects of a sclerostin antibody, romosozumab, and hPTH(1-34) in rat vertebral cancellous bone. Both are bone-forming agents but have different modes of action at the tissue level in cancellous bone. Romosozumab caused a rapid marked increase in bone formation that attenuated or self-regulated with continued treatment. In contrast, hPTH (1-34) caused a sustained increase in bone formation and resorption. To understand the consequences of these different modes of action on the osteoblast lineage, selected subpopulations of the osteoblast lineage were targeted for estimation of total number using the stereological methods. These methods are detailed in Ominsky et al. (2015) and have demonstrated subtle but significant differences in the effects of these two bone-forming agents on the mature osteoblast and osteoprogenitor populations. These analyses demonstrated that sclerostin antibody maintains vertebral cancellous bone formation rates comparable to hPTH (1-34) with fewer osteoblasts and lower osteoblast density associated with significantly lower numbers of osteoprogenitors due to decreased demand for osteoblasts on the active bone-forming surfaces. Furthermore, they provided insight into the potential cellular mechanism for the self-regulation of bone formation with romosozumab. These data would only have been obtainable through the use of stereological methods.

8.7 Cortical Bone Dynamics

8.7.1 Periosteal and Endocortical Bone Apposition

It is necessary to measure the mineralizing perimeter and the mineral apposition rate separately at periosteal and endocortical surfaces for the quantification of periosteal and endocortical bone formation rates. Similar to cancellous bone histomorphometry, the double-labeled perimeter is usually used for the calculation of the mineralizing perimeter (M.Pm/B.Pm = Db.Lb.Pm/B.Pm) in rat and mouse studies. In non-rodent species, the double-labeled + one half of the single-labeled perimeter (M.Pm = Db.Lb.Pm+0.5 * S.Lb.Pm) is used to calculate the mineralizing perimeter (M.Pm/B.Pm = MS/BS). Periosteal bone apposition slows down with age in all species. Therefore, when planning a study in non- or slowly growing animals, it may be advisable to use separate pairs of fluorochrome markers for cancellous/endocortical and periosteal bone labeling, because the marker intervals used for cancellous/endocortical bone dynamic histomorphometry may result in inseparable double labels at periosteal surfaces.

8.7.2 Endocortical Bone Resorption

Osteoclasts are only found at low numbers at endocortical bone surfaces. Therefore, osteoclast numbers are usually not quantified at endocortical bone surfaces. Rather, to assess osteoclastic bone resorption at endocortical surfaces, the endocortical eroded perimeter (Ec.E.Pm/Ec.Pm) can be measured (Reim et al. 2008).

8.7.3 Intracortical Bone Remodeling

Longitudinal assessment of cortical bone mass in a chronic study with a test article can be easily accomplished using several in vivo densitometric techniques such as pQCT or HRpQCT. Changes in cortical bone mineral density (BMD) across the treatment period may suggest changes in intracortical remodeling in response to the test article, because as cutting cones are created and subsequently refilled with bone, there are reversible changes in BMD, known as remodeling space. Confirming this finding requires determination of the histologic correlate of changes in cortical remodeling or activation frequency, the number of new BMUs created per unit area of cortical bone.

For the evaluation of intracortical bone remodeling parameters, it is necessary to measure Haversian wall width (H.W.Wi). H.W.Wi is an average measurement on complete and quiescent osteons per bone site. Often 40 measurements are sufficient for a stable mean. At each of these osteons, the wall width is the distance measured from orthogonal intersects drawn from the cement line to the Haversian canal along the shortest axis of the osteons. In addition, the Haversian perimeter (H.Pm) can be measured, which includes perimeters traced at the edges of cavities seen in forming, resorbing, or resting osteons but excludes Volkmann's canals. It is also possible to derive the Haversian activation frequency (H.Ac.f; No./year), using the measured bone formation rate (H.BFR/BS) divided by H.W.Th. H.BFR is calculated by multiplying MAR and M.Pm/B.Pm within cortical bone osteons [H.BFR/BS = (Db. Lb.Pm+0.5 * S.Lb.Pm) * MAR/H.Pm]. The evaluation of Harversian remodeling is of limited benefit in juvenile animals as this process becomes only common in skeletally mature non-rodents. It is recommended to use mature animals if effects on Haversian systems are important considerations under study. Rats and mice lack true Haversian remodeling. Therefore, rats and mice are inappropriate models to study the effects of a test article on intracortical, Haversian remodeling.

Addressing temporal changes by correlative histomorphometry has previously required sequential cortical bone biopsies. To avoid the need for sequential biopsies, a technique was developed by the late Prof. Robert Schenk that takes advantage of the use of a unique sequence of fluorochrome labels administered at uniform intervals across the study (e.g., a single administration of a fluorochrome label every 3 or 4 weeks) (Boyce et al. 1996; Ominsky et al. 2016). The sequence of labels requires that each label (e.g., calcein green, alizarin complexone, xylenol orange, tetracycline) be given in a pattern such that the timing of administration is known and that each label can be recognized individually. Examples of sequences are given in Boyce et al. (1996) and Ominsky et al. (2016). At the end of the study, a cortical bone sample is collected, and undecalcified cross sections are prepared for analysis. Cortical osteons are temporally categorized by their specific sequence of fluorochrome labels contained within the osteon and tallied or aged according to their outermost (earliest) label (Fig. 8.7). The number of labeled osteons for each temporal category expressed as the number of labeled osteons per square millimeter of cortical bone provides an estimate of cortical activation that occurred approximately 3 weeks earlier than the timing of administration of the outermost (earliest) label. The 3-week interval corresponds to the average time required for the completion of the cutting cone and initiation of bone formation (High 1987). This method has been used to characterize the changes in cortical activation frequency in response to chronic intermittent hPTH(1-34) treatment in the dog (Boyce et al. 1996) and romosozumab treatment in the cynomolgus monkey (Ominsky et al. 2016).

8.8 Longitudinal Bone Growth and Epiphyseal Plate Width

General and skeletal toxicity often involves suppression of longitudinal bone growth in juvenile animals. The most sensitive assessment of longitudinal bone growth is the direct measurement of longitudinal bone growth rate (Lo.B.G.R), using fluorochrome labels.

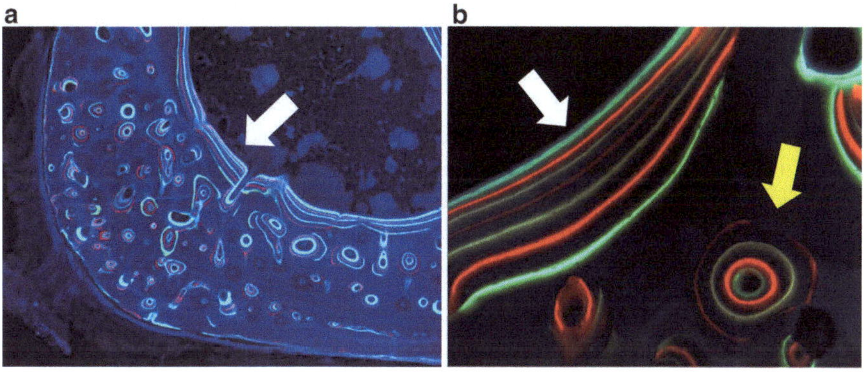

Fig. 8.7 Cortical long bone from an ovariectomized cynomolgus monkey with high intracortical remodeling administered with a sequence of fluorochrome labels over a 6-month period. (**a**) At low magnification, extensive incorporation of labels in the majority of cortical osteons is evident. Stacked labels are present on the endocortical surface (*white arrow*) indicating active bone formation over the period beginning at the time of administration of the deepest label sustained to the time of the label closest to the endocortical surface (original magnification 40×). (**b**) Higher magnification of the endocortical surface (*white arrow*) with stacked labels each corresponding to a specific time in the study, allowing for the temporal reconstruction of changes in endocortical mineral apposition rate. An osteon (*yellow arrow*) contains four labels; the sequence of the labels places the formation of the osteon within a specific time frame in the 6-month period (original magnification 200×). Temporal assignment of all labeled osteons in the cortex allows for the reconstruction of cortical remodeling rates over the 6-month period

Lo.B.G.R can be measured using single or double fluorochrome labels (Erben 2003). When a single label is used, the mean distance between the most proximal traces of the label in the metaphyseal cancellous bone and the first calcified cartilage septa in the growth plate is measured. The Lo.B.G.R is then given by the measured mean distance divided by the time interval between the administration of the label and the necropsy. When a double label is used for this type of measurement, the Lo.B.G.R is given by the mean distance between the most proximal traces of the two labels in the spongiosa divided by the time interval between the labels. It is important to consider that fluorochrome labels in cancellous bone will disappear as the bone elongates, because metaphyseal cancellous bone is eventually resorbed toward the diaphyseal marrow cavity in a growing long bone. Disappearance of such bone label could occur within 2 weeks after exposure considering that in young 6-week-old rats, the tibial longitudinal bone growth may attain 300 μm/day (Hansson et al. 1972). High bone turnover, such as induced by estrogen deficiency in rats or mice, accelerates loss of the metaphyseal label. Therefore, the time period that can be covered for measuring longitudinal bone growth is limited and depends on the growth rate and on the bone turnover in the secondary spongiosa.

Specific drugs may either increase or decrease the epiphyseal plate width (Ep.Pl.Wi) (Hall et al. 2006; Van Leeuwen et al. 2003). This parameter can be measured reliably at the tibia in growing animals due to its relatively linear profile by taking about 10–12 equidistant widths along the entire physis to derive a mean width (μm).

The measurement of this age-dependent and bone site-specific parameter is a futile exercise if performed in a cohort of animals with various ages or approaching the time of age-related physeal closure. Measurement of the cartilaginous plate can be obtained from toluidine blue- or Goldner-stained undecalcified bone sections or from decalcified H&E-stained sections and is made by taking widths from the end of the zone of calcification abutting with the bone marrow of the primary spongiosa to the junction of the resting zone and bone/marrow (Fig. 8.1b). If an effect is detected upon this parameter, further dissection of the effect can be done by measurement of the sub-compartments of the growth plate such as the resting, proliferative, and hypertrophic zones which may provide insight into the mechanism for test article-mediated changes in growth plate width (Wu et al. 2012).

8.9 Conclusions

Bone histomorphometry is one of the many tools that can be employed to understand the cellular mechanisms responsible for test article-related structural changes in the bone or to prospectively address a potential target- or class-related theoretical bone liability. The methods of bone histomorphometry are sensitive, detecting effects of test articles on bone resorption, formation, mineralization, remodeling rates, and growth. Effects can be detected in short-term studies before changes in bone mass, the long-term consequence of alterations in bone metabolism, occur.

References

Bagi CM, Berryman E, Moalli MR. Comparative bone anatomy of commonly used laboratory animals: implications for drug discovery. Comp Med. 2011;61:76–85.

Bonucci E, Ballanti P. Osteoporosis-bone remodeling and animal models. Toxicol Pathol. 2014;42:957–69.

Boyce RW, Ebert DC, Youngs TA, Paddock CL, Mosekilde L, Stevens ML, et al. Unbiased estimation of vertebral trabecular connectivity in calcium-restricted ovariectomized minipigs. Bone. 1995a;16:637–42.

Boyce RW, Paddock CL, Gleason JR, Sietsema WK, Eriksen EF. The effects of risedronate on canine cancellous bone remodeling: three-dimensional kinetic reconstruction of the remodeling site. J Bone Miner Res. 1995b;10:211–21.

Boyce RW, Paddock CL, Franks AF, Jankowsky ML, Eriksen EF. Effects of intermittent hPTH(1-34) alone and in combination with 1,25$(OH)_2D_3$ or risedronate on endosteal bone remodeling in canine cancellous and cortical bone. J Bone Miner Res. 1996;11:600–13.

Boyce RW, Dorph-Petersen KA, Lyck L, Gundersen HJ. Design-based stereology: introduction to basic concepts and practical approaches for estimation of cell number. Toxicol Pathol. 2010;38:1011–25.

Boyce RW, Niu QT, Ominsky MS. Kinetic reconstruction reveals time-dependent effects of romosozumab on bone formation and osteoblast function in vertebral cancellous and cortical bone in cynomolgus monkeys. Bone. 2017;101:77–87.

Cruz-Orive LM, Weibel ER. Recent stereological methods for cell biology: a brief survey. Am J Phys. 1990;258:L148–56.
Dempster DW, Compston JE, Drezner MK, Glorieux FH, Kanis JA, Malluche H, et al. Standardized nomenclature, symbols, and units for bone histomorphometry: a 2012 update of the report of the ASBMR Histomorphometry Nomenclature Committee. J Bone Miner Res. 2013;28:2–17.
Erben RG. Embedding of bone samples in methylmethacrylate: an improved method suitable for bone histomorphometry, histochemistry, and immunohistochemistry. J Histochem Cytochem. 1997;45:307–13.
Erben RG. Bone labeling techniques. In: An YH, Martin KL, editors. Handbook of histology methods for bone and cartilage. Totowa: Humana Press; 2003. p. 99–117.
Erben RG, Glosmann M. Histomorphometry in rodents. Methods Mol Biol. 2012;816:279–303.
Eriksen EF. Normal and pathological remodeling of human trabecular bone: three dimensional reconstruction of the remodeling sequence in normals and in metabolic bone disease. Endocr Rev. 1986;7:379–408.
Eriksen EF, Gundersen HJ, Melsen F, Mosekilde L. Reconstruction of the formative site in iliac trabecular bone in 20 normal individuals employing a kinetic model for matrix and mineral apposition. Metab Bone Dis Relat Res. 1984a;5:243–52.
Eriksen EF, Melsen F, Mosekilde L. Reconstruction of the resorptive site in iliac trabecular bone: a kinetic model for bone resorption in 20 normal individuals. Metab Bone Dis Relat Res. 1984b;5:235–42.
Eriksen EF, Langdahl B, Vesterby A, Rungby J, Kassem M. Hormone replacement therapy prevents osteoclastic hyperactivity: a histomorphometric study in early postmenopausal women. J Bone Miner Res. 1999;14:1217–21.
Frost HM. Bone histomorphometry: analysis of trabecular bone dynamics. In: Recker RR, editor. Bone histomorphometry: techniques and interpretation. Boca Raton: CRC Press; 1983a. p. 109–31.
Frost HM. Bone histomorphometry: choice of marking agent and labeling schedule. In: Recker RR, editor. Bone histomorphometry: techniques and interpretation. Boca Raton: CRC Press; 1983b. p. 37–52.
Frost HM. Bone histomorphometry: correction of labeling 'escape error'. In: Recker RR, editor. Bone histomorphometry: techniques and interpretation. Boca Raton: CRC Press; 1983c. p. 133–42.
Hall AP, Westwood FR, Wadsworth PF. Review of the effects of anti-angiogenic compounds on the epiphyseal growth plate. Toxicol Pathol. 2006;34:131–47.
Hansson LI, Menander-Sellman K, Stenstrom A, Thorngren KG. Rate of normal longitudinal bone growth in the rat. Calcif Tissue Res. 1972;10:238–51.
High WB. Effects of orally administered prostaglandin E-2 on cortical bone turnover in adult dogs: a histomorphometric study. Bone. 1987;8:363–73.
Howard CV, Reed MG. Unbiased stereology, three-dimensional measurements in microscopy. Abingdon: Garland Science/Bios Scientific Publishers; 2005.
Kilborn SH, Trudel G, Uhthoff H. Review of growth plate closure compared with age at sexual maturity and lifespan in laboratory animals. Contemp Top Lab Anim Sci. 2002;41:21–6.
Muhlfeld C, Nyengaard JR, Mayhew TM. A review of state-of-the-art stereology for better quantitative 3D morphology in cardiac research. Cardiovasc Pathol. 2010;19:65–82.
Nyengaard JR. Stereologic methods and their application in kidney research. J Am Soc Nephrol. 1999;10:1100–23.
Odgaard A, Gundersen HJG. Quantification of connectivity in cancellous bone, with special emphasis on 3-D reconstructions. Bone. 1993;14:173–82.
Ominsky MS, Vlasseros F, Jolette J, Smith SY, Stouch B, Doellgast G, et al. Two doses of sclerostin antibody in cynomolgus monkeys increases bone formation, bone mineral density, and bone strength. J Bone Miner Res. 2010;25:948–59.
Ominsky MS, Niu QT, Li C, Li X, Ke HZ. Tissue-level mechanisms responsible for the increase in bone formation and bone volume by sclerostin antibody. J Bone Miner Res. 2014;29:1424–30.

Ominsky MS, Brown DL, Van G, Cordover D, Pacheco E, Frazier E, et al. Differential temporal effects of sclerostin antibody and parathyroid hormone on cancellous and cortical bone and quantitative differences in effects on the osteoblast lineage in young intact rats. Bone. 2015;81:380–91.

Ominsky MS, Boyd SK, Varela A, Jolette J, Felx M, Doyle N, et al. Romosozumab improves bone mass and strength while maintaining bone quality in Ovariectomized Cynomolgus monkeys. J Bone Miner Res. 2016;32:788–801.

Paddock C, Youngs T, Eriksen E, Boyce R. Validation of wall thickness estimates obtained with polarized light microscopy using multiple fluorochrome labels: correlation with erosion depth estimates obtained by lamellar counting. Bone. 1995;16:381–3.

Parfitt AM. Stereologic basis of bone histomorphometry: theory of quantitative microscopy and reconstruction of the third dimension. In: Recker RR, editor. Bone histomorphometry: techniques and interpretation. Boca Raton: CRC Press; 1983. p. 53–87.

Parfitt AM, Mathews CH, Villanueva AR, Kleerekoper M, Frame B, Rao DS. Relationships between surface, volume, and thickness of iliac trabecular bone in aging and in osteoporosis. Implications for the microanatomic and cellular mechanisms of bone loss. J Clin Invest. 1983;72:1396–409.

Parfitt AM, Drezner MK, Glorieux FH, Kanis JA, Malluche H, Meunier PJ, et al. Bone histomorphometry: standardization of nomenclature, symbols, and units. Report of the ASBMR Histomorphometry Nomenclature Committee. J Bone Miner Res. 1987;2:595–610.

Reim NS, Breig B, Stahr K, Eberle J, Hoeflich A, Wolf E, et al. Cortical bone loss in androgen-deficient aged male rats is mainly caused by increased endocortical bone remodeling. J Bone Miner Res. 2008;23:694–704.

Roach HI, Mehta G, Oreffo RO, Clarke NM, Cooper C. Temporal analysis of rat growth plates: cessation of growth with age despite presence of a physis. J Histochem Cytochem. 2003;51:373–83.

Schenk RK, Olah AJ, Herrmann W. Preparation of calcified tissues for light microscopy. In: Dickson GR, editor. Methods of calcified tissue preparation. Amsterdam: Elsevier; 1984. p. 1–56.

Smith SY, Varela A, Jolette J. Nonhuman primate models of osteoporosis. In: Duque G, Watanabe K, editors. Osteoporosis research – animal models. London: Springer Verlag; 2011. p. 135–57.

Steiniche T, Eriksen EF, Kudsk H, Mosekilde L, Melsen F. Reconstruction of the formative site in trabecular bone by a new, quick, and easy method. Bone. 1992;13:147–52.

Sterchi DL, Callis GM. Staining techniques for undecalcified bone embedded in methylmethacrylate. National Society for Histotechnology Symposium, Philadelphia, Workshop No. 63; 1993.

Storm T, Steiniche T, Thamsborg G, Melsen F. Changes in bone histomorphometry after long-term treatment with intermittent, cyclic etidronate for postmenopausal osteoporosis. J Bone Miner Res. 1993;8:199–208.

Van Leeuwen BL, Hartel RM, Jansen HW, Kamps WA, Hoekstra HJ. The effect of chemotherapy on the morphology of the growth plate and metaphysis of the growing skeleton. Eur J Surg Oncol. 2003;29:49–58.

Vesterby A. Star volume in bone research. A histomorphometric analysis of trabecular bone structure using vertical sections. Anat Rec. 1993;235:325–34.

Weibel ER, Hsia CC, Ochs M. How much is there really? Why stereology is essential in lung morphometry. J Appl Physiol (1985). 2007;102:459–67.

Wessler S. Introduction: what is a model? In: Committee on animal models for thrombosis and hemorrhagic diseases, Institute of Laboratory Animal Resources, editor. Animal models of thrombosis and hemorrhagic diseases. Bethesda: National Institute of Health; 1976. p. xi–xvi.

Wu S, Levenson A, Kharitonenkov A, De LF. Fibroblast growth factor 21 (FGF21) inhibits chondrocyte function and growth hormone action directly at the growth plate. J Biol Chem. 2012;287:26060–7.

Part III
Bone as an Endocrine Organ

Chapter 9
Bone and Muscle

Chenglin Mo, Zhiying Wang, Lynda Bonewald, and Marco Brotto

Abstract Muscle and bone are anatomically integrated with each other to form the musculoskeletal system. Muscle has been considered as an endocrine organ to affect other tissues, including bone, via secreting factors within the microenvironment to modulate their biological functions. Thus muscle is essential for bone development, modeling, and remodeling. Similar to muscle, bone can also be considered an endocrine organ targeting other tissues such as muscle. Therefore, the aim of this chapter is to summarize endocrine factors derived from either bone or muscle and their potential benefits for the therapy of simultaneous bone-muscle dysfunction, such as osteoporosis and sarcopenia in the elderly population. Herein, we first describe the synchronous development of the musculoskeletal system during embryogenesis and how this close relationship between bone and muscle continues in postnatal life. Then the factors regulating both bone and muscle, including factors like myogenic regulatory factors (MRFs) and myostatin from muscle, bone-derived factors like bone morphogenetic proteins (BMPs) and prostaglandin E_2 (PGE_2), as well as other secretory factors, are discussed to provide a better understanding of the mechanisms underlying the parallel development of musculoskeletal system diseases. Finally, we discuss some meaningful and innovative treatment strategies, which preferentially strengthen the musculoskeletal unit as a whole.

Keywords Endocrine • Myogenic • Osteocalcin • Embryogenesis • Extracellular matrix (ECM) • Transdifferentiation • Sarcopenia

---

C. Mo • Z. Wang • M. Brotto (✉)
Bone-Muscle Collaborative Sciences, School of Nursing & Health Innovation, The University of Texas at Arlington, 411 S. Nedderman Drive, Pickard Hall 525, Arlington, TX 76019, USA
e-mail: marco.brotto@uta.edu

L. Bonewald
Indiana Center for Musculoskeletal Health, Department of Anatomy and Cell Biology, Orthopaedic Surgery, School of Medicine, Indiana University, Indianapolis, IN 46202, USA

S.Y. Smith et al. (eds.), *Bone Toxicology*, Molecular and Integrative Toxicology,
DOI 10.1007/978-3-319-56192-9_9

9.1 Introduction

Bone and muscle are two integrated components in the musculoskeletal system. In an average adult, bones make up about 15% of overall body mass, and skeletal muscle accounts for 35–45% body mass (Janssen et al. 2000). One of the fundamental functions of bone and muscle is to produce controlled, precise movements. To accomplish this, skeletal muscle contracts in response to action potentials, which shortens and pulls the tendon. The tendon, in turn, pulls the bone to make us sit, stand, walk, or run. In addition to these commonly identified actions, bone and muscle together serve to protect soft tissues and organs including the brain, heart, and lung, to maintain body temperature, and to play critical metabolic roles in our lives. Bone serves as an internal reservoir for calcium to ensure the proper function of nerves and muscle, and skeletal muscle is responsible for over 80% of carbohydrate storage (DeFronzo et al. 1985; Kovacs 2011). Moreover, the skeleton also contributes to glucose homeostasis, further intertwining the actions of bone and muscle beyond locomotion (Motyl et al. 2010). Recently, bone and muscle have been considered to be endocrine organs. It has been discovered that muscles, especially contracting muscles, are able to release peptides or proteins termed "myokines", including interleukin (IL)-6, IL-8, and IL-15 that appear to provide a reservoir of muscle-derived factors for communication with other organs in the human body via autocrine, paracrine, or endocrine mechanisms (Argiles et al. 2009; Gray and Kamolrat 2011; Pedersen and Fischer 2007). The importance of these myokines for human fitness and their roles in diseases, such as diabetes and obesity, is beginning to emerge and be recognized by the scientific community. On the other hand, in bone, osteocytes are the cells in skeleton that sense and respond to mechanical loads to regulate bone remodeling (Bonewald 2011; Javaheri et al. 2014). Upon sensing load, factors (a major one being prostaglandin E_2, PGE_2) important for regulation of bone formation, bone mass, and resorption are produced and secreted by osteocytes and travel through their lacuna-canalicular system to the circulation, thereby reaching and regulating the functions of other tissues and organs (Bonewald 2011; Dallas et al. 2013; Kitase et al. 2010). The integration of bone and muscle continues in aging. Aging results in the progressive and parallel loss of bone (osteopenia) and skeletal muscle (sarcopenia) with profound consequences for quality of life (He et al. 2016; Isaacson and Brotto 2014; Karasik and Kiel 2008; Pereira et al. 2015). Age-associated sarcopenia results in reduced endurance, poor balance, and reduced mobility that predispose elderly individuals to falls, which frequently result in fracture because of concomitant osteoporosis (Ferrucci et al. 2014; Frisoli et al. 2011; Hong et al. 2015; Szulc et al. 2005). Thus, a better understanding of the mechanisms underlying the parallel development and involution of these tissues is critical to developing new and more effective means to combat osteoporosis and sarcopenia in our increasingly greater population of aged individuals. Overall, bone and muscle integrate with each other and work as a unit beginning with the embryo and throughout life, and their communication relies on both physical and biochemical interactions.

9.1.1 Bone and Muscle in Embryonic Development

Mammalian bone and skeletal muscle are involved with locomotion and develop in close association from the somitic mesoderm. The skeletal system develops from the mesodermal germ layer, which can be divided into three basic regions, according to their positions from medial to lateral: (1) paraxial mesoderm, (2) intermediate mesoderm, and (3) lateral plate mesoderm (Fig. 9.1). The paraxial mesoderm plays a major role in the formation of the musculoskeletal system. Somitogenesis is a critical step that occurs in the paraxial mesoderm where cells divide into blocklike structures called somites. Each somite contains specific precursor populations for the development of bones (sclerotome), tendons (syndotome), skeletal muscles (myotome), and dermis (dermatome). Cells within the sclerotome go through the stages of precartilage, chondrification, and ossification to form the vertebrae, the rib, and part of the occipital bone. Meanwhile, cells of the myotome committed to muscle generation become myoblasts characterized by an elongated, spindle shape. Myoblasts will undergo proliferation, differentiation, and fusion to form multinucleated cylindrical structures called myotubes. Eventually, fusion of myotubes give rise to muscle fibers, which then group into bundles and the bundles come together to form the tissue. Simultaneously, while limb bones are developing from the somatic layer of lateral plate mesoderm, limb muscle is developing from in situ myoblasts around the developing limb bones and myoblasts migrating from myotome in the paraxial mesoderm (Fig. 9.1) (Brent and Tabin 2002; Pourquie 2001).

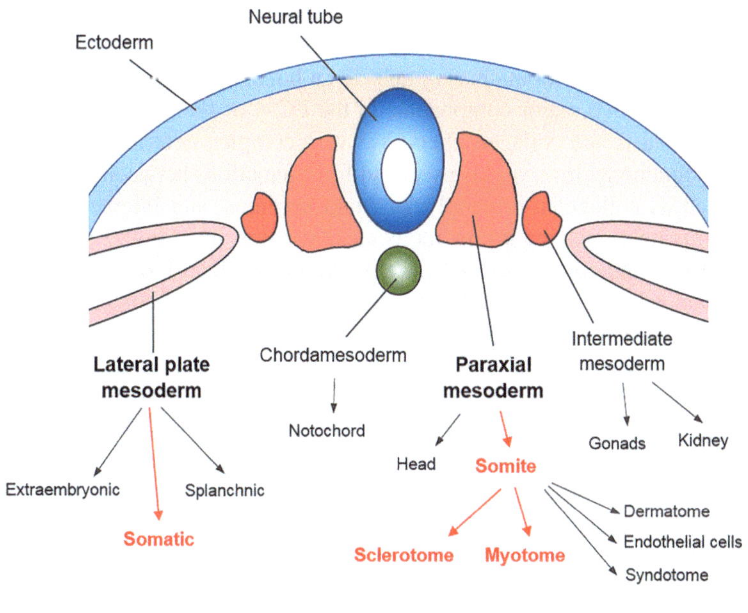

Fig. 9.1 Embryonic development of bone and muscle

Bone and muscle develop synchronously during embryogenesis, and this process is precisely controlled by many signaling pathways, such as Wnt and BMP signaling, and dysregulation of these pathways will lead to defects in bone/muscle growth and patterning (Edom-Vovard and Duprez 2004; Winnier et al. 1995).

9.1.2 The Postnatal Connection Between Bone and Muscle

The close relationship between bone and muscle during embryogenesis continues in postnatal life. The two tissues work as a functional unit in locomotion and have important functions in nutrition storage, regulation of metabolism, and other aspects. In terms of development, bone and muscle reach their peak tissue mass synchronously, which is determined by genetic background, nutrition, and environmental stimuli. Mechanical loading has been shown to be able to stimulate and strengthen both bone and muscle simultaneously. The extracellular matrix (ECM) of bone and muscle plays an essential role in converting mechanical stimulation to intracellular biochemical signaling in both tissues.

Endomysium, perimysium, and epimysium are the three compartments of muscle ECM. Endomysium locates between muscle fibers, and perimysium and epimysium encase fascicle and entire muscle, respectively. Epimysium, the outermost layer of muscle ECM, acts as a buffer to protect muscles from strain and friction between muscles or between bone and muscle to promote smooth movement during locomotion. Due to its important protective function, epimysium is ubiquitous in muscles. For attachment of skeletal muscle to bone, epimysium fuses with the periosteum, the outer layer of ECM of bone, and this is called direct or fleshy attachment. At both ends of the muscle, epimysium converges with endomysium and perimysium on tendon for anchoring muscle on bone.

Although the two major components in the ECM of skeletal muscle are tightly packed fibroblasts and collagen fibers, it is very complex in terms of structure and biological functions. It serves as a scaffold for interactions between components of ECM, such as collagens, proteoglycans, and laminins, and interactions between ECM components and adhesion molecules, such as integrins. The interactions between ECM and adhesion receptors activate intracellular biochemical signaling and cytoskeletal rearrangement. Furthermore, by sensing mechanical loading, ECM can regulate the activities of growth factors. Enzymes have also been identified in the skeletal muscle ECM. Tyrosine kinase receptors play important signaling roles. Mechanical stress has been shown to activate members of the mitogen-activated protein kinase (MAPK) family, including extracellular signal-regulated kinases 1 and 2 (MAPK-erk1+2, p44), stress-activated protein kinases p38 (MAPK-p38), c-jun NH_2-terminal kinase (MAPK-jnk, p54), and extracellular signal-regulated kinase 5 (MAPK-erk5). It is not unexpected that metalloproteinases (MMPs) and tissue inhibitors of metalloproteinases (TIMPs) also have been identified in ECM due to their critical functions in ECM remodeling (Kjaer et al. 2006; Gillies and Lieber 2011).

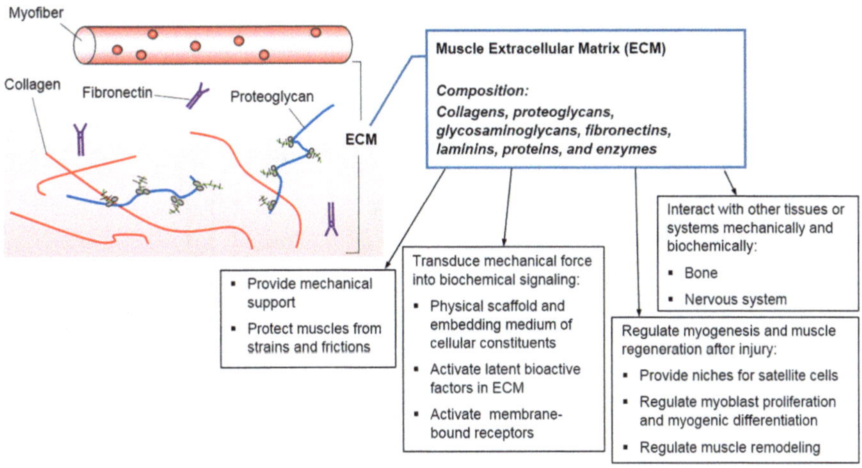

Fig. 9.2 Composition of muscle ECM and its functions

ECM not only protects muscle from damage induced by mechanic stimuli, but also regulates intracellular signaling for muscle function and myogenesis (Fig. 9.2). For instance, the physical properties of muscle ECM are important for maintaining the satellite cell niche.

As mentioned above, the muscle ECM makes contact with the periosteum of bone. Periosteum is a thin sheath of ECM, which is further divided into two distinct layers, an outer fibrous connective layer and an inner layer with osteoblastic potential for bone growth and fracture healing (Fig. 9.3).

Outer layer The outer layer of periosteum can be separated into two zones. The superficial zone mainly consists of collagenous matrix and elongated fibroblasts. Collagen fibrils contribute almost half of the volume of this tissue to generate a dense layer also characterized by few blood vessels. The deeper zone of the outer layer is also composed of collagen fibers and fibroblasts, and each of them contributes 25% of the volume of this zone. Most vessels, predominantly capillaries, locate in the deeper zone of the outer layer (Squier et al. 1990). The outer layer fuses with muscle epimysium and connects joint capsule and tendon through continuous collagen fibrils. Vessels inside periosteum are not only important for blood supply of bone and even skeletal muscle but also could play an essential role for systemic regulation for bone and muscle as endocrine organs.

Inner cambium layer Compared with the outer layer, the cambium layer is relatively thin. The thickness of cambium layer changes during development. It is thickest in the fetus, but becomes indistinguishable from the outer layer in adults. This change is accompanied by decreases in cell number in this layer, especially the number of mesenchymal progenitor cells. Mesenchymal progenitor cells can differentiate into osteoblasts and are critically important for bone formation in the

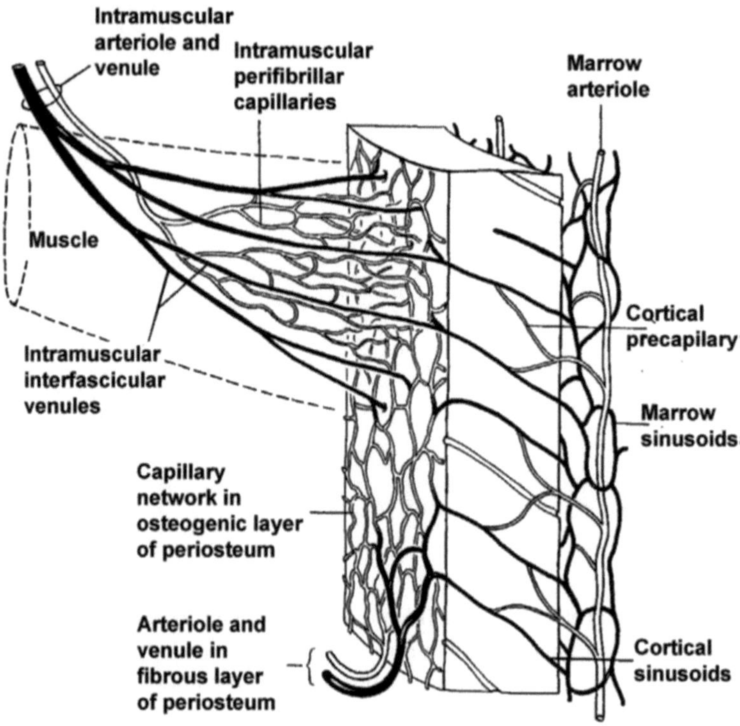

Fig. 9.3 Structure of the periosteum and the vascular connection between bone and muscle in the periosteum (Adapted from Fig. 9.37 of Brooks and Revell 1998)

fetus. This osteogenic capacity significantly decreases during childhood, but can be revived by conditions and challenges, such as bone fracture (Ito et al. 2001).

In vitro studies revealed that cells derived from periosteum proliferate rapidly and respond to physiological levels of mechanical strain by increasing cell growth and PGE_2 production (Jones et al. 1991). Moreover, treatment with parathyroid hormone, PTH, a potent bone anabolic hormone, significantly increases bone matrix protein production in periosteal cells compared to untreated control cells (Midura et al. 2003).

Similar to muscle, bone also contains protein factors in the ECM, and some of these factors previously were considered to be expressed only in cells of mineralized tissue such as bone and teeth. These include dentin matrix protein 1 (DMP1), dentin phosphophoryn (DPP), and osteopontin (Ravindran and George 2014). The presence of these proteins in ECM suggests that mechanical loading can also induce biological signaling through factors in the ECM in both bone and muscle. Based on the fact that bone and muscle are integrated through ECM, ECM could play an important role in bone-muscle interaction and remodeling. An example of this is the activation of transforming growth factor-beta (TGF-β) in ECM by mechanical loading. TGF-β is synthesized by cells and then deposited into ECM as a large latent complex associated with the latency-associated peptide (LAP), which binds to

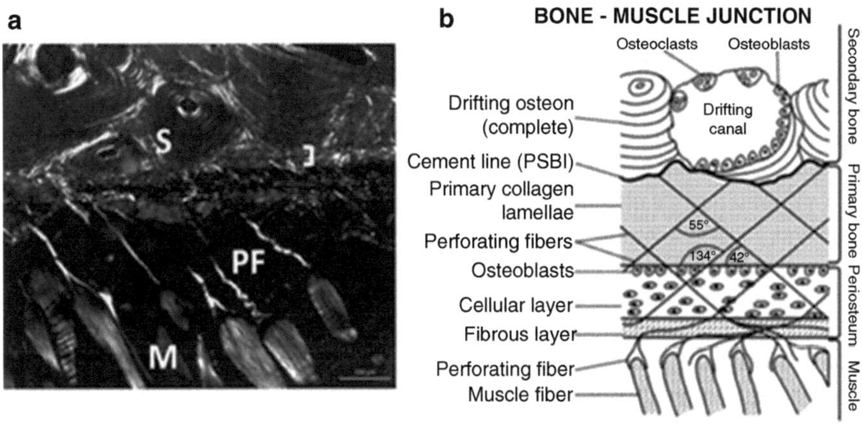

Fig. 9.4 (**a**) Photomicrograph of the bone-muscle junction (periosteal surface facing down, female aged 23 years). Perforating fibers (PF) connect muscle fibers (M) with the interface of primary (narrow at this age) and secondary (S) bone. (**b**) Diagram of the bone-muscle junction (Modified from Schitzler (2015))

latent TGF-beta binding protein (LTBP). Mechanical loading induces a conformational change of the latent complex leading to the release of TGF-β (Hinz 2009). There are three identified isoforms of TGF-β: TGF-β1, TGF-β2, and TGF-β3. Although all isoforms of TGF-β can be detected in the periosteum, only deletion of TGF-β2 results in severe abnormalities in bone at birth in animal models (Sanford et al. 1997; Kulkarni et al. 1993; Kaartinen et al. 1995). Further studies revealed that TGF-β1 plays an essential role in bone remodeling, which is associated with promoting the migration of bone mesenchymal stem cells (Tang et al. 2009), mineralization of osteoblasts (Ota et al. 2013), and apoptosis of osteoclasts (Murakami et al. 1998). TGF-β1 has been considered as the most effective TGF-β isoform in muscle. TGF-β1 signaling plays an important role in the determination of muscle fiber types and the development of muscle fibrosis (Burks and Cohn 2011). In addition to LAP and LTBP, decorin, a leucine-rich proteoglycan in ECM, also can bind to TGF-β to inactivate TGF-β signaling (Yamaguchi et al. 1990). In vitro data also suggest that decorin can act as a negative regulator of myostatin (Miura et al. 2006).

In a recent study performed by Dr. Christine Schnitzler, perforating fibers originating from muscle were discovered (Schnitzler 2015). These fibers connect muscle fibers with the primary-secondary bone interface during bone development, but their function in the transmission of bone-muscle signaling, including mechanical and biochemical signals, has not been explored (Fig. 9.4).

9.1.3 Role of ECM in Bone and Muscle Diseases

Based on their important functions, it is reasonable to expect that the ECMs of muscle and bone play important roles in the development of muscle and bone diseases and could be important therapeutic targets for the treatments of muscle and bone diseases.

Accumulating data have shown that dysfunction within the components of muscle ECM is an important cause of muscular dystrophy. Heterotrimeric laminins are important components in basement membrane of skeletal muscle for transducing mechanical signal into intracellular signaling by interacting with integrins. During muscle development, laminin-111 composed of α1, β1, and γ1 chains is the major isoform of laminin in embryonic muscle, but it is replaced by laminin-211 (α2, β1, and γ1, previously called merosin) in adulthood. Completed laminin α2 chain deficiency or mutation in *LAMA2* gene results in merosin-deficient congenital muscular dystrophy 1A (MDC1A), which accounts for approximately 40% cases of congenital muscular dystrophy in Europe. MDC1A leads to muscle weakness and can be lethal due to respiratory failure and feeding difficulties (Holmberg and Durbeej 2013; Gawlik and Durbeej 2011). Multiple mouse models with laminin α2 chain deficiency have been developed and effectively mimic MDC1A development in human. Knowledge obtained from these models has greatly promoted the understanding of mechanisms behind MDCIA and the development of potential therapeutic interventions (Holmberg and Durbeej 2013). Interestingly, supplement with laminin-111 improves mobility and muscle strength in MDC1A mice, which is associated with the prevention of fibrosis and apoptosis of muscle cells. Antiapoptotic effects of laminin-111 also can be observed in primary myoblasts isolated from MDC1A patients (Rooney et al. 2012). In addition, treatment with laminin-111 prevents the development of muscle diseases in *mdx* mice, an animal model for Duchenne muscular dystrophy (Rooney et al. 2009). The activation of the proteasome could play a role in the development of muscle atrophy in MDC1A mice. Systemic injection of the proteasome inhibitor MG-132 prevents muscle atrophy in quadriceps and improves the locomotion and lifespan of MDC1A mice (Carmignac et al. 2011).

In the muscle ECM, the collagen network is formed outside the basement membrane. Collagen VI has been correlated with the development of myopathy (Tagliavini et al. 2013; Lampe and Bushby 2005). Collagen VI mainly is composed of α1, α2, and α3 chains. Animal models with *COL6A1* deletion or *COL6A3* mutation show similar symptoms as Bethlem myopathy in humans (Shah et al. 1986; Bonaldo et al. 1998). Data from recent studies indicate that α6 chain, a newly identified collagen VI component in human, may also be involved in the progress of collagen VI-related myopathies (Bonaldo et al. 1998; Sabatelli et al. 2012; Vevers 1970). Moreover, collagen VI affects mitochondrial function and satellite cell self-renewal, which are important for muscle function and disease development in muscle (Urciuolo et al. 2013; Bernardi and Bonaldo 2013).

Compared with the studies of muscle ECM in muscle diseases, less is known about the role of the periosteum in osteoporosis and fracture healing. Current treatments for osteoporosis mainly focus on enhancing bone strength through increasing bone mineral density (Kling et al. 2014). However, bone size is another important factor determining bone strength and fracture risk (Vanderschueren et al. 2006; Allen et al. 2004). During puberty, periosteal bone formation is greater in men than in women, which results in larger bone size and enhanced bone strength (Vanderschueren et al. 2006). Traditionally, this difference is attributed to the stimulatory effect of androgen and inhibitory effect of estrogen on periosteal expansion (Vanderschueren et al. 2006). Recent data from animal models and humans have revealed that sex steroids have complex functions in regulating periosteal bone formation. Inactivation of estrogen receptor in male mice leads to reduced radial bone growth and thinner bones (Vidal et al. 2000). Treatment with estrogen (estradiol), but not testosterone, significantly recovered attenuated periosteal expansion in a young man suffering from aromatase deficiency, which is responsible for estrogen deficiency (Bouillon et al. 2004). On the other hand, in preclinical studies, the administration of letrozole, an inhibitor of aromatase, did not show significant effect on bone but caused delayed sex maturation in both male and female juvenile rats (Pouliot et al. 2013). These data suggest that maintenance of proper levels of sex hormones is critical for periosteal bone formation and bone size.

Parathyroid hormone, PTH, also plays an important role in the regulation of bone size. Treatment with PTH shows increased bone mass and increased resistance to fracture in adult ovariectomized rhesus monkeys (Fox et al. 2007). Consistent with in vitro results using periosteal cells, enhanced modeling of periosteal surface after once daily PTH treatment contributes to increased cortical bone width (Burr et al. 2001). Clearly, PTH is one of the most anabolic osteogenic factors.

The periosteum may also play a role in bone resorption due to osteoporosis. Compared with normal controls, osteoporotic rats have significantly more active osteoclasts in the periosteum, especially in the metaphyseal periosteum. Since the metaphyseal area is a common area where osteoporosis-induced fracture occurs, this finding may have important clinical relevance (Fan et al. 2010). To date, knowledge about the role of periosteum in the development of osteoporosis and how the periosteum responds to therapeutics for the treatment of osteoporosis is limited. To target the periosteum for developing new treatments for bone diseases, future studies are necessary and may offer a very promising new venue of unexplored possibilities.

9.2 Bone-Muscle Biochemical Interactions

9.2.1 Bone and Muscle Interact Through Secretory Factors

In recent years, results from different research groups have supported the concept that muscle functions as an endocrine organ. It regulates systemic metabolism by secretory myokines to interact with other tissues or organs, including the brain, liver, fat, and bone, while also having autocrine effects. The discovered myokines include myostatin, leukemia inhibitory factor (LIF), IL-6, IL-7, brain-derived neurotrophic factor (BDNF), insulin-like growth factor I (IGF-1), fibroblast growth factor 2 (FGF2), and follistatin-like protein 1 (FSTL-1). Some of them have been shown to be important for bone development, such as IGF-1 and FGF2 (Pedersen and Febbraio 2012). Given the proximity of bone and muscle, it is reasonable to postulate that many myokines could impact bone directly through binding to specific receptors. Moreover, indirect effects could also be involved in bone-muscle interaction, since myokines can affect the pancreas and adipose, and other tissues, which have been shown in previous studies to crosstalk with bone biochemically. For example, irisin, a newly identified myokine, can induce the switch of white fat to brown-fat-like tissue, which could subsequently exert effects on bone (Zhang et al. 2014). In addition, recent data also suggest that irisin can directly regulate osteogenesis in osteoblast and bone marrow stromal cell cultures (Qiao et al. 2016; Colaianni et al. 2015).

Similar to muscle, bone also has been considered as an endocrine organ. Klein-Nulend and her team observed that bone cells secreted prostaglandins in response to mechanical stress induced in pulsating fluid flow experiments (Klein-Nulend et al. 1997). Research findings continue to support the significant role of prostaglandins in bone homeostasis, particularly the E and F series of prostaglandins (Agas et al. 2013; Hartke and Lundy 2001; Mo et al. 2012; Raisz 1999; Yoshida et al. 2002). Research performed in the early part of this century provided evidence that in addition to their function as a sensory and responsive cell, osteocytes also serve to help regulate bone density through the secretion of sclerostin, a protein that inhibits bone formation (Winkler et al. 2003). This work tested the hypothesis that the dysregulation in bone formation resulted from phenotypes observed in osteosclerosis patients. This hypothesis was further supported through genetic testing and the development of transgenic mice with increased sclerostin production and low bone mass. Another factor shown to mainly be produced by osteocytes in bone is fibroblast growth factor 23, FGF23, a bone-derived protein that has been identified as integral to vitamin D metabolism and the regulation of systemic phosphate levels by targeting the kidney. The impressive list of bone-derived factors continues to grow and includes: adenosine triphosphate (ATP), calcium, Dickkopf-1 (DKK1), DMP1, FGF23, matrix extracellular phosphoglycoprotein (MEPE), nitric oxide, osteoprotegerin (OPG), osteocalcin, prostaglandins (particularly PGE_2), receptor activator of nuclear factor-kappaB ligand (RANKL), and sclerostin. These factors represent a myriad of biochemical structures ranging from simple organic molecules to complex proteins, all of which help illustrate the diversity and far-reaching impact of bone as an endocrine organ.

As described above, conditions are predetermined in utero for the connection between bones and muscles, as they share a common mesenchymal precursor and experience organogenesis through a tightly orchestrated network of genes during intrauterine development. Their anatomic proximity lends credence to the hypothesis that bones and muscles influence each other in a paracrine nature. Evidence of such a relationship exists in pathologic conditions, for example, some of the bone stress syndromes where inflammation localizes to the muscle area underneath the periosteal region spreads into the bone itself. These situations support the paracrine relationship hypothesis, suggesting inflammatory molecules from adjacent muscle fibers may penetrate into this region of the bone. Another powerful clinical example of this paracrine relationship is the muscle flap application to compounded bone fractures. The effect of this therapeutic approach provides significantly faster healing for fractured bones. Although the specific molecular mechanism of action is not completely understood, the introduction of muscle flaps has been used as a successful therapeutic approach to treat chronic osteomyelitis and to accelerate the healing of bone fractures (Chan et al. 2012). These mechanisms might display further importance for bone and muscle healing after musculoskeletal injury. Our experiments performed using osteocyte and muscle cell lines have revealed that PGE_2 secretion from osteocytes is more than 100 times greater than PGE_2 secretion from muscle cells. This excess amount of PGE_2 from osteocytes could interplay with injured muscles, which would aid in muscle regeneration and repair. Intriguingly, recent in vitro studies have provided support for a role of osteocyte secreted PGE_2 in aiding with the process of myogenesis (Mo et al. 2012). While these studies were originally performed with the myogenic cell line C2C12, PGE_2 signaling is also a potent stimulator of myogenic differentiation in primary myoblasts/myotubes (Mo et al. 2015).

Myostatin was discovered in the late 1990s to be a potent inhibitor of muscle growth. It is expressed during development and in adult skeletal muscle, serving as an important negative regulator of skeletal muscle growth (Joulia-Ekaza and Cabello 2007; McPherron and Lee 1997). Myostatin appears to decrease myoblast proliferation. The myostatin-deficient mouse model has increased muscle size and strength, with individual muscles weighing significantly more than wild-type mice (Zimmers et al. 2002). Hamrick used this myostatin-deficient mouse model to investigate the effects of increased muscle mass on bone mineral content and density. He and his team found that although a consistent correlation was not found in all regions of the skeletal system, there was increased cortical bone mineral density in the distal femur and an increased periosteal circumference along the humerus (Elkasrawy and Hamrick 2010; Hamrick 2003; Hamrick et al. 2002, 2005). Another group used the same myostatin-deficient mouse model to look at the impact of the chronic loss of myostatin on multiple organ systems and found that it appeared to preserve bone density (Morissette et al. 2009).

Osteocalcin is a bone-derived factor synthesized by osteoblasts. Carboxylated osteocalcin, generated from a vitamin K-dependent posttranslational carboxylation, presents at high concentrations in bone ECM and stimulates bone mineral maturation (Hauschka et al. 1989). On the other hand, undercarboxylated osteocalcin (ucOC) binds to the GPRC6A receptor, affecting distant adipocytes and pancreatic β cells in mice (Pi et al. 2011). However, elevated serum ucOC has been considered as a biomarker for high risk of hip fracture in institutionalized elderly women (Szulc et al. 1993, 1996). Interestingly, osteoblasts also naturally express the osteotesticular phosphatase gene (Esp), which inhibits the function of osteocalcin (Coiro et al. 2012). With this information in mind, it is of specific interest to the discussion of bone-muscle crosstalk that the Gprc6a knockout mouse displays the phenotype of decreased muscle mass, while the Esp knockout mouse has increased muscle mass. Through these observations, it can be proposed that osteocalcin, a known bone cell factor, may play a role in the regulation of muscle mass. Recent data from mouse and human indicate that aerobic exercise increases circulating ucOC levels and induces osteocalcin signaling in muscle leading to myokine IL-6 production.

Moreover, supplement with ucOC is able to restore reduced exercise capacity in mice (Mera et al. 2016).

Due to the endocrine properties of factors derived from bone and muscle, it is reasonable to expect that structural components important for the release of factors also affect bone-muscle biochemical crosstalk. Connexin 43 is an integral membrane protein in gap junctions in bone and has been shown to play an essential role in the release of PGE_2 from osteocytes in response to loading (Cherian et al. 2005). Targeted knockdown of connexin 43 in osteoblast/osteocyte not only decreases cortical bone thickness but also causes a defective muscle phenotype in fast twitch extensor digitorum longus (EDL) muscle (Shen et al. 2015). Interestingly, in a study performed by Gorski et al., conditional deletion of membrane-bound transcription factor peptidase site 1 (MBTPS1) in bone did not induce significant changes in bone, except for a 25% increase in femoral stiffness. However, MBTPS1 deletion in bone results in a 12% increase in muscle mass and a 30% increase in contractile force of soleus, a slow-twitch muscle, pointing to the signaling role of bone cells to muscle cells (Gorski et al. 2016).

The endocrine communication that continues to be revealed through effects of factors derived from these tissues is not limited merely to a bone-muscle connection. These insights into bone-muscle crosstalk will begin to be translated into meaningful and innovative therapeutic approaches, and unprecedented advances will be achieved in the fight against chronic diseases such as obesity, diabetes, osteoporosis, and sarcopenia.

9.2.2 Factors Regulating Both Bone and Muscle

Myf5, Myod, myogenin, and myogenic regulatory factor 4 (MRF4) are myogenic regulatory factors (MRFs), a group of basic helix-loop-helix (bHLH) transcription factors binding to the promoters or the enhancers of muscle-specific genes (Weintraub et al. 1991). Traditionally, Myf5 and Myod are considered as early MRFs required for myogenic commitment of satellite cells, while myogenin and MRF4 are late end differentiation factors. Although the development of myotome is delayed until Myod is expressed in *Myf5*$^{-/-}$ mice, the morphology of skeletal muscle is normal in the newborns (Braun et al. 1992). *Myod*$^{-/-}$ mice have a similar skeletal muscle phenotype as seen in *Myf5*$^{-/-}$ mice, and functionally, the mobility of *Myod* mutants is normal. The expression of myogenin and MRF4 is not altered, but elevated and prolonged expression of Myf5 was observed (Rudnicki et al. 1992). However, the double mutant of *Myf5* and *Myod* shows a drastic skeletal muscle phenotype represented by skeletal muscle development does not occur (Rudnicki et al. 1993). These results suggest that Myf5 and Myod have distinct but also partially overlapping functions in skeletal muscle development.

Different from *Myf5* and *Myod* single mutants, *myogenin*$^{-/-}$ mice show severe defective skeletal muscle development (Hasty et al. 1993). Dramatic muscle mass loss also is observed. Further analysis revealed that the population of myoblasts and the expression of Myf5 and Myod were not affected, but they could not proceed to myogenic differentiation (Hasty et al. 1993; Nabeshima et al. 1993). Expression of MRF4 is complex during muscle development compared with other MRFs. It is detectable in somite myotome at embryonic day 9.0 (E9.0), but diminishes at E11.5. MRF4 expression resumes at the muscle fiber differentiation stage at E16.0 (Rhodes and Konieczny 1989). This expression profile suggests that MRF4 may regulate both the commitment and differentiation phases. Knocking out MRF4 results in a similar muscle phenotype as *Myf5* homozygous deletion. Consistent with this phenotype, Myf5 expression is significantly inhibited, whereas the expression of Myod and myogenin is not affected (Braun and Arnold 1995).

Besides the defects observed in skeletal muscle, *MyoD*, *Myf5*, *myogenin*, and *MRF4* mutations also result in bone anomalies. In *Myf5* mutant fetuses, the most notable change in bone is the defective development of the rib cage. When the *MyoD* and *Myf5* mutations were combined, anomalies were observed in the cervical vertebrae, scapula, humerus and other bones (Rot-Nikcevic et al. 2006). These changes led to significant postural anomalies. *MRF4*-null mice also show deformation in the ribs, characterized by bifurcated and fused ribs (Zhang et al. 1995). Similar to the *MRF4* mutants, the rib cages in *myogenin*$^{-/-}$ mice were malformed with significant shorter sterna (Vivian et al. 2000). Double mutant mice lacking both *MRF4* and *myogenin* show an exaggerated bone phenotype represented by severely malformed ribs and sterna (Vivian et al. 2000). However, *MRF4/MyoD* double mutations did not lead to rib defects distinguishable from those resulted from *MRF4* mutation (Rawls et al. 1998).

The bone phenotypes in the mutants of MRFs could result from the loss of contraction and mechanical loading from skeletal muscle in embryonic bone development. Recently, several studies have indicated that MRFs also regulate bone formation postnatally. MyoD promotes osteogenic differentiation of MC3T3 preosteoblast cells through binding to the bHLH domain in the osterix promoter (Hewitt et al. 2008) and strengthens the effect of BMP7 on osteogenesis of C3H10T1/2 cells (Komaki et al. 2004). Postnatal knockdown of myogenin does not affect the development of skeletal muscle in mice, but the body size of this mouse model is significantly smaller than wild-type control, suggesting myogenin regulates postnatal development of the skeleton (Knapp et al. 2006). Moreover, the result from analysis of myogenin polymorphism revealed that a haplotype of myogenin is associated with bone loss in the hip of postmenopausal women (Harslof et al. 2013). These findings suggest that MRFs are critical for mechanical stimuli on bone and interact with signaling important for bone formation, such as BMPs and β-catenin, during bone and muscle development or regeneration.

BMPs are a group of factors having crucial effects on bone formation. Most BMPs have osteogenic capacity, especially BMP2, BMP4, BMP5, BMP6, BMP7, and BMP9. The loss of the functions of BMP2 and BMP4 causes severe phenotypes of defective osteogenesis (Bandyopadhyay et al. 2006), but BMP2 has a stronger effect in chondrocyte proliferation and maturation and fracture healing than BMP4 (Shu et al. 2011; Tsuji et al. 2006, 2008). BMP5 and BMP7 could have redundant function in the regulation of development of sclerotome and myotome (Solloway and Robertson 1999). BMP9 can induce osteogenic differentiation from mesenchymal stem cells, including stem cells derived from adipose tissue and muscle (Lu et al. 2016; Xiang et al. 2012). In contrast, BMP3 and BMP13 act as negative regulators of bone formation (Daluiski et al. 2001; Shen et al. 2009). BMP3 is a non-signaling ligand for activin receptor type 2b (ActRllb). By binding to ActRllb, BMP3 inhibits signaling induced by osteogenic BMPs, such as BMP2 (Kokabu et al. 2012). Notably, ActRllb has been linked to the function of myostatin in muscle. Inhibitors of ActRllb, or soluble ActRllb, which bind to and neutralize the signaling of myostatin significantly increase muscle mass in mice and humans (Lee et al. 2005; Lach-Trifilieff et al. 2014). Besides bone, BMPs also regulate the development and diseases in other organs or tissues, such as muscle, adipose tissue, and reproductive system (Duprez et al. 1996; Shimasaki et al. 1999; Wang et al. 2014).

BMP2 and BMP4 are the two most studied BMPs in muscle. During embryonic development, administration of BMP2 inhibits myogenesis while promoting osteogenesis in limb bud cultures (Duprez et al. 1996). BMPs, especially BMP4, work in concert with Noggin, an antagonist of BMP signaling, to control the expression of Pax3 and MRFs, such as MyoD and Myf5, in somites (Reshef et al. 1998). Localization, concentration, and timing are crucial factors to determine the effect of BMPs on myogenesis. Myogenesis is first initiated in muscle progenitor cells in the medial somites, which is closely related with the expression of Noggin in the dorsomedial lip. Overexpression of Noggin can induce the expression of MyoD and formation of myotome in adjacent areas in the somite. In contrast, high level of BMP4 in the somite can inhibit myotome formation through the downregulation of MyoD expression. Moreover, the activation of BMP signaling in the paraxial mesoderm can induce the expression of lateral plate genes (Reshef et al. 1998; Epperlein et al. 2007).

In postnatal muscle development or muscle regeneration, BMP signaling plays an important role in controlling the fate of muscle satellite cells. Upon activation by injury or other stimuli, satellite cells become active and proliferate. BMP signaling is important for maintaining satellite cells in their proliferative stage to enlarge the pool of satellite cells and their descendants, but high levels of BMP signaling inhibit cell cycle exit and initiation of myogenic differentiation, which results in attenuated myotube development in vitro (Friedrichs et al. 2011; Ono et al. 2011). On the other hand, inhibition of BMP signaling by Noggin enhances myogenic differentiation in the studies using C2C12 cells or satellite cells. However, the inhibition of BMP signaling in mice leads to muscle atrophy and smaller regenerative muscle fibers after injury due to premature differentiation of satellite cells. In addition, BMP signaling can interact with myostatin in regulation of muscle mass (Sartori et al. 2013). The results further confirm that timing and concentration are critical for the effect of BMPs on skeletal muscle.

The levels of BMPs in muscle are associated with the activation of the transcriptional factor β-catenin. β-catenin is a key downstream co-activator of Wnt signaling. The Wnt/β-catenin pathway has been shown to be critical for normal bone development and postnatally is responsible for the response of bone to mechanical loading (Bonewald 2011). Deletion of β-catenin in osteocytes abolishes the anabolic response to loading (Javaheri et al. 2014) and enhances the negative effects of unloading in male mice (Maurel et al. 2016). In muscle, overexpression of β-catenin activators Wnt1 and Wnt3a in isolated primary myoblasts and mice significantly inhibits myogenesis (Kuroda et al. 2013).This effect is mediated by the upregulation of BMP4. β-catenin has been shown to be able to directly interact with MyoD or increase MyoD expression to promote myogenesis (Kim et al. 2008; Pan et al. 2015). During muscle regeneration, it is strongly upregulated in satellite cells. Moreover, activation of the Wnt/β-catenin signaling pathway causes changes in the structure of chromatin at the MyoD and Myf5 promoters, leading to increased expression of these two MRFs in satellite cells, which subsequently promotes myoblast proliferation in skeletal muscle (Fujimaki et al. 2014). All of these findings suggest that the Wnt/β-catenin signaling plays a critical role in the regulation of satellite cells in adult muscle regeneration. These results imply that similar to BMP signaling, transient activation of β-catenin is necessary for muscle development and regeneration, but constant activation or overexpression of its activator will lead to the inhibition of myogenesis. Data from our group demonstrated that factors released by C2C12 myotubes and contracting muscle fibers protect MLO-Y4 osteocyte-like cells from dexamethasone-induced apoptosis through the activation of Wnt/β-catenin signaling (Jahn et al. 2012).

Further studies are necessary to dissect the interactions among MRFs, BMPs, and Wnt/β-catenin in bone-muscle development and biochemical crosstalk, and response to mechanical loading, which will provide new insights for understanding mechanical and biochemical interactions between bone and muscle in development, regeneration, and diseases.

In addition to factors abovementioned, additional soluble factors are produced by other tissues or organs and also have functions on both bone and muscle, which are briefly summarized in Table 9.1.

9.2.3 Transdifferentiation of Myoblasts into the Osteoblast Lineage

Transdifferentiation of myoblasts into osteogenic cells serves as additional evidence for bone-muscle cellular and biochemical interaction. Myogenesis of C2C12 myoblasts was inhibited after treatment with high-dose (300 ng/ml) BMP2. At the same time, the osteoblastic phenotype was induced (Katagiri et al. 1994). The transdifferentiation of C2C12 cells induced by BMP2 is associated with the inhibition of MyoD through the activation of downstream factors Smad1/Smad5 (Yamamoto et al. 1997). BMP2 also can induce the expression of core-binding factor α1 (Cbfa1, also called runt-related transcription factor 2 (RUNX2)), a transcriptional factor important for osteogenesis. Cbfa1 can transdifferentiate C2C12 cells with or without interacting with Smad1/Smad5 (Nishimura et al. 2002). Overexpression of Runx2/Cbfa1 in mouse primary myoblast inhibits the expression of MyoD and

Table 9.1 Soluble factors showing effects on bone/muscle

Factors	Effects on bone/muscle	
Vitamin D	Bone	Vitamin D stimulates bone formation and mineralization in osteoblasts (Fretz et al. 2007; Shi et al. 2007)
	Muscle	Vitamin D signaling regulates the proliferation and differentiation of C2C12 skeletal muscle cells and shows growth-promoting effects on size of myotube (Girgis et al. 2014)
		Vitamin D supplements can improve muscle strength and reduce risk of falls (Murad et al. 2011; Snijder et al. 2006)
		Annual oral administration of high-dose vitamin D results in an increased risk of falls and fractures among older community-dwelling female individuals (Sanders et al. 2010)
	Musculoskeletal system	Vitamin D deficiency is positively associated with musculoskeletal pain (Heidari et al. 2010; Plotnikoff and Quigley 2003; Wepner et al. 2014)
IGF-I	Bone	IGF-I stimulates survival, proliferation, differentiation, and matrix production in cultured osteoblast cells (Hill et al. 1997; Hock et al. 1988)
		Higher osteocyte lacunae occupancy, increased bone formation rate (BFR), bone volume, and bone mineral density (BMD) are obtained in IGF-I overexpressed transgenic mice (Zhao et al. 2000)
		PTH that promotes bone anabolism via IGF-I signaling pathway has been reported (Bonjour 2016; Esen et al. 2015)
	Muscle	IGF-I promotes myoblast proliferation, differentiation, and fiber formation during normal growth as well as during regeneration after injury (Sharples et al. 2015; Nindl 2010; James et al. 1996; Foulstone et al. 2004; Bikle et al. 2015)
Sex steroids like estrogens, testosterone, etc.	Bone	Estrogens attenuate the generation rate of osteoclast and osteoblast progenitors in the bone marrow to exert a proapoptotic effect on osteoclasts and an antiapoptotic effect on osteoblasts and osteocytes, consequently help to maintain a balance between bone formation and resorption (Riggs et al. 2002; Manolagas 2000, 2010; Almeida et al. 2013)
		Testosterone stimulates osteoblast proliferation through the estrogen receptor (ER) and has direct and indirect inhibitory effects on osteoclast formation and bone resorption through androgen receptor (AR) (Nakano et al. 1994; Michael et al. 2005; Damien et al. 2000)
		Both estrogens and testosterone contribute to the maintenance of bone homeostasis and strength, building the skeleton and preventing the bone loss. Sharp decline of estrogen levels at menopause in females and testosterone deficiency in elderly men accelerate the age-dependent skeleton involution and lead to the development of osteoporosis (Vanderschueren et al. 2014; Carson and Manolagas 2015)
	Muscle	Sex steroids promote the growth and maintenance of muscle mass and strength, improve the metabolic function of skeletal muscles, and thus stimulate their recovery after injury (Carson and Manolagas 2015; Spangenburg et al. 2012; McClung et al. 2006)
Glucocorticoids	Bone	Glucocorticoid excess induces impaired osteoblastogenesis and osteoclastogenesis, diminished bone formation and turnover, thus leads to secondary osteoporosis (Ziegler and Kasperk 1998; Weinstein et al. 1998)
	Muscle	Glucocorticoids can cause muscle wasting and stimulate skeletal muscle atrophy (Rahnert et al. 2016; Fry et al. 2016; Bodine and Furlow 2015)

myogenin and leads to functional osteogenesis (Gersbach et al. 2004). Other osteogenic factors, such as osteoactivin, show similar capacity in transdifferentiating C2C12 cells. The effect of osteoactivin is mediated by upregulating the level of phosphorylated focal adhesion kinase (Sondag et al. 2014).

In the bone fracture-healing model, satellite cells from muscle express chondrogenic factors Nkx3.2 and Sox9, downstream factors of BMP2, and transdifferentiate into chondrocytes (Cairns et al. 2012). This may not be the final destination of satellite cells, since chondrocytes can continue to transdifferentiate into osteoblasts in bone fracture healing (Zhou et al. 2014). Factors released or induced by fractured bone also promote transdifferentiation of cells derived from muscle in vitro. TNF-alpha plays a central role in this process. Low concentrations of TNF-alpha enhance osteogenic transdifferentiation of muscle-derived stromal cells and fracture healing (Glass et al. 2011).

9.3 Potential Therapeutic Approaches Targeting the Muscle-Bone Unit

9.3.1 Development of Sarcopenia and Osteoporosis During Aging

Age is an important factor inducing mass loss in both skeletal muscle and bone. Sarcopenia, an age-related muscle atrophy, is characterized by the loss of both strength and skeletal muscle mass. Muscle mass starts to gradually decrease when age reaches around 30. Generally, it reduces by 3–8% per decade, but this process accelerates in people over 60 (English and Paddon-Jones 2010). As a result of aging, the number of muscle satellite cells decreases, leading to lower capacity of muscle regeneration and loss of muscle fibers (Shefer et al. 2010). Signaling from motor neurons also diminishes during aging due to the reduced number of neurons and their increased dysfunction (Manini et al. 2013). This contributes to impaired contractile function and reduced muscle strength in sarcopenia. Similarly, osteoporosis is the most common metabolic bone disorder and is characterized by the progressive loss of bone mass. Bones are in a constant state of destruction and rebuilding, and in young, healthy individuals, the balance between bone formation and resorption is maintained. The decrease in bone density that has come to be known as osteoporosis appears to be the result of a growing imbalance of these two processes. The body of knowledge surrounding osteoporosis is growing as it is now recognized as the most common metabolic bone disease in the United States.

Sarcopenia and osteoporosis are often present in the same patients, but the mechanisms behind this concomitant expression are not clear at present. Changes in the muscle ECM contribute to the development of sarcopenia. During aging, structural changes in the ECM of muscle include an increase in the amount of collagen, reduced viscoelastic properties in the elastic fiber system, and higher levels of fat infiltration in skeletal muscle (Kragstrup et al. 2011). These changes alter the niche for satellite cells, which is critical for maintaining the proper number of satellite cells. In both fast- and slow-twitch muscles, the abundance of resident satellite cells, but not their inherent myogenic capacity, declines during aging (Shefer et al. 2006). Due to the reduced protective function of ECM in aging, subtle myofiber injuries routinely occur during normal muscle activities, and the need for muscle regeneration increases. Results from multiple studies have confirmed that the levels of MyoD and myogenin in aged muscles are significantly higher than young muscles, suggesting muscle regeneration is active in aged muscles (Musaro et al. 1995; Dedkov et al. 2003). This continuous demand for functional satellite cells eventually reduces or exhausts satellite cell pool, which subsequently limits myofiber repair and contributes to sarcopenia.

As discussed previously, changes in the muscle ECM will affect mechanical stimulation of bone and biochemical interaction within ECM. This may affect the regenerative responses in bone progenitor cells. In the aged periosteum, progenitor cells show diminished response to PTH 1–34 treatment, demonstrated by impeded proliferation, differentiation, and fracture healing (Yukata et al. 2014). Taken together, reduced regenerative capacity could be a shared mechanism for sarcopenia and osteoporosis.

Since bone and muscle are endocrine organs, the release of cytokines and other factors could change with aging. Signaling molecules, such as IGF-1, FGF2, IL-6, myostatin, and BMPs, could change. Modified biochemical crosstalk between bone and muscle could lead to the development of sarcopenia and osteoporosis concomitantly.

9.3.2 Current Therapeutic Approaches for Sarcopenia and Osteoporosis

The Food and Drug Administration (FDA)-approved medications for the prevention and treatment of osteoporosis include the following two main types: antiresorptive drugs and osteoanabolic agents. Antiresorptive agents, which act by inhibiting bone resorption, include a wide drug armamentarium: bisphosphonates such as alendronate and risedronate; selective estrogen receptor modulators (SERMs) such as raloxifene and bazedoxifene; monoclonal antibodies such as denosumab, romosozumab, and blosozumab; calcitonin; estrogens; and strontium ranelate (Cantudo-Cuenca et al. 2016). Bisphosphonates, which reduce both vertebral and nonvertebral fracture risks (a primary indicator of treatment efficacy for osteoporosis), are the most widely used first-line antiresorptive therapy for patients with osteoporosis, a prior fragility fracture, or osteopenia, as well as individuals with other metabolic bone diseases such as Paget's disease, multiple myeloma, bone metastases, and cancer-induced hypercalcemia (Body et al. 2012; Chen and Sambrook 2011; Milat and Ebeling 2016). Some cases of prolonged musculoskeletal pain have been reported in up to 20–25% of patients on alendronate and risedronate, the majority of the patients

(66%) experienced relief after alendronate therapy was discontinued, and a few patients (11%) redeveloped pain following readministration of the drug (Bock et al. 2007; Wysowski and Chang 2005). Denosumab (Prolia®, Xgeva®) is a fully human monoclonal antibody that inhibits the activity of RANKL, the main stimulator of osteoclastogenesis and of osteoclast activity (Kong et al. 1999). It has been reported that the safety of denosumab and its efficacy in reducing fractures is not significantly different from bisphosphonates after 12–24 months of treatment. Nevertheless, denosumab was found to be more effective in increasing bone mineral density (Beaudoin et al. 2016). However, unlike the bisphosphonates, discontinuance of denosumab leads to bone loss; therefore additional therapy is recommended after discontinuance. After 5–10 years of use, denosumab is proving to have similar or better safety outcomes than the bisphosphonates (Anastasilakis et al. 2012).

Recombinant human PTH drug teriparatide (Forteo®) is the only FDA-approved osteoanabolic drug recommended for both men and postmenopausal women with osteoporosis at high fracture risk. In contrast to antiresorptive agents, which preserve bone microarchitecture and inhibit bone loss, teriparatide acts primarily by stimulating osteoblast recruitment and activity, as well as osteoblastic bone formation. Teriparatide can significantly reduce risks for vertebral and nonvertebral fractures by 65% and 53%, respectively, in postmenopausal women who have previously suffered a vertebral fracture (Neer et al. 2001). However, teriparatide cannot be administered longer than 2 years due to the observation of an increased risk of osteosarcoma in the rats treated with teriparatide, even though the very high doses were applied in the rat's short life cycle in these preclinical studies, suggesting that it may not be predictive of the long-term risk-benefit profile for teriparatide in humans (Vahle et al. 2002; Sikon and Batur 2010). In addition, the administration of teriparatide could be significantly limited due to its low patient compliance/tolerance (daily-based subcutaneous injection) and higher cost as compared with other antiresorptive drugs (Levine 2006). Newer anabolic therapeutics are being developed to target an osteocyte factor, sclerostin, showing even greater bone formation than Forteo®, denosumab, or the bisphosphonates (Genant et al. 2016; Recknor et al. 2015). Results of clinical studies demonstrated that therapeutic potential of these anti-sclerostin monoclonal antibodies, such as romosozumab/AMG785 from Amgen, blosozumab/LY2541546 from Eli Lilly, BPS804 from Novartis, etc., in human clinical trials is eagerly awaited (Genant et al. 2016; Recknor et al. 2015; Lewiecki 2014; MacNabb et al. 2016; Recker et al. 2015; Reichert 2015). Accordingly, it has been suggested that sclerostin antibody therapy will achieve FDA approval by 2017 and eventually be set as the gold standard for osteoporosis treatment by 2021 (MacNabb et al. 2016).

Indeed, falls-related fracture is a marker of simultaneous dysfunction of both bones and muscles: osteosarcopenia (Binkley and Cooper 2015; Drey et al. 2016; Landi et al. 2012; Yu et al. 2014). Fracture risk has been reported to be increased 3.5-fold in the elderly male osteosarcopenic patients, which is significantly higher than that in the patients with only osteoporosis (2.4-fold) or sarcopenia (2.3-fold) (Yu et al. 2014). All currently available antiresorptive and osteoanabolic medications for osteoporosis still do not sufficiently increase bone strength and reduce concurrent falls risk or resolve underlying sarcopenia and frailty. Actually, at present, the FDA has approved about 49 drugs for the treatment of musculoskeletal system-related diseases from 2005, but none of them targets both muscle and bone at the same time. Thus, more potent therapeutic approaches are needed, which preferentially strengthen the musculoskeletal unit as a whole and simultaneously address coexisting sarcopenia, frailty, and falls risk, especially in older adults.

9.3.2.1 Exercise

Exercise could be an effective treatment strategy targeting bone-muscle unit for osteoporosis/sarcopenia. Physical activity refers to the body movement that is produced by skeletal muscle contractions and that increases energy expenditure and contributes to a healthy energy balance and increases muscle function and bone mass (American College of Sports Medicine 2009). Exercise training for health and function contains different components, i.e., strength or power training, aerobic exercise, and balance and flexibility training (Goisser et al. 2015; Montero-Fernandez and Serra-Rexach 2013). It is evident that a multicomponent training, including aerobic activity and other types of training (resistance and/or strength exercises), improves bone mass and bone metabolism in older adults and especially osteopenic and osteoporotic women (Castrogiovanni et al. 2016). In contrast to aerobic activity, resistance exercise training appears to have a larger effect on augmenting muscle mass, increasing skeletal muscle protein synthesis and specific major histocompatibility complex synthesis, thereby attenuating the development of sarcopenia (Hughes et al. 2004; Sipila and Suominen 1995; Yarasheski et al. 1993). In combination with dietary weight loss interventions, strength exercise demonstrated positive effects on sarcopenic obesity in the elderly (Goisser et al. 2015). Strong evidence exists based on exercise interventions in osteoporosis/sarcopenia that physical activity, whatever the adopted training, always has beneficial effects on patients suffering from osteoporosis and not only on bone homeostasis but also on the whole skeletal muscle system.

9.3.2.2 Medications

In recent years, there is increasing awareness that a concerted treatment approach would be ideal for the treatment of osteoporosis and sarcopenia because bone loss shares common causes with muscle wasting, instead of only coexisting (Huo et al. 2015; Verschueren et al. 2013). Muscle and bone share many endocrine, paracrine, and autocrine signaling pathways, such as growth hormone (GH)/IGF signaling pathways, vitamin D signaling pathways, glucocorticoid receptor (GR) signaling pathways, androgen and estrogen receptor signaling pathways, etc. Related endocrine regulators, including GH/IGF, vitamin D receptor (VDR), GR, as well as myostatin and follistatin, are affected by aging and associated with osteoporosis and sarcopenia (Laurent et al. 2016).

9.3.2.2.1 Hormone Replacement Therapies (HRT)

Application of hormone supplementation has been widely accepted as an effective strategy to augment muscle and bone mass (Burton and Sumukadas 2010). Hormonal abnormalities, primarily the decreased secretion of testosterone, estrogen, GH/IGF-1, and dehydroepiandrosterone (DHEA), are associated with loss of muscle strength and mass, profound bone mineralization disorders, and thereby increased risk of fracture following falls (Burton and Sumukadas 2010; Jagielska et al. 2016). Therefore, it is easy to hypothesize that HRT may play a role in the treatment of osteoporosis/sarcopenia. However, the effect and extent of hormone supplementation is found to be largely dependent on gender, age, reproductive status of the individual, type of hormone administered, duration of use, and route of administration (Southmayd and De Souza 2016). Three years of uninterrupted administration of GH in younger GH-deficient adults shows significantly improved muscle physical function (Jorgensen et al. 1994), but the results of GH supplementation in healthy non-GH-deficient older people are more controversial. Some studies have shown an increase in muscle mass but no improvement in muscle strength (Papadakis et al. 1996), whereas others have shown an increase in both muscle mass and strength after administration of GH supplementation (Welle et al. 1996). It is possible that comorbidities, for example, or different levels of activity, or even genetic factors could have direct influence in the discrepancy of these results.

HRT is expensive and frequently associated with serious adverse effects, even though lean muscle mass and bone density can be significantly increased by these therapeutic strategies. Snyder et al. investigated the effects of physiological testosterone replacement for 3 years in previously untreated hypogonadal men. Their results indicate that bone mineral density (BMD) of the spine and hip and fat-free muscle mass were significantly increased by testosterone replacements, but the size of prostate gland also markedly increased by about twofold during the first 6 months (Snyder et al. 2000). This increment in prostate volume might be detrimental to the men older than 60, in which the prevalence of early stage prostate cancer is already high (Jemal et al. 2002). Estrogen replacement therapy exhibits modest efficacy to attenuate the loss of muscle mass among perimenopausal women, but the risk for breast cancer caused by HRT has led to it not being recommended for the treatment of sarcopenia (Taaffe et al. 2005; Chlebowski et al. 2003). Therefore, administration of these hormone medications can be a complex and uncertain process.

9.3.2.2.2 Calcium and Vitamin D Supplementation

Calcium and vitamin D are essential components for bone health. Calcium has a key role in many physiological processes including skeletal mineralization, muscle contraction, nerve impulse transmission, blood clotting, and hormone secretion. More than 99% of calcium in the body is stored in the skeleton as hydroxyapatite, which provides skeletal strength and is a source of calcium for multiple calcium-mediated functions as well as for the maintenance of serum calcium within the normal range. The importance of vitamin D on musculoskeletal system has been well established. Often considered more of a steroid hormone than a vitamin, vitamin D signaling is beneficial for bone health via its regulation on calcium absorption and homeostasis by its biologically active form in circulation: 1, 25-dihydroxyvitamin D_3 (1,25$(OH)_2D_3$) (Cantorna et al. 2004). It promotes intestinal calcium intakes and enables mineralization of newly formed osteoid tissue in bone. It is well recognized that prolonged vitamin D deficiency is detrimental to the skeleton, resulting in rickets in children and osteomalacia in adults (Thacher and Clarke 2011). However, during the aging process, a decline in intestinal calcium absorption and serum 1,25$(OH)_2D_3$ concentration, combined with an increase in catabolism of 1,25$(OH)_2D_3$ by cytochrome P24A1, results in age-related bone loss (Matkovits and Christakos 1995). Thus calcium and vitamin D supplementation is an important strategy to ensure bone health among adults older than 65 years. Generally, a serum 25-hydroxy-vitamin D level greater than 30 ng/mL is recommended for the management and prevention of osteoporosis and fractures (Kim et al. 2015).

Numerous observational studies have demonstrated an association between vitamin D deficiency and not only osteoporosis but also sarcopenia (Lappe and Binkley 2015; Okuno et al. 2009; Tanner and Harwell 2015; Wintermeyer et al. 2016). Vitamin D, primarily 1,25-dihydroxyvitamin D (1,25$(OH)_2$D), binding to the VDR expressed in skeletal muscle promotes muscle protein synthesis and enhances calcium uptake across the cell membrane (Bischoff et al. 2001; Pojednic et al. 2015). Low vitamin D levels result in atrophy predominantly of the type 2 (fast twitch) muscle fibers in common with sarcopenia (Ziambaras and Dagogo-Jack 1997). Low vitamin D levels in older people may associate with reduced muscle function such as proximal muscle weakness as well as increased physical disability (Mowe et al. 1999). However, benefit of administration of calcium/vitamin D supplements in physical performance is controversial. Recent meta-analyses of randomized trials suggest that vitamin D supplementation increases muscle strength but not mass or power in elderly persons (Beaudart et al. 2014). The reduction in the risk of falls from vitamin D supplementation, with or without calcium, was only observed in the most severely vitamin D deficient population, i.e., institutionalized elderly, but not in community-dwelling elderly (Bolland et al. 2014; Cameron et al. 2012; Gillespie et al. 2001). Moreover, potentially harmful effects of excessive calcium or vitamin D supplementation on health are recently suggested. The life-threatening acute or chronic toxicity, including primarily kidney injures and hypercalcemia, has been reported due to overdose of vitamin D supplemented with calcium (Kaur et al. 2015; Rocha et al. 2011). Therefore, avoiding vitamin D deficiency may have musculoskeletal benefits, although a longer follow-up period is required in order to assess the safety profile of calcium and vitamin D supplementation in older people before it is recommended as a treatment for osteoporosis or sarcopenia in clinical practice.

9.3.3 *Other Innovative Treatment Strategies to Target Muscle-Bone Unit*

9.3.3.1 Myostatin-Activin Pathway-Associated Protein Therapeutics

Myostatin, a member of the TGFβ superfamily of growth factors, is a highly conserved negative regulator of skeletal muscle mass that is upregulated in many conditions of muscle wasting. It binds to and activates the activin type II receptors ActRIIA and ActRIIB (mostly ActRIIB) and exhibits strong effects on muscle cell growth and differentiation as well as intracellular catabolic and anabolic signaling pathways (Han et al. 2007; Smith and Lin 2013; Tsuchida et al. 2008; White and LeBrasseur 2014). Many myostatin inhibitors/blockers, either by binding directly to myostatin itself or by binding to components of activin type II receptor complex, are now under development for the treatment of muscle wasting disorders, muscular dystrophy, sarcopenia, and other musculoskeletal diseases. These promising therapeutic proteins include LY2495655 (myostatin antibody) from Eli Lilly, PF-06252616 (myostatin antibody) from Pfizer, MYO-029 (myostatin antibody) from Wyeth, BYM388/bimagrumab (ActRIIB antibody) from Novartis, REGN1033 (myostatin antibody) from Regeneron, ACE-083 (myostatin antibody) and ACE-011/sotatercept (ActRIIA-IgG1-Fc antibody) from Acceleron Pharma, etc. (Woodhouse et al. 2016; Wagner et al. 2008; Sherman et al. 2013; Salzler et al. 2016; Latres et al. 2015; Evans et al. 1982; Cohen et al. 2015; Pearsall et al. 2015).

Similar to "bone" signaling pathways affecting muscle, the possibility of muscle-to-bone effects is also gaining ground. Animal studies revealed that myostatin has direct effects on the proliferation and differentiation of osteoprogenitor cells, and inhibition of the myostatin pathway increases bone turnover and leads to bone mass accrual (Elkasrawy and Hamrick 2010; Buehring and Binkley 2013; Hamrick et al. 2006). Blocking myostatin exhibits potential activity to improve bone quality in diet-induced obese rats through a mechanotransduction pathway from skeletal muscle, suggesting that administration of myostatin antagonists may be a promising therapy for treating obesity and obesity-induced bone loss (Tang et al. 2016). In addition, inhibition of activin type II receptor ActRIIB, which also is present in osteoblasts and chondroblasts, leads to a 30% increased bone strength in mice (Hamrick et al. 2006). Another activating type II receptor ActRIIA has been found to be involved in bone remodeling regulation in that inhibiting the ActRIIA signaling cascade increases bone formation and decreases bone resorption (Hittel et al. 2010; Pearsall et al. 2008). A dramatic increase in bone formation and bone mass and strength has been observed in both normal mice and ovariectomized mice with established bone loss after administration of ActRIIA-mFc fusion protein (ActRIIA fused to a murine IgG2a-Fc) to inhibit activin signaling (Pearsall et al. 2008). ACE-011/sotatercept, a soluble ActRIIA fusion protein, was developed by Acceleron Pharma for the treatment of chemotherapy-induced anemia (CIA) in patients with metastatic non-small cell lung cancer (NSCLC). Anemia is one of the most common and debilitating complications associated with cancer chemotherapy leading to muscle fatigue and weakness. Results from animal studies such as cynomolgus monkeys suggest that sotatercept improves cancellous bone volume by promoting bone formation and inhibiting bone resorption through a dual anabolic and antiresorptive mechanism in nonhuman primates (Lotinun et al. 2010). In clinical trials of sotatercept, besides hemoglobin and hematocrit, the levels of bone formation biomarkers and BMD also increased in both healthy volunteers and patients with multiple myeloma. These potential antitumor effects and anabolic bone activity in early clinical studies make sotatercept a promising therapeutic candidate for cancer-related anemia and malignant bone diseases (Sherman et al. 2013; Fields et al. 2013).

9.3.3.2 Gene Therapies

In comparison with traditional protein delivery, gene therapy has the potential to achieve a directed, sustained, and regulated expression of authentically processed, nascent proteins (Legre et al. 1989). Skeletal muscle is a practical target organ for gene therapy because it is easily accessible and has a large capacity for protein production. Muscle cells can be easily isolated and cultivated, and numerous reports have indicated the existence of inducible osteoprogenitor cells in skeletal muscle (Dawson et al. 2014; Mastrogiacomo et al. 2005; Matthews et al. 2016). Thus muscle cells could serve as an efficient gene delivery vehicle (vector) to muscle or bone. Recombinant human BMP2 (rhBMP2), an osteogenic protein which can promote bone healing and regeneration, is now approved by the US FDA for restricted clinical use in open tibial fractures and anterior spinal fusion. Since a large quantity of the rhBMP2 is needed to enhance the bone healing potential during treatment, gene therapy in the bone defect could be a promising option to obtain a sustained expression of the osteogenic proteins and further enhancement of bone healing (Kawai et al. 2003; Osawa et al. 2009). Lee et al. genetically engineered freshly isolated human skeletal muscle cells with adenovirus and retrovirus to express human BMP2, and then these cells were implanted into nonhealing bone defects in mice. Their results indicate that myoblast-mediated gene delivery can enhance bone healing by delivering BMP2 during the 4- to 8-week period, and a fraction of muscle cells differentiate into osteogenic cells (Lee et al. 2002). These muscle-derived cells coupled with ex vivo gene therapy were also reported successfully by other groups using different vectors (Day et al. 1999).

Hypophosphatasia (HPP) is heterogeneous disease due to a mutation in the gene encoding tissue-nonspecific alkaline phosphatase (TNALP) that results in severely impaired bone and tooth mineralization, seizures, and hypercalcemia defects (Linglart and Biosse-Duplan 2016). Recently, a novel muscle-directed gene therapy has been reported for the treatment of a severe infantile form of HPP (Nakamura-Takahashi et al. 2016). Ferreira et al. constructed a self-complementary adeno-associated virus type 8 (scAAV8) vector that expresses bone-targeted TNALP with deca-aspartates at the C-terminus (TNALP-D_{10}) via the muscle-specific creatine kinase (MCK) promoter (scAAV8-MCK-TNALP-D_{10}). Direct injection of AAV vector into muscle has been widely used for the treatment of Duchenne muscular dystrophy (DMD) and systemic metabolic disease like lipoprotein lipase deficiency (Ferreira et al. 2014; Okada and Takeda 2013). Efficacy and safety profiles of this muscle-directed gene therapy for HPP were assessed using neonatal TNALP knockout (*Akp2*$^{-/-}$) mice via injection into the bilateral quadriceps femoris muscle. Results indicated that plasma ALP activities in scAAV8-MCK-TNALP-D_{10}-treated *Akp2*$^{-/-}$ mice were increased by more than tenfold over those in wild-type mice. Moreover, a significantly sustained TNALP-D_{10} expression and a prolonged survival and normal physical activities were observed in lethal HPP model during a 90-day experiment. Improved bone architecture in *Akp2*$^{-/-}$ mice was also demonstrated on X-ray images after treatment. These findings suggest that this muscle-directed gene therapy approach could be a safe and practical strategy to delivery bone-targeted therapeutic regents for the treatment of bone diseases.

In summary, bone and muscle have a lifetime interaction from embryogenesis to aging. This interaction was considered to mainly be mechanical, but biochemical crosstalk between these two tissues is gradually being accepted by the scientific community. Based on the fact that bone and muscle work as a unit in many aspects, it is expected that some signaling pathways or factors are critical for the regulation of development, function, and diseases in both bone and muscle. By identifying these pathways and factors, treatments targeting both bone and muscle diseases simultaneously will be developed. Identification of both exercise and diet interventions that promote optimal bone-muscle crosstalk could also prove very fruitful for the treatment of musculoskeletal diseases.

References

Agas D, et al. Prostaglandin F2alpha: a bone remodeling mediator. J Cell Physiol. 2013;228(1):25–9.

Allen MR, Hock JM, Burr DB. Periosteum: biology, regulation, and response to osteoporosis therapies. Bone. 2004;35(5):1003–12.

Almeida M, et al. Estrogen receptor-alpha signaling in osteoblast progenitors stimulates cortical bone accrual. J Clin Invest. 2013;123(1):394–404.

American College of Sports Medicine, et al. American College of Sports Medicine position stand. Exercise and physical activity for older adults. Med Sci Sports Exerc. 2009;41(7):1510–30.

Anastasilakis AD, et al. Long-term treatment of osteoporosis: safety and efficacy appraisal of denosumab. Ther Clin Risk Manag. 2012;8:295–306.

Argiles JM, Lopez-Soriano FJ, Busquets S. Therapeutic potential of interleukin-15: a myokine involved in muscle wasting and adiposity. Drug Discov Today. 2009;14(3–4):208–13.

Bandyopadhyay A, et al. Genetic analysis of the roles of BMP2, BMP4, and BMP7 in limb patterning and skeletogenesis. PLoS Genet. 2006;2(12):e216.

Beaudart C, et al. The effects of vitamin D on skeletal muscle strength, muscle mass, and muscle power: a systematic review and meta-analysis of randomized controlled trials. J Clin Endocrinol Metab. 2014;99(11):4336–45.

Beaudoin C, et al. Denosumab compared to other treatments to prevent or treat osteoporosis in individuals at risk of fracture: a systematic review and meta-analysis. Osteoporos Int. 2016;27(9):2835–44.

Bernardi P, Bonaldo P. Mitochondrial dysfunction and defective autophagy in the pathogenesis of collagen VI muscular dystrophies. Cold Spring Harb Perspect Biol. 2013;5(5):a011387.

Bikle DD, et al. Role of IGF-I signaling in muscle bone interactions. Bone. 2015;80:79–88.

Binkley N, Cooper C. Sarcopenia, the next frontier in fracture prevention: introduction from the guest editors. J Clin Densitom. 2015;18(4):459–60.

Bischoff HA, et al. In situ detection of 1,25-dihydroxyvitamin D3 receptor in human skeletal muscle tissue. Histochem J. 2001;33(1):19–24.

Bock O, et al. Common musculoskeletal adverse effects of oral treatment with once weekly alendronate and risedronate in patients with osteoporosis and ways for their prevention. J Musculoskelet Neuronal Interact. 2007;7(2):144–8.

Bodine SC, Furlow JD. Glucocorticoids and skeletal muscle. Adv Exp Med Biol. 2015;872:145–76.

Body JJ, et al. Extraskeletal benefits and risks of calcium, vitamin D and anti-osteoporosis medications. Osteoporos Int. 2012;23(Suppl 1):S1–23.

Bolland MJ, et al. The effect of vitamin D supplementation on skeletal, vascular, or cancer outcomes: a trial sequential meta-analysis. Lancet Diabetes Endocrinol. 2014;2(4):307–20.

Bonaldo P, et al. Collagen VI deficiency induces early onset myopathy in the mouse: an animal model for Bethlem myopathy. Hum Mol Genet. 1998;7(13):2135–40.

Bonewald LF. The amazing osteocyte. J Bone Miner Res. 2011;26(2):229–38.

Bonjour JP. The dietary protein, IGF-I, skeletal health axis. Horm Mol Biol Clin Investig. 2016;28(1):39–53.

Bouillon R, et al. Estrogens are essential for male pubertal periosteal bone expansion. J Clin Endocrinol Metab. 2004;89(12):6025–9.

Braun T, Arnold HH. Inactivation of Myf-6 and Myf-5 genes in mice leads to alterations in skeletal muscle development. EMBO J. 1995;14(6):1176–86.

Braun T, et al. Targeted inactivation of the muscle regulatory gene Myf-5 results in abnormal rib development and perinatal death. Cell. 1992;71(3):369–82.

Brent AE, Tabin CJ. Developmental regulation of somite derivatives: muscle, cartilage and tendon. Curr Opin Genet Dev. 2002;12(5):548–57.

Brooks M, Revell WJ. Blood supply of bone. London: Springer; 1998.

Buehring B, Binkley N. Myostatin – the holy grail for muscle, bone, and fat? Curr Osteoporos Rep. 2013;11(4):407–14.

Burks TN, Cohn RD. Role of TGF-beta signaling in inherited and acquired myopathies. Skelet Muscle. 2011;1(1):19.

Burr DB, et al. Intermittently administered human parathyroid hormone(1–34) treatment increases intracortical bone turnover and porosity without reducing bone strength in the humerus of ovariectomized cynomolgus monkeys. J Bone Miner Res. 2001;16(1):157–65.

Burton LA, Sumukadas D. Optimal management of sarcopenia. Clin Interv Aging. 2010;5:217–28.

Cairns DM, et al. Interplay of Nkx3.2, Sox9 and Pax3 regulates chondrogenic differentiation of muscle progenitor cells. PLoS One. 2012;7(7):e39642.

Cameron ID, et al. Interventions for preventing falls in older people in care facilities and hospitals. Cochrane Database Syst Rev. 2012;12:CD005465.

Cantorna MT, et al. Vitamin D status, 1,25-dihydroxyvitamin D3, and the immune system. Am J Clin Nutr. 2004;80(6 Suppl):1717S–20S.

Cantudo-Cuenca MR, et al. Suitability of teriparatide and level of acceptance of pharmacotherapeutic recommendations in a healthcare management area. Farm Hosp. 2016;40(4):237–45.

Carmignac V, Quere R, Durbeej M. Proteasome inhibition improves the muscle of laminin alpha2 chain-deficient mice. Hum Mol Genet. 2011;20(3):541–52.

Carson JA, Manolagas SC. Effects of sex steroids on bones and muscles: similarities, parallels, and putative interactions in health and disease. Bone. 2015;80:67–78.

Castrogiovanni P, et al. The importance of physical activity in osteoporosis. From the molecular pathways to the clinical evidence. Histol Histopathol. 2016;31(11):1183–94.

Chan JK, et al. Soft-tissue reconstruction of open fractures of the lower limb: muscle versus fasciocutaneous flaps. Plast Reconstr Surg. 2012;130(2):284e–95e.

Chen JS, Sambrook PN. Antiresorptive therapies for osteoporosis: a clinical overview. Nat Rev Endocrinol. 2011;8(2):81–91.

Cherian PP, et al. Mechanical strain opens connexin 43 hemichannels in osteocytes: a novel mechanism for the release of prostaglandin. Mol Biol Cell. 2005;16(7):3100–6.

Chlebowski RT, et al. Influence of estrogen plus progestin on breast cancer and mammography in healthy postmenopausal women: the Women's Health Initiative randomized trial. JAMA. 2003;289(24):3243–53.

Cohen S, Nathan JA, Goldberg AL. Muscle wasting in disease: molecular mechanisms and promising therapies. Nat Rev Drug Discov. 2015;14(1):58–74.

Coiro V, et al. Effect of physiological exercise on osteocalcin levels in subjects with adrenal incidentaloma. J Endocrinol Investig. 2012;35(4):357–8.

Colaianni G, et al. The myokine irisin increases cortical bone mass. Proc Natl Acad Sci U S A. 2015;112(39):12157–62.

Dallas SL, Prideaux M, Bonewald LF. The osteocyte: an endocrine cell … and more. Endocr Rev. 2013;34(5):658–90.

Daluiski A, et al. Bone morphogenetic protein-3 is a negative regulator of bone density. Nat Genet. 2001;27(1):84–8.

Damien E, Price JS, Lanyon LE. Mechanical strain stimulates osteoblast proliferation through the estrogen receptor in males as well as females. J Bone Miner Res. 2000;15(11):2169–77.

Dawson JI, et al. Concise review: bridging the gap: bone regeneration using skeletal stem cell-based strategies – where are we now? Stem Cells. 2014;32(1):35–44.

Day CS, et al. Use of muscle cells to mediate gene transfer to the bone defect. Tissue Eng. 1999;5(2):119–25.

Dedkov EI, et al. MyoD and myogenin protein expression in skeletal muscles of senile rats. Cell Tissue Res. 2003;311(3):401–16.

DeFronzo RA, et al. Effects of insulin on peripheral and splanchnic glucose metabolism in noninsulin-dependent (type II) diabetes mellitus. J Clin Invest. 1985;76(1):149–55.

Drey M, et al. Osteosarcopenia is more than sarcopenia and osteopenia alone. Aging Clin Exp Res. 2016;28(5):895–9.

Duprez DM, et al. Bone morphogenetic protein-2 (BMP-2) inhibits muscle development and promotes cartilage formation in chick limb bud cultures. Dev Biol. 1996;174(2):448–52.

Edom-Vovard F, Duprez D. Signals regulating tendon formation during chick embryonic development. Dev Dyn. 2004;229(3):449–57

Elkasrawy MN, Hamrick MW. Myostatin (GDF-8) as a key factor linking muscle mass and bone structure. J Musculoskelet Neuronal Interact. 2010;10(1):56–63.

English KL, Paddon-Jones D. Protecting muscle mass and function in older adults during bed rest. Curr Opin Clin Nutr Metab Care. 2010;13(1):34–9.

Epperlein HH, et al. BMP-4 and Noggin signaling modulate dorsal fin and somite development in the axolotl trunk. Dev Dyn. 2007;236(9):2464–74.

Esen E, et al. PTH promotes bone anabolism by stimulating aerobic glycolysis via IGF signaling. J Bone Miner Res. 2015;30(11):1959–68.

Evans RA, et al. Long-term experience with a calcium-thiazide treatment for Paget's disease of bone. Miner Electrolyte Metab. 1982;8(6):325–33.

Fan W, et al. Structural and cellular features in metaphyseal and diaphyseal periosteum of osteoporotic rats. J Mol Histol. 2010;41(1):51–60.

Ferreira V, et al. Immune responses to intramuscular administration of alipogene tiparvovec (AAV1-LPL(S447X)) in a phase II clinical trial of lipoprotein lipase deficiency gene therapy. Hum Gene Ther. 2014;25(3):180–8.

Ferrucci L, et al. Interaction between bone and muscle in older persons with mobility limitations. Curr Pharm Des. 2014;20(19):3178–97.

Fields SZ, et al. Activin receptor antagonists for cancer-related anemia and bone disease. Expert Opin Investig Drugs. 2013;22(1):87–101.

Foulstone EJ, et al. Differential signalling mechanisms predisposing primary human skeletal muscle cells to altered proliferation and differentiation: roles of IGF-I and TNFalpha. Exp Cell Res. 2004;294(1):223–35.
Fox J, et al. Effects of daily treatment with parathyroid hormone 1–84 for 16 months on density, architecture and biomechanical properties of cortical bone in adult ovariectomized rhesus monkeys. Bone. 2007;41(3):321–30.
Fretz JA, et al. 1,25-dihydroxyvitamin D3 induces expression of the Wnt signaling co-regulator LRP5 via regulatory elements located significantly downstream of the gene's transcriptional start site. J Steroid Biochem Mol Biol. 2007;103(3–5):440–5.
Friedrichs M, et al. BMP signaling balances proliferation and differentiation of muscle satellite cell descendants. BMC Cell Biol. 2011;12:26.
Frisoli A Jr, et al. Severe osteopenia and osteoporosis, sarcopenia, and frailty status in community-dwelling older women: results from the Women's Health and Aging Study (WHAS) II. Bone. 2011;48(4):952–7.
Fry CS, et al. Glucocorticoids increase skeletal muscle NF-kappaB inducing kinase (NIK): links to muscle atrophy. Physiol Rep. 2016;4(21):e13014.
Fujimaki S, et al. Wnt protein-mediated satellite cell conversion in adult and aged mice following voluntary wheel running. J Biol Chem. 2014;289(11):7399–412.
Gawlik KI, Durbeej M. Skeletal muscle laminin and MDC1A: pathogenesis and treatment strategies. Skelet Muscle. 2011;1(1):9.
Genant HK, et al. Effects of romosozumab compared with teriparatide on bone density and mass at the spine and hip in postmenopausal women with low bone mass. J Bone Miner Res. 2016;32(1):181–7.
Gersbach CA, et al. Runx2/Cbfa1 stimulates transdifferentiation of primary skeletal myoblasts into a mineralizing osteoblastic phenotype. Exp Cell Res. 2004;300(2):406–17.
Gillespie LD, et al. Interventions for preventing falls in elderly people. Cochrane Database Syst Rev. 2001;3:CD000340.
Gillies AR, Lieber RL. Structure and function of the skeletal muscle extracellular matrix. Muscle Nerve. 2011;44(3):318–31.
Girgis CM, et al. Vitamin D signaling regulates proliferation, differentiation, and myotube size in C2C12 skeletal muscle cells. Endocrinology. 2014;155(2):347–57.
Glass GE, et al. TNF-alpha promotes fracture repair by augmenting the recruitment and differentiation of muscle-derived stromal cells. Proc Natl Acad Sci U S A. 2011;108(4):1585–90.
Goisser S, et al. Sarcopenic obesity and complex interventions with nutrition and exercise in community-dwelling older persons – a narrative review. Clin Interv Aging. 2015;10:1267–82.
Gorski JP, et al. Deletion of Mbtps1 (Pcsk8, S1p, ski-1) gene in osteocytes stimulates soleus muscle regeneration and increased size and contractile force with age. J Biol Chem. 2016;291(9):4308–22.
Gray SR, Kamolrat T. The effect of exercise induced cytokines on insulin stimulated glucose transport in C2C12 cells. Cytokine. 2011;55(2):221–8.
Hamrick MW. Increased bone mineral density in the femora of GDF8 knockout mice. Anat Rec A Discov Mol Cell Evol Biol. 2003;272(1):388–91.
Hamrick MW, McPherron AC, Lovejoy CO. Bone mineral content and density in the humerus of adult myostatin-deficient mice. Calcif Tissue Int. 2002;71(1):63–8.
Hamrick MW, et al. Leptin treatment induces loss of bone marrow adipocytes and increases bone formation in leptin-deficient ob/ob mice. J Bone Miner Res. 2005;20(6):994–1001.
Hamrick MW, et al. Increased muscle mass with myostatin deficiency improves gains in bone strength with exercise. J Bone Miner Res. 2006;21(3):477–83.
Han S, et al. Crystal structure of activin receptor type IIB kinase domain from human at 2.0 Angstrom resolution. Protein Sci. 2007;16(10):2272–7.
Harslof T, et al. Polymorphisms of muscle genes are associated with bone mass and incident osteoporotic fractures in Caucasians. Calcif Tissue Int. 2013;92(5):467–76.

Hartke JR, Lundy MW. Bone anabolic therapy with selective prostaglandin analogs. J Musculoskelet Neuronal Interact. 2001;2(1):25–31.
Hasty P, et al. Muscle deficiency and neonatal death in mice with a targeted mutation in the myogenin gene. Nature. 1993;364(6437):501–6.
Hauschka PV, et al. Osteocalcin and matrix Gla protein: vitamin K-dependent proteins in bone. Physiol Rev. 1989;69(3):990–1047.
He H, et al. Relationship of sarcopenia and body composition with osteoporosis. Osteoporos Int. 2016;27(2):473–82.
Heidari B, et al. Association between nonspecific skeletal pain and vitamin D deficiency. Int J Rheum Dis. 2010;13(4):340–6.
Hewitt J, et al. The muscle transcription factor MyoD promotes osteoblast differentiation by stimulation of the Osterix promoter. Endocrinology. 2008;149(7):3698–707.
Hill PA, Tumber A, Meikle MC. Multiple extracellular signals promote osteoblast survival and apoptosis. Endocrinology. 1997;138(9):3849–58.
Hinz B. Tissue stiffness, latent TGF-beta1 activation, and mechanical signal transduction: implications for the pathogenesis and treatment of fibrosis. Curr Rheumatol Rep. 2009;11(2):120–6.
Hittel DS, et al. Myostatin decreases with aerobic exercise and associates with insulin resistance. Med Sci Sports Exerc. 2010;42(11):2023–9.
Hock JM, Centrella M, Canalis E. Insulin-like growth factor I has independent effects on bone matrix formation and cell replication. Endocrinology. 1988;122(1):254–60.
Holmberg J, Durbeej M. Laminin-211 in skeletal muscle function. Cell Adhes Migr. 2013;7(1):111–21.
Hong W, et al. Prevalence of sarcopenia and its relationship with sites of fragility fractures in elderly Chinese men and women. PLoS One. 2015;10(9):e0138102.
Hughes VA, et al. Anthropometric assessment of 10-y changes in body composition in the elderly. Am J Clin Nutr. 2004;80(2):475–82.
Huo YR, et al. Phenotype of osteosarcopenia in older individuals with a history of falling. J Am Med Dir Assoc. 2015;16(4):290–5.
Isaacson J, Brotto M. Physiology of mechanotransduction: how do muscle and bone "Talk" to one another? Clin Rev Bone Miner Metab. 2014;12(2):77–85.
Ito Y, et al. Localization of chondrocyte precursors in periosteum. Osteoarthr Cartil. 2001;9(3):215–23.
Jagielska GW, et al. Bone mineralization disorders as a complication of anorexia nervosa – etiology, prevalence, course and treatment. Psychiatr Pol. 2016;50(3):509–20.
Jahn K, et al. Skeletal muscle secreted factors prevent glucocorticoid-induced osteocyte apoptosis through activation of beta-catenin. Eur Cell Mater. 2012;24:197–209. discussion 209–10
James PL, Stewart CE, Rotwein P. Insulin-like growth factor binding protein-5 modulates muscle differentiation through an insulin-like growth factor-dependent mechanism. J Cell Biol. 1996;133(3):683–93.
Janssen I, et al. Skeletal muscle mass and distribution in 468 men and women aged 18–88 yr. J Appl Physiol (1985). 2000;89(1):81–8.
Javaheri B, et al. Deletion of a single beta-catenin allele in osteocytes abolishes the bone anabolic response to loading. J Bone Miner Res. 2014;29(3):705–15.
Jemal A, et al. Cancer statistics, 2002. CA Cancer J Clin. 2002;52(1):23–47.
Jones DB, et al. Biochemical signal transduction of mechanical strain in osteoblast-like cells. Biomaterials. 1991;12(2):101–10.
Jorgensen JO, et al. Three years of growth hormone treatment in growth hormone-deficient adults: near normalization of body composition and physical performance. Eur J Endocrinol. 1994;130(3):224–8.
Joulia-Ekaza D, Cabello G. The myostatin gene: physiology and pharmacological relevance. Curr Opin Pharmacol. 2007;7(3):310–5.
Kaartinen V, et al. Abnormal lung development and cleft palate in mice lacking TGF-beta 3 indicates defects of epithelial-mesenchymal interaction. Nat Genet. 1995;11(4):415–21.

Karasik D, Kiel DP. Genetics of the musculoskeletal system: a pleiotropic approach. J Bone Miner Res. 2008;23(6):788–802.
Katagiri T, et al. Bone morphogenetic protein-2 converts the differentiation pathway of C2C12 myoblasts into the osteoblast lineage. J Cell Biol. 1994;127(6 Pt 1):1755–66.
Kaur P, Mishra SK, Mithal A. Vitamin D toxicity resulting from overzealous correction of vitamin D deficiency. Clin Endocrinol. 2015;83(3):327–31.
Kawai M, et al. Ectopic bone formation by human bone morphogenetic protein-2 gene transfer to skeletal muscle using transcutaneous electroporation. Hum Gene Ther. 2003;14(16):1547–56.
Kim CH, et al. Beta-catenin interacts with MyoD and regulates its transcription activity. Mol Cell Biol. 2008;28(9):2941–51.
Kim KM, et al. Calcium and vitamin D supplementations: 2015 position statement of the Korean Society for Bone and Mineral Research. J Bone Metab. 2015;22(4):143–9.
Kitase Y, et al. Mechanical induction of PGE2 in osteocytes blocks glucocorticoid-induced apoptosis through both the beta-catenin and PKA pathways. J Bone Miner Res. 2010;25(12):2657–68.
Kjaer M, et al. Extracellular matrix adaptation of tendon and skeletal muscle to exercise. J Anat. 2006;208(4):445–50.
Klein-Nulend J, et al. Pulsating fluid flow stimulates prostaglandin release and inducible prostaglandin G/H synthase mRNA expression in primary mouse bone cells. J Bone Miner Res. 1997;12(1):45–51.
Kling JM, Clarke BL, Sandhu NP. Osteoporosis prevention, screening, and treatment: a review. J Women's Health (Larchmt). 2014;23(7):563–72.
Knapp JR, et al. Loss of myogenin in postnatal life leads to normal skeletal muscle but reduced body size. Development. 2006;133(4):601–10.
Kokabu S, et al. BMP3 suppresses osteoblast differentiation of bone marrow stromal cells via interaction with Acvr2b. Mol Endocrinol. 2012;26(1):87–94.
Komaki M, et al. MyoD enhances BMP7-induced osteogenic differentiation of myogenic cell cultures. J Cell Sci. 2004;117(Pt 8):1457–68.
Kong YY, et al. OPGL is a key regulator of osteoclastogenesis, lymphocyte development and lymph-node organogenesis. Nature. 1999;397(6717):315–23.
Kovacs CS. Calcium and bone metabolism disorders during pregnancy and lactation. Endocrinol Metab Clin N Am. 2011;40(4):795–826.
Kragstrup TW, Kjaer M, Mackey AL. Structural, biochemical, cellular, and functional changes in skeletal muscle extracellular matrix with aging. Scand J Med Sci Sports. 2011;21(6):749–57.
Kulkarni AB, et al. Transforming growth factor beta 1 null mutation in mice causes excessive inflammatory response and early death. Proc Natl Acad Sci U S A. 1993;90(2):770–4.
Kuroda K, et al. Canonical Wnt signaling induces BMP-4 to specify slow myofibrogenesis of fetal myoblasts. Skelet Muscle. 2013;3(1):5.
Lach-Trifilieff E, et al. An antibody blocking activin type II receptors induces strong skeletal muscle hypertrophy and protects from atrophy. Mol Cell Biol. 2014;34(4):606–18.
Lampe AK, Bushby KM. Collagen VI related muscle disorders. J Med Genet. 2005;42(9):673–85.
Landi F, et al. Sarcopenia as a risk factor for falls in elderly individuals: results from the ilSIRENTE study. Clin Nutr. 2012;31(5):652–8.
Lappe JM, Binkley N. Vitamin D and sarcopenia/falls. J Clin Densitom. 2015;18(4):478–82.
Latres E, et al. Myostatin blockade with a fully human monoclonal antibody induces muscle hypertrophy and reverses muscle atrophy in young and aged mice. Skelet Muscle. 2015;5:34.
Laurent MR, et al. Muscle-bone interactions: from experimental models to the clinic? A critical update. Mol Cell Endocrinol. 2016;432:14–36.
Lee JY, et al. Enhancement of bone healing based on ex vivo gene therapy using human muscle-derived cells expressing bone morphogenetic protein 2. Hum Gene Ther. 2002;13(10):1201–11.
Lee SJ, et al. Regulation of muscle growth by multiple ligands signaling through activin type II receptors. Proc Natl Acad Sci U S A. 2005;102(50):18117–22.
Legre R, et al. Evaluation of 106 free flaps. Analysis of the failures and the indications. Ann Chir Plast Esthet. 1989;34(5):385–91.

Levine JP. Pharmacologic and nonpharmacologic management of osteoporosis. Clin Cornerstone. 2006;8(1):40–53.

Lewiecki EM. Role of sclerostin in bone and cartilage and its potential as a therapeutic target in bone diseases. Ther Adv Musculoskelet Dis. 2014;6(2):48–57.

Linglart A, Biosse-Duplan M. Hypophosphatasia. Curr Osteoporos Rep. 2016;14(3):95–105.

Lotinun S, et al. A soluble activin receptor type IIA fusion protein (ACE-011) increases bone mass via a dual anabolic-antiresorptive effect in Cynomolgus monkeys. Bone. 2010;46(4):1082–8.

Lu S, et al. Bone morphogenetic protein 9 (BMP9) induces effective bone formation from reversibly immortalized multipotent adipose-derived (iMAD) mesenchymal stem cells. Am J Transl Res. 2016;8(9):3710–30.

MacNabb C, Patton D, Hayes JS. Sclerostin antibody therapy for the treatment of osteoporosis: clinical prospects and challenges. J Osteoporos. 2016;2016:6217286.

Manini TM, Hong SL, Clark BC. Aging and muscle: a neuron's perspective. Curr Opin Clin Nutr Metab Care. 2013;16(1):21–6.

Manolagas SC. Birth and death of bone cells: basic regulatory mechanisms and implications for the pathogenesis and treatment of osteoporosis. Endocr Rev. 2000;21(2):115–37.

Manolagas SC. From estrogen-centric to aging and oxidative stress: a revised perspective of the pathogenesis of osteoporosis. Endocr Rev. 2010;31(3):266–300.

Mastrogiacomo M, Derubeis AR, Cancedda R. Bone and cartilage formation by skeletal muscle derived cells. J Cell Physiol. 2005;204(2):594–603.

Matkovits T, Christakos S. Variable in vivo regulation of rat vitamin D-dependent genes (osteopontin, Ca, Mg-adenosine triphosphatase, and 25-hydroxyvitamin D3 24-hydroxylase): implications for differing mechanisms of regulation and involvement of multiple factors. Endocrinology. 1995;136(9):3971–82.

Matthews BG, et al. Osteogenic potential of alpha smooth muscle actin expressing muscle resident progenitor cells. Bone. 2016;84:69–77.

Maurel DB, et al. Beta-catenin haplo insufficient male mice do not lose bone in response to hindlimb unloading. PLoS One. 2016;11(7):e0158381.

McClung JM, et al. Estrogen status and skeletal muscle recovery from disuse atrophy. J Appl Physiol (1985). 2006;100(6):2012–23.

McPherron AC, Lee SJ. Double muscling in cattle due to mutations in the myostatin gene. Proc Natl Acad Sci U S A. 1997;94(23):12457–61.

Mera P, et al. Osteocalcin signaling in myofibers is necessary and sufficient for optimum adaptation to exercise. Cell Metab. 2016;23(6):1078–92.

Michael H, et al. Estrogen and testosterone use different cellular pathways to inhibit osteoclastogenesis and bone resorption. J Bone Miner Res. 2005;20(12):2224–32.

Midura RJ, et al. Parathyroid hormone rapidly stimulates hyaluronan synthesis by periosteal osteoblasts in the tibial diaphysis of the growing rat. J Biol Chem. 2003;278(51):51462–8.

Milat F, Ebeling PR. Osteoporosis treatment: a missed opportunity. Med J Aust. 2016;205(4):185–90.

Miura T, et al. Decorin binds myostatin and modulates its activity to muscle cells. Biochem Biophys Res Commun. 2006;340(2):675–80.

Mo C, et al. Prostaglandin E2: from clinical applications to its potential role in bone- muscle crosstalk and myogenic differentiation. Recent Pat Biotechnol. 2012;6(3):223–9.

Mo C, et al. Prostaglandin E2 promotes proliferation of skeletal muscle myoblasts via EP4 receptor activation. Cell Cycle. 2015;14(10):1507–16.

Montero-Fernandez N, Serra-Rexach JA. Role of exercise on sarcopenia in the elderly. Eur J Phys Rehabil Med. 2013;49(1):131–43.

Morissette MR, et al. Effects of myostatin deletion in aging mice. Aging Cell. 2009;8(5):573–83.

Motyl KJ, McCabe LR, Schwartz AV. Bone and glucose metabolism: a two-way street. Arch Biochem Biophys. 2010;503(1):2–10.

Mowe M, Haug E, Bohmer T. Low serum calcidiol concentration in older adults with reduced muscular function. J Am Geriatr Soc. 1999;47(2):220–6.

Murad MH, et al. Clinical review: the effect of vitamin D on falls: a systematic review and meta-analysis. J Clin Endocrinol Metab. 2011;96(10):2997–3006.
Murakami T, et al. Transforming growth factor-beta1 increases mRNA levels of osteoclastogenesis inhibitory factor in osteoblastic/stromal cells and inhibits the survival of murine osteoclast-like cells. Biochem Biophys Res Commun. 1998;252(3):747–52.
Musaro A, et al. Enhanced expression of myogenic regulatory genes in aging skeletal muscle. Exp Cell Res. 1995;221(1):241–8.
Nabeshima Y, et al. Myogenin gene disruption results in perinatal lethality because of severe muscle defect. Nature. 1993;364(6437):532–5.
Nakamura-Takahashi A, et al. Treatment of hypophosphatasia by muscle-directed expression of bone-targeted alkaline phosphatase via self-complementary AAV8 vector. Mol Ther Methods Clin Dev. 2016;3:15059.
Nakano Y, et al. The receptor, metabolism and effects of androgen in osteoblastic MC3T3-E1 cells. Bone Miner. 1994;26(3):245–59.
Neer RM, et al. Effect of parathyroid hormone (1–34) on fractures and bone mineral density in postmenopausal women with osteoporosis. N Engl J Med. 2001;344(19):1434–41.
Nindl BC. Insulin-like growth factor-I, physical activity, and control of cellular anabolism. Med Sci Sports Exerc. 2010;42(1):35–8.
Nishimura R, et al. Core-binding factor alpha 1 (Cbfa1) induces osteoblastic differentiation of C2C12 cells without interactions with Smad1 and Smad5. Bone. 2002;31(2):303–12.
Okada T, Takeda S. Current challenges and future directions in recombinant AAV-mediated gene therapy of Duchenne muscular dystrophy. Pharmaceuticals (Basel). 2013;6(7):813–36.
Okuno J, et al. Evaluation of the association between impaired renal function and physical function among community-dwelling Japanese frail elderly based on the estimated glomerular filtration rate (eGFR). Nihon Ronen Igakkai Zasshi. 2009;46(1):63–70.
Ono Y, et al. BMP signalling permits population expansion by preventing premature myogenic differentiation in muscle satellite cells. Cell Death Differ. 2011;18(2):222–34.
Osawa K, et al. Osteoinduction by microbubble-enhanced transcutaneous sonoporation of human bone morphogenetic protein-2. J Gene Med. 2009;11(7):633–41.
Ota K, et al. TGF-beta induces Wnt10b in osteoclasts from female mice to enhance coupling to osteoblasts. Endocrinology. 2013;154(10):3745–52.
Pan YC, et al. Wnt3a signal pathways activate MyoD expression by targeting cis-elements inside and outside its distal enhancer. Biosci Rep. 2015;35(2):e00180.
Papadakis MA, et al. Growth hormone replacement in healthy older men improves body composition but not functional ability. Ann Intern Med. 1996;124(8):708–16.
Pearsall RS, et al. A soluble activin type IIA receptor induces bone formation and improves skeletal integrity. Proc Natl Acad Sci U S A. 2008;105(19):7082–7.
Pearsall R, et al. ACE-083 increases muscle hypertrophy and strength in C57BL/6 mice. Neuromuscul Disord. 2015;25:S218.
Pedersen BK, Febbraio MA. Muscles, exercise and obesity: skeletal muscle as a secretory organ. Nat Rev Endocrinol. 2012;8(8):457–65.
Pedersen BK, Fischer CP. Beneficial health effects of exercise – the role of IL-6 as a myokine. Trends Pharmacol Sci. 2007;28(4):152–6.
Pereira FB, Leite AF, de Paula AP. Relationship between pre-sarcopenia, sarcopenia and bone mineral density in elderly men. Arch Endocrinol Metab. 2015;59(1):59–65.
Pi M, Wu Y, Quarles LD. GPRC6A mediates responses to osteocalcin in beta-cells in vitro and pancreas in vivo. J Bone Miner Res. 2011;26(7):1680–3.
Plotnikoff GA, Quigley JM. Prevalence of severe hypovitaminosis D in patients with persistent, nonspecific musculoskeletal pain. Mayo Clin Proc. 2003;78(12):1463–70.
Pojednic RM, et al. Effects of 1,25-dihydroxyvitamin D3 and vitamin D3 on the expression of the vitamin d receptor in human skeletal muscle cells. Calcif Tissue Int. 2015;96(3):256–63.
Pouliot L, et al. Assessment of a nonsteroidal aromatase inhibitor, letrozole, in juvenile rats. Birth Defects Res B Dev Reprod Toxicol. 2013;98(5):374–90.

Pourquie O. Vertebrate somitogenesis. Annu Rev Cell Dev Biol. 2001;17:311–50.
Qiao X, et al. Irisin promotes osteoblast proliferation and differentiation via activating the MAP kinase signaling pathways. Sci Rep. 2016;6:18732.
Rahnert JA, et al. Glucocorticoids Alter CRTC-CREB signaling in muscle cells: impact on PGC-1alpha expression and atrophy markers. PLoS One. 2016;11(7):e0159181.
Raisz LG. Prostaglandins and bone: physiology and pathophysiology. Osteoarthr Cartil. 1999;7(4):419–21.
Ravindran S, George A. Multifunctional ECM proteins in bone and teeth. Exp Cell Res. 2014;325(2):148–54.
Rawls A, et al. Overlapping functions of the myogenic bHLH genes MRF4 and MyoD revealed in double mutant mice. Development. 1998;125(13):2349–58.
Recker RR, et al. A randomized, double-blind phase 2 clinical trial of blosozumab, a sclerostin antibody, in postmenopausal women with low bone mineral density. J Bone Miner Res. 2015;30(2):216–24.
Recknor CP, et al. The effect of discontinuing treatment with blosozumab: follow-up results of a phase 2 randomized clinical trial in postmenopausal women with low bone mineral density. J Bone Miner Res. 2015;30(9):1717–25.
Reichert JM. Antibodies to watch in 2015. MAbs. 2015;7(1):1–8.
Reshef R, Maroto M, Lassar AB. Regulation of dorsal somitic cell fates: BMPs and Noggin control the timing and pattern of myogenic regulator expression. Genes Dev. 1998;12(3):290–303.
Rhodes SJ, Konieczny SF. Identification of MRF4: a new member of the muscle regulatory factor gene family. Genes Dev. 1989;3(12B):2050–61.
Riggs BL, Khosla S, Melton LJ 3rd. Sex steroids and the construction and conservation of the adult skeleton. Endocr Rev. 2002;23(3):279–302.
Rocha PN, et al. Hypercalcemia and acute kidney injury caused by abuse of a parenteral veterinary compound containing vitamins A, D, and E. J Bras Nefrol. 2011;33(4):467–71.
Rooney JE, Gurpur PB, Burkin DJ. Laminin-111 protein therapy prevents muscle disease in the mdx mouse model for Duchenne muscular dystrophy. Proc Natl Acad Sci U S A. 2009;106(19):7991–6.
Rooney JE, et al. Laminin-111 protein therapy reduces muscle pathology and improves viability of a mouse model of merosin-deficient congenital muscular dystrophy. Am J Pathol. 2012;180(4):1593–602.
Rot-Nikcevic I, et al. Myf5−/−: MyoD−/− amyogenic fetuses reveal the importance of early contraction and static loading by striated muscle in mouse skeletogenesis. Dev Genes Evol. 2006;216(1):1–9.
Rudnicki MA, et al. Inactivation of MyoD in mice leads to up-regulation of the myogenic HLH gene Myf-5 and results in apparently normal muscle development. Cell. 1992;71(3):383–90.
Rudnicki MA, et al. MyoD or Myf-5 is required for the formation of skeletal muscle. Cell. 1993;75(7):1351–9.
Sabatelli P, et al. Expression of collagen VI alpha5 and alpha6 chains in human muscle and in Duchenne muscular dystrophy-related muscle fibrosis. Matrix Biol. 2012;31(3):187–96.
Salzler RR, et al. Myostatin deficiency but not anti-myostatin blockade induces marked proteomic changes in mouse skeletal muscle. Proteomics. 2016;16(14):2019–27.
Sanders KM, et al. Annual high-dose oral vitamin D and falls and fractures in older women: a randomized controlled trial. JAMA. 2010;303(18):1815–22.
Sanford LP, et al. TGFbeta2 knockout mice have multiple developmental defects that are non-overlapping with other TGFbeta knockout phenotypes. Development. 1997;124(13):2659–70.
Sartori R, et al. BMP signaling controls muscle mass. Nat Genet. 2013;45(11):1309–18.
Schnitzler CM. Childhood cortical porosity is related to microstructural properties of the bone-muscle junction. J Bone Miner Res. 2015;30(1):144–55.
Shah D, Spencer H, Bhoorasingh P. Leiomyoma of the trachea. West Indian Med J. 1986;35(4):327–9.
Sharples AP, et al. Longevity and skeletal muscle mass: the role of IGF signalling, the sirtuins, dietary restriction and protein intake. Aging Cell. 2015;14(4):511–23.

Shefer G, et al. Satellite-cell pool size does matter: defining the myogenic potency of aging skeletal muscle. Dev Biol. 2006;294(1):50–66.
Shefer G, et al. Reduced satellite cell numbers and myogenic capacity in aging can be alleviated by endurance exercise. PLoS One. 2010;5(10):e13307.
Shen B, et al. BMP-13 emerges as a potential inhibitor of bone formation. Int J Biol Sci. 2009;5(2):192–200.
Shen H, et al. Deletion of connexin43 in osteoblasts/osteocytes leads to impaired muscle formation in mice. J Bone Miner Res. 2015;30(4):596–605.
Sherman ML, et al. Multiple-dose, safety, pharmacokinetic, and pharmacodynamic study of sotatercept (ActRIIA-IgG1), a novel erythropoietic agent, in healthy postmenopausal women. J Clin Pharmacol. 2013;53(11):1121–30.
Shi YC, et al. Effects of continuous activation of vitamin D and Wnt response pathways on osteoblastic proliferation and differentiation. Bone. 2007;41(1):87–96.
Shimasaki S, et al. A functional bone morphogenetic protein system in the ovary. Proc Natl Acad Sci U S A. 1999;96(13):7282–7.
Shu B, et al. BMP2, but not BMP4, is crucial for chondrocyte proliferation and maturation during endochondral bone development. J Cell Sci. 2011;124(Pt 20):3428–40.
Sikon A, Batur P. Profile of teriparatide in the management of postmenopausal osteoporosis. Int J Womens Health. 2010;2:37–44.
Sipila S, Suominen H. Effects of strength and endurance training on thigh and leg muscle mass and composition in elderly women. J Appl Physiol (1985). 1995;78(1):334–40.
Smith RC, Lin BK. Myostatin inhibitors as therapies for muscle wasting associated with cancer and other disorders. Curr Opin Support Palliat Care. 2013;7(4):352–60.
Snijder MB, et al. Vitamin D status in relation to one-year risk of recurrent falling in older men and women. J Clin Endocrinol Metab. 2006;91(8):2980–5.
Snyder PJ, et al. Effects of testosterone replacement in hypogonadal men. J Clin Endocrinol Metab. 2000;85(8):2670–7.
Solloway MJ, Robertson EJ. Early embryonic lethality in Bmp5;Bmp7 double mutant mice suggests functional redundancy within the 60A subgroup. Development. 1999;126(8):1753–68.
Sondag GR, et al. Osteoactivin induces transdifferentiation of C2C12 myoblasts into osteoblasts. J Cell Physiol. 2014;229(7):955–66.
Southmayd EA, De Souza MJ. A summary of the influence of exogenous estrogen administration across the lifespan on the GH/IGF-1 axis and implications for bone health. Growth Horm IGF Res. 2016;32(1):181–7.
Spangenburg EE, et al. Regulation of physiological and metabolic function of muscle by female sex steroids. Med Sci Sports Exerc. 2012;44(9):1653–62.
Squier CA, Ghoneim S, Kremenak CR. Ultrastructure of the periosteum from membrane bone. J Anat. 1990;171:233–9.
Szulc P, et al. Serum undercarboxylated osteocalcin is a marker of the risk of hip fracture in elderly women. J Clin Invest. 1993;91(4):1769–74.
Szulc P, et al. Serum undercarboxylated osteocalcin is a marker of the risk of hip fracture: a three year follow-up study. Bone. 1996;18(5):487–8.
Szulc P, et al. Low skeletal muscle mass is associated with poor structural parameters of bone and impaired balance in elderly men – the MINOS study. J Bone Miner Res. 2005;20(5):721–9.
Taaffe DR, et al. Estrogen replacement, muscle composition, and physical function: the Health ABC Study. Med Sci Sports Exerc. 2005;37(10):1741–7.
Tagliavini F, et al. Ultrastructural changes in muscle cells of patients with collagen VI-related myopathies. Muscle Ligaments Tendons J. 2013;3(4):281–6.
Tang Y, et al. TGF-beta1-induced migration of bone mesenchymal stem cells couples bone resorption with formation. Nat Med. 2009;15(7):757–65.
Tang L, et al. Inhibiting myostatin signaling prevents femoral trabecular bone loss and microarchitecture deterioration in diet-induced obese rats. Exp Biol Med (Maywood). 2016;241(3):308–16.

Tanner SB, Harwell SA. More than healthy bones: a review of vitamin D in muscle health. Ther Adv Musculoskelet Dis. 2015;7(4):152–9.
Thacher TD, Clarke BL. Vitamin D insufficiency. Mayo Clin Proc. 2011;86(1):50–60.
Tsuchida K, et al. Signal transduction pathway through activin receptors as a therapeutic target of musculoskeletal diseases and cancer. Endocr J. 2008;55(1):11–21.
Tsuji K, et al. BMP2 activity, although dispensable for bone formation, is required for the initiation of fracture healing. Nat Genet. 2006;38(12):1424–9.
Tsuji K, et al. BMP4 is dispensable for skeletogenesis and fracture-healing in the limb. J Bone Joint Surg Am. 2008;90(Suppl 1):14–8.
Urciuolo A, et al. Collagen VI regulates satellite cell self-renewal and muscle regeneration. Nat Commun. 2013;4:1964.
Vahle JL, et al. Skeletal changes in rats given daily subcutaneous injections of recombinant human parathyroid hormone (1–34) for 2 years and relevance to human safety. Toxicol Pathol. 2002;30(3):312–21.
Vanderschueren D, et al. Clinical review: sex steroids and the periosteum – reconsidering the roles of androgens and estrogens in periosteal expansion. J Clin Endocrinol Metab. 2006;91(2):378–82.
Vanderschueren D, et al. Sex steroid actions in male bone. Endocr Rev. 2014;35(6):906–60.
Verschueren S, et al. Sarcopenia and its relationship with bone mineral density in middle-aged and elderly European men. Osteoporos Int. 2013;24(1):87–98.
Vevers GM. Nature. 1970;226(5240):89.
Vidal O, et al. Estrogen receptor specificity in the regulation of skeletal growth and maturation in male mice. Proc Natl Acad Sci U S A. 2000;97(10):5474–9.
Vivian JL, Olson EN, Klein WH. Thoracic skeletal defects in myogenin- and MRF4-deficient mice correlate with early defects in myotome and intercostal musculature. Dev Biol. 2000;224(1):29–41.
Wagner KR, et al. A phase I/II trial of MYO-029 in adult subjects with muscular dystrophy. Ann Neurol. 2008;63(5):561–71.
Wang RN, et al. Bone Morphogenetic Protein (BMP) signaling in development and human diseases. Genes Dis. 2014;1(1):87–105.
Weinstein RS, et al. Inhibition of osteoblastogenesis and promotion of apoptosis of osteoblasts and osteocytes by glucocorticoids. Potential mechanisms of their deleterious effects on bone. J Clin Invest. 1998;102(2):274–82.
Weintraub H, et al. The myoD gene family: nodal point during specification of the muscle cell lineage. Science. 1991;251(4995):761–6.
Welle S, et al. Growth hormone increases muscle mass and strength but does not rejuvenate myofibrillar protein synthesis in healthy subjects over 60 years old. J Clin Endocrinol Metab. 1996;81(9):3239–43.
Wepner F, et al. Effects of vitamin D on patients with fibromyalgia syndrome: a randomized placebo-controlled trial. Pain. 2014;155(2):261–8.
White TA, LeBrasseur NK. Myostatin and sarcopenia: opportunities and challenges – a mini-review. Gerontology. 2014;60(4):289–93.
Winkler DG, et al. Osteocyte control of bone formation via sclerostin, a novel BMP antagonist. EMBO J. 2003;22(23):6267–76.
Winnier G, et al. Bone morphogenetic protein-4 is required for mesoderm formation and patterning in the mouse. Genes Dev. 1995;9(17):2105–16.
Wintermeyer E, et al. Crucial role of vitamin D in the musculoskeletal system. Nutrients. 2016;8(6):319.
Woodhouse L, et al. A phase 2 randomized study investigating the efficacy and safety of myostatin antibody LY2495655 versus placebo in patients undergoing elective total hip arthroplasty. J Frailty Aging. 2016;5(1):62–70.
Wysowski DK, Chang JT. Alendronate and risedronate: reports of severe bone, joint, and muscle pain. Arch Intern Med. 2005;165(3):346–7.

Xiang L, et al. BMP9-induced osteogenetic differentiation and bone formation of muscle-derived stem cells. J Biomed Biotechnol. 2012;2012:610952.

Yamaguchi Y, Mann DM, Ruoslahti E. Negative regulation of transforming growth factor-beta by the proteoglycan decorin. Nature. 1990;346(6281):281–4.

Yamamoto N, et al. Smad1 and smad5 act downstream of intracellular signalings of BMP-2 that inhibits myogenic differentiation and induces osteoblast differentiation in C2C12 myoblasts. Biochem Biophys Res Commun. 1997;238(2):574–80.

Yarasheski KE, Zachwieja JJ, Bier DM. Acute effects of resistance exercise on muscle protein synthesis rate in young and elderly men and women. Am J Phys. 1993;265(2 Pt 1):E210–4.

Yoshida K, et al. Stimulation of bone formation and prevention of bone loss by prostaglandin E EP4 receptor activation. Proc Natl Acad Sci U S A. 2002;99(7):4580–5.

Yu R, Leung J, Woo J. Incremental predictive value of sarcopenia for incident fracture in an elderly Chinese cohort: results from the Osteoporotic Fractures in Men (MrOs) study. J Am Med Dir Assoc. 2014;15(8):551–8.

Yukata K, et al. Aging periosteal progenitor cells have reduced regenerative responsiveness to bone injury and to the anabolic actions of PTH 1-34 treatment. Bone. 2014;62:79–89.

Zhang W, Behringer RR, Olson EN. Inactivation of the myogenic bHLH gene MRF4 results in up-regulation of myogenin and rib anomalies. Genes Dev. 1995;9(11):1388–99.

Zhang Y, et al. Irisin stimulates browning of white adipocytes through mitogen-activated protein kinase p38 MAP kinase and ERK MAP kinase signaling. Diabetes. 2014;63(2):514–25.

Zhao G, et al. Targeted overexpression of insulin-like growth factor I to osteoblasts of transgenic mice: increased trabecular bone volume without increased osteoblast proliferation. Endocrinology. 2000;141(7):2674–82.

Zhou X, et al. Chondrocytes transdifferentiate into osteoblasts in endochondral bone during development, postnatal growth and fracture healing in mice. PLoS Genet. 2014;10(12):e1004820.

Ziambaras K, Dagogo-Jack S. Reversible muscle weakness in patients with vitamin D deficiency. West J Med. 1997;167(6):435–9.

Ziegler R, Kasperk C. Glucocorticoid-induced osteoporosis: prevention and treatment. Steroids. 1998;63(5–6):344–8.

Zimmers TA, et al. Induction of cachexia in mice by systemically administered myostatin. Science. 2002;296(5572):1486–8.

Chapter 10
Pituitary Hormone-Driven Mechanism for Skeletal Loss

Tony Yuen, Li Sun, Wahid Abu-Amer, Peng Liu, Terry F. Davies, Harry C. Blair, Maria New, Alberta Zallone, and Mone Zaidi

Abstract A single specific function has been traditionally ascribed to each anterior and posterior pituitary glycoprotein hormone. However, it has become clear over the past decade that these hormones and their receptors have more ubiquitous functions. Organs, such as the skeleton, are regulated by and respond to pituitary hormones, particularly when circulating levels are perturbed in disease. Additionally, certain pituitary hormones are also expressed in bone cells, underscoring paracrine regulation.

The function of pituitary glycoprotein receptors in skeletal control appears evolutionarily more distant than effects on primary endocrine tissues (Blair et al. 2011). Growth hormone (GH), follicle-stimulating hormone (FSH), thyroid-stimulating hormone (TSH), adrenocorticotrophic hormone (ACTH), prolactin, oxytocin, and vasopressin all affect bone, and in mice, the haploinsufficiency of either the ligand and/or receptor yields a skeletal phenotype with the primary target organ remaining unaffected. Recognition and in-depth analysis of the mechanism of action of each pituitary hormone has improved our understanding of bone pathophysiology and opens new avenues for therapy. Here we discuss the interaction of each pituitary hormone with bone and the potential it holds in understanding and treating osteoporosis.

Keywords Pituitary-bone axis • Pituitary hormones • GPCR's • Anti-resorptive • Anabolic

---

T. Yuen (✉) • L. Sun • W. Abu-Amer • P. Liu • T.F. Davies • M. New • M. Zaidi
The Mount Sinai Bone Program, Department of Medicine, and Department of Pediatrics, Icahn School of Medicine at Mount Sinai, New York, NY, USA
e-mail: tony.yuen@mssm.edu

H.C. Blair
Departments of Pathology and of Cell Biology, University of Pittsburgh School of Medicine and the Pittsburgh VA Medical Center, Pittsburgh, PA, USA

A. Zallone
Department of Histology, University of Bari, Bari, Italy

S.Y. Smith et al. (eds.), *Bone Toxicology*, Molecular and Integrative Toxicology,
DOI 10.1007/978-3-319-56192-9_10

10.1 Evolution of Pituitary Hormone Receptor Signaling

ACTH is the best example of a pituitary hormone being part of a widely distributed G protein-coupled receptor (GPCR) system, which is known to participate in local cell differentiation. However, this local function seems to pale in comparison to the pituitary-adrenal signaling axis. There are five melanocortin receptors, including the ACTH receptor MC2R, which regulate various cellular functions, such as pigment production, appetite, and sexual function. All are controlled by ligands processed from a single large prohormone, pro-opiomelanocorticin (POMC). Hormone production occurs by tissue-specific regulated proteolysis, with ACTH being the predominant product in the anterior pituitary. At other sites, pro-opiomelanocorticin, three melanotropins, and β-endorphin are synthesized from the same precursor. ACTH is produced by human macrophage/monocyte cells (Pallinger and Csaba 2008), making it possible that MC2Rs in bone may also be activated by local, instead of pituitary-derived, ACTH. This level of decentralized control is also exemplified by corticotropin-releasing factor (CRF), which stimulates pituitary ACTH production in the adult while stimulating cortisol synthesis directly in the fetus (Sirianni et al. 2005).

TSH and FSH, along with chorionic gonadotropin (hCG) and luteinizing hormone (LH), belong to the subfamily of heterodimeric protein ligands that share a common α-chain, which specificity depending on their distinct β-chains. These hormones have primitive functions. Notably, in coelenterates for example, a TSHR family gene is identifiable, widely expressed, and shows the intron-exon structure found in mammals (Vibede et al. 1998). In lower vertebrates, such as in bony fish, there is abundant TSHR expression in the thyroid, but the receptor is also detectable in the ovaries, heart, muscle, and brain (Kumar et al. 2000). In fish, the gonadal expression of LHR and FSHR is well documented. In fact, multiple differently processed forms of the FSHR occur in fish (Kobayashi and Andersen 2008); this may reflect isoforms with differing functions. Further, in fish, the FSHR binds both FSH and LH, whereas the LHR recognizes only LH (Bogerd et al. 2005). While high-level FSHR expression is restricted to gonads, low-level expression is seen in spleen (Kumar et al. 2001), quite similarly to findings in human cells.

We and others have recently reported the production of a TSHβ variant by bone marrow macrophages (Vincent et al. 2009; Baliram et al. 2013); this splice variant activates the TSHR and may do so locally in bone. Lymphocytes also express TSH (Smith et al. 1983; Harbour et al. 1989), but such production is unlikely to affect circulating levels. There is no evidence, however, for bone or marrow cell production of FSH, although coproduction of TSHβ and FSH is noted in CD11β cells from mouse thyroid (Klein and Wang 2004). Overall, therefore, the presence of GPCRs in tissues other than traditional endocrine targets, such as the skeleton, and in cases, coexistence of their ligands, comes as no surprise. What does come as a surprise, however, is that the skeleton appears to be more sensitive to GPCR stimulation than the primary target organs, at least in mouse genetic and limited human studies.

10.2 Growth Hormone and IGF-1 Action on Bone

Growth hormone (GH), a single-chain polypeptide, plays a vital role in bone growth, modeling, and remodeling. It directly acts through a GPCR, but its primary action occurs via the release of insulin-like growth factors (IGFs). IGF-1 is synthesized mainly in the liver, and ~80 % circulates bound to IGF-binding protein-3 (IGFBP3) and the acid labile subunit (ALS). In GHR-deficient mice, both growth retardation and osteoporosis are rescued by IGF-1 overexpression (De Jesus et al. 2009), attesting to the relative importance of IGF-1 over GH. Furthermore, despite elevated GH levels, mice lacking both liver IGF-1 (LID) and ALS, with depleted serum IGF-1, show reduced bone growth and bone strength (Yakar et al. 2002). These results suggest that the skeletal effects of GH require IGF-1. In addition, GH-induced osteoclastic activity appears to require IGF-1 released from bone, which then activates bone resorption by acting on osteoclastic receptors, as well as by altering RANKL expression (Guicheux et al. 1998; Hou et al. 1997; Rubin et al. 2002).

Nonetheless, evidence to suggest that GH can act independently of IGF is somewhat limited and, in some instances, contradictory. GH replacement reverses the increased adiposity in hypophysectomized rats, while IGF-1 replacement does not (Menagh et al. 2010). Furthermore, GH reverses osteopenia in ovariectomized LID mice (Fritton et al. 2010). These results suggest that GH also directly acts on bone.

10.3 FSH Facilitates Bone Loss Along with Estrogen Deprivation

We discovered that FSH directly stimulates bone resorption by osteoclasts (Sun et al. 2006; Iqbal et al. 2006). Several studies have confirmed direct effects of FSH on the skeleton in rodents and humans. For example, amenorrheic women with a higher mean serum FSH (~35 IU/L) have greater bone loss than those with lower levels (~8 IU/L) in the face of near-equal estrogen levels (Devleta et al. 2004). Patients with functional hypothalamic amenorrhea, in whom both FSH and estrogen were low, show slight to moderate skeletal defects (Podfigurna-Stopa et al. 2012). Women harboring an activating FSHR-N680S polymorphism, rs6166, have lower bone mass and high resorption markers (Rendina et al. 2010); this attests to a role for FSHRs in human physiology. The exogenous administration of FSH to rats augments ovariectomy-induced bone loss, and a FSH antagonist reduces bone loss after ovariectomy or FSH injection (Liu et al. 2010a, b).

FSH increases osteoclast formation, function, and survival through a distinct FSHR isoform (Sun et al. 2006, 2010, Robinson et al. 2010; Wu et al. 2007). Two groups have, however, failed to identify FSHRs on osteoclasts, having likely used primers targeted to the ovarian isoform (Allan et al. 2010; Ritter et al. 2008). We very consistently find FSHR in human $CD14^+$ cells and osteoclasts using nested primers and sequencing to verify the specificity of the reaction and amplifying regions that contain an intron to avoid the pitfall of genomic DNA contamination (Robinson et al. 2010).

Wu et al. show that the osteoclastogenic response to FSH was abolished in mice lacking ITAM adapter signaling molecules (Wu et al. 2007). This suggests an interaction between FSH and immune receptor complexes, although the significance of such an interaction remains unclear. In a separate study, FSHR activation was shown to enhance RANK receptor expression (Cannon et al. 2011). In addition, FSH indirectly stimulates osteoclast formation by releasing osteoclastogenic cytokines, namely, IL-1β, TNF-α, and IL-6, in proportion to the surface expression of FSHRs (Iqbal et al. 2006; Cannon et al. 2010). In a study of 36 women between the ages of 20 and 50, serum FSH concentrations correlated with circulating cytokine concentrations (Cannon et al. 2010; Gertz et al. 2010).

Correlations between bone mineral density (BMD) declines and serum FSH levels have been documented extensively. The Study of Women's Health Across The Nations (SWAN), a longitudinal cohort of 2,375 perimenopausal women, showed a strong correlation between serum FSH levels and markers of bone resorption, as well as an equally strong correlation between changes in FSH levels over 4 years and decrements in BMD (Sowers et al. 2003). Analyses of data from Chinese women showed similar trends: a significant association between bone loss and high serum FSH (Xu et al. 2009; Wu et al. 2010). In a group of southern Chinese women aged between 45 and 55 years, those in the highest quartile of serum FSH lost bone at a 1.3- to 2.3-fold higher rate than those in the lowest quartile (Cheung et al. 2011). Furthermore, analysis of a National Health and Nutrition Examination Survey (NHANES) III cohort of women between the ages of 42 and 60 years showed a strong correlation between serum FSH and femoral neck BMD (Gallagher et al. 2010). Likewise, a cross-sectional analysis of 92 postmenopausal women found that serum osteocalcin and C-telopeptide levels were both positively correlated with FSH, but not with estradiol (Garcia-Martin et al. 2012).

The BONTURNO Study group showed that women with serum FSH levels of >30 IU/mL had significantly higher bone turnover markers than age-matched women, despite having normal menses (Adami et al. 2008). In contrast, Gourlay et al. fail to show a strong relationship between bone mass and FSH or indeed estrogen (Gourlay et al. 2011). Interestingly, however, the same authors show an independent correlation between FSH and lean mass (Gourlay et al. 2012). This latter association makes biological sense inasmuch as FSHRs are present on mesenchymal stem cells (Sun et al. 2006), known to have the propensity for adipocyte and myocyte differentiation. Notwithstanding whether there is a cause-effect relationship between a rising FSH and BMD changes, there is clear evidence favoring the use of FSH as a serum marker for identifying "fast bone losers" during the early phases of the menopausal transition, most notably the late perimenopause (Zaidi et al. 2009).

It has been difficult to tease out the action of FSH from that of estrogen in vivo as FSH releases estrogen and the actions of FSH and estrogen on the osteoclast are opposed. The injection of FSH into mice with intact ovaries (Ritter et al. 2008) or its transgenic overexpression (Allan et al. 2010), even in *hpg* mice, is unlikely to reveal pro-resorptive actions of FSH. This is because direct effects of FSH on the osteoclast will invariably be masked by the anti-resorptive and anabolic actions of the ovarian estrogen so released in response to FSH.

Clinically, there is evidence that women with low FSH levels undergo less bone loss than women with higher FSH levels, estrogen levels being nearly equal (Devleta et al. 2004), and importantly, that the effectiveness of estrogen therapy is related to the degree of FSH suppression (Kawai et al. 2004). With that said, patients with pituitary hypogonadism do lose bone. Luperide treatment, and hence the lowering of FSH, has not been shown to prevent hypogonadal hyper-resorption (Drake et al. 2010). While this proves that low estrogen is a *cause of* acute hypogonadal bone loss, it does not exclude a role for FSH in human skeletal homeostasis (Drake et al. 2010). Rather than blocking FSH in acute hypogonadism, where the effect of low estrogen is likely to be overwhelming, FSH inhibition during the late perimenopause, particularly when estrogen levels are normal and FSH is high, could potentially be of therapeutic significance.

We have recently developed a polyclonal antibody against a 13-amino acid long conserved peptide sequence of the computationally defined FSHR-binding domain of FSHβ (Zhu et al. 2012a, b). We found that this antibody not only attenuated osteoclastogenesis induced by FSH in vitro, but also prevented the loss of bone post-ovariectomy (Zhu et al. 2012a, b). Most importantly, we noted through a detailed histomorphometric analysis that while bone resorption was reduced, bone formation was surprisingly elevated significantly (Zhu et al. 2012a). This decoupling likely underscored the osteoprotective action of the antibody. Further studies documented FSHRs on mesenchymal stem cells. Additionally, bone marrow stromal cells from $FSHR^{-/-}$ mice showed significantly greater colony-forming (Cfu-f) potential, suggesting a profound increase in osteoblastogenesis in the absence of the FSHR. Importantly, antibody injections did not affect serum estrogen levels in wild-type mice (Zhu et al. 2012a). Collectively, our studies provide clear evidence that lowering FSH in a hypogonadal state prevents bone loss.

10.4 TSH Exerts Osteoprotection Through a Dual Action

TSH directly inhibits osteoclasts (Abe et al. 2003). TSHR haploinsufficiency in heterozygotic $TSHR^{+/-}$ mice results in osteoporosis, while thyroid hormone levels remain unaffected (Abe et al. 2003). Moreover, $TSHR^{-/-}$ mice are osteoporotic, a phenotype that cannot be explained by the known pro-osteoclastic action of thyroid hormones, particularly as $TSHR^{-/-}$ mice are hypothyroid (Novack 2003). Furthermore, thyroid hormone replacement that renders $TSHR^{-/-}$ mice euthyroid reverses skeletal runting but not the osteoporotic phenotype (Abe et al. 2003). Thus, TSH acts on bone independently of thyroid hormones. These findings also suggest that the osteoporosis of hyperthyroidism may, in part, be due to low TSH (Zaidi et al. 2006; Baliram et al. 2012). At least 20 clinical studies have since documented tight and highly reproducible correlations between low TSH levels, bone loss, bone geometry, and fracture risk in patient cohorts across the globe [e.g., (Mazziotti et al. 2010; Moon et al. 2015; Morris 2007; Grimnes et al. 2008; Frohlich and Wahl 2015; Kim et al. 2015; Blum et al. 2015; Noh et al. 2015; Tournis et al. 2015; Leader et al. 2014; Abrahamsen et al. 2014; Christy et al. 2014; Taylor et al. 2013; Chin et al. 2013; Kim et al. 2006, 2010; Baqi et al. 2010a, b; Svare et al. 2009; Heemstra et al. 2008; Mikosch et al. 2008)]. Evidence also shows that TSH protects the skeleton by exerting anti-resorptive and anabolic actions in rodent models and in people (Abe et al. 2003; Ma et al. 2011; Dumic-Cule et al. 2014; Hase et al. 2006; Zhang et al. 2014; Bagriacik et al. 2012; Baliram et al. 2011; Boutin et al. 2014; Sampath et al. 2007; Martini et al. 2008; Mazziotti et al. 2005).

The osteoporosis of TSHR deficiency is of the high-turnover variety. Osteoclastic activity is increased in $TSHR^{-/-}$ mice, similar to *hyt/hyt* mice in which TSHR signaling is defective (Britto et al. 1994; Sun et al. 2008). Recombinant TSH has been shown to attenuate the genesis, function, and survival of osteoclasts in ex vivo murine bone marrow (Abe et al. 2003) and in vitro ES cell cultures (Ma et al. 2009). The latter suggested a role early in bone development (Ma et al. 2009). In contrast, the overexpression of constitutively activated TSHR in osteoclast precursor cells (Hase et al. 2006) or transgenically, in mouse precursors (Sun et al. 2008), inhibited osteoclastogenesis. In postmenopausal women, a single subcutaneous injection of TSH drastically lowered serum C-telopeptide to premenopausal levels within 2 days, with recovery at day 7 (Mazziotti et al. 2005). In none of the studies with TSH replacement did thyroid hormones increase, exemplifying again that the pituitary-bone axis is more primitive than the pituitary-thyroid axis.

This anti-osteoclastogenic action of TSH is mediated by reduced NF-κB and JNK signaling and TNFα production (Abe et al. 2003; Hase et al. 2006). The effect of TSH on TNFα synthesis is mediated transcriptionally by binding of two high mobility group box proteins, HMGB1 and HMGB2, to TNFα gene promoter (Yamoah et al. 2008). TNFα production is expectedly upregulated in osteoporotic $TSHR^{-/-}$ mice (Abe et al. 2003), and the genetic deletion of TNFα in these mice reverses the osteoporosis, as well as the bone formation and resorption defects, proving that the $TSHR^{-/-}$ phenotype is mediated by TNFα, at least in part (Hase et al. 2006; Sun et al. 2013). The osteoporosis, low bone formation, and hyperresorption that accompany TSH deficiency were fully rescued in compound mouse mutants in which TNFα is genetically deleted on a homozygote $TSHR^{-/-}$ or heterozygote $TSHR^{+/-}$ background (Sun et al. 2013). Studies using ex vivo bone marrow cell cultures show that TSH inhibits and stimulates TNFα production from macrophages and osteoblasts, respectively (Sun et al. 2013). TNFα, in turn, stimulates osteoclastogenesis but also enhances the production of a variant TSHβ in bone marrow (Baliram et al. 2013; Sun et al. 2013). This locally produced TSH suppresses osteoclast formation in a negative feedback loop.

The role of TSH in osteoblast regulation is less defined. While it inhibits osteoblastogenesis in bone marrow-derived cell cultures, TSH stimulates differentiation and mineralization in murine cell cultures through a Wnt5a-dependent mechanism

(Baliram et al. 2011). Likewise, in vivo, intermittently administered TSH is anabolic in both rats and mice (Sampath et al. 2007; Sun et al. 2008). Thus, the effect of TSH may be differentiation stage dependent. However, in vivo, in rats, TSH, injected up to once every 2 weeks, inhibits ovariectomy-induced bone loss 28 weeks following ovariectomy (Sun et al. 2008). Calcein labeling provides evidence for a direct anabolic action of intermittent TSH (Sampath et al. 2007). In humans, Martini et al. (2008) show an increase in PINP, a marker of bone formation, validating the conclusion that bolus doses of TSH are indeed anabolic.

Consistent with our work and that of other labs, epidemiologic studies show 4.5-fold increase in the risk of vertebral fractures, and 3.2-fold increase in the risk of non-vertebral fractures is seen at TSH levels <0.1 IU/L (Bauer et al. 2001). There is also a strong negative correlation between low serum TSH and high C-telopeptide levels, without an association with thyroid hormone (Zofkova and Hill 2008). In patients on L-thyroxine, greater bone loss has been noted in those with a suppressed TSH than in those without suppression (Baqi et al. 2010b; La Vignera et al. 2008; Flynn et al. 2010). The Tromso study supports this: participants with serum TSH below 2 SD had a significantly lower BMD, those with TSH above 2 SD had a significantly increased BMD, whereas there was no association between TSH and BMD at normal TSH levels (Grimnes et al. 2008). In patients taking suppressive doses of thyroxine for thyroid cancer, the serum level of cathepsin K, a surrogate resorption marker, was elevated (Mikosch et al. 2008), and the HUNT 2 study found a positive correlation between TSH and BMD at the distal forearm (Svare et al. 2009). Furthermore, evaluation of the NHANES data has shown that the odds ratio for correlations between TSH and bone mass ranged between 2 and 3.4 (Morris 2007). Euthyroid women with serum TSH in the lower tertile of normal displayed a higher incidence of vertebral fractures, independent of age, BMD, and thyroid hormones (Mazziotti et al. 2010). Finally, patients harboring the TSHR-D727E polymorphism had high bone mass (Heemstra et al. 2008); similar allelic associations have been reported from the United Kingdom and in the Rotterdam study (van der Deure et al. 2008; Albagha et al. 2005).

Physiologically, therefore, TSH uncouples bone remodeling by inhibiting osteoclastic bone resorption and stimulating osteoblastic bone formation, particularly when given intermittently. Furthermore, absent TSH signaling stimulates bone remodeling directly and through TNFα production, causing net bone loss. We compared the effect of inducing hyperthyroidism through the implantation of T_4 pellets in wild-type and $TSHR^{-/-}$ mice to determine whether mice with absent TSHR signaling lost more bone (Baliram et al. 2012). Whereas wild-type hyperthyroid mice lost bone, expectedly from the pro-resorptive actions of thyroid hormones, the loss was greater in hyperthyroid $TSHR^{-/-}$ mice – clearly demonstrating a direct action of TSH signaling on bone that was independent of thyroid hormone levels (Baliram et al. 2012). Low TSH levels may thus contribute to the pathophysiology of osteoporosis of hyperthyroidism, which has traditionally been attributed to high thyroid hormone levels alone.

10.5 ACTH Directly Affects VEGF Expression in Bone

Glucocorticoids, under natural regulation mainly by ACTH, are important co-regulators of many processes including vascular tone, central metabolism, and immune response. At higher, pharmacological levels, they exert anti-inflammatory and immunosuppressant effects, with incident complications including diabetes, osteoporosis, and osteonecrosis. Osteonecrosis, in particular, is a painful debilitating condition that affects metabolically active bone, typically the femoral head (Mankin 1992), and invariably requires surgical treatment. The underlying mechanisms of glucocorticoid-induced osteonecrosis are poorly understood, although a key finding is that osteonecrosis occurs prior to macroscopic vascular changes (Eberhardt et al. 2001).

Isales et al. (2010) discovered that bone-forming units strongly express MC2Rs. We showed that as with the adrenal cortex, ACTH induces VEGF production in osteoblasts through its action on MC2Rs (Zaidi et al. 2010). This likely translates into the protection by ACTH of glucocorticoid-induced osteonecrosis in a rabbit model (Zaidi et al. 2010). An independent report with consistent findings was recently published (Wang et al. 2010). We speculate that VEGF suppression secondary to ACTH suppression may contribute to the bone damage with long-term glucocorticoid therapy. Much needs to be done to validate this idea toward a therapeutic advantage, considering that ACTH analogs are already approved for human use.

10.6 Effects of Prolactin on Bone

Prolactin (PRL), a peptide hormone secreted by the anterior pituitary, acts to induce and maintain lactation, as well as to suppress folliculogenesis and libido. During pregnancy, it increases the calcium bioavailability for milk production and fetal skeletalogenesis by promoting intestinal calcium absorption and skeletal mobilization (Lotinun et al. 1998). Accelerated bone turnover and bone loss is noted in hyperprolactinemic adults (Naylor et al. 2000). Antagonism of PRL by bromocriptine, a dopamine agonist, reverses the bone loss (Lotinun et al. 2003).

This osteoclastic action of PRL is traditionally thought to arise from the accompanying hypoestrogenemia (Meaney et al. 2004). However, it has been shown that osteoblasts express PRLRs (Coss et al. 2000), suggesting a direct interaction between PRL and the osteoblast. In fact, the pattern of bone loss is distinct in PRL-exposed and ovariectomized rats (Seriwatanachai et al. 2008a). Ex vivo, PRL decreases osteoblast differentiation markers (Seriwatanachai et al. 2008a), in part, through the PI3K signaling pathway (Seriwatanachai et al. 2008b). PRL injected into adult mice accelerates bone resorption (Seriwatanachai et al. 2008a), notably by increasing the RANK/OPG ratio (Seriwatanachai et al. 2008a). Osteoclasts themselves do not possess PRLRs (Coss et al. 2000). In contrast, in infant rats, PRL causes net bone gain (Krishnamra and Seemoung 1996), increasing osteocalcin expression. Likewise, in human fetal osteoblast cells, PRL decreases the RANKL/OPG ratio (Seriwatanachai et al. 2008b). It appears therefore that the net effect of PLR on bone depends on the stage of development. In the fetus, it promotes bone growth and mineralization while accelerating bone resorption in the mother to make nutrients available.

10.7 Opposing Actions of Oxytocin and Vasopressin on Bone

Oxytocin (OXT) is a nonapeptide synthesized in the hypothalamus and released into circulation via the posterior pituitary. Its primary function is to mediate the milk ejection reflex in nursing mammals. It also stimulates uterine contraction during parturition; however OXT is not a requirement for this function. Thus, OXT null mice can deliver normally but are unable to nurse. Subcutaneous OXT injection completely rescues the milk ejection phenotype, attesting to this being a peripheral, as opposed to a central action (Nishimori et al. 1996). Central actions of OXT include the regulation of social behavior, including sexual and maternal behavior, affiliation, social memory, as well as penile erection and ejaculation (Young et al. 1998; Insel and Harbaugh 1989; Mantella et al. 2003; Argiolas et al. 1988). It also controls food, predominantly carbohydrate, intake centrally (Sclafani et al. 2007). Thus, the social amnesia, aggressive behavior, and overfeeding observed in $OXT^{-/-}$ and $OXTR^{-/-}$ mice are reversed on intracerebroventricular OXT injection (Ferguson et al. 2000).

OXT acts on a GPCR, present in abundance on osteoblasts (Copland et al. 1999), osteoclasts, and their precursors (Colucci et al. 2002). In line with the ubiquitous distribution of OXTRs, cells of bone marrow also synthesize OXT, suggesting the existence of autocrine and paracrine interactions (Colaianni et al. 2011). In vitro, OXT stimulates osteoblast differentiation and bone formation. Thus, $OXT^{-/-}$ and $OXTR^{-/-}$ mice, including the haploinsufficient heterozygotes with normal lactation, display severe osteoporosis due to a bone-forming defect (Tamma et al. 2009). This not only indicates that the osteoblast is the target for OXT but also that bone is more sensitive to OXT than the breast, hitherto considered its primary target. Once again, the finding emphasizes a relatively primitive pituitary-bone axis. Effects of OXT on bone resorption in vivo appear minimal, as OXT stimulates osteoclastogenesis but inhibits the activity of mature osteoclasts, with a net zero effect on resorption.

In vivo gain-of-function studies document a direct effect of OXT on bone. Intraperitoneal OXT injections result in increased BMD and ex vivo osteoblast formation (Tamma et al. 2009). In contrast, short-term intracerebroventricular OXT does not affect bone turnover markers. OXT injections in wild-type rats alter the RANKL/OPG ratio in favor on bone formation, again attesting to an anabolic action (Elabd et al. 2007).

Although unproven, OXT may have a critical role in bone anabolism during pregnancy and lactation. Both are characterized by excessive bone resorption in favor of fetal and postpartum bone growth, respectively (Wysolmerski 2002). This bone loss is, however, completely reversed upon weaning by a yet unidentified mechanism (Sowers et al. 1995). OXT peaks in blood during late pregnancy and lactation, and while its proosteoclastogenic action may contribute to intergenerational calcium transfer, its anabolic action could enable the restoration of the maternal skeleton. That $OXT^{-/-}$ pups show hypomineralized skeletons, and $OXT^{-/-}$ moms display reduced bone formation markers are suggestive of such actions. The question whether estrogen, via its positive regulation of osteoblastic OXT production, can synergize this action through a local feed-forward loop remains to be determined. These studies nonetheless pave the way for a greater understanding of pregnancy and lactation-associated osteoporosis, as well as new potential therapeutic options.

Arginine-vasopressin (AVP), another posterior pituitary hormone, is a nonapeptide that differs from OXT only by two amino acids. Its primary functions are to retain water in the body by increasing water reabsorption in the kidney and to constrict blood vessels, thereby increasing arterial blood pressure. AVP exerts its actions through G protein-coupled receptors. AVPR1A is present in the kidney, liver, peripheral vasculature, and brain. AVPR2 is expressed predominantly in the kidney. Expression of AVPR1A and AVPR2 is also found in osteoblasts and osteoclasts (Tamma et al. 2013). AVPR1B, also known as vasopressin 3 receptor, is expressed in the anterior pituitary and the brain, but not in the skeleton.

AVP can affect the skeleton directly and is yet another component of the pituitary-bone axis. For example, mice injected with AVP display reduced osteoblast formation and increased osteoclast formation (Tamma et al. 2013). Conversely, mice injected with the AVPR1A antagonist SR49059 had enhanced bone mass due to increased osteoblastogenesis and reduced osteoclast formation and bone resorption (Tamma et al. 2013). This high bone mass phenotype was also observed in AVPR1A-deficient mice (Tamma et al. 2013). In contrast, AVPR2 does not have a significant role in bone remodeling, since the specific AVPR2 inhibitor tolvaptan does not affect bone formation or bone mass (Sun et al. 2016). Collectively, the data establish a primary role for AVP signaling in bone mass regulation. Of note is that hyponatremic patients have elevated circulating AVP levels and a high fracture risk, prompting for further studies on the skeletal actions of AVPR inhibitors used commonly in these patients.

The highly homologous AVP and OXT can share their receptors in the regulation of bone formation by osteoblasts. AVPR1A and OXTR have opposing effects on bone mass. Notably, the deletion of OXTR in $OXTR^{-/-}$:$AVPR1A^{-/-}$ double-mutant mice reversed the high bone mass phenotype in $AVPR1A^{-/-}$ mice (Sun et al. 2016). While AVP-stimulated gene expression is inhibited when the OXTR is deleted in $AVPR1A^{-/-}$ cells, OXTR is dispensable for AVP action in inhibiting osteoblastogenesis and gene expression (Sun et al. 2016). Furthermore, OXT does not interact with AVPRs in vivo in a model of lactation-induced bone loss in which OXT levels are high (Sun et al. 2016).

10.8 Conclusion

The direct regulation of bone by glycoprotein hormones, since their discovery, helps explain some of the inconsistencies of older models that assumed that pituitary signaling was mediated entirely via endocrine organs through steroid-family signals. Important direct responses include actions of TSH, FSH, ACTH, oxytocin, and vasopressin in bone. It is important, in evaluating these new signaling mechanisms, to consider that the skeletal responses may or may not have similar mechanisms to the responses of the traditional endocrine targets and that the signals may vary in importance due to secondary endocrine and paracrine control. Nevertheless, the discovery of direct skeletal responses of pituitary hormones offers a new set of therapeutic opportunities.

Disclosures M.Z. consults for Merck, Novartis, and Roche and is a named inventor of a patent related to osteoclastic bone resorption filed by the Icahn School of Medicine at Mount Sinai (ISMMS). In the event the patent is licensed, he would be entitled to a share of any proceeds ISMMS receives from the licensee. All other authors have nothing to disclose.

References

Abe E, Marians RC, Yu W, Wu XB, Ando T, Li Y, Iqbal J, Eldeiry L, Rajendren G, Blair HC, Davies TF, Zaidi M. TSH is a negative regulator of skeletal remodeling. Cell. 2003;115:151–62.

Abrahamsen B, Jorgensen HL, Laulund AS, Nybo M, Brix TH, Hegedus L. Low serum thyrotropin level and duration of suppression as a predictor of major osteoporotic fractures-the OPENTHYRO register cohort. J Bone Miner Res. 2014;29:2040–50.

Adami S, Bianchi G, Brandi ML, Giannini S, Ortolani S, DiMunno O, Frediani B, Rossini M, group Bs, et al. Calcif Tissue Int. 2008;82:341–7.

Albagha OME, Natarajan R, Reid DM, Ralston SH. The D727E polymorphism of the human thyroid stimulating hormone receptor is associated with bone mineral density and bone loss in women from the UK. J Bone Miner Res. 2005;20(Suppl 1):S341.

Allan CM, Kalak R, Dunstan CR, McTavish KJ, Zhou H, Handelsman DJ, Seibel MJ. Follicle-stimulating hormone increases bone mass in female mice. Proc Natl Acad Sci U S A. 2010;107:22629–34.

Argiolas A, Collu M, Gessa GL, Melis MR, Serra G. The oxytocin antagonist d(CH2)5Tyr(Me)-Orn8-vasotocin inhibits male copulatory behaviour in rats. Eur J Pharmacol. 1988;149:389–92.

Bagriacik EU, Yaman M, Haznedar R, Sucak G, Delibasi T. TSH-induced gene expression involves regulation of self-renewal and differentiation-related genes in human bone marrow-derived mesenchymal stem cells. J Endocrinol. 2012;212:169–78.

Baliram R, Chow A, Huber AK, Collier L, Ali MR, Morshed SA, Latif R, Teixeira A, Merad M, Liu L, Sun L, Blair HC, Zaidi M, Davies TF. Thyroid and bone: macrophage-derived TSH-beta splice variant increases murine osteoblastogenesis. Endocrinology. 2013;154:4919–26.

Baliram R, Latif R, Berkowitz J, Frid S, Colaianni G, Sun L, Zaidi M, Davies TF. Thyroid-stimulating hormone induces a Wnt-dependent, feed-forward loop for osteoblastogenesis in embryonic stem cell cultures. Proc Natl Acad Sci U S A. 2011;108:16277–82.

Baliram R, Sun L, Cao J, Li J, Latif R, Huber AK, Yuen T, Blair HC, Zaidi M, Davies TF. Hyperthyroid-associated osteoporosis is exacerbated by the loss of TSH signaling. J Clin Invest. 2012;122:3737–41.

Baqi L, Payer J, Killinger Z, Hruzikova P, Cierny D, Susienkova K, Langer P. Thyrotropin versus thyroid hormone in regulating bone density and turnover in premenopausal women. Endocr Regul. 2010;44:57–63.

Baqi L, Payer J, Killinger Z, Susienkova K, Jackuliak P, Cierny D, Langer P. The level of TSH appeared favourable in maintaining bone mineral density in postmenopausal women. Endocr Regul. 2010;44:9–15.

Bauer DC, Ettinger B, Nevitt MC, Stone KL. Study of osteoporotic fractures research G: risk for fracture in women with low serum levels of thyroid-stimulating hormone. Ann Intern Med. 2001;134:561–8.

Blair HC, Robinson LJ, Sun L, Isales C, Davies TF, Zaidi M. Skeletal receptors for steroid-family regulating glycoprotein hormones: a multilevel, integrated physiological control system. Ann N Y Acad Sci. 2011;1240:26–31.

Blum MR, Bauer DC, Collet TH, Fink HA, Cappola AR, da Costa BR, Wirth CD, Peeters RP, Asvold BO, den Elzen WP, Luben RN, Imaizumi M, Bremner AP, Gogakos A, Eastell R, Kearney PM, Strotmeyer ES, Wallace ER, Hoff M, Ceresini G, Rivadeneira F, Uitterlinden AG, Stott DJ, Westendorp RG, Khaw KT, Langhammer A, Ferrucci L, Gussekloo J, Williams GR, Walsh JP, Juni P, Aujesky D, Rodondi N, Thyroid Studies C. Subclinical thyroid dysfunction and fracture risk: a meta-analysis. JAMA. 2015;313:2055–65.

Bogerd J, Granneman JC, Schulz RW, Vischer HF. Fish FSH receptors bind LH: how to make the human FSH receptor to be more fishy? Gen Comp Endocrinol. 2005;142:34–43.

Boutin A, Eliseeva E, Gershengorn MC. Neumann S: beta-Arrestin-1 mediates thyrotropin-enhanced osteoblast differentiation. FASEB J. 2014;28:3446–55.

Britto JM, Fenton AJ, Holloway WR, Nicholson GC. Osteoblasts mediate thyroid hormone stimulation of osteoclastic bone resorption. Endocrinology. 1994;134:169–76.

Cannon JG, Cortez-Cooper M, Meaders E, Stallings J, Haddow S, Kraj B, Sloan G, Mulloy A. Follicle-stimulating hormone, interleukin-1, and bone density in adult women. Am J Phys Regul Integr Comp Phys. 2010;298:R790–8.

Cannon JG, Kraj B, Sloan G. Follicle-stimulating hormone promotes RANK expression on human monocytes. Cytokine. 2011;53:141–4.

Cheung E, Tsang S, Bow C, Soong C, Yeung S, Loong C, Cheung CL, Kan A, Lo S, Tam S, Tang G, Kung A. Bone loss during menopausal transition among southern Chinese women. Maturitas. 2011;69:50–6.

Chin KY, Ima-Nirwana S, Mohamed IN, Aminuddin A, Johari MH, Ngah WZ. Thyroid-stimulating hormone is significantly associated with bone health status in men. Int J Med Sci. 2013;10:857–63.

Christy AL, D'Souza V, Babu RP, Takodara S, Manjrekar P, Hegde A, Rukmini MS. Utility of C-terminal telopeptide in evaluating levothyroxine replacement therapy-induced bone loss. Biomark Insights. 2014;9:1–6.

Colaianni G, Di Benedetto A, Zhu LL, Tamma R, Li J, Greco G, Peng Y, Dell'Endice S, Zhu G, Cuscito C, Grano M, Colucci S, Iqbal J, Yuen T, Sun L, Zaidi M, Zallone A. Regulated production of the pituitary hormone oxytocin from murine and human osteoblasts. Biochem Biophys Res Commun. 2011;411:512–5.

Colucci S, Colaianni G, Mori G, Grano M, Zallone A. Human osteoclasts express oxytocin receptor. Biochem Biophys Res Commun. 2002;297:442–5.

Copland JA, Ives KL, Simmons DJ, Soloff MS. Functional oxytocin receptors discovered in human osteoblasts. Endocrinology. 1999;140:4371–4.

Coss D, Yang L, Kuo CB, Xu X, Luben RA, Walker AM. Effects of prolactin on osteoblast alkaline phosphatase and bone formation in the developing rat. Am J Physiol Endocrinol Metab. 2000;279:E1216–25.

De Jesus K, Wang X, Liu JL. A general IGF-I overexpression effectively rescued somatic growth and bone deficiency in mice caused by growth hormone receptor knockout. Growth Factors. 2009;27:438–47.

van der Deure WM, Uitterlinden AG, Hofman A, Rivadeneira F, Pols HA, Peeters RP, Visser TJ. Effects of serum TSH and FT4 levels and the TSHR-Asp727Glu polymorphism on bone: the Rotterdam study. Clin Endocrinol. 2008;68:175–81.

Devleta B, Adem B, Senada S. Hypergonadotropic amenorrhea and bone density: new approach to an old problem. J Bone Miner Metab. 2004;22:360–4.

Drake MT, McCready LK, Hoey KA, Atkinson EJ, Khosla S. Effects of suppression of follicle-stimulating hormone secretion on bone resorption markers in postmenopausal women. J Clin Endocrinol Metab. 2010;95:5063–8.

Dumic-Cule I, Draca N, Luetic AT, Jezek D, Rogic D, Grgurevic L, Vukicevic S. TSH prevents bone resorption and with calcitriol synergistically stimulates bone formation in rats with low levels of calciotropic hormones. Horm Metab Res. 2014;46:305–12.

Eberhardt AW, Yeager-Jones A, Blair HC. Regional trabecular bone matrix degeneration and osteocyte death in femora of glucocorticoid- treated rabbits. Endocrinology. 2001;142:1333–40.

Elabd SK, Sabry I, Hassan WB, Nour H, Zaky K. Possible neuroendocrine role for oxytocin in bone remodeling. Endocr Regul. 2007;41:131–41.

Ferguson JN, Young LJ, Hearn EF, Matzuk MM, Insel TR, Winslow JT. Social amnesia in mice lacking the oxytocin gene. Nat Genet. 2000;25:284–8.

Flynn RW, Bonellie SR, Jung RT, MacDonald TM, Morris AD, Leese GP. Serum thyroid-stimulating hormone concentration and morbidity from cardiovascular disease and fractures in patients on long-term thyroxine therapy. J Clin Endocrinol Metab. 2010;95:186–93.

Fritton JC, Emerton KB, Sun H, Kawashima Y, Mejia W, Wu Y, Rosen CJ, Panus D, Bouxsein M, Majeska RJ, Schaffler MB, Yakar S. Growth hormone protects against ovariectomy-induced bone loss in states of low circulating insulin-like growth factor (IGF-1). J Bone Miner Res. 2010;25:235–46.

Frohlich E, Wahl R. Mechanisms in endocrinology: impact of isolated TSH levels in and out of normal range on different tissues. Eur J Endocrinol. 2016;174:R29–41.

Gallagher CM, Moonga BS, Kovach JS. Cadmium, follicle-stimulating hormone, and effects on bone in women age 42–60 years, NHANES III. Environ Res. 2010;110:105–11.

Garcia-Martin A, Reyes-Garcia R, Garcia-Castro JM, Rozas-Moreno P, Escobar-Jimenez F, Munoz-Torres M. Role of serum FSH measurement on bone resorption in postmenopausal women. Endocrine. 2012;41:302–8.

Gertz ER, Silverman NE, Wise KS, Hanson KB, Alekel DL, Stewart JW, Perry CD, Bhupathiraju SN, Kohut ML, Van Loan MD. Contribution of serum inflammatory markers to changes in bone mineral content and density in postmenopausal women: a 1-year investigation. J Clin Densitom. 2010;13:277–82.

Gourlay ML, Preisser JS, Hammett-Stabler CA, Renner JB, Rubin J. Follicle-stimulating hormone and bioavailable estradiol are less important than weight and race in determining bone density in younger postmenopausal women. Osteoporos Int. 2011;22:2699–708.

Gourlay ML, Specker BL, Li C, Hammett-Stabler CA, Renner JB, Rubin JE. Follicle-stimulating hormone is independently associated with lean mass but not BMD in younger postmenopausal women. Bone. 2012;50:311–6.

Grimnes G, Emaus N, Joakimsen RM, Figenschau Y, Jorde R. The relationship between serum TSH and bone mineral density in men and postmenopausal women: the Tromso study. Thyroid. 2008;18:1147–55.

Guicheux J, Heymann D, Rousselle AV, Gouin F, Pilet P, Yamada S, Daculsi G. Growth hormone stimulatory effects on osteoclastic resorption are partly mediated by insulin-like growth factor I: an in vitro study. Bone. 1998;22:25–31.

Harbour DV, Kruger TE, Coppenhaver D, Smith EM, Meyer WJ 3rd. Differential expression and regulation of thyrotropin (TSH) in T cell lines. Mol Cell Endocrinol. 1989;64:229–41.

Hase H, Ando T, Eldeiry L, Brebene A, Peng Y, Liu L, Amano H, Davies TF, Sun L, Zaidi M, Abe E. TNFalpha mediates the skeletal effects of thyroid-stimulating hormone. Proc Natl Acad Sci U S A. 2006;103:12849–54.

Heemstra KA, van der Deure WM, Peeters RP, Hamdy NA, Stokkel MP, Corssmit EP, Romijn JA, Visser TJ, Smit JW. Thyroid hormone independent associations between serum TSH levels and indicators of bone turnover in cured patients with differentiated thyroid carcinoma. Eur J Endocrinol. 2008;159:69–76.

Hou P, Sato T, Hofstetter W, Foged NT. Identification and characterization of the insulin-like growth factor I receptor in mature rabbit osteoclasts. J Bone Miner Res. 1997;12:534–40.

Insel TR, Harbaugh CR. Lesions of the hypothalamic paraventricular nucleus disrupt the initiation of maternal behavior. Physiol Behav. 1989;45:1033–41.

Iqbal J, Sun L, Kumar TR, Blair HC, Zaidi M. Follicle-stimulating hormone stimulates TNF production from immune cells to enhance osteoblast and osteoclast formation. Proc Natl Acad Sci U S A. 2006;103:14925–30.

Isales CM, Zaidi M, Blair HC. ACTH is a novel regulator of bone mass. Ann N Y Acad Sci. 2010;1192:110–6.

Kawai H, Furuhashi M, Suganuma N. Serum follicle-stimulating hormone level is a predictor of bone mineral density in patients with hormone replacement therapy. Arch Gynecol Obstet. 2004;269:192–5.

Kim S, Jung J, Jung JH, Kim SK, Kim RB, Hahm JR. Risk factors of bone mass loss at the lumbar spine: a longitudinal study in healthy Korean pre- and Perimenopausal women older than 40 years. PLoS One. 2015;10:e0136283.

Kim DJ, Khang YH, Koh JM, Shong YK, Kim GS. Low normal TSH levels are associated with low bone mineral density in healthy postmenopausal women. Clin Endocrinol. 2006;64:86–90.

Kim BJ, Lee SH, Bae SJ, Kim HK, Choe JW, Kim HY, Koh JM, Kim GS. The association between serum thyrotropin (TSH) levels and bone mineral density in healthy euthyroid men. Clin Endocrinol. 2010;73:396–403.

Klein JR, Wang HC. Characterization of a novel set of resident intrathyroidal bone marrow-derived hematopoietic cells: potential for immune-endocrine interactions in thyroid homeostasis. J Exp Biol. 2004;207:55–65.

Kobayashi T, Andersen O. The gonadotropin receptors FSH-R and LH-R of Atlantic halibut (*Hippoglossus hippoglossus*), 1: isolation of multiple transcripts encoding full-length and truncated variants of FSH-R. Gen Comp Endocrinol. 2008;156:584–94.

Krishnamra N, Seemoung J. Effects of acute and long-term administration of prolactin on bone 45Ca uptake, calcium deposit, and calcium resorption in weaned, young, and mature rats. Can J Physiol Pharmacol. 1996;74:1157–65.

Kumar RS, Ijiri S, Kight K, Swanson P, Dittman A, Alok D, Zohar Y, Trant JM. Cloning and functional expression of a thyrotropin receptor from the gonads of a vertebrate (bony fish): potential thyroid-independent role for thyrotropin in reproduction. Mol Cell Endocrinol. 2000;167:1–9.

Kumar RS, Ijiri S, Trant JM. Molecular biology of the channel catfish gonadotropin receptors: 2. Complementary DNA cloning, functional expression, and seasonal gene expression of the follicle-stimulating hormone receptor. Biol Reprod. 2001;65:710–7.

La Vignera S, Vicari E, Tumino S, Ciotta L, Condorelli R, Vicari LO, Calogero AE. L-thyroxin treatment and post-menopausal osteoporosis: relevance of the risk profile present in clinical history. Minerva Ginecol. 2008;60:475–84.

Leader A, Ayzenfeld RH, Lishner M, Cohen E, Segev D, Hermoni D. Thyrotropin levels within the lower normal range are associated with an increased risk of hip fractures in euthyroid women, but not men, over the age of 65 years. J Clin Endocrinol Metab. 2014;99:2665–73.

Liu S, Cheng Y, Fan M, Chen D, Bian Z. FSH aggravates periodontitis-related bone loss in ovariectomized rats. J Dent Res. 2010;89:366–71.

Liu S, Cheng Y, Xu W, Bian Z. Protective effects of follicle-stimulating hormone inhibitor on alveolar bone loss resulting from experimental periapical lesions in ovariectomized rats. J Endod. 2010;36:658–63.

Lotinun S, Limlomwongse L, Krishnamra N. The study of a physiological significance of prolactin in the regulation of calcium metabolism during pregnancy and lactation in rats. Can J Physiol Pharmacol. 1998;76:218–28.

Lotinun S, Limlomwongse L, Sirikulchayanonta V, Krishnamra N. Bone calcium turnover, formation, and resorption in bromocriptine- and prolactin-treated lactating rats. Endocrine. 2003;20:163–70.
Ma R, Latif R, Davies TF. Thyrotropin-independent induction of thyroid endoderm from embryonic stem cells by activin A. Endocrinology. 2009;150:1970–5.
Ma R, Morshed S, Latif R, Zaidi M, Davies TF. The influence of thyroid-stimulating hormone and thyroid-stimulating hormone receptor antibodies on osteoclastogenesis. Thyroid. 2011;21:897–906.
Mankin HJ. Nontraumatic necrosis of bone (osteonecrosis). N Engl J Med. 1992;326:1473–9.
Mantella RC, Vollmer RR, Li X, Amico JA. Female oxytocin-deficient mice display enhanced anxiety-related behavior. Endocrinology. 2003;144:2291–6.
Martini G, Gennari L, De Paola V, Pilli T, Salvadori S, Merlotti D, Valleggi F, Campagna S, Franci B, Avanzati A, Nuti R, Pacini F. The effects of recombinant TSH on bone turnover markers and serum osteoprotegerin and RANKL levels. Thyroid. 2008;18:455–60.
Mazziotti G, Porcelli T, Patelli I, Vescovi PP, Giustina A. Serum TSH values and risk of vertebral fractures in euthyroid post-menopausal women with low bone mineral density. Bone. 2010;46:747–51.
Mazziotti G, Sorvillo F, Piscopo M, Cioffi M, Pilla P, Biondi B, Iorio S, Giustina A, Amato G, Carella C. Recombinant human TSH modulates in vivo C-telopeptides of type-1 collagen and bone alkaline phosphatase, but not osteoprotegerin production in postmenopausal women monitored for differentiated thyroid carcinoma. J Bone Miner Res. 2005;20:480–6.
Meaney AM, Smith S, Howes OD, O'Brien M, Murray RM, O'Keane V. Effects of long-term prolactin-raising antipsychotic medication on bone mineral density in patients with schizophrenia. Br J Psychiatry. 2004;184:503–8.
Menagh PJ, Turner RT, Jump DB, Wong CP, Lowry MB, Yakar S, Rosen CJ, Iwaniec UT. Growth hormone regulates the balance between bone formation and bone marrow adiposity. J Bone Miner Res. 2010;25:757–68.
Mikosch P, Kerschan-Schindl K, Woloszczuk W, Stettner H, Kudlacek S, Kresnik E, Gallowitsch HJ, Lind P, Pietschmann P. High cathepsin K levels in men with differentiated thyroid cancer on suppressive L-thyroxine therapy. Thyroid. 2008;18:27–33.
Moon JH, Jung KY, Kim KM, Choi SH, Lim S, Park YJ, Park DJ, Jang HC. The effect of thyroid stimulating hormone suppressive therapy on bone geometry in the hip area of patients with differentiated thyroid carcinoma. Bone. 2015;83:104–10.
Morris MS. The association between serum thyroid-stimulating hormone in its reference range and bone status in postmenopausal American women. Bone. 2007;40:1128–34.
Naylor KE, Iqbal P, Fledelius C, Fraser RB, Eastell R. The effect of pregnancy on bone density and bone turnover. J Bone Miner Res. 2000;15:129–37.
Nishimori K, Young LJ, Guo Q, Wang Z, Insel TR, Matzuk MM. Oxytocin is required for nursing but is not essential for parturition or reproductive behavior. Proc Natl Acad Sci U S A. 1996;93:11699–704.
Noh HM, Park YS, Lee J, Lee W. A cross-sectional study to examine the correlation between serum TSH levels and the osteoporosis of the lumbar spine in healthy women with normal thyroid function. Osteoporos Int. 2015;26:997–1003.
Novack DV. TSH, the bone suppressing hormone. Cell. 2003;115:129–30.
Pallinger E, Csaba G. A hormone map of human immune cells showing the presence of adrenocorticotropic hormone, triiodothyronine and endorphin in immunophenotyped white blood cells. Immunology. 2008;123:584–9.
Podfigurna-Stopa A, Pludowski P, Jaworski M, Lorenc R, Genazzani AR, Meczekalski B. Skeletal status and body composition in young women with functional hypothalamic amenorrhea. Gynecol Endocrinol. 2012;28:299–304.
Rendina D, Gianfrancesco F, De Filippo G, Merlotti D, Esposito T, Mingione A, Nuti R, Strazzullo P, Mossetti G, Gennari L. FSHR gene polymorphisms influence bone mineral density and bone turnover in postmenopausal women. Eur J Endocrinol. 2010;163:165–72.

Ritter V, Thuering B, Saint Mezard P, Luong-Nguyen NH, Seltenmeyer Y, Junker U, Fournier B, Susa M, Morvan F. Follicle-stimulating hormone does not impact male bone mass in vivo or human male osteoclasts in vitro. Calcif Tissue Int. 2008;82:383–91.

Robinson LJ, Tourkova I, Wang Y, Sharrow AC, Landau MS, Yaroslavskiy BB, Sun L, Zaidi M, Blair HC. FSH-receptor isoforms and FSH-dependent gene transcription in human monocytes and osteoclasts. Biochem Biophys Res Commun. 2010;394:12–7.

Rubin J, Ackert-Bicknell CL, Zhu L, Fan X, Murphy TC, Nanes MS, Marcus R, Holloway L, Beamer WG, Rosen CJ. IGF-I regulates osteoprotegerin (OPG) and receptor activator of nuclear factor-kappaB ligand in vitro and OPG in vivo. J Clin Endocrinol Metab. 2002;87:4273–9.

Sampath TK, Simic P, Sendak R, Draca N, Bowe AE, O'Brien S, Schiavi SC, McPherson JM, Vukicevic S. Thyroid-stimulating hormone restores bone volume, microarchitecture, and strength in aged ovariectomized rats. J Bone Miner Res. 2007;22:849–59.

Sclafani A, Rinaman L, Vollmer RR, Amico JA. Oxytocin knockout mice demonstrate enhanced intake of sweet and nonsweet carbohydrate solutions. Am J Phys Regul Integr Comp Phys. 2007;292:R1828–33.

Seriwatanachai D, Thongchote K, Charoenphandhu N, Pandaranandaka J, Tudpor K, Teerapornpuntakit J, Suthiphongchai T, Krishnamra N. Prolactin directly enhances bone turnover by raising osteoblast-expressed receptor activator of nuclear factor kappaB ligand/osteoprotegerin ratio. Bone. 2008;42:535–46.

Seriwatanachai D, Charoenphandhu N, Suthiphongchai T, Krishnamra N. Prolactin decreases the expression ratio of receptor activator of nuclear factor kappaB ligand/osteoprotegerin in human fetal osteoblast cells. Cell Biol Int. 2008;32:1126–35.

Sirianni R, Rehman KS, Carr BR, Parker CR Jr, Rainey WE. Corticotropin-releasing hormone directly stimulates cortisol and the cortisol biosynthetic pathway in human fetal adrenal cells. J Clin Endocrinol Metab. 2005;90:279–85.

Smith EM, Phan M, Kruger TE, Coppenhaver DH, Blalock JE. Human lymphocyte production of immunoreactive thyrotropin. Proc Natl Acad Sci U S A. 1983;80:6010–3.

Sowers M, Eyre D, Hollis BW, Randolph JF, Shapiro B, Jannausch ML, Crutchfield M. Biochemical markers of bone turnover in lactating and nonlactating postpartum women. J Clin Endocrinol Metab. 1995;80:2210–6.

Sowers MR, Greendale GA, Bondarenko I, Finkelstein JS, Cauley JA, Neer RM, Ettinger B. Endogenous hormones and bone turnover markers in pre- and perimenopausal women: SWAN. Osteoporos Int. 2003;14:191–7.

Sun L, Peng Y, Sharrow AC, Iqbal J, Zhang Z, Papachristou DJ, Zaidi S, Zhu LL, Yaroslavskiy BB, Zhou H, Zallone A, Sairam MR, Kumar TR, Bo W, Braun J, Cardoso-Landa L, Schaffler MB, Moonga BS, Blair HC, Zaidi M. FSH directly regulates bone mass. Cell. 2006;125:247–60.

Sun L, Tamma R, Yuen T, Colaianni G, Ji Y, Cuscito C, Bailey J, Dhawan S, Lu P, Calvano CD, Zhu LL, Zambonin CG, Di Benedetto A, Stachnik A, Liu P, Grano M, Colucci S, Davies TF, New MI, Zallone A, Zaidi M. Functions of vasopressin and oxytocin in bone mass regulation. Proc Natl Acad Sci U S A. 2016;113:164–9.

Sun L, Vukicevic S, Baliram R, Yang G, Sendak R, McPherson J, Zhu LL, Iqbal J, Latif R, Natrajan A, Arabi A, Yamoah K, Moonga BS, Gabet Y, Davies TF, Bab I, Abe E, Sampath K, Zaidi M. Intermittent recombinant TSH injections prevent ovariectomy-induced bone loss. Proc Natl Acad Sci U S A. 2008;105:4289–94.

Sun L, Zhang Z, Zhu LL, Peng Y, Liu X, Li J, Agrawal M, Robinson LJ, Iqbal J, Blair HC, Zaidi M. Further evidence for direct pro-resorptive actions of FSH. Biochem Biophys Res Commun. 2010;394:6–11.

Sun L, Zhu LL, Lu P, Yuen T, Li J, Ma R, Baliram R, Moonga SS, Liu P, Zallone A, New MI, Davies TF, Zaidi M. Genetic confirmation for a central role for TNFalpha in the direct action of thyroid stimulating hormone on the skeleton. Proc Natl Acad Sci U S A. 2013;110:9891–6.

Svare A, Nilsen TI, Bjoro T, Forsmo S, Schei B, Langhammer A. Hyperthyroid levels of TSH correlate with low bone mineral density: the HUNT 2 study. Eur J Endocrinol. 2009;161:779–86.

Tamma R, Colaianni G, Zhu LL, DiBenedetto A, Greco G, Montemurro G, Patano N, Strippoli M, Vergari R, Mancini L, Colucci S, Grano M, Faccio R, Liu X, Li J, Usmani S, Bachar M, Bab I, Nishimori K, Young LJ, Buettner C, Iqbal J, Sun L, Zaidi M, Zallone A. Oxytocin is an anabolic bone hormone. Proc Natl Acad Sci U S A. 2009;106:7149–54.

Tamma R, Sun L, Cuscito C, Lu P, Corcelli M, Li J, Colaianni G, Moonga SS, Di Benedetto A, Grano M, Colucci S, Yuen T, New MI, Zallone A, Zaidi M. Regulation of bone remodeling by vasopressin explains the bone loss in hyponatremia. Proc Natl Acad Sci U S A. 2013;110:18644–9.

Taylor PN, Razvi S, Pearce SH, Dayan CM. Clinical review: a review of the clinical consequences of variation in thyroid function within the reference range. J Clin Endocrinol Metab. 2013;98:3562–71.

Tournis S, Antoniou JD, Liakou CG, Christodoulou J, Papakitsou E, Galanos A, Makris K, Marketos H, Nikopoulou S, Tzavara I, Triantafyllopoulos IK, Dontas I, Papaioannou N, Lyritis GP, Alevizaki M. Volumetric bone mineral density and bone geometry assessed by peripheral quantitative computed tomography in women with differentiated thyroid cancer under TSH suppression. Clin Endocrinol. 2015;82:197–204.

Vibede N, Hauser F, Williamson M, Grimmelikhuijzen CJ. Genomic organization of a receptor from sea anemones, structurally and evolutionarily related to glycoprotein hormone receptors from mammals. Biochem Biophys Res Commun. 1998;252:497–501.

Vincent BH, Montufar-Solis D, Teng BB, Amendt BA, Schaefer J, Klein JR. Bone marrow cells produce a novel TSHbeta splice variant that is upregulated in the thyroid following systemic virus infection. Genes Immun. 2009;10:18–26.

Wang G, Zhang CQ, Sun Y, Feng Y, Chen SB, Cheng XG, Zeng BF. Changes in femoral head blood supply and vascular endothelial growth factor in rabbits with steroid-induced osteonecrosis. J Int Med Res. 2010;38:1060–9.

Wu Y, Torchia J, Yao W, Lane NE, Lanier LL, Nakamura MC, Humphrey MB. Bone microenvironment specific roles of ITAM adapter signaling during bone remodeling induced by acute estrogen-deficiency. PLoS One. 2007;2:e586.

Wu XY, Wu XP, Xie H, Zhang H, Peng YQ, Yuan LQ, Su X, Luo XH, Liao EY. Age-related changes in biochemical markers of bone turnover and gonadotropin levels and their relationship among Chinese adult women. Osteoporos Int. 2010;21:275–85.

Wysolmerski JJ. The evolutionary origins of maternal calcium and bone metabolism during lactation. J Mammary Gland Biol Neoplasia. 2002;7:267–76.

Xu ZR, Wang AH, Wu XP, Zhang H, Sheng ZF, Wu XY, Xie H, Luo XH, Liao EY. Relationship of age-related concentrations of serum FSH and LH with bone mineral density, prevalence of osteoporosis in native Chinese women. Clin Chim Acta. 2009;400:8–13.

Yakar S, Rosen CJ, Beamer WG, Ackert-Bicknell CL, Wu Y, Liu JL, Ooi GT, Setser J, Frystyk J, Boisclair YR, LeRoith D. Circulating levels of IGF-1 directly regulate bone growth and density. J Clin Invest. 2002;110:771–81.

Yamoah K, Brebene A, Baliram R, Inagaki K, Dolios G, Arabi A, Majeed R, Amano H, Wang R, Yanagisawa R, Abe E. High-mobility group box proteins modulate tumor necrosis factor-alpha expression in osteoclastogenesis via a novel deoxyribonucleic acid sequence. Mol Endocrinol. 2008;22:1141–53.

Young WS 3rd, Shepard E, DeVries AC, Zimmer A, LaMarca ME, Ginns EI, Amico J, Nelson RJ, Hennighausen L, Wagner KU. Targeted reduction of oxytocin expression provides insights into its physiological roles. Adv Exp Med Biol. 1998;449:231–40.

Zaidi M, Sun L, Davies TF, Abe E. Low TSH triggers bone loss: fact or fiction? Thyroid. 2006;16:1075–6.

Zaidi M, Sun L, Robinson LJ, Tourkova IL, Liu L, Wang Y, Zhu LL, Liu X, Li J, Peng Y, Yang G, Shi X, Levine A, Iqbal J, Yaroslavskiy BB, Isales C, Blair HC. ACTH protects against glucocorticoid-induced osteonecrosis of bone. Proc Natl Acad Sci U S A. 2010;107:8782–7.

Zaidi M, Turner CH, Canalis E, Pacifici R, Sun L, Iqbal J, Guo XE, Silverman S, Epstein S, Rosen CJ. Bone loss or lost bone: rationale and recommendations for the diagnosis and treatment of early postmenopausal bone loss. Curr Osteoporos Rep. 2009;7:118–26.

Zhang W, Zhang Y, Liu Y, Wang J, Gao L, Yu C, Yan H, Zhao J, Xu J. Thyroid-stimulating hormone maintains bone mass and strength by suppressing osteoclast differentiation. J Biomech. 2014;47:1307–14.

Zhu LL, Blair H, Cao J, Yuen T, Latif R, Guo L, Tourkova IL, Li J, Davies TF, Sun L, Bian Z, Rosen C, Zallone A, New MI, Zaidi M. Blocking antibody to the beta-subunit of FSH prevents bone loss by inhibiting bone resorption and stimulating bone synthesis. Proc Natl Acad Sci U S A. 2012;109:14574–9.

Zhu LL, Tourkova I, Yuen T, Robinson LJ, Bian Z, Zaidi M, Blair HC. Blocking FSH action attenuates osteoclastogenesis. Biochem Biophys Res Commun. 2012;422:54–8.

Zofkova I, Hill M. Biochemical markers of bone remodeling correlate negatively with circulating TSH in postmenopausal women. Endocr Regul. 2008;42:121–7.

Chapter 11
Kidney–Bone: Interaction

Olena V. Andrukhova and Reinhold G. Erben

Abstract Kidney–bone interaction is a key controlling mechanism for the physiologic electrolyte balance in the body important for many physiological functions from the intracellular cell signaling to the energy metabolism and skeletal mineralization. Kidney, as a primary avenue for the regulating extracellular fluid and electrolyte homeostasis, participates to the maintenance of the continuous bone turnover and mineralization. At the same time, bone is the permanent physiological mineral store which could be activated for the normalizing electrolyte shift in the blood. This functional co-regulation in the kidney–bone interaction axis is accomplished by the tight feedback regulation between both organs and requires the input from the nervous and endocrine systems. The impact on the mineral homeostasis is achieved through the action of the (1) kidney-derived factors like vitamin D, acid–base metabolites, as well as blood electrolyte status and (2) bone-derived factors including FGF23, MEPE, and osteocalcin. Dysfunction in any of these factors alone or in combination can disrupt a multitude of the mineral balance in the cross-organ communication between the kidneys, intestine, bones, and parathyroid gland. Indeed, two essential bone-acting hormones vitamin D and parathyroid hormone (PTH) influence mineral balances by affecting the rate of intestinal absorption, renal reabsorption, and bone resorption and closing the key endocrine regulatory loop in mineral homeostasis.

Keywords Kidney–bone • Mineral homeostasis • Energy metabolism • Mineralization • Vitamin D • FGF23 • MEPE • Osteocalcin • Calcitonin • Acid–base • Sodium • Magnesium

O.V. Andrukhova, PhD (✉) • R.G. Erben, MD, DVM
Department of Biomedical Sciences, University of Veterinary Medicine Vienna, Veterinärplatz 1, 1210 Vienna, Austria
e-mail: Olena.Andrukhova@vetmeduni.ac.at; Reinhold.Erben@vetmeduni.ac.at

S.Y. Smith et al. (eds.), *Bone Toxicology*, Molecular and Integrative Toxicology,
DOI 10.1007/978-3-319-56192-9_11

11.1 Renal Regulation of Electrolyte Homeostasis, Extracellular Volume, and Acid–Base Balance

The kidney contains two populations of nephrons, cortical (or superficial), and juxtamedullary (or deep) with the same basic structures, but different locations and lengths of specific segments. Blood enters the capillary from the afferent arteriole and exits the capillary by the efferent arteriole. Blood is filtered at the glomerulus, a capillary system, from which an ultrafiltrate of plasma enters into Bowman's space. The glomerular filtration barrier consists of capillaries fenestrated endothelium, basement membrane, and epithelial cells (podocytes), preventing filtration of blood cells, proteins, and most macromolecules into the glomerular ultrafiltrate. Mesangial cells in the glomerulus are able to contract, decreasing the surface area for filtration.

Another important structural and functional aspect is the juxtaglomerular apparatus, consisting of the macula densa, an area in the distal convoluted tubular epithelium, and specialized renin-producing cells in the afferent arteriole. The macula densa cells of the juxtaglomerular apparatus are important in sensing tubular fluid flow and sodium delivery to the distal nephron. Since the macula densa is located in the direct proximity to the afferent arteriole, it can regulate the release of the kidney-derived hormone renin which in turn regulates renal plasma flow and glomerular filtration rate (GFR) (autoregulation).

11.1.1 Renal Transport Processes

Plasma filtered into Bowman's space enters the proximal tubule where the majority of the fluid and the solutes are reabsorbed through passive transport. In the following, fine-tuning of the reabsorptive processes takes place at distal tubular sites. The proximal tubule is composed of S1, S2, and S3 segments, which differ in the depth of the brush border and the amount of mitochondria in the cells. As the filtrate is reabsorbed, and less is present in the tubule in subsequent segments, the deeper brush border increases surface area, which enhances continued reabsorption. The high number of mitochondria in S1 is consistent with a high rate of active transport in that segment. From S1 to S3 segments, the brush border becomes progressively larger and the number of mitochondria decreases.

11.1.1.1 Sodium-Driven Solute Transport

Sodium is the major extracellular cation necessary for maintenance of general fluid and electrolyte homeostasis. Sodium is also a major driving force for the renal reabsorption of fluid, electrolytes, and a variety of nutrients. As sodium transporters carry sodium and other solutes, they generate the driving force for water reabsorption.

Approximately 65–70% of the water in tubular fluid is reabsorbed from the proximal tubule primarily following sodium reabsorption. In the proximal convoluted tubule (S1 and S2), the flow occurs by secondary active sodium cotransport with glucose, amino acids, phosphate, and organic acids. The proximal tubule also has Na^+/H^+ antiporters, which allow H^+ secretion into the proximal renal tubular fluid.

In the proximal convoluted tubule (S3), the electrochemical gradient facilitates chloride reabsorption paracellularly or through antiporters, resulting in apical secretion of anions such as OH^-, HCO_3^-, SO_4^-, and oxalate. The thin descending limb of Henle segment is important for tubular fluid concentration and is impermeable to sodium but permeable to water and is sensitive for arginine vasopressin (AVP) action. The thick ascending limb of Henle (TALH) is impermeable to water, but has specialized apical Na^+–K^+–$2Cl^-$ 1 and 2 cotransporters (NKCC1/2) and Na^+/H^+ antiporters facilitating reabsorption of electrolytes. In the early distal tubule, the Na^+–Cl^- cotransporters are targets for the sodium reabsorption regulating hormone aldosterone, resulting in increased water retention. Na^+–Cl^- cotransporters can be inhibited by thiazide diuretics. The epithelial sodium channel (ENaC) is another aldosterone-sensitive channel which can be blocked by the diuretic amiloride. The collecting tubule is also a place for aldosterone signaling, acting on the ENaC and NKCC1/2 channels located there (see Fig. 11.1).

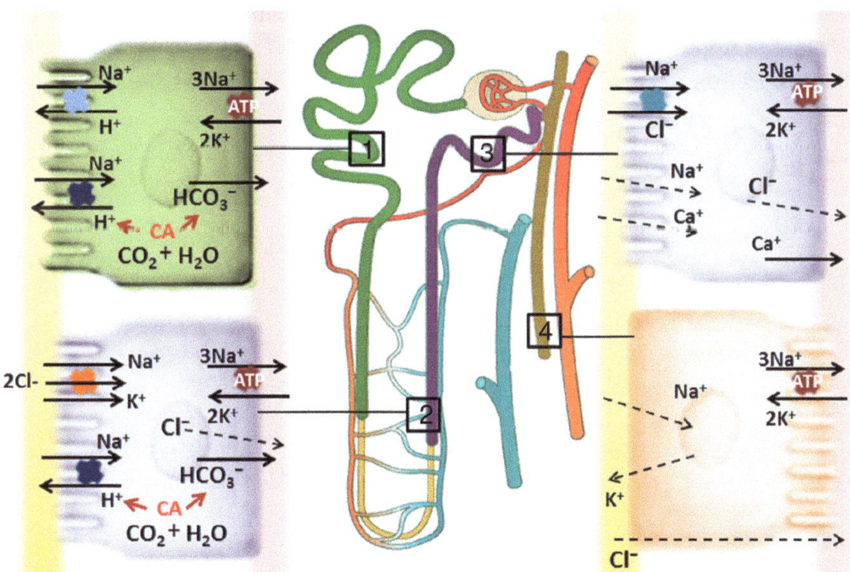

Fig. 11.1 Nephron sites of sodium reabsorption. Sodium reabsorption is critical for fluid and electrolyte homeostasis. Reabsorption of the filtered sodium load in the proximal tubule is 67% (1), in the loop of Henle 25% (2), in the distal tubule 4% (3), and in the collecting duct 4% (4). Sodium reabsorption can be stimulated in the proximal tubule by angiotensin II and in the distal tubule and collecting duct by aldosterone action. Sodium reabsorption in the loop of Henle can be stimulated by sympathetic nerves. The gradient for sodium transport into the cell is maintained by the basolateral Na^+/K^+ ATPase

11.1.1.2 Bicarbonate and Potassium Handling

Plasma *bicarbonate* is necessary for acid–base balance. HCO_3^- is effectively reabsorbed through a cation exchange and active H^+ pumps. In the tubular lumen, filtered HCO_3^- and secreted H^+ form CO_2 and H_2O, catalyzed by carbonic anhydrase. CO_2 diffuses into the cell. Intracellular carbonic anhydrase converts it back to carbonic acid and then HCO_3^- is transported out of the cell via basolateral HCO_3^-/Cl^- exchangers or Na^+–HCO_3^- cotransporters, and H^+ is secreted back into the tubular lumen and can be excreted or reused (see Fig. 11.2).

Potassium is a major electrolyte important to overall homeostasis. Potassium is pumped into cells via Na^+/K^+ ATPase, which is stimulated by insulin and epinephrine. Potassium handling in the proximal tubule occurs by paracellular diffusion down the electrochemical gradient. In the thick ascending limb of Henle Na^+–K^+–$2Cl^-$, cotransporters generate sodium and chloride gradients to facilitate transport of K^+. Potassium can be secreted into the late distal tubules and in collecting ducts via aldosterone-sensitive K^+ channels. K^+ is also secreted into the collecting ducts by α-intercalated cells, in exchange for H^+. Renal potassium handling is influenced by dietary potassium intake. Intestinal K^+ absorption regulates plasma K^+ concentration. When dietary K^+ intake varies, K^+ secretion from the collecting duct in α-intercalated and in principal cells is mutually regulated to normalize total K^+ reabsorption.

In addition to responding to increased plasma K^+ concentrations, aldosterone is also released in response to decreased plasma volume, as part of the renin–angiotensin–aldosterone system (RAAS). Aldosterone increases the Na^+/K^+ ATPase, Na^+/H^+ anti-

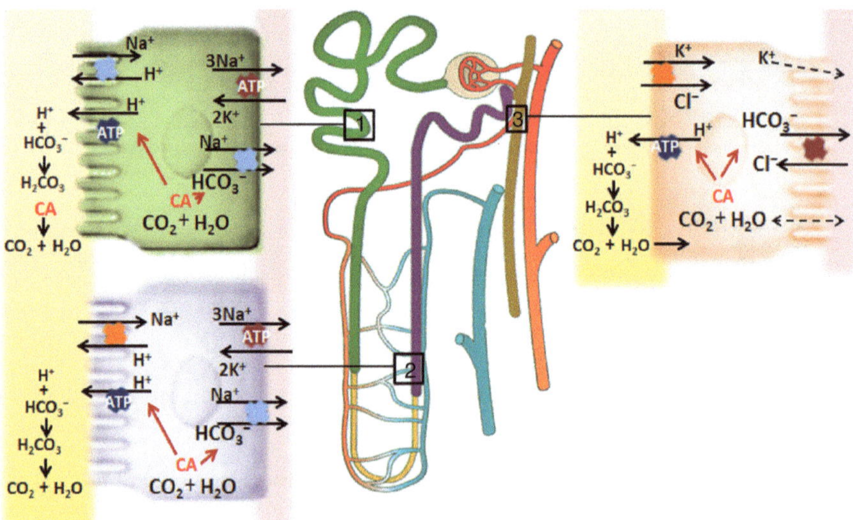

Fig. 11.2 Renal HCO_3^- reabsorption. Bicarbonate is freely filtrated at the glomerulus and is reabsorbed through a process involving secretion of the H^+. Bicarbonate is reabsorbed in the proximal tubule by 80% (1), loop of Henle 5% (2), and collecting duct 5% (3). Under physiological conditions 100% of the bicarbonate is reabsorbed

porters, and Na^+–Cl^- cotransporters in the late distal tubules and collecting duct, independent of plasma potassium levels. This increases K^+ secretion into the renal tubules. For maintaining the acid–base status, H^+ can be secreted from the principal cells or retained in α-intercalated cells into the collecting ducts using the K^+ gradient.

11.1.1.3 Calcium and Phosphate Transport

Calcium and phosphate are important during fetal and postnatal development for bone and tissue growth and continue to be important in the adult for bone health. The kidneys control plasma levels of calcium and phosphate by altering their rate of reabsorption. Most of the calcium and phosphate in the body (99% and 85%, respectively) is found in bone matrix. Renal phosphate and calcium reabsorption are regulated by PTH, the vitamin D hormone, and FGF23.

About 40% of plasma Ca^{2+} is bound to proteins, leaving 60% free for filtration at the glomeruli. The kidneys reabsorb ~99% of the filtered Ca^{2+} at sites throughout the nephron. In the proximal tubule and thick ascending limb of Henle, Ca^{2+} is reabsorbed paracellularly in parallel with Na^+ reabsorption. Since Ca^{2+} follows sodium reabsorption, reductions in sodium reabsorption (such as those induced by NKCC inhibition) will also reduce Ca^{2+} reabsorption. The distal tubule is the site of hormonally controlled Ca^{2+} reabsorption. Transport is transcellular and is facilitated by the high electrochemical gradient from the tubule into the cell. Once in the cell, transport into the interstitium is through active Ca^{2+} ATPase and Na^+/Ca^{2+} exchangers on the basolateral membrane. Renal handling of Ca^{2+} is regulated by the effects of PTH and FGF23 on calcium transporters and by the effect of the vitamin D hormone on intracellular calcium-binding proteins in the distal tubule. Low plasma Ca^{2+} directly stimulates PTH release from the parathyroid glands. PTH activates apical calcium channels and stimulates basolateral Ca^{2+} transporters, whereas FGF23 stimulates the membrane abundance of calcium transporters in the distal tubule (see below).

Phosphates are required for bone matrix formation as well as for intracellular energy homeostasis. The majority of plasma phosphate (Pi) (more than 90%) is available for filtration, and Pi reabsorption and excretion are highly dependent on diet and age. Under normal dietary conditions, transporters are present only in the proximal tubules, and ~75% of the filtered phosphate is reabsorbed by apical Na^+–Pi cotransporters type 2a and 2c. The remaining Pi can be used to buffer H^+, forming titratable acid, or excreted. In growing children and with low Pi diets, Na^+–Pi cotransporters are also present in the proximal straight tubules and distal tubules, facilitating reabsorption. Renal Pi reabsorption is primarily controlled by FGF23 and PTH, both of which affect the number of Na^+–Pi cotransporters in the apical membrane of the proximla tubule. High dietary Pi increases PTH and FGF23 secretion which both cause a reduction in the number of Na^+–Pi cotransporters in the apical membrane, increasing Pi excretion. PTH is secreted in response to high plasma Pi concentrations or low plasma calcium concentrations and decreases apical Na^+–Pi cotransporters, reducing reabsorption and increasing urinary excretion of Pi.

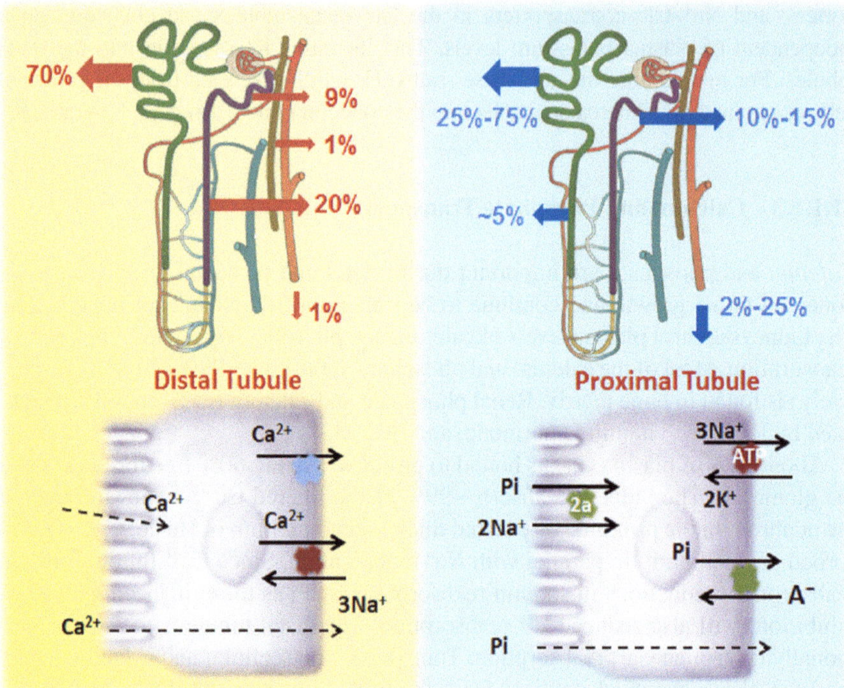

Fig. 11.3 Renal calcium and phosphate handling. Calcium is reabsorbed along the nephron and very little is excreted. Calcium reabsorption takes place in the distal nephron under the control of PTH and FGF23 which regulate abundance and activity of transient receptor potential cation channel (TRPV5) calcium channels. Phosphate is reabsorbed in the proximal tubule where PTH and FGF23 suppress membrane expression of NaPi2a and Napi2c phosphate cotransporters

Plasma Ca^{2+} and phosphate regulation are connected because of the constant bone resorption and deposition. In response to low plasma Ca^{2+}, vitamin D increases intestinal calcium and phosphate absorption, and PTH induces bone resorption. Both actions increase blood Ca^{2+} and phosphate. In the kidneys, PTH increases calcium reabsorption, but decreases phosphate transporters and increases urinary phosphate excretion. Similar to PTH, FGF23 increases calcium reabsorption and decreases urinary phosphate reabsorption (see Fig. 11.3).

11.1.2 Regulation of Extracellular Fluid

Renal sodium handling is closely regulated as part of the important process of extracellular fluid (ECF) homeostasis. A number of intrarenal factors can alter sodium (and thus, fluid) reabsorption in response to changes in glomerular filtration rate (GFR), tubular fluid flow rate, baroreceptor response to changes in blood pressure, and medullary blood flow in the *vasa recta*.

11.1.2.1 Neurohumoral Control of Renal Sodium Reabsorption

The kidneys constantly respond to changes in blood pressure and ECF volume status. When plasma volume is contracted, fluid and sodium conservation systems are activated. Sympathetic nerves increase sodium reabsorption through several mechanisms. Sympathetic nerves innervate afferent and efferent arterioles (via α-adrenergic receptors). During sympathetic stimulation, arterioles constrict, decreasing GFR and thus decreasing sodium excretion. There is also direct sympathetic innervation of the proximal tubule and loop of Henle, which, when activated, stimulates sodium reabsorption.

The renin–angiotensin–aldosterone system (RAAS) has an important role in regulation of the renal retention of sodium and water and, thus, in the regulation of ECF volume. Low tubular sodium concentration and low tubular fluid flow lead to increased renin secretion in the juxtaglomerular apparatus. Renin cleaves angiotensinogen (a plasma protein secreted by the liver) to angiotensin (Ang) I, which is converted to Ang II by angiotensin-converting enzyme (ACE) in the lung and other tissues; Ang II stimulates adrenal release of the mineralocorticoid aldosterone. In the kidney, Ang II has dual effects: it directly stimulates sodium (and thus water) reabsorption in the proximal tubule and has vasoconstrictor effects on afferent and efferent arterioles, which result in a lower GFR and subsequent sodium retention. Aldosterone binds to the cytoplasmic mineralocorticoid receptors in the late distal tubules and collecting tubules and stimulates Na^+ and water retention as well as K^+ secretion. Aldosterone increases apical Na^+ channels and Na^+/H^+ antiporters and increases basolateral Na^+/K^+ ATPase activity. This further increases sodium and water retention and limits urinary losses of sodium. Atrial natriuretic peptide (ANP) is primarily produced by myocytes in the right atrium of the heart and is released into the blood in response to atrial stretch, as with increased blood volume. ANP increases GFR and inhibits collecting tubule Na^+ reabsorption, causing natriuresis and diuresis, reducing ECF volume. Urodilatin is a natriuretic peptide related to ANP that is produced in the renal tubular cells and secreted into the tubules and acts at the collecting tubules to decrease Na^+ reabsorption, resulting in natriuresis/diuresis. AVP increases water channels in the collecting duct, enhancing solute-free water reabsorption.

11.1.2.2 Renal Regulation of Plasma Volume and Osmolarity

Control of ECF is a continual process, with changes in plasma osmolarity and volume stimulating multiple neural and hormonal systems to regulate the renal concentration and dilution of the urine. When plasma volume is contracted, fluid and sodium conservation systems are activated. When extracellular volume decreases, sympathetic nervous system (SNS) activity increases renal vascular resistance and decreases GFR. Activation of the RAAS, which increases Ang II and aldosterone, enhances sodium (and thus water) reabsorption in the proximal tubules and collecting duct. The increase in AVP increases water channels (aquaporins) in the collecting duct, enhancing water absorption. To maintain homeostasis, almost all of the filtered sodium must be reabsorbed—loss of even a few percent of the filtered sodium can result in severe sodium deficiency. Although aldosterone increases sodium reabsorption by only about 2–3%, urinary losses of sodium in aldosterone insufficiency (Addison's disease) can lead to severe sodium depletion, ECF volume contraction, and lowered blood pressure.

11.1.3 Regulation of Acid–Base Balance by the Kidney

Physiologic pH range corresponds to a narrow range in H^+ concentration. The kidneys play an important role in the maintenance of a narrow physiologic window for H^+ concentration. The daily acid load is always buffered in the plasma and is then dissociated from the buffers in the kidneys in the process of bicarbonate metabolism. To control urinary pH, the free H^+ is secreted into the renal tubular lumen, where it is incorporated into either phosphates (titratable acid) or ammonia.

11.1.3.1 Phosphate Buffering

In the normal adult organism, ~75% of the filtered phosphate is reabsorbed and therefore unavailable for generating titratable acid. Thus, only 25% of the filtered HPO_4^{2-} can be used. Formation of titratable acids occurs in the collecting duct at the α-intercalated cells. At this point of the nephron, phosphate reabsorption is complete and the final reabsorption of HCO_3^- is occurring. Thus, the excess H^+ that is actively secreted can bind to the HPO_4^{2-}, creating the titratable acid $H_2PO_4^-$, which is then excreted. Under normal conditions, titratable acids are the main source of acid excretion, but their maximal rate of excretion is fixed because titratable acid formation depends on the amount of phosphate reabsorption occurring, and this is not enough to eliminate the daily acid load. The remaining acid is buffered by NH_3, and when the acid load increases, ammoniagenesis will be further stimulated to handle the load.

11.1.3.2 Ammoniagenesis

The proximal tubule cells are capable of producing ammonia from glutamine, extracted from the tubular fluid as well as from peritubular capillary blood. The ammonia can buffer H^+ by forming ammonium (NH_4^+) that can be excreted in the urine. A key aspect of this reaction is that the excretion of NH_4^+ produces new HCO_3^- that is reabsorbed into the plasma. In the proximal tubular cells, the glutamine is hydrolyzed to produce two NH_3 which immediately bind two H^+, forming two NH_4^+. Additional α-ketoglutarate metabolism yields two HCO_3^- which are secreted into the tubular fluid. The NH_3 stays in the interstitial fluid, increasing the medullary interstitial concentration of NH_3. The dissociated H^+ is secreted into Henle's loop in exchange for Na^+. The NH_3 gradient promotes secretion of NH_3 into the tubular lumen of the collecting ducts where it binds free H^+ reforming NH_4^+, that is excreted in the urine.

11.1.3.3 Acidosis and Alkalosis

Acidosis can result from either gain of acid or loss of bicarbonate. Net acid gain can arise from either decreased respiration (increasing CO_2) or from the accumulation of acids from metabolic sources (metabolic acidosis). *Metabolic acidosis* can result from the gain of the following acids: (1) keto acids, from β-oxidation of fatty acids during starvation and poorly controlled diabetes mellitus; (2) phosphoric acid, in renal failure, as a result of the decreased renal function; and (3) lactic acid, which is released from tissues during hypoxia or heart failure. Metabolic acidosis can also result from the loss of bicarbonate by (1) increased fecal elimination by prolonged diarrhea, (2) increased urinary excretion because of insufficient HCO_3^- reabsorption in the proximal tubule, or (3) elevated H^+ secretion into the collecting duct, causing enhanced HCO_3^- secretion from the β-intercalated cells of the collecting ducts to buffer the urine (distal renal tubular acidosis). Whether due to addition of acid or loss of bicarbonate, the resulting acid load must be buffered systemically and then excreted by the kidneys or compensated by the respiratory system in order to reduce pCO_2 and, therefore, acid load. Chronically, the main compensation will occur in the kidneys through reaction of H^+ with phosphates and ammonia to form titratable acids and ammonium for excretion.

Alkalosis results from the loss of fixed acid or gain of HCO_3^-. Hyperventilation will cause the loss of CO_2 and result in respiratory alkalosis. *Metabolic alkalosis* can result from (1) vomiting, which results in the loss of significant amounts of H^+ in the form of HCl and volume contraction; (2) sodium bicarbonate overdose; and (3) chronic diuretic usage, which can cause hypokalemia and volume depletion, causing in turn a counter-regulatory aldosterone excess with increased renal tubular H^+ secretion. During alkalosis, HCO_3^-/Cl^- exchange in β-intercalated cells of the renal collecting ducts is activated, allowing normalization of the urinary pH.

11.2 Kidney–Bone Endocrine Axis

A normal kidney–bone interaction is important for bone remodeling, a continuous process involving deposition and resorption of minerals from the exchangeable calcium-phosphate mineral pool. In the bone, the activities of osteoblasts and osteoclasts are normally in balance, with bone deposition matching bone resorption (Teitelbaum 2000). Continual remodeling of the bone is important and contributes to maintaining bone strength when the skeleton and muscles are stressed or exercised. When the remodeling balance is negative, as seen in the elderly population, osteoporotic bone can result (Ducy et al. 2000).

Several hormones are either synthesized in the kidney or act on the kidney, including parathyroid hormone (PTH), 1,25-dihydroxyvitamin D_3 [1,25(OH)2D], and calcitonin. These hormones are known to regulate anabolic and catabolic processes in bone development, growth, and maintenance. In addition, bone-derived factors, including fibroblast growth factor 23(FGF23), MEPE, and osteocalcin, can influence systemic calcium and phosphate homeostasis (Quarles 2003; Baum 2014; Rowe 2004a). In addition, these hormones regulate the serum concentration of each other. Thus, dysregulation of one hormone in this axis can lead to the disbalance of the other hormones. However, the optimal ratio in systemic calcium and phosphate balance is variable and changes during life. Growing organisms are in positive balance for both calcium and phosphate for bone growth. A disbalance in systemic calcium and phosphate homeostasis can result in the accumulation of unmineralized osteoid, causing rickets in growing organisms and osteomalacia in adults. The current understanding of the cross talk between bone and kidney has provided the basis for a novel endocrine network, regulating calcium and phosphate metabolism.

11.2.1 Endocrine Control of Calcium and Phosphate Homeostasis

Calcium and phosphate are substrates for mineralization of bone and hence the formation of the bone. The kidneys play a central role in the homeostasis of these ions. Under physiological conditions, the circulating levels of calcium and phosphate are in balance in the way that the amount of calcium and phosphate absorbed from the intestine is equal to that excreted by the kidneys. When circulating levels of these ions decline, gastrointestinal absorption, bone resorption, and renal tubular reabsorption increase to normalize their levels. Renal regulation of the circulating levels of calcium and phosphate occurs through glomerular filtration and tubular reabsorption. Calcium and phosphate levels are tightly regulated by several hormones including 1,25 dihydroxyvitamin D3, parathyroid hormone (PTH), and FGF23, and these hormones also regulate the serum concentration of each other (Rowe 2015b; Penido and Alon 2012).

1,25-dihydroxyvitamin D3 acts on the intestine to increase the absorption of calcium and phosphate. 1,25-dihydroxyvitamin D3 also inhibits PTH secretion while stimulating the skeletal secretion of FGF23. PTH secretion is increased by hypocalcemia and acts to increase serum calcium by stimulating osteoclasts to resorb bone, increasing 1,25 dihydroxyvitamin D3 production and increasing renal calcium reabsorption. All of these processes are directed to increase serum calcium. PTH inhibits renal phosphate reabsorption and increases FGF23. On the other hand, FGF23 inhibits PTH secretion; decreases 1α-hydroxylase; the enzyme responsible for synthesis

Table 11.1 Endocrine renal calcium and phosphate regulation

Modulation of Ca^{2+} transport (low serum calcium)			Modulation of Pi transport (high serum phosphate)		
Factor	Action site	Mechanism	Factor	Nephron site	Mechanism
↑ PTH	Kidney DCT	Activate TRPV5	↑ PTH	Kidney PT	Inhibit NaPi2a
↑ PTH	Kidney PT	Activate vitamin D production	↑ PTH	Kidney PT	Activate Vitamin D production
↑ PTH	Bone	Activate bone resorption	↑ PTH	Bone	Activate bone resorption
↑ FGH23	Kidney DCT	Activate TRPV5	↑ FGH23	Kidney PT	Inhibit NaPi2a
↑ Vitamin D	Intestine	Activate Ca^{2+} absorption	↓ Vitamin D	Intestine	Inhibit Pi absorption

of 1,25-dihydroxyvitamin D_3; and increases inactivation of 1,25-dihydroxyvitamin D_3 by stimulating 24-hydroxylase (Portale et al. 1989; Schiavi and Kumar 2004; Yu and White 2005). The net effect of all abovementioned bone- and kidney-derived regulators as well as PTH and calcitonin will lead to stable serum calcium and phosphate levels while also providing calcium and phosphate pools for bone mineralization (see Table 11.1).

11.2.1.1 Bone-Derived Regulators: FGF23, MEPE, and Osteocalcin

11.2.1.1.1 FGF23

FGF23 is a 251 amino acid protein primarily produced in osteocytes and osteoblasts (Yoshiko et al. 2007). Its molecular action requires binding to a family of FGF receptors 1, 2, 3, and 4 (FGFR 1–4) which is facilitated by the transmembrane co-receptor αKlotho (Razzaque and Lanske 2007). αKlotho is a type 1 membrane protein that exists in a transmembrane form, important for the FGF23 co-receptor function, and in a shed, soluble form that can be released from the plasma membrane by the activity of disintegrin and metalloproteinases (ADAM-10 and ADAM-17). The physiological role of soluble αKlotho is currently not entirely clear. Several studies suggested a role of soluble αKlotho in the regulation of renal calcium and phosphate reabsorption (Urakawa et al. 2006; Olauson et al. 2014; Nabeshima 2008). Although FGF receptors are ubiquitously expressed, the downstream effects of FGF23 have been mainly described in the kidney, parathyroid gland, and heart (Urakawa et al. 2006). It is still not completely clear which FGFRs are responsible for the actions of FGF23 in different cell types. αKlotho expression is predominantly detected in the kidney, the parathyroid gland, and the brain (choroid plexus). The site-specific actions of FGF23 are thought to be related to the expression of αKlotho. However, recent studies have suggested a αKlotho-independent

FGF23–FGFR4 interaction in the heart, activating pro-hypertrophic downstream signaling networks in cardiomyocytes (Faul 2012; Grabner et al. 2015). Solid evidence has been provided that FGF23 signals through the FGFR1c/Klotho complex to fulfill its well-described phosphaturic effects in the kidney. Acting on the basolateral membrane in renal proximal tubular epithelial cells, it downregulates the membrane abundance of the type IIa sodium-phosphate cotransporter (NaPi-2a), by phosphorylation of the scaffolding protein NHE exchange regulatory cofactor (NHERF)-1 through ERK1/2 and serum/glucocorticoid-regulated kinase-1 (SGK1) signaling in a FGFR1c/αKlotho-dependent fashion, decreasing phosphate reabsorption in renal proximal tubules (Shimada et al. 2004; Andrukhova et al. 2012; Bowe et al. 2001). It was also suggested that FGF23 suppresses the apical membrane abundance of other phosphate cotransporter in renal proximal tubules such as type IIc sodium-phosphate cotransporter (NaPi-2c) and thereby additionally regulates apical phosphate entry into the epithelial cells. However, because recent evidence revealed that the role of NaPi-2c is minor in phosphate homeostasis under physiological conditions, it is plausible that FGF23 regulates NaPi-2a and 2c in parallel, but that the phosphaturic action of FGF23 is mainly determined by downregulation of the apical membrane abundance of NaPi-2a, at least in mice.

Another important renal effect of FGF23 is to suppresses renal 1α-hydroxylase expression and to stimulate 24-hydroxylase expression, downregulating 1,25-dihydroxyvitamin D3 production by the kidney (Perwad et al. 2007). Both FGFR3 and 4 were suggested to be involved in the FGF23-mediated regulation of 1,25-dihydroxyvitamin D3 production. The net effect of a decrease in renal production of active vitamin D is a lower active absorption of calcium and phosphate from the intestinal lumen of the gut. The secretion of FGF23 in bone is stimulated by 1,25-dihydroxyvitamin D3 hormone and by increased extracellular phosphate, forming a feedback loop between bone and kidney (Fig. 11.4).

The parathyroid gland is another possible target organ of FGF23 action with the presence of both αKlotho and FGF receptors. It was reported in several studies that circulating levels of FGF23 and PTH are controlled by each other by a negative feedback loop (Ben-Dov et al. 2007; Krajisnik et al. 2007). However, the data are still controversial and need further clarification.

FGF23 can be cleaved into carboxyl-terminal (aa 180–251) and amino-terminal portions (aa 1–24). The family with sequence similarity 20, member C (Fam20C), has been shown to phosphorylate FGF23, and such phosphorylation promotes FGF23 proteolysis by furin through blocking O-glycosylation by polypeptide N-acetylgalactosaminyltransferase 3 (GalNAc-T3). The C-terminal fragment of FGF23 is able to interact with FGFRs and blocks binding of the intact FGF23 molecule. The exact role of FGF23 cleavage is still unclear, but it is likely that the interplay between phosphorylation and O-glycosylation of FGF23 may determine its bioactivity.

The physiological effects of FGF23 can explain the symptoms of several human genetic diseases associated with gain or loss of function of FGF23. Tumor-induced osteomalacia, X-linked hypophosphatemia, and autosomal dominant hypophosphatemic rickets are characterized by markedly elevated FGF23 levels

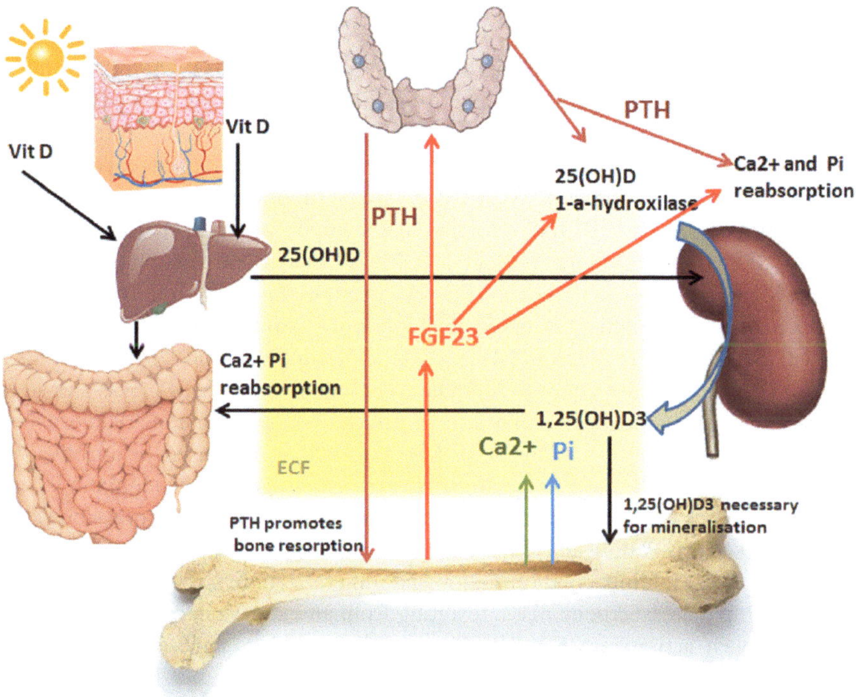

Fig. 11.4 Endocrine control of calcium and phosphate homeostasis

and are associated with urinary phosphate wasting, hypophosphatemia, reduced 1,25-dihydroxyvitamin D3 levels, and osteomalacia (Liu et al. 2007a, b; White et al. 2001). Conversely, loss-of-function mutations in FGF23 or αKlotho (Ichikawa et al. 2007; Araya et al. 2005) are characterized by hyperphosphatemia, increased 1,25-dihydroxyvitamin D3 levels, and ectopic calcification. Moreover, elevated serum FGF23 is one of the earliest markers in chronic kidney disease, and it correlates with disease progression. Since serum phosphate was shown to stimulate FGF23 expression, hyperphosphatemia observed in chronic kidney disease was believed to drive bony FGF23 secretion. However, serum FGF23 levels increase with mild kidney dysfunction even before there is an increase in serum phosphate, suggesting some other phosphate-independent regulator(s) of FGF23 expression. The increase in serum phosphate results in a decrease in ionized calcium, low 1,25-dihydroxyvitamin D3 levels, and an increase in serum PTH. The low levels of 1,25 dihydroxyvitamin D3 and chronic hyperparathyroidism will result in a bone mineralization defect and increased bone resorption. These syndromes underscore the importance of FGF23 in maintaining normal mineral homeostasis (Fig. 11.3).

11.2.1.1.2 MEPE

Matrix extracellular phosphoglycoprotein MEPE (also called OF45 in the rat) was identified in differential gene-expression profiling in tumors of TIO patients (De Beur et al. 2002; Shimada et al. 2001). MEPE is exclusively expressed in osteoblasts, osteocytes, and odontoblasts. MEPE belongs to the small integrin-binding ligand N-linked glycoprotein (SIBLING) family of proteins. Together with the other members of the SIBLING proteins like sialophosphoprotein (DSPP), osteopontin (SPP1), dentin-matrix-protein-1 (DMP-1), and bone sialophosphoprotein (BSP), MEPE is involved in the regulation of the mineralization of the extracellular matrix in bone. The highly conserved acidic serine–aspartate-rich MEPE-associated motif (ASARM motif) found in MEPE, DSPP, osteopontin, and DMP-1 is important for the role of these genes in modulating mineralization. MEPE was shown to be markedly upregulated in patients with X-linked hypophosphatemic rickets and in a murine homologue of human XLH, *Hyp* mice. XLH is caused by inactivating mutations in the cell surface metalloprotease PHEX, phosphate-regulating gene with homologies to endopeptidases on the X chromosome. MEPE expression could be suppressed by PHEX and 1,25-dihydroxyvitamin-D3. Current evidence indicates that MEPE inhibits bone formation and mineralization, since MEPE-deficient mice develop increased bone mass, resulting from an increase in osteoblast number and osteoblast activity, with unaltered osteoclast number and osteoclast surface (Gowen et al. 2003).

It has been postulated that MEPE have phosphaturic actions in the kidney (Quarles and Drezner 2001). Several studies confirmed that MEPE (or MEPE peptides) inhibit renal phosphate handling in vivo and in vitro. A micropuncture study showed that MEPE infusion inhibited fractional phosphate reabsorption, in the absence of any significant change in glomerular filtration rate (Shirley et al. 2010). A suppressive effect on the proximal tubule NaPi-2a cotransporter was proposed as molecular mechanism underlying the phosphaturic action of MEPE. However, deletion of MEPE failed to rescue the hypophosphatemia in Hyp mice, suggesting that MEPE is not the cause of phosphaturia observed in Hyp mice (Liu et al. 2005).

Cathepsin B was shown to cleave MEPE in vitro. However, at present it is not clear whether degradation fragments of MEPE's can be functionally active. Low-molecular-weight fragments have been identified in conditioned media from cells secreting recombinant MEPE (Guo et al. 2002). Moreover, evidence indicates that MEPE is a substrate of, or forms, a complex with PHEX (Campos et al. 2003). The biological significance of the later finding is uncertain, but it may suggest a MEPE-associated regulation of mineralization through a PHEX-dependent mechanism.

11.2.1.1.3 Osteocalcin

Osteocalcin has routinely been used as a serum marker of osteoblastic bone formation and believed to act in the bone matrix to regulate mineralization. New genetic and pharmacologic evidence now points to a hormonal role of osteocalcin.

Osteocalcin, also referred to as bone γ-carboxyglutamic acid (Gla) protein or BGP, is a 46–50 amino acid, 5.6 kDa secreted protein that is produced primarily by osteoblasts (Razzaque 2011). Smaller amounts are also produced by odontoblasts of the teeth and hypertrophic chondrocytes.

Osteocalcin together with the other Gla-protein MGP belongs to a distinct subgroup of the larger vitamin K-dependent protein family, the constituents of which are primarily involved in blood coagulation. Mature osteocalcin is secreted into the bone microenvironment and then undergoes a conformational change that aligns its calcium-binding Gla residues with the calcium ions in hydroxyapatite. This property was proposed as a mechanism that enables osteocalcin to initiate the formation of hydroxyapatite crystals (Price and Williamson 1985). Circulating osteocalcin is present either as carboxylated or as undercarboxylated forms. The carboxylation of osteocalcin is a complex process, and the Esp gene, encoding the osteotesticular protein tyrosine phosphatase (OST-PTP), has been implicated in the carboxylation process.

Osteocalcin is also involved in regulation of glucose metabolism (Clemens and Karsenty 2011). These experimental observations suggest that osteoblastic insulin signaling can influence systemic glucose homeostasis through osteocalcin secretion. Experimental studies have reported that insulin signaling in osteoblasts is not only important for normal bone acquisition but also stimulates osteocalcin production. On the other hand, osteocalcin infusion improves insulin sensitivity in insulin receptor mutant mice. It is believed that the acidic conditions (low pH) created by the osteoclastic resorption process in the bone activate osteocalcin within the bone which is then released from the bone into the circulation and stimulates pancreatic insulin secretion which in turn facilitates bone formation. Human cross-sectional studies examining the association of circulating levels of total and/or undercarboxylated osteocalcin with glucose metabolism indicate that the levels of osteocalcin are negatively correlated with fasting glucose, fasting insulin, HOMA-IR (a representation of insulin resistance), body mass index, and hyperlipidemia (Kindblom et al. 2009; Yeap et al. 2010; Chin et al. 2014).

Serum osteocalcin levels are high in patients with CKD. The accumulation of osteocalcin in the serum of patients with CKD can be related to decreased renal clearance, increased bone metabolism, or a combination of both. In patients with CKD, the progressive increase in serum osteocalcin levels closely corresponds with intact PTH and alkaline phosphatase levels. The progression of vascular calcification has similarities to bone formation, and the serum undercarboxylated osteocalcin levels are higher in CKD patients with carotid artery calcification, suggesting that undercarboxylated osteocalcin levels may be a biomarker for vascular calcification. The production of osteocalcin by human adipose tissue adds additional complexity in energy metabolism (Foresta et al. 2010). Experimental studies have shown that undercarboxylated osteocalcin levels are negatively correlated with fat mass, fasting plasma glucose levels in male type 2 diabetic patients independent of age, duration of diabetes, body stature, renal function, and glucose or fat metabolism. Further studies are necessary to determine how undercarboxylated osteocalcin is involved in CKD pathology.

11.2.1.2 Kidney-Derived Regulators: Vitamin D and Acid-Based Metabolites

11.2.1.2.1 Vitamin D

Vitamin D (cholecalciferol) is available from dietary sources and from the action of ultraviolet light on 7-dehydroxycholesterol in the skin. The synthesis of vitamin D in the skin is the most important source of vitamin D in humans and depends on the season and latitude which determine the intensity of ultraviolet irradiation. Vitamin D is transported by vitamin D-binding protein (DBP) in the blood to the liver where it is hydroxylated at C-25 to produce 25-hydroxyvitamin D3 (25(OH)D3), the major circulating form of vitamin D. The serum level of 25(OH)D3 is considered to be one of the most reliable biomarkers of vitamin D status (Bikle et al. 2013; Heaney et al. 2009). 25(OH)D3 is transported by DBP to the kidney. DBP is partially filtered and taken up again in proximal tubular epithelial cells by endocytic internalization through binding of the DBP to megalin, an epithelial transmembrane protein and a member of the low-density lipoprotein receptor superfamily (Chun et al. 2014).

In the proximal renal tubule, 25(OH)D3 is hydroxylated by the renal 25(OH)D 1α-hydroxylase (mitochondrial CYP27B1), resulting in the formation of 1,25(OH)2D3, the hormonally active form of vitamin D responsible for its biologic actions. At the same time, as a feedback mechanism in the kidney 24-hydroxylase (CYP24A1) limits the amount of circulating 1,25(OH)2D3 by catalyzing the conversion of 25(OH)D3 and 1,25(OH)2D3 into 24-hydroxylated products which are targeted for excretion and decrease the pool of 25(OH)D3 available for 1α-hydroxylation by producing 24,25(OH)2D3. The balance in the CYP27B1 and CYP24A1 expression in the proximal renal tubule is tightly regulated. PTH was shown to be one factor mediating this regulation by activating CYP27B1 transcription (Bikle 2014; Henry 2011). 1,25(OH)2D3 in turn suppresses PTH production in the parathyroid gland and is regulating its own production by inhibiting CYP27B1 (Canaff and Hendy 2002; Brenza and DeLuca 2000).

The biological actions of 1,25(OH)2D3 are mediated by the vitamin D receptor, acting as a transcriptional factor. The genomic mechanism of 1,25(OH)2D3 action involves the direct binding of the ligand-activated vitamin D receptor/retinoic X receptor (VDR/RXR) heterodimeric complex to specific DNA sequences. Numerous VDR co-regulatory proteins involved in different physiological processes have been identified. 1,25(OH)2D3 is involved in several processes in order to control plasma calcium levels. The main actions of the active form of vitamin D are to increase the plasma calcium concentration and to promote bone mineralization.

When the circulating levels of calcium decrease, the reduced activation of the calcium sensing receptor (CaSR) on the parathyroid gland results in increased PTH secretion. This stimulates the renal 1α-hydroxylase, increasing the formation of 1,25(OH)2D3. In parallel, PTH will also increase resorption of calcium and phosphate from bone and increase renal calcium reabsorption by acting on the distal tubular epithelial calcium channel TRPV5. The total effect will be increasing plasma calcium levels.

The principal action of 1,25(OH)2D3 and the VDR is to promote intestinal calcium absorption. 1,25(OH)2D3 controls intestinal transcellular calcium transport by regulation of calcium entry through the apical membrane calcium channel TRPV6, regulation of the expression of the calcium-binding protein calbindin-D9k, and regulation of calcium extrusion by stimulating basolateral membrane PMCA1b (Christakos et al. 2014).

In the kidney, calcium absorption is regulated by 1,25(OH)2D3 and PTH which both increase calcium reabsorption in the distal tubules acting on calbindin-D28k and TRPV5, respectively. The proximal tubules in the kidney are not only the major site of 1,25(OH)2D3 synthesis but are also important for phosphate reabsorption. Phosphate reabsorption in the proximal tubules is regulated by several factors including FGF23, PTH, and 1,25(OH)2D3. PTH and FGF23 promote renal phosphate loss by decreasing the NaPi-2a/2c at the apical membrane. 1,25(OH)2D3 indirectly regulates renal phosphate handling by increasing FGF23 expression in osteocytes (Forster et al. 2011). Thus, 1,25(OH)2D3 stimulates calcium reabsorption and indirectly promotes urinary phosphate loss.

The main action of 1,25(OH)2D3 in bone is to enhance bone resorption by increasing the expression of RANKL, an important osteoclastogenic factor, in osteoprogenitor cells and osteoblasts (Kim et al. 2006). RANKL binds to its cognate receptor RANK in osteoclast precursors and increases osteoclast formation and finally bone resorption. Besides stimulating bone resorption during a negative calcium balance, 1,25(OH)2D3 also inhibits bone matrix mineralization by increasing the pyrophosphate (PPi) levels and osteopontin expression, both potent mineralization inhibitors. On the other hand, in mature osteoblasts 1,25(OH)2D3 has anabolic and anticatabolic activity by activating low-density lipoprotein receptor-related protein 5 (LRP5), a co-receptor in wingless (Wnt) signaling, a pathway known to mediate anabolic effects in osteoblasts (Gee et al. 2013).

1,25(OH)2D3 levels in the blood are tightly regulated by a feedback control loop involving PTH and FGF23. This overall control of 1,25(OH)2D3 levels is important, since overproduction of 1,25(OH)2D3 can cause hypercalcemia. Rickets is a disease caused by insufficient 1,25(OH)2D3 levels, resulting in stunted bone mineralization and growth. In adults, prolonged vitamin D deficiency will decrease bone mineralization, resulting in osteomalacia. Also renal rickets due to chronic renal damage limits the production of 1,25(OH)2D3 and will cause osteomalacia.

11.2.1.2.2 Acid-Based Metabolites

In addition to the hormonal regulation by PTH, 1,25(OH)2D3, and FGF23, hormone-independent changes in plasma calcium and phosphate concentration can be caused by acid–base disorders. Chronic acidosis was shown to activate bone resorption (Lemann et al. 2003). Moreover, metabolic acidosis observed in CKD patients correlates with bone resorption and low systemic calcium levels. Metabolic acidosis stimulates osteoclastic bone resorption and reduces osteoblastic bone synthesis (Bushinski and Nilsson 1995). Reversing the metabolic acidosis ameliorates bone disease in renal osteodystrophy. Metabolic acidosis may also stimulate PTH secretion and may negatively impact 1,25(OH)2D3 production. However, these effects are not clear-cut, and some results have been conflicting. Chronic metabolic acidosis impairs skeletal growth. This is believed to be not only due to the direct skeletal effect of chronic metabolic acidosis but may also be due to the effect of acidosis on the release and effectiveness of growth hormone (Lemann et al. 2003).

11.2.1.3 PTH and Calcitonin

11.2.1.3.1 PTH

Parathyroid hormone (PTH) is the primary hormone controlling the short-term regulating of the plasma calcium concentration. PTH is synthesized by the chief cells of the parathyroid glands as a 110 amino acid pre-prohormone and is cleaved to a prohormone in the endoplasmic reticulum and to the active 84 amino acid peptide hormone in the Golgi apparatus. The secretory vesicles with the PTH are packaged in the cytoplasm of the chief cells and can be released in the circulation when the circulating calcium level drops. The changes in the plasma calcium concentration are recognized by the chief cells with the help of the calcium sensing receptor (CaSR). When the plasma calcium concentration is decreased, changes in the CaSR activity cause a rapid intracellular signaling response, resulting in PTH release into the circulation. PTH binds to the PTH receptor 1, a seven transmembrane G-protein coupled receptor, to activate the PKA, PKC, and MAPK pathways in the kidney and bone. In addition to serum calcium, vitamin D can also suppress PTH expression and parathyroid gland hyperplasia.

The primary response to PTH in the kidney is the stimulation of calcium reabsorption in the distal tubule by activating specific ion channels such as TRPV5 (van Abel et al. 2005) and the increase in phosphate excretion in the proximal tubule by downregulating NaPi-2a via both PKA and PKC dependent pathways. It also indirectly enhances intestinal calcium and phosphate uptake by stimulating the production of 1,25(OH)2D3 by transcriptional activation of 1α-hydroxylase. Because the kidney is the main site of vitamin D activation, loss of PTH or loss of kidney function impairs vitamin D metabolism. The PTH actions in the kidney are rapid and are a main avenue for the rapid restoration of plasma calcium and phosphate levels.

PTH regulates renal phosphate reabsorption in the proximal tubule by downregulating the apical expression and promoting the degradation of the sodium-phosphate cotransporters NaPi-2a/2c via NHERFs protein phosphorylation by PKA and PKC dependent pathways (Segawa et al. 2007; Traebert et al. 2000). NHERF1 colocalizes with NaPi-IIa and forms a complex through its PDZ domain, stabilizing the expression of NaPi-2a at the apical membrane. PTH treatment disrupts this complex by stimulating the phosphorylation of NHERF1, which leads to the dissociation and subsequent endocytosis of NaPi-2a (Weinman et al. 2007).

In bone, PTH is able to act both as an anabolic as well as a catabolic agent. Recent investigations have markedly advanced our understanding of the cellular and molecular mechanisms behind the contrasting effects of the two PTH regimens on bone. Studies in mice indicate that chronic and intermittent PTH administration increase osteoblast number by distinct mechanisms. The anabolic effect of intermittent PTH may partially be accounted for by attenuation of osteoblast apoptosis, whereas chronic elevation of PTH has no effect on osteoblast survival (Jilka 2007). Instead, the osteoblastogenic action of continuous elevation of PTH results from direct actions of the hormone on osteocytes inhibiting the expression of the sclerostin (Sost) gene which through the binding to the LRP5, LPR6, and LRP4 preventing activation of Wnt signaling critical for osteoblastogenesis (Bellido et al. 2005). The expression and activity of Runx2, an osteoblast-specific transcription factor essential for osteogenesis, is also augmented by PTH, resulting in increased bone formation (Selvamurugan et al. 2009).

Continuous PTH administration shifts gene-expression patterns in osteoblasts to participate in the bone resorption process by secreting various cytokines involved in the regulation of osteoclast differentiation and activation. PTH increases osteoclast formation by upregulating RANKL and the RANKL/OPG ratio through the PKA pathway. Although it is recognized that osteoblastic lineage cells mediate osteoclastogenesis and bone resorption mediated by PTH, the precise stage of differentiation of the PTH target cell is unknown. Recent evidence demonstrates that osteocytes are a major source of RANKL, because mice lacking RANKL in osteocytes exhibit osteopetrosis. Alternatively, activation of the PTH receptor in osteocytes may induce the production and secretion of factors, such as IL-6-type cytokines, which in turn may stimulate RANKL expression in other cells, such as stromal–osteoblastic cells (Raggatt et al. 2008; Lee 2002; Horwood et al. 1998). The bone resorptive effect of PTH also involves activation of the transcription factor/cAMP-responsive-element-binding protein (ATF/CREB) family of leucine zipper transcription factors, which bind to cAMP response elements (CRE). Another member of the ATF/CREB family, CRE modulator (CREM), has been shown to be implicated in PTH signaling, as a typical anabolic response (Liu 2007).

Excessive PTH secretion results in hypercalcemia and hypophosphatemia. Primary hyperparathyroidism is usually caused by benign parathyroid tumors that produce PTH but do not have feedback regulation. The elevated PTH increases the renal activation of vitamin D, increases plasma calcium through increasing renal calcium reabsorption and bone resorption, and decreases plasma phosphate by reducing renal phosphate reabsorption. Although renal calcium reabsorption is elevated, the significantly increased filtered calcium load results in elevated urinary calcium and phosphate excretion, increasing the chance of kidney stone formation from calcium-phosphate and calcium-oxalate formation in urine. Secondary hyperparathyroidism is caused in response to the primary problem of low plasma calcium levels. In secondary hyperparathyroidism, the low plasma calcium levels are usually caused by vitamin D deficiency and/or chronic renal failure, and the parathyroid glands secrete PTH to correct this problem. Thus, PTH levels will be high, but plasma calcium will be low or normal.

Parathyroid hormone-related protein (PTHrp) is a protein that mimics the calcium-mobilizing effects of PTH. It is present in a variety of tissues including smooth muscle, breast, and the central nervous system and is believed to act in a paracrine manner. Although its normal actions are still not completely described, PTHrp is also secreted from certain tumors, such as breast cancers and renal cell and squamous cell carcinomas.

11.2.1.3.2 Calcitonin

Calcitonin is a 32 amino acid peptide hormone that is produced in the parafollicular cells of the thyroid gland, and although it appears to have effects on calcium handling, it does not contribute to the rapid regulation of calcium. It is released in response to elevated plasma calcium levels and acts on G-protein-coupled receptors.

The major effect of calcitonin on the kidney is to enhance calcium excretion by inhibition of renal tubular calcium reabsorption (Cochran et al. 1970). In hypercalcemic patients with metastatic bone disease, the administration of calcitonin induces a rapid fall in serum calcium due primarily to inhibition of renal tubular reabsorption by calcitonin.

In bone, the osteoclasts are the major target for the action of calcitonin. Calcitonin acts directly on osteoclast to inhibit osteoclast motility through calcitonin receptor. Calcitonin has also been demonstrated to interfere with osteoclast differentiation from precursor cells and fusion of mononucleated precursors to form multinucleated osteoclasts in bone marrow cultures (Kurihara et al. 1991). While the molecular mechanism(s) by which calcitonin inhibits osteoclast-mediated bone resorption are not fully understood, the activation of G-proteins and a subsequent rise in intracellular cAMP is likely to play an important role in this process. Incubation of osteoclasts in media containing the cAMP analogue (Bu)2cAMP duplicates the effects of calcitonin on osteoclastic motility and bone resorption (Chambers et al. 1985a).

However, the significance of the physiologic effects of calcitonin in mammals is still debatable, since removal of the parafollicular cells in the thyroid gland has no dramatic effect on mineral homeostasis.

11.2.2 Endocrine Control of Sodium Homeostasis

11.2.2.1 Sodium and Bone

Sodium is the major extracellular cation and is the most important regulator of plasma osmolality, normally tightly regulated by the homeostatic process of osmoregulation. Changes in blood osmolality are detected by osmoreceptor cells in the hypothalamus, and this results in changes in the secretion of the hormone arginine vasopressin (AVP), regulating water reabsorption through activation of the aquaporin proteins in the kidney collecting duct. Bones are extremely rich in sodium

(Baylis and Thompson 1988) and are the only tissue in which the sodium concentration is actually higher than in the ECF. Up to 40% of the bone minerals are exchangeable with the circulating pool. Therefore, bone serves as an internal sodium "storage" that could be mobilized during sodium deficiency to control the sodium concentration at the levels needed for maintenance of the blood volume, blood pressure, and tissue perfusion.

Hyponatremia is the most common electrolyte imbalance encountered in clinical practice and is common in the elderly (Hannon and Verbalis 2014). It is generally accepted as a potential new independent risk factor for bone resorption and fractures and was shown to be associated with adverse outcomes independent of the underlying condition (Hoorn et al. 2006). Potential mechanisms through which hyponatremia could affect bone quality are increased oxidative stress and free radical accumulation, decreases in intracellular calcium, and decreases in cellular uptake of ascorbic acid, leading to an increase in osteoclastic activity (Hannon and Verbalis 2014). The actual mechanism by which osteoclasts could sense and then respond to hyponatremia needs further clarifications.

It has been suggested that the relative excess of AVP, rather than the hyponatremia itself, may be the cause of increased bone loss in hyponatremia through inhibition of the Runx2 protein, resulting in decreased osteoblast differentiation and inhibited bone formation rate (Tamma et al. 2013). The data accumulated to date support the hypothesis that osteoclast-mediated bone resorption during hyponatremic conditions occurs to preserve sodium homeostasis.

At the same time, it has been proposed that high dietary sodium intake results in a sodium-mediated increase in renal calcium excretion (calciuria), which is a risk factor for osteoporosis. However, interindividual variation in sodium-induced calciuria in cross-sectional and some salt-loading studies are explained by differences in intake of other dietary factors known to affect urinary calcium excretion, such as protein and potassium (Sellmeyer et al. 2002).

Na^+ reabsorption along the kidney nephron plays a crucial role in the regulation of Na^+ balance. To maintain the extracellular volume constant, the amount of Na^+ excreted needs to match the Na^+ input by intestinal absorption. Nearly 99% of filtered Na^+ is reabsorbed in the kidney. In different segments of the renal tubule, different cotransporters, counter-transporters, or channels on the apical surface of tubule cells mediate Na^+ reabsorption. Most frequently used strategy to control renal Na^+ reabsorption is a use of the diuretics. The most common ones are thiazides targeting the NCC cotransporter and amiloride targeting the ENaC channel in the distal nephron.

11.2.2.2 RAAS System

The principal regulator of renal Na^+ reabsorption is the steroid hormone aldosterone, which is produced in the *zona glomerulosa* of the adrenal cortex. Aldosterone, like all steroid hormones, together with its intracellular mineralocorticoid receptor works as a transcription factor to alter gene expression in cells. Aldosterone belongs to the renin–angiotensin–aldosterone system (RAAS) playing an important role in regulating of renal Na^+ reabsorption and, therefore, in control of the blood volume and systemic vascular resistance, which together influence arterial pressure. Renin, which is primarily released by the kidneys, stimulates the formation of angiotensin in blood and tissues, which in turn stimulates the release of aldosterone from the adrenal cortex. Juxtaglomerular cells associated with the afferent arteriole entering the renal glomerulus are the primary site of renin storage and release. Specialized cells (*macula densa*) of distal tubules lie adjacent to the juxtaglomerular cells of the afferent arteriole. The macula densa senses the concentration of sodium and chloride ions in the tubular fluid and stimulates renin production. Circulating renin acts upon a circulating substrate, angiotensinogen, that undergoes proteolytic cleavage to form the angiotensin I which in vascular endothelium, particularly in the lungs, will be converted to the angiotensin II by the specific angiotensin-converting enzyme (ACE). Angiotensin II directly or through activation of the aldosterone expression in the adrenal gland stimulates sodium reabsorption in the kidney by acting on the distal tubules thiazide-sensitive Na+−Cl− cotransporter (NCC) and epithelial sodium channel (ENaC). It also stimulates release of AVP from the pituitary gland, which increases fluid retention by the kidneys. It was suggested that the molecular action of aldosterone and angiotensin II in the kidney involves activation of the with-no-lysine[K] (WNK1-4) serine/threonine protein kinase family proteins, members of the mitogen-activated protein kinase (MEK) family lacking the invariant catalytic lysine in subdomain II of protein kinases necessary for binding ATP (Xu et al. 2000). It was shown that both WNK1 and WNK4 increase apical abundance and activity of the NCC and Na^+–K^+–$2Cl^-$ cotransporter 1 (NKCC1) in the renal distal convoluted tubule and collecting duct, as well as reduced expression of the renal outer medullary K+ channel (ROMK) (Siew and O'Shaughnessy 2013). WNKs increase the functional activity of NCC through downstream effectors, namely, STE20/sporulation-specific protein 1(SPS1)-related proline/alanine-rich kinase (SPAK) and oxidative stress-responsive 1 (OSR1) (Gagnon and Delpire 2012).

11.2.2.3 Bone-Derived Regulators: FGF23

Clinical studies have provided evidence for an association between thiazide treatment and increased bone mineral density, suggesting a link between renal Na^+ handling and bone homeostasis either directly or via regulation of renal Ca^{2+} homeostasis (Rejnmark et al. 2003). This was further supported by the fact that the renal TRPV5, one of the most important Ca^{2+} transporters, is WNK regulated in the distal tubule (Jiang et al. 2008).

Recently, an additional link between kidney and bone was found for maintenance of sodium homeostasis. FGF23, the bone-derived hormone, was identified as an upstream activator of WNK4 signaling in the renal distal tubular epithelium (Andrukhova et al. 2014a, b). It was shown that FGF23 and αKlotho deficient mice showed decreased membrane expression of NCC in renal distal tubules, leading to renal sodium wasting, reduced plasma volume, and lower blood pressure despite elevated aldosterone secretion. Conversely, recombinant FGF23 increased renal sodium retention, plasma expansion, hypertension, and induced heart hypertrophy by upregulation of renal NCC expression. This action of FGF23 requires the presence of FGF receptor 1c and of the co-receptor αKlotho in order to activate the WNK4 signaling axis.

11.2.3 Magnesium Homeostasis

Magnesium is the second most abundant intracellular divalent cation. Besides the important physiological functions like intracellular signaling, oxidative phosphorylation, cardiovascular tone, neuromuscular excitability, and cofactor for protein and DNA synthesis, magnesium is also important for bone formation. Approximately 10–30% of the filtered magnesium in the kidney is absorbed in the proximal tubule. Although the exact mechanisms are not known, magnesium is believed to be absorbed via a paracellular pathway driven by a chemical gradient generated by Na gradient-driven water transport. A paracellular pathway in the thick ascending limb absorbs 40–70% of filtered magnesium, mostly enhanced by lumen-positive transepithelial voltage, in which claudin-16 and claudin-19 play an important role. The remaining 5–10% of magnesium is reabsorbed in the distal convoluted tubule mainly by active transcellular transport mediated by transient receptor potential cation channel, subfamily 6, member 6 (TRPM6). Mutations in the claudin 16 (paracellin) paracellular protein in the thick ascending limb of Henle's loop and in the TRPM6 magnesium channel expressed in distal tubules found in patients with renal magnesium wasting and hypomagnesemia underscore the importance of these transport proteins in magnesium homeostasis (Wagner 2007; Hou et al. 2007). Moreover, epidermal growth factor (EGF) was also identified as a hormonal regulator of TRPM6 activity which may explain the hypomagnesemia observed in patients with EGF mutations (Groenestege et al. 2007).

At least 50% of the total body magnesium content resides in bone as hydroxyapatite crystals. Although the bone magnesium stores are dynamic, the transporters that mediate magnesium flux in and out of bone have not yet been determined. Bone surface Mg^{2+} levels are related to serum Mg^{2+}. Accordingly, surface bone Mg^{2+} increases with Mg^{2+} loading (Cunningham et al. 2012). Dietary magnesium restriction causes decreased bone magnesium content (Alfrey et al. 1974). Hypomagnesemia in rodents induces osteopenia with accelerated bone turnover, decreased bone volume, and decreased bone strength (Elin 1987). Several studies shown that magnesium deficiency resulted in an increase in circulating substance P, tumor necrosis factor α (TNF-α), and interleukin 1β (IL-1β) as well as in activation of the RANKL/OPG system, favoring an increase in bone resorption. Hypomagnesemia is also correlated in humans and in rodent models with low serum PTH and 1,25(OH)(2)-vitamin D levels which may also contribute to reduced bone formation. Moreover, it was also postulated that Mg^{2+} restriction promotes oxidative stress, partly as a consequence of inflammation and partly because of the reduced antioxidant defenses which occur upon Mg^{2+} restriction. The increased amounts of free radicals potentiate the activity of osteoclasts and depress that of osteoblasts (Castiglioni et al. 2013). An optimal range of Mg^{2+} concentrations might be required to ensure bone homeostasis. However, more studies are required to clarify the effects of high and low Mg^{2+} concentrations on bone metabolism and structure.

References

Alfrey AC, Miller NL, Trow R. Effect of age and magnesium depletion on bone magnesium pools in rats. J Clin Invest. 1974;54:1074–81.

Andrukhova O, Slavic S, Smorodchenko A, Zeitz U, Shalhoub V, Lanske B, Pohl EE, Erben RG. FGF23 regulates renal sodium handling and blood pressure. EMBO Mol Med. 2014a;6(6):744–59.

Andrukhova O, Smorodchenko A, Egerbacher M, Streicher C, Zeitz U, Goetz R, Shalhoub V, Mohammadi M, Pohl EE, Lanske B, Erben RG. FGF23 promotes renal calcium reabsorption through the TRPV5 channel. EMBO J. 2014b;33(3):229–46.

Andrukhova O, Zeitz U, Goetz R, Mohammadi M, Lanske B, Erben RG. FGF23 acts directly on renal proximal tubules to induce phosphaturia through activation of the ERK1/2-SGK1 signaling pathway. Bone. 2012;51(3):621–8.

Araya K, Fukumoto S, Backenroth R, et al. A novel mutation in fibroblast growth factor 23 gene as a cause of tumoral calcinosis. J Clin Endocrinol Metab. 2005;90:5523–7.

Baum M. The bone kidney axis. Curr Opin Pediatr. 2014;26(2):177–9. doi:10.1097/MOP.0000000000000071.

Baylis PH, Thompson CJ. Osmoregulation of vasopressin secretion and thirst in health and disease. Clin Endocrinol. 1988;29:549–76.

Bellido T, Ali AA, Gubrij I, Plotkin LI, Fu Q, O'Brien CA, Manolagas SC, Jilka RL. Chronic elevation of parathyroid hormone in mice reduces expression of sclerostin by osteocytes: a novel mechanism for hormonal control of osteoblastogenesis. Endocrinology. 2005;146(11):4577–83.

Ben-Dov IZ, Galitzer H, Lavi-Moshayoff V, Goetz R, Kuro-o M, Mohammadi M, Sirkis R, Naveh-Many T, Silver J. The parathyroid is a target organ for FGF23 in rats. J Clin Invest. 2007;117:4003–8.

Bikle DD. Vitamin D metabolism, mechanism of action, and clinical applications. Chem Biol. 2014;21:319–29.

Bikle D, Adams J, Christakos S. Vitamin D: production, metabolism and clinical requirements. In: Rosen C, editor. Primer Metab Bone Dis. Hoboken: Wiley; 2013. p. 235–45.

Bowe AE, Finnegan R, Jan de Beur SM, et al. FGF-23 inhibits renal tubular phosphate transport and is aPHEXsubstrate. Biochem Biophys Res Commun. 2001;284:977–81.

Brenza HL, DeLuca HF. Regulation of 25-hydroxyvitamin D3 1alpha-hydroxylase gene expression by parathyroid hormone and 1,25-dihydroxyvitamin D3. Arch Biochem Biophys. 2000;381:143–52.

Bushinski DA1, Nilsson EL. Additive effects of acidosis and parathyroid hormone on mouse osteoblastic and osteoclastic function. Am J Phys. 1995;269(6 Pt 1):C1364–70.

Campos M, Couture C, Hirata IY, Juliano MA, Loisel TP, Crine P, Juliano L, Boileau G, Carmona AK. Human recombinant endopeptidase PHEX has a strict S1′ specificity for acidic residues and cleaves peptides derived from fibroblast growth factor-23 and matrix extracellular phosphoglycoprotein. Biochem J. 2003;373(Pt 1):271–9.

Canaff L, Hendy GN. Human calcium-sensing receptor gene. Vitamin D response elements in promoters P1 and P2 confer transcriptional responsiveness to 1,25- dihydroxyvitamin D. J Biol Chem. 2002;277:30337–50.

Castiglioni S1, Cazzaniga A, Albisetti W, Maier JA. Magnesium and osteoporosis: current state of knowledge and future research directions. Forum Nutr. 2013;5(8):3022–33. doi:10.3390/nu5083022.
Chambers TJ, McSheehy PM, Thomson BM, Fuller K. The effect of calcium-regulating hormones and prostaglandins on bone resorption by osteoclasts disaggregated from neonatal rabbit bones. Endocrinology. 1985a;116(1):234–9.
Chin KY, Ima-Nirwana S, Mohamed IN, Ahmad F, Ramli ES, Aminuddin A, et al. Serum osteocalcin is significantly related to indices of obesity and lipid profile in Malaysian men. Int J Med Sci. 2014;11:151–7.
Christakos S, Lieben L, Masuyama R, Carmeliet G. Vitamin D endocrine system and the intestine. Bone Key Reports. 2014;3:496.
Chun RF, Peercy BE, Orwoll ES, Nielson CM, Adams JS, Hewison M. VitaminDand DBP: the free hormone hypothesis revisited. J Steroid Biochem Mol Biol. 2014;144:132–7.
Clemens TL, Karsenty G. The osteoblast: an insulin target cell controlling glucose homeostasis. J Bone Miner Res. 2011;26:677–80.
Cochran M, Peacock M, Sachs G, Nordin BE. Renal effects of calcitonin. Br Med J. 1970;1(5689):135–7.
Cunningham J, Rodriguez M, Messa P. Magnesium in chronic kidney disease stages 3 and 4 and in dialysis patients. Clin Kidney J. 2012;5:i39–51.
De Beur SM, et al. Tumors associated with oncogenic osteomalacia express genes important in bone and mineral metabolism. J Bone Miner Res. 2002;17:1102–10.
Ducy P, Schinke T, Karsenty G. The osteoblast: a sophisticated fibroblast under central surveillance. Science. 2000;289(5484):1501–4.
Elin RJ. Assessment of magnesium status. Clin Chem. 1987;33:1965–70.
Faul C. Fibroblast growth factor 23 and the heart. Curr Opin Nephrol Hypertens. 2012;21(4):369–75.
Foresta C, Strapazzon G, De Toni L, et al. Evidence for osteocalcin production by adipose tissue and its role in human metabolism. J Clin Endocrinol Metab. 2010;95:3502–6.
Forster RE, Jurutka PW, Hsieh JC, Haussler CA, Lowmiller CL, Kaneko I, Haussler MR, Kerr Whitfield G. Vitamin D receptor controls expression of the anti-aging klotho gene in mouse and human renal cells. Biochem Biophys Res Commun. 2011;414:557–62.
Gagnon KB, Delpire E. Molecular physiology of SPAK and OSR1: two Ste20-related protein kinases regulating ion transport. Physiol Rev. 2012;92:1577–617.
Gee J, Bailey H, Kim K, Kolesar J, Havighurst T, Tutsch KD, See W, Cohen MB, Street N, Levan L, Jarrard D, Wilding G. Phase II open label, multi-center clinical trial of modulation of intermediate endpoint biomarkers by 1alpha-hydroxyvitamin D2 in patients with clinically localized prostate cancer and high grade pin. Prostate. 2013;73:970–8.
Gowen LC, Petersen DN, Mansolf AL, Qi H, Stock JL, Tkalcevic GT, Simmons HA, Crawford DT, Chidsey-Frink KL, Ke HZ, McNeish JD, Brown TA. Targeted disruption of the osteoblast/osteocyte factor 45 gene (OF45) results in increased bone formation and bone mass. J Biol Chem. 2003;278:1998–2007.
Grabner A, Amaral AP, Schramm K, Singh S, Sloan A, Yanucil C, Li J, Shehadeh LA, Hare JM, David V, Martin A, Fornoni A, Di Marco GS, Kentrup D, Reuter S, Mayer AB, Pavenstädt H, Stypmann J, Kuhn C, Hille S, Frey N, Leifheit-Nestler M, Richter B, Haffner D, Abraham R, Bange J, Sperl B, Ullrich A, Brand M, Wolf M, Faul C. Activation of Cardiac Fibroblast Growth Factor Receptor 4 Causes Left Ventricular Hypertrophy. Cell Metab. 2015;22(6):1020–32.
Groenestege WM, Thébault S, van der Wijst J, van den Berg D, Janssen R, Tejpar S, van den Heuvel LP, van Cutsem E, Hoenderop JG, Knoers NV, Bindels RJ. Impaired basolateral sorting of pro-EGF causes isolated recessive renal hypomagnesemia. J Clin Invest. 2007;117(8):2260–7.
Guo R, Rowe PS, Liu S, Simpson LG, Xiao ZS, Quarles LD. Inhibition of MEPE cleavage by PHEX. Biochem Biophys Res Commun. 2002;297:38–45.
Hannon MJ1, Verbalis JG. Sodium homeostasis and bone. Curr Opin Nephrol Hypertens. 2014;23(4):370–6. doi:10.1097/01.mnh.0000447022.51722.f4.

Heaney RP, Horst RL, Cullen DM, Armas LA. Vitamin D3 distribution and status in the body. J Am Coll Nutr. 2009;28:252–6.
Henry HL. Regulation of vitamin D metabolism. Best practice & research. Clin Endocrinol Metab. 2011;25:531–41.
Hoorn EJ, Lindemans J, Zietse R. Development of severe hyponatraemia in hospitalized patients: treatment-related risk factors and inadequate management. Nephrol Dial Transplant. 2006;21:70–6.
Horwood NJ, Udagawa N, Elliott J, et al. Interleukin 18 inhibits osteoclast formation via T cell production of granulocyte macrophage colony-stimulating factor. J Clin Invest. 1998;
Hou J, Shan Q, Wang T, Gomes AS, Yan Q, Paul DL, Bleich M, Goodenough DA. Transgenic RNAi depletion of claudin-16 and the renal handling of magnesium. J Biol Chem. 2007;282(23):17114–22.
Ichikawa S, Imel EA, Kreiter ML, et al. A homozygous missense mutation in human KLOTHO causes severe tumoral calcinosis. J Clin Invest. 2007;117:2684–91.
Jiang Y, Cong P, Williams SR, et al. WNK4 regulates the secretory pathway via which TRPV5 is targeted to the plasma membrane. Biochem Biophys Res Commun. 2008;375:225–9.
Jilka RL. Molecular and cellular mechanisms of the anabolic effect of intermittent PTH. Bone. 2007;40(6):1434–46.
Kim S, Yamazaki M, Zella LA, Shevde NK, Pike JW. Activation of receptor activator of NF-kappaB ligand gene expression by 1,25-dihydroxyvitamin D3 is mediated through multiple long-range enhancers. Mol Cell Biol. 2006;26:6469–86.
Kindblom JM, Ohlsson C, Ljunggren O, Karlsson MK, Tivesten A, Smith U, et al. Plasma osteocalcin is inversely related to fat mass and plasma glucose in elderly Swedish men. J Bone Miner Res Off J Am Soc Bone Miner Res. 2009;24:785–91.
Krajisnik T, Bjorklund P, Marsell R, et al. Fibroblast growth factor-23 regulates parathyroid hormone and 1alpha-hydroxylase expression in cultured bovine parathyroid cells. J Endocrinol. 2007;195:125–31.
Kurihara N, Civin C, Roodman GD. Osteotropic factor responsiveness of highly purified populations of early and late precursors for human multinucleated cells expressing the osteoclast phenotype. J Bone Miner Res. 1991;6(3):257–61.
Lee SK, Lorenzo JA. Regulation of receptor activator of nuclear factor-kappa B ligand and osteoprotegerin mRNA expression by parathyroid hormone is predominantly mediated by the protein kinase a pathway in murine bone marrow culture. Bone. 2002;31:252–9.
Lemann J Jr, Bushinsky DA, Hamm LL. Bone buffering of acid and base in humans. Am J Physiol Ren Physiol. 2003;285(5):F811–32.
Liu S1, Brown TA, Zhou J, Xiao ZS, Awad H, Guilak F, Quarles LD. Role of matrix extracellular phosphoglycoprotein in the pathogenesis of X-linked hypophosphatemia. J Am Soc Nephrol. 2005;16(6):1645–53. Epub 2005 Apr 20.
Liu S, Gupta A, Quarles LD. Emerging role of fibroblast growth factor 23 in a bone-kidney axis regulating systemic phosphate homeostasis and extracellular matrix mineralization. Curr Opin Nephrol Hypertens. 2007a;16:329–35.
Liu F, Lee SK, Adams DJ, et al. CREM deficiency in mice alters the response of bone to intermittent parathyroid hormone treatment. Bone. 2007b;40:1135–43.
Nabeshima Y. The discovery of α-Klotho and FGF23 unveiled new insight into calcium and phosphate homeostasis cell. Mol Life Sci. 2008;65:3218–30.
Olauson H, Vervloet MG, Cozzolino M, Massy ZA, Ureña Torres P, Larsson TE. New insights into the FGF23-Klotho axis. Semin Nephrol. 2014;34(6):586–97.
Penido MG, Alon US. Phosphate homeostasis and its role in bone health. Pediatr Nephrol. 2012;27:2039–48.
Perwad F, Zhang MY, Tenenhouse HS, Portale AA. Fibroblast growth factor 23 impairs phosphorus and vitamin D metabolism in vivo and suppresses 25-hydroxyvitamin D-1alpha-hydroxylase expression in vitro. Am J Physiol Ren Physiol. 2007;293:F1577–83.

Portale AA, Halloran BP, Morris RC Jr. Physiologic regulation of the serum concentration of 1,25-dihydroxyvitamin D by phosphorus in normal men. J Clin Invest. 1989;83:1494–9.
Price PA, Williamson MK. Primary structure of bovine matrix Gla protein, a new vitamin K-dependent bone protein. J Biol Chem. 1985;260:14971–5.
Quarles LD. FGF23, PHEX, and MEPE regulation of phosphate homeostasis and skeletal mineralization. Am J Physiol Endocrinol Metab. 2003;285(1):E1–9.
Quarles LD, Drezner MK. Pathophysiology of X-linked hypophosphatemia, tumor-induced osteomalacia, and autosomal dominant hypophosphatemia: a perPHEXing problem. J Clin Endocrinol Metab. 2001;86:494–6.
Raggatt LJ, Qin L, Tamasi J, et al. Interleukin-18 is regulated by parathyroid hormone and is required for its bone anabolic actions. J Biol Chem. 2008;283:6790–8. SFX Studies using mice lacking IL-18 demonstrate a role for IL-18 in the anabolic actions of PTH in bone.
Razzaque MS. Osteocalcin: a pivotal mediator or an innocent bystander in energy metabolism? Nephrol Dial Transplant. 2011;26(1):42–5.
Razzaque MS, Lanske B. The emerging role of the fibroblast growth factor-23-klotho axis in renal regulation of phosphate homeostasis. J Endocrinol. 2007;194:1–10.
Rejnmark L, Vestergaard P, Pedersen AR, Heickendorff L, Andreasen F, Mosekilde L. Dose–effect relations of loop- and thiazide-diuretics on calcium homeostasis: a randomized, double-blinded Latin-square multiple cross-over study in postmenopausal osteopenic women. Eur J Clin Investig. 2003;33:41–50.
Rowe PS. The wrickkened pathways of FGF23, MEPE and PHEX. Crit Rev Oral Biol Med. 2004;15(5):264–81.
Rowe PS. A unified model for bone-renal mineral and energy metabolism. Curr Opin Pharmacol. 2015;22:64–71. doi:10.1016/j.coph.2015.03.006.
Schiavi SC, Kumar R. The phosphatonin pathway: new insights in phosphate homeostasis. Kidney Int. 2004;65(1):1–14.
Segawa H, Yamanaka S, Onitsuka A, et al. Parathyroid hormone-dependent endocytosis of renal type IIc Na-Pi cotransporter. Am J Physiol Ren Physiol. 2007;292:F395–403.
Sellmeyer DE, Schloetter M, Sebastian A. Potassium citrate prevents increased urine calcium excretion and bone resorption induced by a high sodium chloride diet. J Clin Endocrinol Metab. 2002;87:2008–12.
Selvamurugan N, Shimizu E, Lee M, et al. Identification and characterization of Runx2 phosphorylation sites involved in matrix metalloproteinase-13 promoter activation. FEBS Lett. 2009;583:1141–6.
Shimada T, Hasegawa H, Yamazaki Y, et al. FGF-23 is a potent regulator of vitamin D metabolism and phosphate homeostasis. J Bone Miner Res. 2004;19:429–35.
Shimada T, Mizutani S, Muto T, et al. Cloning and characterization of FGF23 as a causative factor of tumor-induced osteomalacia. Proc Natl Acad Sci U S A. 2001;98:6500–5.
Shirley DG, Faria NJ, Unwin RJ, et al. Direct micropuncture evidence that matrix extracellular phosphoglycoprotein inhibits proximal tubular phosphate reabsorption. Nephrol Dial Transplant Advance. Access published on May 14, 2010. 2010; doi:10.1093/ndt/gfq263.
Siew K, O'Shaughnessy KM. Extrarenal roles of the with-no-lysine[K] kinases (WNKs). Clin Exp Pharmacol Physiol. 2013;40(12):885–94.
Tamma R, Sun L, Cuscito C, et al. Regulation of bone remodeling by vasopressin explains the bone loss in hyponatremia. Proc Natl Acad Sci U S A. 2013;110:18644–9.
Teitelbaum SL. Bone resorption by osteoclasts. Science. 2000;289(5484):1504–8.
Traebert M, Volkl H, Biber J, et al. Luminal and contraluminal action of 1–34 and 3–34 PTH peptides on renal type IIa Na-P(i) cotransporter. Am J Physiol Ren Physiol. 2000;278:F792–8.
Urakawa I, Yamazaki Y, Shimada T, et al. Klotho converts canonical FGF receptor into a specific receptor for FGF23. Nature. 2006;444:770–4.
van Abel M, Hoenderop JG, van der Kemp AW, et al. Coordinated control of renal Ca(2+) transport proteins by parathyroid hormone. Kidney Int. 2005;68:1708–21.

Wagner CA. Disorders of renal magnesium handling explain renal magnesium transport. J Nephrol. 2007;20(5):507–10.

Weinman EJ, Biswas RS, Peng G, et al. Parathyroid hormone inhibits renal phosphate transport by phosphorylation of serine 77 of sodium–hydrogen exchanger regulatory factor-1. J Clin Invest. 2007;117:3412–20.

White KE, Carn G, Lorenz-Depiereux B, Benet-Pages A, Strom TM, Econs MJ. Autosomal-dominant hypophosphatemic rickets (ADHR) mutations stabilize FGF-23. Kidney Int. 2001;60(6):2079–86.

Xu B, English JM, Wilsbacher JL, Stippec S, Goldsmith EJ, Cobb MH. WNK1, a novel mammalian serine/threonine protein kinase lacking the catalytic lysine in subdomain II. J Biol Chem. 2000;275:16795–801.

Yeap BB, Chubb SA, Flicker L, McCaul KA, Ebeling PR, Beilby JP, et al. Reduced serum total osteocalcin is associated with metabolic syndrome in older men via waist circumference, hyperglycemia, and triglyceride levels. Eur J Endocrinol. 2010;163:265–72.

Yoshiko Y, Wang H, Minamizaki T, et al. Mineralized tissue cells are a principal source of FGF23. Bone. 2007;40:1565–73.

Yu X, White KE. FGF23 and disorders of phosphate homeostasis. Cytokine Growth Factor Rev. 2005;16(2):221–32.

Chapter 12
Bone and the Immune System

M. Neale Weitzmann

Abstract The mechanisms that control physiological and pathological bone mass and turnover have been extensively investigated over the past several decades. One surprising discovery that emerged during this journey is the existence of an immuno-skeletal interface (ISI), a centralization of shared cells and cytokine effectors serving dual, but separate, functions within both the immune and skeletal systems. The existence of this ISI has profound consequences for normal bone turnover and homeostasis and also appears to be key to the etiologies of multiple diverse osteoporotic conditions relating to rheumatoid arthritis, periodontal infection, postmenopausal osteoporosis, hyperparathyroidism, HIV infection, and antiretroviral therapy. Although most ISI activities elucidated to date converge on the process of bone resorption, recent studies have identified unexpected anabolic roles of the ISI in bone formation. Specifically, the bone anabolic actions of parathyroid hormone and the pharmaceutical costimulation inhibitor CTLA4-Ig (abatacept) are mediated though mechanisms involving T-cells. This chapter documents some of the key discoveries in the field of osteoimmunology in the context of both physiological and pathological bone turnover.

Keywords Bone • Osteoporosis • T-cells • B-cells • Immuno-skeletal interface • ISI • Immune system • Osteoimmunology

12.1 Introduction

A surprising finding in bone biology is how tightly the skeletal system is intertwined with the immune response. This immuno-skeletal interface (ISI) comprising common cell and cytokine effectors mediates responses within both immune and

M.N. Weitzmann, PhD (✉)
The Atlanta U.S. Department of Veterans Affairs Medical Center, Decatur, GA, USA

Department of Medicine, Division of Endocrinology and Metabolism and Lipids, Emory University School of Medicine, 101 Woodruff Circle, 1305 WMB, Atlanta, GA 30322-0001, USA
e-mail: mweitzm@emory.edu

S.Y. Smith et al. (eds.), *Bone Toxicology*, Molecular and Integrative Toxicology, DOI 10.1007/978-3-319-56192-9_12

skeletal systems and has spawned an independent subspecialty of bone biology referred to as "osteoimmunology".

The ISI encompasses major components of both innate and adaptive immunities including lymphocytes (both B-cells and T-cells), monocytes/macrophages, and dendritic cells and is now recognized to affect both bone resorption and bone formation.

The notion of T-cells impacting bone turnover is however not a new concept, and as far back as the early 1970s, reports were already emerging to suggest an association between osteoclastic bone loss and activated immune cells. One of the earliest of these reports was in relation to periodontitis, an inflammatory disease associated with bacterial infection and characterized by lymphocyte infiltration of the periodontal tissues leading to alveolar bone resorption and causing tooth loss. Interestingly, when human peripheral blood leukocytes were treated with lymphocyte antigens derived from dental plaque deposits, or with the T-cell activator, phytohemagglutinin (PHA), the culture supernatants were found to contain an osteoclastogenic substance, referred to as "osteoclast activating factor (OAF)". OAF was observed to stimulate bone resorption and the release of calcium from fetal rat bones in ex vivo organ cultures (Horton et al. 1972, 1974).

Other classical activators of T-cells including pokeweed mitogen and concanavalin-A were also demonstrated to induce OAF, further implicating activated T-cells in the process of osteoclastogenesis and bone resorption (Trummel et al. 1975).

Multiple human lymphoid cell lines derived from both T- and B-cell tumors, but not other neoplasms, were likewise found to secrete OAF, which was speculated to account for the high incidence of bone lesions in affected patients (Mundy et al. 1974). Detailed investigations of the skeletons of T-cell lymphoma patients further revealed complex bone manifestations including osteolytic and osteoblastic lesions and a diffuse osteoporosis. These observations led to a speculation that similar to mitogen and antigen-activated T-cells, T-cell lymphomas may derive from a subset of T-cells capable of producing bone-resorbing factor(s) akin to OAF (Brigham et al. 1982).

In the field of rheumatology, the concept of immune-directed bone and cartilage destruction had long been recognized. Rheumatoid arthritis is a common autoimmune disease characterized by chronic inflammation of the synovial joints and by infiltrating T-cells, macrophages, and plasma cells, all of which show signs of activation (Feldmann et al. 1996). Indeed, enriched populations of B-cells were also found to produce OAF upon activation (Chen et al. 1976), although other studies, using highly purified B-cells stimulated with lipopolysaccharide (LPS), found no OAF-like activity in contrast to cells activated in the presence of macrophages (Horowitz et al. 1984).

It would take another two and a half decades and the discovery of the receptor activator of NF-κB (RANK), its ligand (RANK ligand, RANKL), and its physiological moderator, the RANKL decoy receptor osteoprotegerin (OPG), as well as the development of genetic mouse models, before the mechanisms underlying this interrelationship between immune and skeletal function in physiological and pathological bone turnover would be explained.

Today the scope of the ISI has exploded in unexpected directions to encompass multiple pathologies of the skeleton including immunodeficiency syndromes such as HIV-1 infection, metabolic disturbances such as hyperparathyroidism, and the archetypal bone disease of women, postmenopausal osteoporosis.

Furthermore, recent data has emerged to suggest that the ISI is not only pertinent to osteoclastic bone catabolism but may also regulate the process of osteoblastic bone formation. The mechanisms by which immune cells cause bone anabolism and whether this can be exploited therapeutically to ameliorate bone disease is a new and unexpected direction in osteoimmunology.

It should be pointed out that many of the concepts outlined in this chapter have yet to be incorporated into mainstream bone biology, and in many quarters, immune cells are still considered to be experimental curiosities, rather than key participants in bone physiology and pathology. A common misconception is that osteoimmunology replaces classical bone biology, when in fact the ISI appears to act as an amplifier of bone metabolism, or in a permissive role, rather than working independently of established bone cells such as osteoclasts, osteoblasts, and osteocytes. However, many of the concepts outlined have been developed in animal models and remain to be ratified in the human system, before they can become better accepted. Nonetheless, a strong body of evidence continues to accumulate in support of a key role of the ISI in bone biology and the remainder of this chapter will summarize some of the key discoveries made over the last two decades in support of a role for the ISI in physiological and pathological bone turnover.

12.2 An Immune Origin for Osteoclasts

Osteoclasts are the unique bone-degrading (resorbing) cells of the body. Originally proposed to derive from cartilage cells (Geddes 1913), later studies based on histological evidence suggested a common osteoprogenitor between osteoblasts (the bone-building cells) and osteoclasts, spawning bone-resorbing cells at one stage of the cycle and then bone-depositing cells at the next (Buring 1975).

By demonstrating that transplantation of bone marrow or spleen cells, or parabiotic cross-circulation to normal siblings could rescue bone resorption in osteopetrotic mice, Walker determined that the likely origin or osteoclasts was in fact hematopoietic cells (Walker 1972, 1975). Shortly thereafter, Buring used radioactive thymidine incorporation to track migration of cells from a labeled rat to an unlabeled rat connected though a parabiotic cross-circulation. The resulting multinucleated osteoclasts contained both radioactively labeled nuclei and non-labeled nuclei. Because only mature granulocytes, lymphocytes, and monocytes in the circulation were able to cross to the unlabeled parabiont, Buring surmised that blood-borne mononuclear cells, most likely monocytes, give rise to mononuclear perivascular cells that further develop onto multinuclear phagocytic cells (the osteoclasts) (Buring 1975). Although Buring assumed that the monocytes were achieving multinuclearity though cellular division, rather than fusion as is now accepted, the identification of the monocyte as the origin of the osteoclast was indeed correct.

Monocytes and their activated progeny, the macrophages, are key immune components functioning in both innate and adaptive immunities. Macrophages, along with B-cells and dendritic cells, function as professional antigen-presenting cells (APC); however, when monocytes are exposed to appropriate cytokines, they differentiate into pre-osteoclasts, which in turn fuse together to form multinucleated mature bone-resorbing osteoclasts. This immune origin of the osteoclast provided the first association between the immune and skeletal systems. However, the significance of the immune system to bone biology was not immediately appreciated, and the precise mechanisms driving this conversion of monocytes to osteoclasts until 1998 were unclear and highly debated. The identification of "OAF" itself remained stubbornly elusive, although a number of inflammatory cytokines were proposed as candidates and intensively investigated.

Early support for a role of IL-1 and/or TNFα came from studies of bone loss due to estrogen deficiency as production of these cytokines was found to be significantly elevated from mononuclear cells derived from the blood of surgically ovariectomized women (Pacifici et al. 1991), a condition associated with bone loss. Animal studies revealed that pharmacological suppression of IL-1 in ovariectomized rats, an animal model of postmenopausal osteoporosis, significantly diminished bone loss (Kimble et al. 1994) while bone loss was completely abolished by simultaneous use of IL-1 and TNFα antagonists (Kimble et al. 1995). These studies provided powerful evidence for a role of IL-1 and TNFα in ovariectomy-induced bone loss.

Other studies, however, invoked a key role of IL-6 in ovariectomy-induced bone loss (Passeri et al. 1993; Manolagas et al. 1995), as well as in Paget's disease of bone (Roodman et al. 1992), and other cytokines were added to the list of candidates for OAF including GM-CSF (Pacifici et al. 1991), M-CSF (Sarma and Flanagan 1996; Srivastava et al. 1998), VEGF (Niida et al. 1999), IL-7 (Weitzmann et al. 2000a, 2002), TGFβ (Hattersley and Chambers 1991; Shinar and Rodan 1990), and others.

An embarrassing truth, however, was that treatment of purified monocytes with inflammatory cytokines, either alone or in combination, failed to produce significant formation of functional osteoclasts.

12.3 Osteoclast Differentiation and the RANK/RANKL/OPG Pathway

In 1998, the field of osteoclast biology was changed forever with the near simultaneous and independent identification of osteoprotegerin ligand (OPGL) (Lacey et al. 1998) and of osteoclast differentiation factor (ODF), (Matsuzaki et al. 1998). OPGL/ODF was a single cytokine, which in the presence of permissive concentrations of the monocyte/macrophage survival factor M-CSF, was capable of inducing the differentiation of monocytes into functional osteoclasts at high frequency.

In fact, the existence of a cytokine structurally identical to OPGL/ODF had been reported by two independent groups of immunologists the previous year and had been named receptor activator of NF-κB (RANK) ligand (RANKL) (Anderson et al. 1997) and TNF-related activation-induced cytokine (TRANCE) (Wong et al. 1997a, b). Interestingly, T-cells were reported to be a dominant source of RANKL/TRANCE expression, and this cytokine was shown to bind to a receptor (RANK (also known as TRANCE-R)) expressed on the surface of dendritic cells (Anderson et al. 1997).

In the context of bone biology, the terminology has now been standardized, and this key osteoclastogenic cytokine is now referred to as RANKL and that of its receptor as RANK.

After extensive investigations, we now understand that what distinguishes an osteoclast precursor from that of a regular monocyte/macrophage is the expression on its surface of the receptor RANK. When the key osteoclastogenic cytokine, RANKL, binds to its receptor, in the presence of permissive concentrations of M-CSF, it induces differentiation of these cells into mononucleated pre-osteoclasts that fuse together to form mature giant multinucleated bone-resorbing cells referred to as osteoclasts (Teitelbaum 2000; Khosla 2001).

Another key discovery just prior to that of RANKL was the identification of a potent inhibitor of osteoclast differentiation by two independent groups, namely, osteoprotegerin (OPG) (Simonet et al. 1997) and osteoclastogenesis inhibitory factor (OCIF) (Tsuda et al. 1997). OPG and OCIF are structurally and functionally identical, and OPG is now the preferred terminology. OPG is a decoy receptor for RANKL and inhibits osteoclastogenesis and bone resorption by binding to RANKL and preventing association with RANK on the osteoclast precursor (Lacey et al. 1998).

The development of genetic models has further allowed for definitive demonstrations of the roles of RANK, RANKL, and OPG in osteoclast differentiation to be made. Genetic ablation of RANKL in mice leads to a near complete failure in osteoclast formation and development of a high bone mass phenotype (osteopetrosis) (Kong et al. 1999a). OPG deletion, by contrast, leads to large numbers of bone-resorbing osteoclasts and low bone mineral density (BMD) (Mizuno et al. 1998; Bucay et al. 1998).

The discovery of the RANK/RANKL/OPG axis has profoundly changed our understanding of the process of osteoclastogenesis and bone resorption. We further understand that the plethora of inflammatory cytokines that were previously associated with bone resorption promote osteoclastogenesis/bone resorption through the following mechanisms: (1) They promote osteoclastogenesis by supporting the survival of monocytes and/or by converting monocytes into osteoclast precursors through upregulation of RANK expression ((e.g., M-CSF (Arai et al. 1999)). (2) They promote production of RANKL by various cell types (e.g., IL-7 promotes RANKL in T-cells (Weitzmann et al. 2000a), and IL-1 and TNFα promote RANKL expression in human osteoblastic cells (Hofbauer et al. 1999a; Wei et al. 2005). (3) They down-modulate expression of the inhibitory factor OPG ((e.g., IL-7 (Weitzmann et al. 2002)). (4) Although TNFα is not generally considered to be osteoclastogenic in the absence of RANKL stimulation, it has a unique property in that it is able to potently amplify the activity of RANKL to superinduce osteoclast formation by interacting with RANK signaling pathways (Cenci et al. 2000; Lam et al. 2000;

Zhang et al. 2001) and stimulating bone resorption through a direct action independent of and strongly synergistic with RANKL (Fuller et al. 2002).

Despite the fact that RANKL was originally identified as a T-cell-produced cytokine, until recently, the RANKL necessary for regulating basal bone turnover has historically been ascribed to the osteoblasts and their precursors, the bone marrow stromal cells (Gori et al. 2000; Suda et al. 1999). However, with the advent of powerful conditional knockout (KO) mouse models, recently a prominent role for additional cell types including osteocytes (Xiong et al. 2011; Nakashima et al. 2011) and hypertrophic chondrocytes (Xiong et al. 2011) has been unveiled.

Interestingly, under basal conditions, the key source of OPG needed to control bone turnover has historically also been ascribed to osteoblast lineage cells (Hofbauer et al. 1999b, 2000). However studies by our lab have now detailed a key role for the immune system in both production and regulation of OPG.

The identification of the RANK/RANKL/OPG pathway has not only opened the door to a wealth of discovery regarding basic bone biology and pathological bone turnover but has provided key insights into the working of the ISI.

The remainder of this chapter will examine in some detail, but non-exhaustively, some of the major research findings over the past two decades into deciphering the interaction between the skeletal and immune systems in physiological bone turnover and in disease.

12.4 Secreted Osteoclastogenic Factor of Activated T-Cells and the Role of the ISI in RANKL-Independent Bone Loss

Since its identification as the key osteoclastogenic cytokine in 1998, RANKL has been considered to be a final downstream effector of physiological and pathological osteoclastogenesis. Although early studies suggested that TNFα, a close family member of RANKL, could also promote RANKL-independent osteoclast formation (Kobayashi et al. 2000), other studies have since concluded that TNFα only augments osteoclastogenesis in the presence of RANKL or if osteoclast precursors have been pre-exposed to RANKL (Lam et al. 2000). Indeed, several studies have reported that the major action of TNFα is to synergize with RANKL (Cenci et al. 2000; Lam et al. 2000) acting downstream of the type 1 (p55) TNF receptor (Zhang et al. 2001) and to amplify osteoclast differentiation and resorptive activity (Fuller et al. 2002).

While TNFα is no longer generally considered to be a direct osteoclastogenic cytokine, we have in fact reported the existence of RANKL-independent osteoclastogenic activity in media derived from activated T-cells that induced osteoclast formation independently of RANKL and in the presence of saturating concentrations of OPG (Weitzmann et al. 2001). Employing biochemical purification, mass spectroscopy, and recombinant DNA technologies, we were indeed able to isolate, identify, clone, and express a recombinant protein for a factor we termed secreted osteoclastogenic factor of activated T-cells (SOFAT). SOFAT was derived from an unusual mRNA splice variant coded by the threonine synthase-like 2 (THNSL2) gene homolog. Like RANKL, SOFAT-induced osteoclastogenic activity is potently augmented by TNFα, and in addition to its osteoclastogenic activity, SOFAT further promotes IL-6 production by osteoblasts (Rifas and Weitzmann 2009).

Although unlikely to contribute significantly to normal physiological osteoclastogenesis, given that a key source of SOFAT is activated T-cells, we hypothesized a role for SOFAT in the exacerbation of bone loss in the context of inflammatory diseases. Indeed, recent independent studies of SOFAT have revealed significantly elevated levels of SOFAT mRNA and protein expression in the gingival tissue of periodontitis patients compared to non-periodontitis controls. This group further demonstrated for the first time the capacity of recombinant SOFAT administration to stimulate formation of osteoclast-like cells in the periodontal ligament in mice in vivo. These studies suggested an important role for SOFAT in alveolar bone loss associated with periodontitis (Jarry et al. 2013). New studies further suggest a broader tissue expression of SOFAT and now include activated B-cells among the key sources of SOFAT (Jarry et al. 2016).

The role of SOFAT in other inflammatory states, however, remains to be studied, as does its molecular mechanisms of action.

12.5 The Immuno-Skeletal Interface in Physiological Osteoclastogenesis and Bone Remodeling

Although Horowitz et al. hypothesized a role for T-cells in physiological bone remodeling as far back as 1984 (Horowitz et al. 1984), the vast majority of investigations into the ISI have been conducted in the context of pathological bone turnover. In fact, evidence suggesting a putative role for T-cells in basal osteoclastogenesis in vivo had already been reported, but the significance thereof was not appreciated at the time. These observations stemmed from studies of nude mice, a strain deficient in T-cells, that displayed significantly increased osteoclast activity. The authors attributed these alterations in part to a putative increase in production of the as yet unidentified cytokine OAF, possibly by B lymphocytes or macrophages, whose normal function was disrupted by the absence of T-cells (Gyarmati et al. 1983).

Almost two decades later, supporting evidence of an inhibitory role for T-cells on osteoclast formation came from studies in which T-cell subsets ($CD4^+$ and $CD8^+$) were depleted in wild-type (WT) mice using neutralizing antibodies. The data revealed that T-cell-depleted bone marrow when treated with vitamin D3 produced significantly enhanced osteoclast formation in vitro (Grcevic et al. 2000). This outcome was unexpected given the accumulating evidence for a pro-osteoclastogenic role for T-cells. Mechanistic examinations further suggested a role for prostaglandin synthesis as indomethacin treatment was observed to abolish the stimulatory effect.

Importantly, T-cell depletion was also reported to induce a complete suppression of OPG production (Grcevic et al. 2000).

Interestingly, while investigating the effects of IL-7 on T-cell-mediated bone loss, our laboratory was able to confirm the earlier report of increased osteoclast number and activity in nude mice and further reported a significantly diminished basal BMD and bone structure indices (Toraldo et al. 2003). Furthermore, we identified a significant decline in OPG, but not of RANKL expression. Importantly, these changes were not directly related to the loss of T-cells but due to a significant decline in OPG production by B-cells (Li et al. 2007a).

12.6 B-Cells: A Dominant Source of OPG Production and a Key Regulator of Physiological Osteoclastogenesis and Bone Remodeling

In fact, almost a decade before, human tonsil B-cells had been reported to secrete copious concentrations of OPG, and in vitro activation of the CD40 costimulatory molecule on the B-cells was found to further upregulate OPG production (Yun et al. 1998). These data were consistent with a murine study involving depletion of B-cell in vivo leading to intensified bone loss in a model of periodontitis and suggesting that B-cell-produced factors can function to limit bone resorption (Klausen et al. 1989).

Studies from our lab involving in vitro osteoclastogenesis experiments with human cells also identified inhibitory activities of peripheral blood B-cells on osteoclast formation, partly mediated through copious secretion by B-cells of TGFβ, promoting osteoclast apoptosis (Weitzmann et al. 2000b). The discovery that TGFβ is also a potent inducer of osteoblastic OPG production (Thirunavukkarasu et al. 2001) led us to further investigate the role of B-cells as a source of OPG.

Using immunomagnetic depletion, we demonstrated that removal of mouse B-cells from whole bone marrow led to a dramatic reduction in total OPG expression and secretion. Further resolving bone marrow B-cells into their major subpopulations, we demonstrated that mature B-cells contribute 40% of the total bone marrow OPG, with additional OPG coming from B-cell precursors, immature B-cell populations, and plasma cells with the total OPG contribution by the B-lineage estimated to be 64% and making the B-lineage the dominant source of OPG in mouse bone marrow (Li et al. 2007a). Consistent with these observations, B-cell-deficient mice displayed significant osteoclastic bone loss driven by a significant OPG deficit in the bone marrow. Adoptive transfer of B-cells into B-cell-deficient mice corrected OPG production and bone loss (Li et al. 2007a).

Similar to B-cell deficiency, loss of T-cells, the CD40 costimulatory receptor on APC and of its ligand CD40 ligand (CD40L) on T-cells, all significantly increased bone resorption and bone loss in these mouse models concurrent with significantly diminished total and B-cell OPG production (Li et al. 2007a). Taken together, the data suggested that lymphocytes (both B-cells and T-cells) function to moderate physiological bone resorption, diminishing bone turnover and contributing to the maintenance of peak BMD. Involvement of the CD40/CD40L system in basal bone turnover is further supported by studies in which polymorphisms in the Kozak sequence of the CD40 gene have been associated with low BMD in postmenopausal women (Pineda et al. 2008). Furthermore, osteoporotic fractures and significantly lower BMD and increase indices of osteoclastic bone resorption are common in patients with X-linked hyper-IgM syndrome, an inherited genetic disease involving mutations in the CD40L gene (Lopez-Granados et al. 2007).

12.7 Mechanisms of Bone Loss in HIV Infection and the Role of the ISI

An important prediction emerging from our studies revealing a protective effect of lymphocytes in basal bone homeostasis was that damage to the adaptive immune response, in particular T-cells and/or B-cells, may have significant repercussions for physiological bone turnover and bone mass.

Evidence in support of this theory were indeed already beginning to emerge in the context of the immunodeficiency disease, acquired immunodeficiency syndrome (AIDS), a state of severe immune compromization resulting from infection by the human immunodeficiency virus (HIV). Early reports suggested increased bone resorption markers concurrent with diminished bone formation in the context of HIV infection (Aukrust et al. 1999) and of significant reductions in BMD in individuals infected with HIV (Tebas et al. 2000). These studies were quickly followed up by a rash of publications reporting high incidence of BMD decline in HIV-infected subjects (Tebas et al. 2000; Jain and Lenhard 2002; Mora et al. 2001; Huang et al. 2001; Mondy and Tebas 2003).

Whether bone loss was a direct effect of HIV-1 infection or related to antiretroviral therapy (ART), also known as highly active antiretroviral therapy (HAART) or combinatorial antiretroviral therapy (cART), was however unclear with evidence both for (Tebas et al. 2000; Jain and Lenhard 2002; Bruera et al. 2003; Dube et al. 2002; Knobel et al. 2001; Brown and Qaqish 2006) and against (Bruera et al. 2003).

In addition, the coexistence of a large number of confounding traditional risk factors for osteoporosis in many HIV-infected subjects further clouded the issue (Ofotokun and Weitzmann 2010; Ofotokun and Weitzmann 2011).

Despite a lack of understanding of the cause, it has been estimated from meta-analysis that nearly two thirds of HIV-infected patients exhibit osteopenia, and full-blown osteoporosis occurs in up to 15% of patients (Brown and Qaqish 2006; Bonjoch et al. 2010; Sharma et al. 2010). Fears were soon raised as to a potential deluge of bone fractures in the HIV-infected community (Amorosa and Tebas 2006; McComsey et al. 2004). As BMD-derived T scores, used to quantify bone status clinically, do not accurately reflect fracture risk in subjects younger than 50 years of age (Hui et al. 1988), the consequences of low BMD in the relatively young HIV-infected population was initially unclear, and early reports of fracture comprised mainly anecdotal case reports (Guaraldi et al. 2001). However, over the next 5 years, a profusion of well-controlled, large population-based, clinical studies emerged demonstrating significant increases in fracture incidence in both men and women and over a wide age range with an average of 2–4% overall increase in fracture and up to ninefold increase in fracture at the hip (Prior et al. 2007; Triant et al. 2008; Young et al. 2011; Womack et al. 2011; Guerri-Fernandez et al. 2013; Prieto-Alhambra et al. 2014). As fracture is still fairly uncommon overall in the relatively younger HIV demographic, not all studies have identified increased fracture incidence including a study in predominantly premenopausal women in the Women's Interagency HIV Study (WIHS) cohort (Yin et al. 2010a). Interestingly, just 5 years later, the same group found in a follow-up study of WIHS women that middle-aged HIV-infected women had a higher adjusted fracture rate than HIV-uninfected women (Sharma et al. 2015). With the average age of HIV demographic now entering the 50s and many subjects now in their 60s, an age when fracture becomes much more common, there is a growing concern as to the skeletal health of these people and the potential for fracture as they continue to age.

The consequence of bone fracture can be dire and is a significant public health concern for both affected individuals and society as a whole (Brainsky et al. 1997). Fractures in the axial skeleton (vertebrae) are associated with deformity and/or chronic back pain, while the clinical management of hip fracture almost always requires invasive surgical intervention with 1-year mortality rates alone ranging from 24% to 33% (Lewis et al. 2006; Bass et al. 2007). In addition, up to 75% of survivors may require long-term rehabilitation in nursing facilities (Kates et al. 2007; Johnell and Kanis 2006) leading to loss of independence, often permanently. Factoring in treatment costs to patients, costs to healthcare systems and Medicare/Medicaid, and the economic costs related to lost productivity of fracture victims, the total annual financial cost of fracture in the US alone is projected to reach an alarming $25 billion by 2025 (Burge et al. 2007). The additional financial drain on healthcare resources associated with fracture in aging HIV-infected subjects has yet to be projected. Despite initial slow recognition of the potential problem, guidelines have finally emerged regarding screening, diagnosis, and treatment of bone disease in the HIV setting (McComsey et al. 2010), and recent revised guidelines now recommend that bone densitometry be performed in men aged ≥50 years, postmenopausal women, patients with a history of fragility fracture, patients receiving chronic glucocorticoid treatment, and patients at high risk of falls (Brown et al. 2015).

12.7.1 B-Cell OPG Production and Bone Loss in HIV Infection

The development of the HIV-1 transgenic rat model of HIV infection through transgenic overexpression of a replication incompetent HIV provirus has proven to be a significant technological advance in the field allowing researchers to differentiate between direct effects of viral action on bone structure and turnover at the molecular level. This model controls for many confounding risk factors for bone loss including smoking, alcohol consumption, drug abuse, vitamin D deficiency, and other lifestyle-associated factors that are common in human populations. Unlike previous mouse HIV models, the transgenic rat leads to effective immunological disruption and development of a constellation of pathologies that in many ways closely resemble that of human AIDS (Reid et al. 2001).

Bone phenotyping studies in HIV transgenic rats by our lab demonstrated a significant reduction in BMD and cortical and trabecular bone micro-architecture driven by a significant elevation in osteoclastic bone resorption. Quantifying OPG and RANKL production from purified B-cells revealed that B-cell OPG production was significantly diminished in the HIV transgenic rat relative to controls. This decline in OPG was further exacerbated by a significant increase in B-cell RANKL production with a net increase in the RANKL/OPG ratio, conditions favorable to enhanced osteoclast formation and bone resorption (Vikulina et al. 2010). Our study demonstrated for the first time a putative role for disruption of the ISI in the bone loss associated with HIV infection.

Although animal models are important mechanistic platforms for exploring hypotheses, the complexity of human pathogenesis ultimately requires the ratification of animal data in a bona fide clinical setting.

To this end, we have recently translated our preclinical HIV model into human HIV-1 infection. In a cross-sectional clinical study involving 58 HIV-negative and 62 ART-naive HIV-positive subjects, we performed an immuno-skeletal profiling to assess bone status in relationship to B-cell production of RANKL and OPG. As expected, we found incidence of bone resorption and osteopenia to be significantly higher in HIV-infected individuals. Importantly, as observed in the preclinical model, expression of RANKL by B-cells was increased significantly, while expression of OPG by B-cells was significantly diminished. B-cell RANKL and OPG remained significantly associated with HIV status in multivariable analysis after adjusting for age, gender, race, BMI, smoking, alcohol consumption, and fracture history. We further demonstrated that the B-cell RANKL/OPG ratio in HIV-infected subjects was skewed in favor of osteoclastic bone resorption and correlated significantly with BMD and derived T- and/or Z-scores at the total hip and femoral neck. For reasons that are not presently understood, there was no significant association between B-cell RANKL/OPG ratio and BMD at the lumbar spine, suggesting that the axial skeleton may be regulated differently (Titanji et al. 2014).

Taken together, our preclinical model in HIV-transgenic rats combined with clinical evidence of inverted B-cell RANKL/OPG ratio in human HIV infection provides the first demonstration that in addition to any other contributing factors, an immunocentric mechanism aligned with B-cell dysfunction may underlie bone loss in HIV-infected subjects. This model is shown diagrammatically in Fig. 12.1.

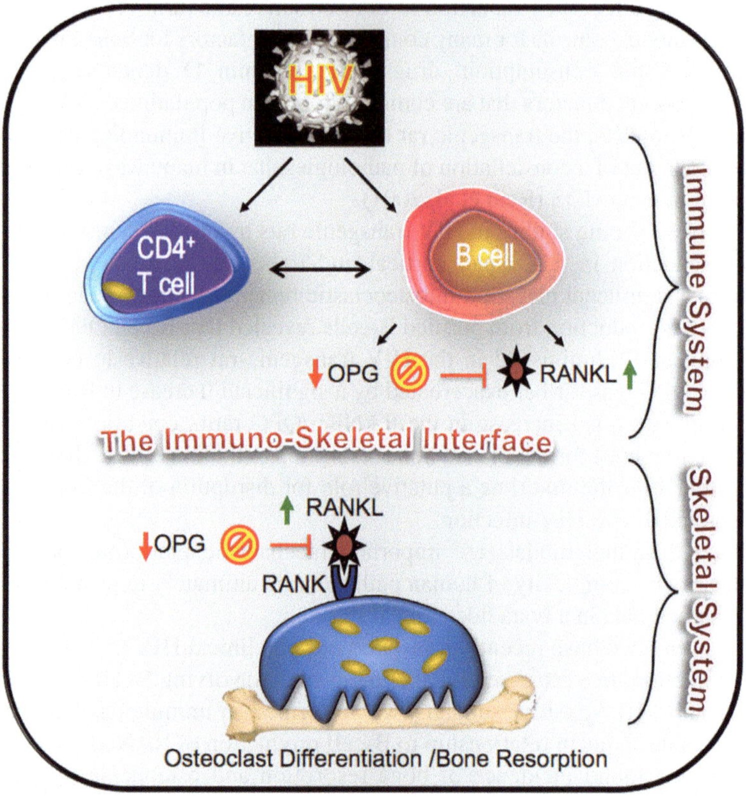

Fig. 12.1 Immunocentric model of HIV-induced bone loss. We propose a model for HIV- induced bone loss whereby HIV infection, either through direct effects on B-cells or though disruption of T-cell costimulation of the humoral system, leads to a decline in the frequency of B-cells secreting OPG, coupled with an increase in the frequency of B-cells secreting RANKL. This disruption of the immuno-skeletal interface results in an increase in the RANKL/OPG ratio that is permissive for osteoclast formation and enhanced bone resorption, which ultimately contributes to bone loss and the elevated bone fracture risk characteristic of HIV-infected individuals (Reproduced with permission from Titanji et al. (2014))

12.7.2 Bone Loss Associated with HIV-1 Antiretroviral Therapy and the Role of the ISI

Although the majority of HIV-infected subjects begin antiretroviral therapy (ART) shortly after being diagnosed with HIV infection, bone loss is not ameliorated and paradoxically may actually intensify (Tebas et al. 2000; Ofotokun and Weitzmann 2010; McComsey et al. 2010; Thomas and Doherty 2003; Yin et al. 2010b; Bolland and Grey 2011). This leads to a 2–6% loss of BMD within the first 2 years of ART initiation (McComsey et al. 2010). This further loss of BMD may significantly compound the bone loss already sustained in patients as a result of years of chronic HIV infection.

Although the mechanisms of ART-induced bone loss are likely to be distinct from those of HIV-induced skeletal decline, resolving the mechanism(s) involved has been frustrated by an inability to accurately model effects of ART on bone cells in vitro (Gibellini et al. 2010; Grigsby et al. 2010a, b) or in animal models in vivo (Wang et al. 2004) as very different outcomes are observed in the absence of viral infection. Furthermore, the multiplicity of ART classes and different agents within classes that are used clinically in combinatorial fashion has made the head-to-head comparison of specific ART constituents difficult. Some ART classes and some specific drugs have been associated with stronger effects on the skeleton (Ofotokun and Weitzmann 2010). In particular tenofovir disoproxil fumarate (TDF) containing regimens have generally been associated with pronounced bone loss; however, even new-generation derivatives such as tenofovir alafenamide (TAF) continue to show bone loss, although declines appear to be smaller than with TDF (Wohl et al. 2016). It is now becoming clear that despite clinical use of many distinct ART drug classes with different modes of action targeting different biochemical pathways, some degree of bone loss appears to be inevitable and common to all ART regimen (Brown et al. 2009; Piso et al. 2011). Consequently, several studies have concluded that at least a core of bone loss is independent of antiretroviral regimen (Bruera et al. 2003; Brown et al. 2009). We have recently examined the early kinetics of bone turnover in HIV-infected subjects initiating ART and found a rapid (within 2 weeks) increase in osteoclastogenic and inflammatory cytokines including RANKL and TNFα, leading to a significant increase in markers of bone resorption that peak by 12 weeks and remain elevated for at least 24 weeks of follow-up (Ofotokun et al. 2016a). Importantly, the rise in bone resorption was significantly correlated with the magnitude of CD4 T-cell reconstitution following ART. Our data further confirmed a previous study by Grant et al., suggesting that low baseline CD4 T-cell count is associated with a greater loss of bone mass after initiation of ART (Grant et al. 2013).

It should be pointed out that changes in serum cytokines were highly polymorphic between subjects with significant differences in the temporal expression of factors and in magnitude. Serum factors were not observed to increase in all patients and in fact declines occurred in some (Ofotokun et al. 2016a). In fact, other studies have reported a significant decline in levels of serum RANKL despite elevated bone resorption (Brown et al. 2011). The data suggest that cytokines quantified in serum may not always be representative of conditions at sites of bone turnover.

Despite discrepancies between studies in the changes in serum cytokines following ART initiation, the data overall suggest that events related to immune regeneration might account for the component of bone loss that is independent of antiretroviral regimen.

In an effort to derive a unifying hypothesis to explain a common route to bone loss for all ART regimes, we consequently proposed that T-cell repopulation and immune reactivation, a direct result of ART-induced HIV disease reversal, leads to the activation of T-cells and production of inflammatory cytokines including RANKL and TNFα. In addition, with the repopulation of $CD4^+$ T-cells (the lynchpin of adaptive immunity), it is further likely that initiation of basal immune function will further lead to production of RANKL and/or TNFα by other immune components though APC and costimulatory pathways creating a chronic inflammatory environment favorable to persistent bone degradation. Indeed chronic inflammation is well established to occur in patients on ART and may underlie a multiplicity of end organ damage associated with HIV/ART including end-stage renal and liver disease, cardiovascular disease, neurocognitive disorders, and indeed bone disease (Deeks et al. 2013).

To test this hypothesis, we reconstituted T-cell-deficient TCRβ knockout (KO) mice by adoptive transfer of syngeneic T-cells to mimic immune reconstitution in the context of immunodeficiency. Our studies (Ofotokun et al. 2015) revealed a significant loss of BMD and bone microarchitecture concurrent with elevated indices of bone resorption that were sustained for at least 3 months after T-cell reconstitution. As predicted, T-cell, B-cell, and monocyte production of RANKL and/or TNFα was significantly and chronically upregulated following T-cell reconstitution.

Although a role for the ISI in ART-associated bone loss has only recently begun to be investigated, our studies suggest that besides direct effects of ART on bone cells, at least some bone loss is ART independent and likely associated with inflammatory effects surrounding immune regeneration and reactivation.

Although previous clinical studies have clearly demonstrated the effectiveness of antiosteoporotic pharmaceuticals, such as zoledronic acid, in ameliorating bone loss caused by ART (Bolland et al. 2007, 2012), these agents remain underutilized, and few subjects are actually treated for their bone loss, especially if loss of BMD does not actually reach clinical definitions of osteoporosis. In reality most patients are likely not even screened for bone loss following ART. In an attempt to develop a simple, reliable prophylaxis for all patients initiating ART, we performed a double-blind, placebo-controlled phase IIb clinical trial in which a single administration of zoledronic acid was infused at time of ART initiation (Ofotokun et al. 2016b). The data revealed that the rise in bone resorption markers and loss of BMD was completely prevented in the zoledronic acid arm through the first 48 weeks of ART, the period when bone loss is the most pronounced (Ofotokun et al. 2016b). Although many patients have already undergone significant bone loss prior to HIV diagnosis, our study suggests that a simple antiresorptive prophylaxis can likely block further loss of BMD associated with ART.

12.8 Role of the ISI in Bone Loss Associated with Hyperparathyroidism

Parathyroid hormone (PTH) is a critical regulator of calcium metabolism and defends the body from hypocalcemia. Small decrements in serum-ionized calcium lead to PTH secretion, stimulating the release of calcium from the skeleton by inducing bone resorption. However, sustained overproduction of PTH in humans (primary or secondary hyperparathyroidism) or chronic infusion of PTH in animal models is a cause of significant osteoclastic bone loss and skeletal deterioration (Pacifici 2010).

Although it is established that PTH promotes bone resorption by increasing the production of RANKL and decreasing the production of OPG in cells of the osteoblast lineage, recent data show that T-cells are permissive and necessary for this catabolic response of PTH. The first evidence of a role for T-cells in PTH-induced bone resorption came from studies involving T-cell-deficient nude mice that were protected from the catabolic effects of transplanted human parathyroid glands from patients with hyperparathyroidism (Hory et al. 2000).

Extensive recent investigations have now confirmed the importance of T-cells in the catabolic activity of PTH by demonstrating that continuous administration of PTH fails to elicit significant cortical bone loss in TCRβ KO mice, a strain devoid of αβ T-cells, but undergo bone loss following adoptive transfer of T-cells (Gao et al. 2008). Furthermore conditional deletion of the PTH receptor on T-cells ablated bone loss supporting a direct action of PTH on T-cells in mediating bone resorption (Bedi et al. 2012). Additional mechanistic studies support a role for TNFα production from T-cells in response to PTH that, in addition to its amplificatory role in RANKL signaling, also upregulates the costimulatory receptor CD40, on bone marrow stromal cells, the osteoblast precursors. Engagement of CD40L produced by activated T-cells to its receptor on the osteoblast precursors promotes proliferative and survival cues contributing to the catabolic action of PTH by stimulating production of RANKL and diminishing OPG production by bone marrow stromal cells (Gao et al. 2008; Bedi et al. 2010; Tawfeek et al. 2010).

12.9 Bone Loss in Rheumatoid Arthritis and the Role of ISI

Although the notion of inflammatory or immune-mediated bone loss in ART is a relatively new concept, in the context of RA and inflammatory autoimmune disease, a key role for the ISI is well established. RA leads to loss of daily function due to chronic pain and fatigue resulting from chronic inflammation in the synovial membrane of affected joints. Almost 2% of adults suffer from RA with the majority of patients also exhibiting deterioration of cartilage and bone in the affected joints that ultimately precipitates permanent disability and increased mortality (Edwards et al. 2004).

Bone loss in RA is not just limited to inflamed joints, but systemic bone loss resulting from disruption of the systemic control of bone remodeling is common (Schett and David 2010). The inflammatory state characteristic of RA results from the activation of both T and B-cells that infiltrate the synovial membrane, a process key for both the initiation and progression of the inflammatory state. These activated T-cells further are key participants in driving the bone loss associated with RA (Edwards et al. 2004; Fournier 2005; Kong et al. 1999b; He et al. 2001; Yanaba et al. 2007) in large measure though secretion of RANKL as well as TNFα. TNFα has been shown to be a key protagonist of both inflammation and bone loss in RA, and transgenic overexpression of TNFα in mice closely mimics the bone and joint destruction associated with human RA (Li and Schwarz 2003; Schett et al. 2003). As a consequence, TNFα antagonists effectively alleviate both inflammation and bone loss in TNFα transgenic mice (Redlich et al. 2004) and are FDA approved for treatment of severe RA in humans (Chopin et al. 2008). A recent study in mice has further demonstrated that antibody-mediated suppression of IL-17 significantly ameliorates TNFα-induced local and systemic bone loss by blocking osteoclast differentiation in vivo (Zwerina et al. 2012). This data places Th17 T-cells and their key product IL-17, downstream of TNFα events on bone. IL-17 has previously been associated with bone erosions in the collagen-induced arthritis model by disrupting the RANKL/OPG balance through induction of RANKL production (Lubberts et al. 2003).

Recent studies have further implicated changes in regulatory T-cells (Tregs), a subset of predominantly CD4+ T-cells (but can also be CD8+) that express CD25 and FOXP3 markers. Tregs are potently immunosuppressive and down-modulate effector T-cell functions (Buchwald et al. 2013). Recent studies suggest that Tregs can suppress bone loss in arthritis in both humans and animal models (Glowacki et al. 2013; Zaiss et al. 2010), and the number of circulating Tregs has been inversely correlated with bone resorption markers in RA patients, suggesting that Tregs might moderate RA bone destruction in vivo (Zaiss et al. 2010).

Interestingly, studies using both human (Kim et al. 2007) and murine (Zaiss et al. 2007) Tregs have demonstrated direct anti-osteoclastogenic activity on RANKL-induced osteoclastogenesis as a possible consequence of production of the anti-osteoclastogenic cytokines TGF-β, IL-4, and IL-10 (Kim et al. 2007; Zaiss et al. 2007) and/or by CTLA4 (Zaiss et al. 2007). In vivo, Tregs potently block bone destruction in TNF transgenic mice, an animal model of RA, and stimulate basal bone mass in Foxp3 transgenic mice and T-cell-deficient RAG-1 null mice reconstituted with Tregs (Zaiss et al. 2010) suggesting direct effect of Tregs on bone cells as the latter models are physiological without inflammatory bone loss.

Because of the involvement of immune cells in the etiology of RA, in recent years, powerful immunomodulatory drugs that block T-cell costimulation are under development or in use for treating inflammatory diseases including otherwise intractable cases of RA (Kuek et al. 2007).

12.10 T-Cells: The Cornerstone of Cell-Mediated Immunity and the Dual-Signal Hypothesis

The ability of T-cells to react to a vast repertoire of unique antigens is a consequence of the T-cell receptor (TCR). Each T-cell clone expresses a unique TCR that recognizes one specific antigen. $CD4^+$ T-cells, also called effector T-cells, express the CD4 surface glycoprotein and respond to antigens presented by major histocompatibility class (MHC) II expressing APCs. B-cells, macrophages, and dendritic cells are termed "professional antigen-presenting cells" and are the dominant APCs in the body. By contrast, T-cells expressing the CD8 surface glycoprotein are termed cytotoxic T-cells and associate with MHC class I antigenic stimuli expressed by most nucleated cells (Abbas et al. 1996). To achieve T-cell activation and effector or cytotoxic function, a specific sequence of stimulatory events is required. In the "dual-signal hypothesis" of T-cell activation, two signals are required, with the first signal being generated by the engagement of the TCR with MHC bearing APCs. This signal alone does not initiate T-cell activation and in fact, through a series of signal transduction responses including initiation of cAMP in the cell, leads to T-cell anergy, a state of being unresponsive to further antigenic stimuli. For T-cell activation, a second costimulatory signal is necessary and is provided by engagement of the CD28 T-cell receptor with B7 molecules (B7-1 (CD80) and B7-2 (CD86)) on professional APC. Following this confirmatory signal, a phosphodiesterase is activated that cleaves cAMP and reverses the anergic block leading to full T-cell activation, cytokine production, clonal expansion, and prevention of anergy (Sayegh 1999). To prevent runaway immune activation and to calm the immune response following resolution of infection, activated T-cells and regulatory T-cells (Tregs) secrete a soluble receptor to CD80 and CD86. This decoy receptor referred to as cytotoxic T-lymphocyte-associated protein 4 (CTLA4) is highly homologous to CD28 and competes for binding of CD28 to CD80 and CD86 eliminating the second signal by preventing costimulation and thus terminating immune responses (Najafian and Sayegh 2000).

Recently a novel pharmaceutical agent has been constructed by generating a fusion protein of CTLA4 with a human immunoglobulin (Ig) chain creating a pharmacological costimulation inhibitor CTLA4-Ig (abatacept) that potently suppresses immune responses and restores immune tolerance in transplant rejection and in inflammatory conditions such as RA (Ruderman and Pope 2005; Vital and Emery 2006).

12.11 The ISI in Bone Formation

12.11.1 Bone Anabolic Effects of the Immunomodulatory Pharmaceutical CTLA4-Ig

Because of the role of the ISI in basal bone homeostasis, as well as in inflammatory bone loss, it was unclear as to whether the action of CTLA4-Ig in bone turnover is likely to be beneficial or harmful in the context of RA. Although CTLA4-Ig may blunt inflammation that drives bone loss, the direct effect of CTLA4-Ig on the regulation of basal bone homeostasis by T-cells and B-cells may have negative consequences. Whether the net effect of CTLA4-Ig treatment is bone gain or bone loss or neutral was unclear. To investigate this issue, we recently performed a study to establish the effect of CTLA4-Ig on basal bone turnover on mice (Roser-Page et al. 2014).

These studies demonstrated that CTLA4-Ig was extremely beneficial for the skeleton leading to a robust gain in BMD and in trabecular and cortical bone mass. These data were consistent with the previous studies by our group demonstrating that CTLA4-Ig prevents ovariectomy-induced bone loss by blunting T-cell activation and downstream inflammatory responses (Grassi et al. 2007) or with another study reporting direct inhibitory effects of CTLA4-Ig on osteoclasts (Axmann et al. 2008). However, analysis of bone turnover markers revealed a perplexing increase in biochemical marker of bone formation, without evidence of any effect on basal bone resorption. This outcome was confirmed by dynamic bone histomorphometry and led us to conclude that CTLA4-Ig may have previously unrecognized bone anabolic activity (Roser-Page et al. 2014).

Additional mechanistic studies revealed that CTLA4Ig-induced bone formation through production of the anabolic Wnt ligand, Wnt10b, and purified T-cells from mice treated with CTLA4-Ig secreted significantly enhanced concentrations of this osteoblast-stimulating factor. Finally, we demonstrated in an in vitro APC assay that antigen-induced activation of T-cells promotes a modest production of Wnt10b that is massively amplified by the addition of CTLA4-Ig (Roser-Page et al. 2014).

Although the exact mechanism driving Wnt10b production in anergic T-cells remains to be elucidated, based on the known presence of cAMP-responsive element-binding protein (CREB) motif sites in the Wnt10b promoter (Fox et al. 2008), we hypothesized that CD3-induced cAMP production following MHC/antigen ligation may induce Wnt10b production. However, in the case of bona fide immune response, the cAMP signal is rapidly counteracted by engagement of CD28 with CD80/CD86 causing activation of phosphodiesterase that cleaves cAMP, terminating signaling. By preventing the upregulation of cAMP, CTLA4-Ig may lead to chronic production of cAMP and hence Wnt10b, promoting osteoblast differentiation and bone formation (Roser-Page et al. 2014). This model is outlined diagrammatically in Fig. 12.2, although it remains to be ratified.

The capacity of CTLA4-Ig to promote bone formation in humans remains unknown; however, if verified, CTLA4-Ig may have potential application as a novel bone anabolic agent for treatment of prevention of osteoporosis.

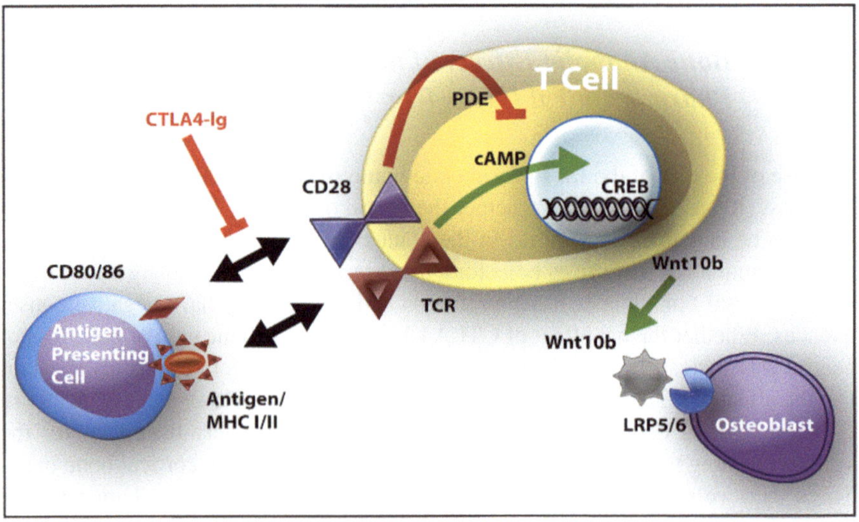

Fig. 12.2 Proposed model of the mechanism for the anabolic response of CTLA4-Ig, involving Wnt10b expression by T-cells. *MHC* major histocompatibility complex, *PDE* phosphodiesterase, *TCR* T-cell receptor, *LRP5/6* low-density lipoprotein receptor-related protein 5/6 (Reproduced with permission from Roser-Page et al. (2014))

12.12 Bone Anabolic Effects of Parathyroid Hormone

As discussed above, chronic exposure to high concentrations of PTH such as what occurs in hyperparathyroidism causes significant bone resorption and skeletal degradation. However, paradoxically when administered in an intermittent manner, PTH leads to a muted catabolic response that is offset completely by a robust induction of bone formation leading to a significant net bone gain. This phenomenon has led to the development of teriparatide, the first FDA-approved bone anabolic agent which is a fragment of human PTH that promotes bone formation in humans (Pacifici 2013).

Interestingly, the anabolic actions of PTH also involve the ISI. The anabolic effect of PTH is a consequence of increased osteoblast proliferation and differentiation; activation of quiescent lining osteoblasts, increasing osteoblast life-span by suppressing apoptosis; and down-modulation of the Wnt receptor antagonist sclerostin in osteocytes (Pacifici 2013). All of these downstream events converge on the anabolic Wnt pathway, but the source of the Wnt ligands driving bone formation following PTH exposure have only recently begun to be identified.

Studies by our group using T-cell-deficient TCRβ KO mice have recently demonstrated a specific role for T-cells in support of the downstream anabolic actions of PTH. Wnt10b was further identified as a key T-cell ligand induced following PTH exposure that supports the anabolic response. Accordingly, reconstitution of WT but not Wnt10b KO T-cells rescues the capacity of PTH to stimulate bone formation in TCRβ KO mice (Terauchi et al. 2009). Although molecular mechanisms are still under investigation, it should be noted that PTH too is a potent inducer of cAMP, and one may thus speculate that PTH induces Wnt10b production in T-cells in a similar manner to that of CTLA4-Ig. PTH however may be able to bypass the normal checks and balances of the adaptive immune response and directly increase cAMP in T-cells, hijacking the T-cell and inducing Wnt10b production independently of APC interactions.

12.13 Role of the ISI in Estrogen Deficiency-Induced Bone Loss

Postmenopausal osteoporosis is the most common form of bone disease and the archetypal osteoporotic condition in women and results from a decline in ovarian function following menopause, in particular the loss of sex steroids, most notably estrogen (Weitzmann and Pacifici 2006).

The etiology of postmenopausal bone loss is extremely complex and has been studied extensively in mouse and rat ovariectomy models in which surgical removal of the ovaries results in a precipitous drop in estrogen leading to a rapid and robust loss of bone mass within a 4–6-week period.

Evidence from early human and animal studies had implicated a role for inflammatory cytokines including IL-1, IL-6, TNFα, and M-CSF in estrogen deficiency bone loss (Pacifici et al. 1991; Passeri et al. 1993; Sarma and Flanagan 1996; Srivastava et al. 1998); however, the mode of action and source of these cytokines remained unclear.

12.13.1 B-cells and Estrogen Deficiency Bone Loss

One of the most interesting findings over the past two decades has been the role of the ISI in estrogen deficiency bone loss. This began with the recognition that B lymphopoiesis is potently upregulated by estrogen deficiency (Masuzawa et al. 1994) and decreased by estrogen supplementation (Erlandsson et al. 2003) which led to speculation of a direct involvement of B-cells in the etiology of postmenopausal bone loss (Miyaura et al. 1997). Studies demonstrating that the administration of IL-7, a potent B-lymphopoietic cytokine, was capable of mimicking the increase in B-lymphopoiesis and indeed the bone loss associated with estrogen deficiency led to the speculation of a direct involvement of B-cells (Miyaura et al. 1997). Additional support was provided by human studies documenting significant increases in RANKL production by both B and T-cells derived from postmenopausal women relative to premenopausal and estrogen-treated postmenopausal controls (Eghbali-Fatourechi et al. 2003).

To further define the specific role of mature B-cells in estrogen deficiency bone loss, our lab performed an ovariectomy on a mouse strain lacking mature B-cells. The outcome was disappointing as the magnitude of bone loss in the ovariectomized B-cell KO mice was not significantly different to that of WT mice, negating a significant role of B-cells in ovariectomy-induced bone loss (Li et al. 2007b). A caveat of these studies however was a significantly diminished baseline BMD and bone volume in the B-cell KO strain, a consequence of an imbalanced RANKL/OPG ratio owing to loss of B-cell OPG (Li et al. 2007a).

However, more recently, studies using a more sophisticated conditional B-cell-specific RANKL ablation model revealed a blunted bone loss following ovariectomy with partial, but significant, protection from vertebral, but not femoral, bone loss. Analysis of femoral bone volume also identified a significant protection from trabecular bone loss (Onal et al. 2012). These data suggest a partial role of B-cell RANKL in the trabecular, but not cortical, bone loss associated with ovariectomy.

12.13.2 T-Cells and Estrogen Deficiency Bone Loss

As previously discussed in detail, activated T-cells are known to be a source of numerous cytokines including TNFα and RANKL, and numerous studies have reported a capability of T-cells to support osteoclast formation. However, the first empirical evidence supporting a role of T-cells in ovariectomy-induced bone loss came from our group with the report that T-cell-deficient nude mice were resistant to bone loss associated with ovariectomy (Cenci et al. 2000). A role for T-cells in ovariectomy-induced bone loss has been independently verified in T-cell-deficient NIH-III beige nude mice that too are protected from bone loss and are rescued by adoptive transfer of T-cells (Yamaza et al. 2008).

Variable outcomes have been reported however, in other immunocompromised mouse strains. Partially, T-cell-deficient Rowett athymic nude rats, for example, were not protected from trabecular bone loss (Sass et al. 1997), while BALBc nude mice (Cho et al. 2012), C57BL6 nude mice, and T-cell receptor α null mice showed cortical, but not trabecular, protection from ovariectomy, and RAG-2 null mice (missing both T-cells and B-cells) lost bone in both compartments (Lee et al. 2006a).

Although differences in experimental conditions including residual T-cells, diets, gut microbiota, environmental factors, and hormonal status may contribute to these varying outcomes, additional work is needed.

Although numerous studies have reported that T-cells are a key source of RANKL, a surprising finding of our study was that T-cells drove bone loss predominantly through upregulation of TNFα production, rather than through RANKL. The importance of TNFα was confirmed by the finding that TNFα KO mice and p55 TNF receptor KO mice were also protected from bone loss. To further demonstrate the key role of T-cells as a source of TNFα in estrogen deficiency bone loss, we adoptively transferred T-cells from TNFα KO mice into nude mice and demonstrated complete protection from ovariectomy-induced bone loss providing strong evidence for T-cell-specific TNFα production in ovariectomy bone loss (Roggia et al. 2001).

Importantly, recent findings using advanced genetic models in which RANKL was specifically ablated in T-cells have confirmed that in the mouse ovariectomy model, T-cell production of RANKL is dispensable for ovariectomy-induced bone loss (Onal et al. 2012).

Dissecting the cellular and molecular mechanisms of how T-cell TNFα is regulated and how it translates into bone loss has been the subject of a decade of intensive research that continues to this day. Mechanistic studies have revealed that the underlying etiology of T-cell TNFα production is the net result of an overall increase in the adaptive immune response brought about by estrogen decline. Estrogen deficiency causes an increase in reactive oxygen species (ROS) that increases expression of the CD80 receptor on dendritic cells (Grassi et al. 2007). In addition, enhanced production of IFNγ upregulates MHCII expression on macrophages though stimulation of the class II transactivator (CIITA) transcription factor (Cenci et al. 2003). This effectively increases the sensitivity of APC to T-cell-mediated costimulatory signals and through bidirectional signaling promotes T-cell activation in response to endogenous antigens leading to TNFα production.

It should be pointed out that while IFNγ is a potent inducer of osteoclastogenesis though immunocentric mechanisms, it has also been reported to mediate a direct suppressive effect on osteoclast differentiation by antagonizing RANKL signal transduction pathways (Takayanagi et al. 2000). The action of IFNγ may thus be the net effect of these two opposing actions, and production of IFNγ from Th1 T-cells further drives up APC activity through actions on CIITA promoting MHC expression (Cenci et al. 2003).

Following estrogen deprivation, T-cells themselves are directly stimulated through a series of signals and cytokine cascades. An upstream event involves increased production of IGF-1 and a concurrent decline in TGFβ, a potent immunosuppressive cytokine that stimulates formation of immunosuppressive Tregs. In addition, the decline in TGFβ, amplified by increased IGF-1, stimulates production of IL-7 (Weitzmann and Pacifici 2006).

IL-7, although potently upregulating B lymphopoiesis, is now also recognized as a master regulator of T-cell differentiation, maturation, and activity (Fry and Mackall 2001). Although early studies had proposed that IL-7-induced bone loss was likely mediated though B-cells (Miyaura et al. 1997), studies in our lab have demonstrated that in vitro osteoclastogenesis driven by IL-7 requires T-cells but B-cells are dispensable (Weitzmann et al. 2000a). We have further demonstrated that IL-7 promotes osteoclastogenesis through direct production of RANKL by T-cells (Weitzmann et al. 2000a; Toraldo et al. 2003) as well as by promoting de novo, bone marrow T-cell differentiation, enhancing thymic function and stimulating peripheral expansion of mature T-cells (Ryan et al. 2005). Finally, IL-7 directly increases T-cell sensitivity to weak antigenic stimuli, amplifying T-cell activation.

Interestingly, as with IFNγ, IL-7 may mediate a suppressive direct signal on osteoclast formation (Lee et al. 2003).

IL-1 has long been associated with inflammatory and estrogen deficiency bone loss and is suggested to function by directly stimulating RANKL production from stromal cells (Wei et al. 2005).

More recently the involvement of IL-17, a cytokine that promotes RANKL production by osteoblasts and has been implicated in the bone loss associated with autoimmune arthritis (Sato et al. 2006), has been added to the cascade of signals required in ovariectomy-induced bone loss. Recent data suggest that estrogen deficiency promotes the differentiation of Th17 CD4 T-cells, the major source of IL-17, promoting bone loss (Tyagi et al. 2012).

Finally, the ability of bone formation to respond and compensate for elevated osteoclastic bone resorption though coupling mechanisms is diminished by the action of cytokines such as TNFα and IL-7 that suppress bone formation further exacerbating bone loss (Weitzmann and Pacifici 2006).

This "vicious cycle" thus perpetuates chronic T-cell activation, TNFα production, and a general inflammatory state propitious for bone loss.

Interestingly, breaking this cycle at any one of these critical nodes is effective in ameliorating ovariectomy-induced bone loss. Pharmacological ablation of IL-1 and TNFα (Kimble et al. 1995) IL-7 (Weitzmann et al. 2002) and IL-17 all reverse bone loss in ovariectomized mice (Tyagi et al. 2012). Similarly, genetic deletion of IFNγ (Gao et al. 2007), TNFα (Roggia et al. 2001), IFNγ (Gao et al. 2007), and IGF-1 (Lindberg et al. 2006) and overexpression of TGFβ by somatic gene therapy (Gao et al. 2004) all effectively prevent bone loss in ovariectomized mice. By contrast IL-7 KO mice show protection from cortical, but not trabecular, bone loss following ovariectomy (Lee et al. 2006b). Interestingly, while systemic overexpression of IL-7 promotes osteolysis (Valenzona et al. 1996), overproduction of IL-7 conditionally in osteoblasts causes a small increase in bone volume in female mice, but a nonsignificant decline in male mice (Aguila et al. 2012). As IL-7 suppresses bone formation (Weitzmann et al. 2002) and IL-7 knockout mice have a marked increase in osteoblast surface (Aguila et al. 2012), some of these discrepancies may relate to context-specific effects on bone formation and resorption.

As antigen-driven APC-mediated immune responses appear to be important in estrogen deficiency bone loss, we abated antigen presentation in mice and reported significant protection from ovariectomy-induced bone loss (Cenci et al. 2003). The nature of the specific antigens involved, however, is still unclear, but resent studies suggest that gut microbiota may be key participants.

Indeed, ovariectomy-induced bone resorption and loss is prevented by administration of the probiotic *Lactobacillus reuteri* by suppressing CD4+ T-cell expansion in the bone marrow (Britton et al. 2014). Similar studies using single strains or mixtures of *Lactobacillus* species also protected mice from cortical bone loss following ovariectomy and was associated with diminished markers of bone resorption and inflammation and prevented an ovariectomy-induced decline in Tregs. Recently, our group reported that germ-free ovariectomized mice are likewise protected from induction of osteoclastogenic cytokines and trabecular bone loss and that sex-steroid-induced bone loss is prevented by probiotics due to beneficial effects on gut permeability (Li et al. 2016).

Although these proof of concept studies in animals suggest that multiple upstream cytokines or cell types may be targeted therapeutically, the potential dangers of targeting the immune system has led to a cautious approach, and therapeutic strategies still focus almost exclusively on downstream mediators including osteoclasts and osteoblasts themselves and antibodies that target RANKL directly (denosumab). Future therapies for postmenopausal osteoporosis may however involve selective targeting of upstream inflammatory cascades.

This inflammatory model of estrogen deficiency-induced bone loss is summarized diagrammatically in Fig. 12.3.

12.13.3 The Role of Antigens in Estrogen Deficiency Bone Loss

One perplexing aspect of an immunocentric basis to ovariectomy-induced bone loss is that adaptive immune responses center on antigen-mediated events.

To attempt to better define the nature of the antigens that support ovariectomy-induced bone loss, we performed a study using a specific strain of mouse (DO11.10) expressing a transgenic T-cell receptor rendering every T-cell responsive to a single antigen (ovalbumin). In the absence of exogenous ovalbumin, basal antigen presentation is completely muted in these mice. In support of a specific requirement for

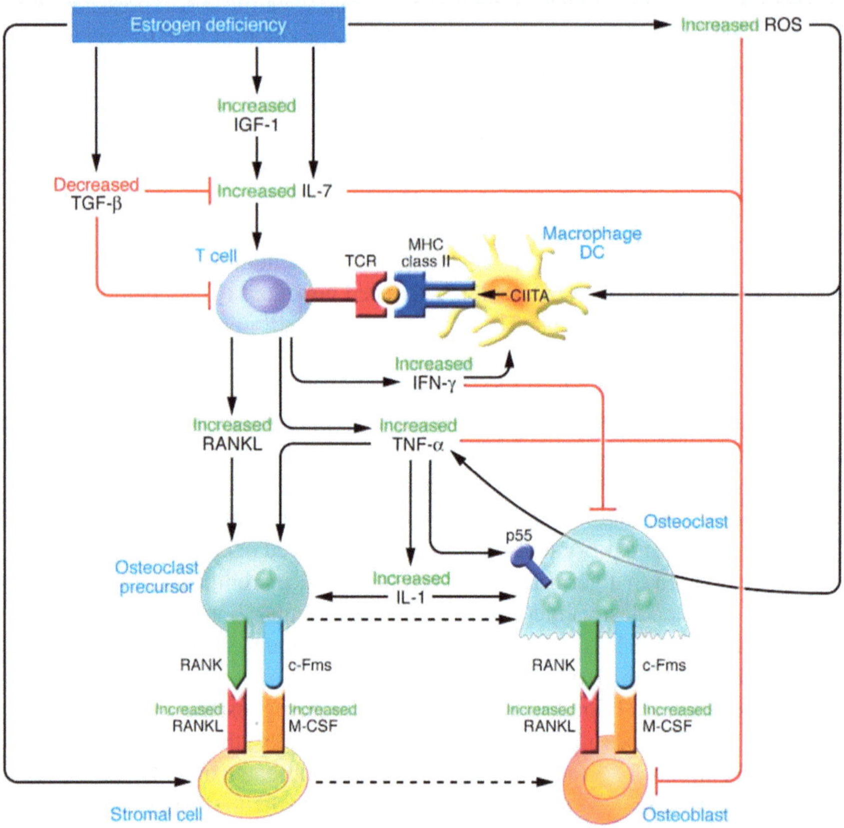

Fig. 12.3 Schematic representation of the major mechanisms and feedback interactions by which estrogen deficiency leads to bone loss. The bone loss induced by estrogen deficiency is due to a complex interplay of hormones and cytokines that converge to disrupt the process of bone remodeling. Estrogen deficiency leads to a global increase in IL-7 production in target organs such as bone, thymus, and spleen, in part through decreases in TGF-β and increased IGF-1 production. This leads to an initial wave of T-cell activation. Activated T-cells release IFNγ, which increases antigen presentation by dendritic cells (DCs) and macrophages (Mϕ) by upregulating MHC class II expression through the transcription factor CIITA. Estrogen deficiency also amplifies T-cell activation and osteoclastogenesis by downregulating antioxidant pathways, leading to an upswing in reactive oxygen species (ROS). The resulting increase in ROS stimulates antigen presentation and the production of TNF by mature osteoclasts. The combined effect of IFNγ and ROS markedly enhances antigen presentation, amplifying T-cell activation and promoting release of the osteoclastogenic factors RANKL and TNFα. TNFα further stimulates stromal cell and osteoblastic RANKL and M-CSF production, in part via IL-1 upregulation, driving osteoclast formation. TNFα and IL-7 further exacerbate bone loss by blunting bone formation through direct repressive effects on osteoblasts (Reproduced with permission from Weitzmann and Pacifici (2006))

antigen presentation, our data revealed that D11.10 mice were completely protected from ovariectomy-induced bone. Furthermore, injection of ovalbumin elicited a robust bone loss in response to ovariectomy allowing us to conclude that the process of antigen presentation was critical; however, any available antigen is sufficient in initiating T-cell-dependent bone loss (Cenci et al. 2003).

It is unknown whether a specific antigen is actually generated in the context of ovariectomy that mediates bone loss; however, one explanation consistent with available data is that as already described above, estrogen deficiency simply leads to an overall heightened response to prevailing antigens, generating an exaggerated immune response with increased reactivity of both APC and T-cells to otherwise weak or tolerogenic antigens.

Low-grade immunological responses are continuously in effect even in the context of otherwise healthy animals and humans (Surh and Sprent 2008). These antigenic reactions are in response to chronic exposure to weak self and foreign antigens derived from multiple sources including antigens inhaled into the lungs, peptide products of digestion, vaginal antigens, and in large measure to bacterial antigens absorbed in the gut. Interestingly, it is now established that persistent antigenic stimulate of T-cells with "self" are essential for long-term homeostatic persistence of peripheral T-cell pools (Surh and Sprent 2008; Tanchot et al. 1997).

An interesting new finding is that osteoclasts themselves are capable of acting as APC and expressing antigens. This activity promotes FoxP3 expression in CD8 T-cells promoting the development of a regulatory T-cell phenotype in CD8 T-cells that then suppress osteoclast activity (Buchwald et al. 2013; Kiesel et al. 2007).

12.13.4 Evidence for an Immunocentric Basis for Postmenopausal Bone Loss in Humans

Although rodent models of postmenopausal osteoporosis are extremely powerful in delineating complex molecular and cellular mechanisms, they may not be fully representative of the human situation. Some of the most compelling human data involve studies quantifying RANKL production by cells of the bone marrow and reveal significant upregulation of RANKL production by both B and T-cells, as well as bone marrow stromal cells derived from postmenopausal women relative to premenopausal and estrogen-treated postmenopausal controls (Eghbali-Fatourechi et al. 2003).

More recently a clinical study examining T-cell activity in women with postmenopausal osteoporosis documented significantly elevated T-cell-produced TNFα and RANKL than in healthy postmenopausal subjects, suggesting that T-cells contribute RANKL and TNFα to the bone loss induced by estrogen deficiency in humans (D'Amelio et al. 2008).

In addition, characterizations of T-cell phenotype and BMD in young patients undergoing elective surgical ovariectomy revealed increased T-cell activation and thymic hypertrophy compared to controls (Adeel et al. 2013). These data are consistent with studies of young animals where thymic function is still intact, showing that thymectomy decreases ovariectomy-induced bone loss by 50% (Ryan et al. 2005).

At this time, data supporting an immunocentric basis for postmenopausal bone loss in humans remains sparse. Clearly, additional clinical studies are needed to ratify the immunocentric model developed in rodents. This will be key for future development of novel anti-inflammatory drugs to ameliorate bone loss in postmenopausal osteoporosis and in other osteoporotic conditions stemming from immune imbalance.

12.14 Conclusion

The identification and elucidation of the ISI has profoundly changed our understanding of how the skeleton is regulated under basal and pathological conditions. The ISI continues to evolve in depth and scope, providing explanations for bone loss in conditions as diverse as inflammation and immunosuppression. The discovery of novel T-cell factors such as SOFAT may have relevance in inflammatory conditions driving osteoclastogenic bone resorption independently of RANKL. In addition to these catabolic actions, recent studies have elucidated novel roles for T-cells in bone formation revealing a deep infiltration of the immune response into all aspects of bone biology.

As basic concepts emerge, the future direction of osteoimmunology is likely to become clinical, ratifying concepts developed in animal models and more translational providing new avenues for therapeutic interventions that target upstream immunological events to allay bone damage by targeting the source of inflammation, rather than downstream effectors of bone resorption, the present standard of care.

Acknowledgments The author gratefully acknowledges research support from the Biomedical Laboratory Research and Development Service of the VA Office of Research and Development (Grant 5I01BX000105) and National Institutes of Health grants from NIAMS (AR059364, AR056090, and AR053607) and NIA (AG040013).

References

Abbas AK, Murphy KM, Sher A. Functional diversity of helper T lymphocytes. Nature. 1996;383:787–93.

Adeel S, Singh K, Vydareny KH, Kumari M, Shah E, Weitzmann MN, Tangpricha V. Bone loss in surgically ovariectomized premenopausal women is associated with T lymphocyte activation and thymic hypertrophy. J Investig Med. 2013;61:1178–83.

Aguila HL, Mun SH, Kalinowski J, Adams DJ, Lorenzo JA, Lee SK. Osteoblast-specific overexpression of human interleukin-7 rescues the bone mass phenotype of interleukin-7-deficient female mice. J Bone Miner Res. 2012;27:1030–42.

Amorosa V, Tebas P. Bone disease and HIV infection. Clin Infect Dis. 2006;42:108–14.

Anderson DM, Maraskovsky E, Billingsley WL, Dougall WC, Tometsko ME, Roux ER, Teepe MC, DuBose RF, Cosman D, Galibert L. A homologue of the TNF receptor and its ligand enhance T-cell growth and dendritic-cell function. Nature. 1997;390:175–9.

Arai F, Miyamoto T, Ohneda O, Inada T, Sudo T, Brasel K, Miyata T, Anderson DM, Suda T. Commitment and differentiation of osteoclast precursor cells by the sequential expression of c-Fms and receptor activator of nuclear factor kappaB (RANK) receptors. J Exp Med. 1999;190:1741–54.

Aukrust P, Haug CJ, Ueland T, Lien E, Muller F, Espevik T, Bollerslev J, Froland SS. Decreased bone formative and enhanced resorptive markers in human immunodeficiency virus infection: indication of normalization of the bone-remodeling process during highly active antiretroviral therapy. J Clin Endocrinol Metab. 1999;84:145–50.

Axmann R, Herman S, Zaiss M, Franz S, Polzer K, Zwerina J, Herrmann M, Smolen J, Schett G. CTLA-4 directly inhibits osteoclast formation. Ann Rheum Dis. 2008;67:1603–9.

Bass E, French DD, Bradham DD, Rubenstein LZ. Risk-adjusted mortality rates of elderly veterans with hip fractures. Ann Epidemiol. 2007;17:514–9.

Bedi B, Li JY, Grassi F, Tawfeek H, Weitzmann MN, Pacifici R. Inhibition of antigen presentation and T-cell costimulation blocks PTH-induced bone loss. Ann N Y Acad Sci. 2010;1192:215–21.

Bedi B, Li JY, Tawfeek H, Baek KH, Adams J, Vangara SS, Chang MK, Kneissel M, Weitzmann MN, Pacifici R. Silencing of parathyroid hormone (PTH) receptor 1 in T-cells blunts the bone anabolic activity of PTH. Proc Natl Acad Sci U S A. 2012;109:E725–33.

Bolland MJ, Grey A. HIV and low bone density: responsible party, or guilty by association? IBMS BoneKey. 2011;8:7–15.

Bolland MJ, Grey AB, Horne AM, Briggs SE, Thomas MG, Ellis-Pegler RB, Woodhouse AF, Gamble GD, Reid IR. Annual zoledronate increases bone density in highly active antiretroviral therapy-treated human immunodeficiency virus-infected men: a randomized controlled trial. J Clin Endocrinol Metab. 2007;92:1283–8.

Bolland MJ, Grey A, Horne AM, Briggs SE, Thomas MG, Ellis-Pegler RB, Gamble GD, Reid IR. Effects of intravenous zoledronate on bone turnover and bone density persist for at least five years in HIV-infected men. J Clin Endocrinol Metab. 2012;97:1922–8.

Bonjoch A, Figueras M, Estany C, Perez-Alvarez N, Rosales J, del Rio L, di Gregorio S, Puig J, Gomez G, Clotet B, Negredo E. High prevalence of and progression to low bone mineral density in HIV-infected patients: a longitudinal cohort study. AIDS. 2010;24:2827–33.

Brainsky A, Glick H, Lydick E, Epstein R, Fox KM, Hawkes W, Kashner TM, Zimmerman SI, Magaziner J. The economic cost of hip fractures in community-dwelling older adults: a prospective study. J Am Geriatr Soc. 1997;45:281–7.

Brigham BA, Bunn PA Jr, Horton JE, Schechter GP, Wahl LM, Bradley EC, Dunnick NR, Matthews MJ. Skeletal manifestations in cutaneous T-cell lymphomas. Arch Dermatol. 1982;118:461–7.

Britton RA, Irwin R, Quach D, Schaefer L, Zhang J, Lee T, Parameswaran N, McCabe LR. Probiotic L. reuteri treatment prevents bone loss in a menopausal ovariectomized mouse model. J Cell Physiol. 2014;229:1822–30.

Brown TT, Qaqish RB. Antiretroviral therapy and the prevalence of osteopenia and osteoporosis: a meta-analytic review. AIDS. 2006;20:2165–74.

Brown TT, McComsey GA, King MS, Qaqish RB, Bernstein BM, da Silva BA. Loss of bone mineral density after antiretroviral therapy initiation, independent of antiretroviral regimen. J Acquir Immune Defic Syndr. 2009;51:554–61.

Brown TT, Ross AC, Storer N, Labbato D, McComsey GA. Bone turnover, osteoprotegerin/RANKL and inflammation with antiretroviral initiation: tenofovir versus non-tenofovir regimens. Antivir Ther. 2011;16:1063–72.

Brown TT, Hoy J, Borderi M, Guaraldi G, Renjifo B, Vescini F, Yin MT, Powderly WG. Recommendations for evaluation and management of bone disease in HIV. Clin Infect Dis. 2015;60:1242–51.

Bruera D, Luna N, David DO, Bergoglio LM, Zamudio J. Decreased bone mineral density in HIV-infected patients is independent of antiretroviral therapy. AIDS. 2003;17:1917–23.

Bucay N, Sarosi I, Dunstan CR, Morony S, Tarpley J, Capparelli C, Scully S, Tan HL, Xu W, Lacey DL, Boyle WJ, Simonet WS. Osteoprotegerin-deficient mice develop early onset osteoporosis and arterial calcification. Genes Dev. 1998;12:1260–8.

Buchwald ZS, Kiesel JR, Yang C, DiPaolo R, Novack DV, Aurora R. Osteoclast-induced Foxp3+ CD8 T-cells limit bone loss in mice. Bone. 2013;56:163–73.

Burge R, Dawson-Hughes B, Solomon DH, Wong JB, King A, Tosteson A. Incidence and economic burden of osteoporosis-related fractures in the United States, 2005–2025. J Bone Miner Res. 2007;22:465–75.

Buring K. On the origin of cells in heterotopic bone formation. Clin Orthop Relat Res. 1975;110:293–301.
Cenci S, Weitzmann MN, Roggia C, Namba N, Novack D, Woodring J, Pacifici R. Estrogen deficiency induces bone loss by enhancing T-cell production of TNF-alpha. J Clin Invest. 2000;106:1229–37.
Cenci S, Toraldo G, Weitzmann MN, Roggia C, Gao Y, Qian WP, Sierra O, Pacifici R. Estrogen deficiency induces bone loss by increasing T-cell proliferation and lifespan through IFN-gamma-induced class II transactivator. Proc Natl Acad Sci U S A. 2003;100:10405–10.
Chen P, Trummel C, Horton J, Baker JJ, Oppenheim JJ. Production of osteoclast-activating factor by normal human peripheral blood rosetting and nonrosetting lymphocytes. Eur J Immunol. 1976;6:732–6.
Cho SW, Sun HJ, Yang JY, Jung JY, Choi HJ, An JH, Kim SW, Kim SY, Park KJ, Shin CS. Human adipose tissue-derived stromal cell therapy prevents bone loss in ovariectomized nude mouse. Tissue Eng Part A. 2012;18:1067–78.
Chopin F, Garnero P, le Henanff A, Debiais F, Daragon A, Roux C, Sany J, Wendling D, Zarnitsky C, Ravaud P, Thomas T. Long-term effects of infliximab on bone and cartilage turnover markers in patients with rheumatoid arthritis. Ann Rheum Dis. 2008;67:353–7.
D'Amelio P, Grimaldi A, Di Bella S, Brianza SZ, Cristofaro MA, Tamone C, Giribaldi G, Ulliers D, Pescarmona GP, Isaia G. Estrogen deficiency increases osteoclastogenesis up-regulating T-cells activity: a key mechanism in osteoporosis. Bone. 2008;43:92–100.
Deeks SG, Tracy R, Douek DC. Systemic effects of inflammation on health during chronic HIV infection. Immunity. 2013;39:633–45.
Dube MP, Qian D, Edmondson-Melancon H, Sattler FR, Goodwin D, Martinez C, Williams V, Johnson D, Buchanan TA. Prospective, intensive study of metabolic changes associated with 48 weeks of amprenavir-based antiretroviral therapy. Clin Infect Dis. 2002;35:475–81.
Edwards JC, Szczepanski L, Szechinski J, Filipowicz-Sosnowska A, Emery P, Close DR, Stevens RM, Shaw T. Efficacy of B-cell-targeted therapy with rituximab in patients with rheumatoid arthritis. N Engl J Med. 2004;350:2572–81.
Eghbali-Fatourechi G, Khosla S, Sanyal A, Boyle WJ, Lacey DL, Riggs BL. Role of RANK ligand in mediating increased bone resorption in early postmenopausal women. J Clin Invest. 2003;111:1221–30.
Erlandsson MC, Jonsson CA, Islander U, Ohlsson C, Carlsten H. Oestrogen receptor specificity in oestradiol-mediated effects on B lymphopoiesis and immunoglobulin production in male mice. Immunology. 2003;108:346–51.
Feldmann M, Brennan FM, Maini R. Rheumatoid arthritis. Cell. 1996;85:307–10.
Fournier C. Where do T-cells stand in rheumatoid arthritis? Joint Bone Spine. 2005;72:527–32.
Fox KE, Colton LA, Erickson PF, Friedman JE, Cha HC, Keller P, MacDougald OA, Klemm DJ. Regulation of cyclin D1 and Wnt10b gene expression by cAMP-responsive element-binding protein during early adipogenesis involves differential promoter methylation. J Biol Chem. 2008;283:35096–105.
Fry TJ, Mackall CL. Interleukin-7: master regulator of peripheral T-cell homeostasis? Trends Immunol. 2001;22:564–71.
Fuller K, Murphy C, Kirstein B, Fox SW, Chambers TJ. TNFalpha potently activates osteoclasts, through a direct action independent of and strongly synergistic with RANKL. Endocrinology. 2002;143:1108–18.
Gao Y, Qian WP, Dark K, Toraldo G, Lin AS, Guldberg RE, Flavell RA, Weitzmann MN, Pacifici R. Estrogen prevents bone loss through transforming growth factor beta signaling in T-cells. Proc Natl Acad Sci U S A. 2004;101:16618–23.
Gao Y, Grassi F, Ryan MR, Terauchi M, Page K, Yang X, Weitzmann MN, Pacifici R. IFN-gamma stimulates osteoclast formation and bone loss in vivo via antigen-driven T-cell activation. J Clin Invest. 2007;117:122–32.
Gao Y, Wu X, Terauchi M, Li JY, Grassi F, Galley S, Yang X, Weitzmann MN, Pacifici R. T-cells potentiate PTH-induced cortical bone loss through CD40L signaling. Cell Metab. 2008;8:132–45.

Geddes AC. The origin of the osteoblast and of the osteoclast. J Anat Physiol. 1913;47:159–76.
Gibellini D, Borderi M, de Crignis E, Clo A, Miserocchi A, Viale P, Re MC. Analysis of the effects of specific protease inhibitors on OPG/RANKL regulation in an osteoblast-like cell line. New Microbiol. 2010;33:109–15.
Glowacki AJ, Yoshizawa S, Jhunjhunwala S, Vieira AE, Garlet GP, Sfeir C, Little SR. Prevention of inflammation-mediated bone loss in murine and canine periodontal disease via recruitment of regulatory lymphocytes. Proc Natl Acad Sci U S A. 2013;110:18525–30.
Gori F, Hofbauer LC, Dunstan CR, Spelsberg TC, Khosla S, Riggs BL. The expression of osteoprotegerin and RANK ligand and the support of osteoclast formation by stromal-osteoblast lineage cells is developmentally regulated. Endocrinology. 2000;141:4768–76.
Grant PM, Kitch D, McComsey GA, Dube MP, Haubrich R, Huang J, Riddler S, Tebas P, Zolopa AR, Collier AC, Brown TT. Low baseline CD4+ count is associated with greater bone mineral density loss after antiretroviral therapy initiation. Clin Infect Dis. 2013;57:1483–8.
Grassi F, Tell G, Robbie-Ryan M, Gao Y, Terauchi M, Yang X, Romanello M, Jones DP, Weitzmann MN, Pacifici R. Oxidative stress causes bone loss in estrogen-deficient mice through enhanced bone marrow dendritic cell activation. Proc Natl Acad Sci U S A. 2007;104:15087–92.
Grcevic D, Lee SK, Marusic A, Lorenzo JA. Depletion of CD4 and CD8 T lymphocytes in mice in vivo enhances 1, 25- dihydroxyvitamin D(3)-stimulated osteoclast-like cell formation in vitro by a mechanism that is dependent on prostaglandin synthesis. J Immunol. 2000;165:4231–8.
Grigsby IF, Pham L, Gopalakrishnan R, Mansky LM, Mansky KC. Downregulation of Gnas, Got2 and Snord32a following tenofovir exposure of primary osteoclasts. Biochem Biophys Res Commun. 2010a;391:1324–9.
Grigsby IF, Pham L, Mansky LM, Gopalakrishnan R, Carlson AE, Mansky KC. Tenofovir treatment of primary osteoblasts alters gene expression profiles: implications for bone mineral density loss. Biochem Biophys Res Commun. 2010b;394:48–53.
Guaraldi G, Ventura P, Albuzza M, Orlando G, Bedini A, Amorico G, Esposito R. Pathological fractures in AIDS patients with osteopenia and osteoporosis induced by antiretroviral therapy. AIDS. 2001;15:137–8.
Guerri-Fernandez R, Vestergaard P, Carbonell C, Knobel H, Aviles FF, Castro AS, Nogues X, Prieto-Alhambra D, Diez-Perez A. HIV infection is strongly associated with hip fracture risk, independently of age, gender, and comorbidities: a population-based cohort study. J Bone Miner Res. 2013;28:1259–63.
Gyarmati J Jr, Mandi B, Fachet J, Varga S, Sikula J. Alterations of the connective tissue in nude mice. Thymus. 1983;5:383–92.
Hattersley G, Chambers TJ. Effects of transforming growth factor beta 1 on the regulation of osteoclastic development and function. J Bone Miner Res. 1991;6:165–72.
He X, Kang AH, Stuart JM. Anti-human type II collagen CD19+ B-cells are present in patients with rheumatoid arthritis and healthy individuals. J Rheumatol. 2001;28:2168–75.
Hofbauer LC, Lacey DL, Dunstan CR, Spelsberg TC, Riggs BL, Khosla S. Interleukin-1beta and tumor necrosis factor-alpha, but not interleukin-6, stimulate osteoprotegerin ligand gene expression in human osteoblastic cells. Bone. 1999a;25:255–9.
Hofbauer LC, Khosla S, Dunstan CR, Lacey DL, Spelsberg TC, Riggs BL. Estrogen stimulates gene expression and protein production of osteoprotegerin in human osteoblastic cells. Endocrinology. 1999b;140:4367–70.
Hofbauer LC, Khosla S, Dunstan CR, Lacey DL, Boyle WJ, Riggs BL. The roles of osteoprotegerin and osteoprotegerin ligand in the paracrine regulation of bone resorption. J Bone Miner Res. 2000;15:2–12.
Horowitz M, Vignery A, Gershon RK, Baron R. Thymus-derived lymphocytes and their interactions with macrophages are required for the production of osteoclast-activating factor in the mouse. Proc Natl Acad Sci U S A. 1984;81:2181–5.
Horton JE, Raisz LG, Simmons HA, Oppenheim JJ, Mergenhagen SE. Bone resorbing activity in supernatant fluid from cultured human peripheral blood leukocytes. Science. 1972;177:793–5.
Horton JE, Oppenheim JJ, Mergenhagen SE, Raisz LG. Macrophage-lymphocyte synergy in the production of osteoclast activating factor. J Immunol. 1974;113:1278–87.

Hory BG, Roussanne MC, Rostand S, Bourdeau A, Drueke TB, Gogusev J. Absence of response to human parathyroid hormone in athymic mice grafted with human parathyroid adenoma, hyperplasia or parathyroid cells maintained in culture. J Endocrinol Investig. 2000;23:273–9.

Huang JS, Wilkie SJ, Sullivan MP, Grinspoon S. Reduced bone density in androgen-deficient women with acquired immune deficiency syndrome wasting. J Clin Endocrinol Metab. 2001;86:3533–9.

Hui SL, Slemenda CW, Johnston CC Jr. Age and bone mass as predictors of fracture in a prospective study. J Clin Invest. 1988;81:1804–9.

Jain RG, Lenhard JM. Select HIV protease inhibitors alter bone and fat metabolism ex vivo. J Biol Chem. 2002;277:19247–50.

Jarry CR, Duarte PM, Freitas FF, de Macedo CG, Clemente-Napimoga JT, Saba-Chujfi E, Passador-Santos F, de Araujo VC, Napimoga MH. Secreted osteoclastogenic factor of activated T-cells (SOFAT), a novel osteoclast activator, in chronic periodontitis. Hum Immunol. 2013;74:861–6.

Jarry CR, Martinez EF, Peruzzo DC, Carregaro V, Sacramento LA, Araujo VC, Weitzmann MN, Napimoga MH. Expression of SOFAT by T- and B-lineage cells may contribute to bone loss. Mol Med Rep. 2016;13:4252–8.

Johnell O, Kanis JA. An estimate of the worldwide prevalence and disability associated with osteoporotic fractures. Osteoporos Int. 2006;17:1726–33.

Kates SL, Kates OS, Mendelson DA. Advances in the medical management of osteoporosis. Injury. 2007;38(Suppl 3):S17–23.

Khosla S. Minireview: the OPG/RANKL/RANK system. Endocrinology. 2001;142:5050–5.

Kiesel J, Miller C, Abu-Amer Y, Aurora R. Systems level analysis of osteoclastogenesis reveals intrinsic and extrinsic regulatory interactions. Dev Dyn. 2007;236:2181–97.

Kim YG, Lee CK, Nah SS, Mun SH, Yoo B, Moon HB. Human CD4+CD25+ regulatory T-cells inhibit the differentiation of osteoclasts from peripheral blood mononuclear cells. Biochem Biophys Res Commun. 2007;357:1046–52.

Kimble RB, Vannice JL, Bloedow DC, Thompson RC, Hopfer W, Kung VT, Brownfield C, Pacifici R. Interleukin-1 receptor antagonist decreases bone loss and bone resorption in ovariectomized rats. J Clin Invest. 1994;93:1959–67.

Kimble RB, Matayoshi AB, Vannice JL, Kung VT, Williams C, Pacifici R. Simultaneous block of interleukin-1 and tumor necrosis factor is required to completely prevent bone loss in the early postovariectomy period. Endocrinology. 1995;136:3054–61.

Klausen B, Hougen HP, Fiehn NE. Increased periodontal bone loss in temporarily B lymphocyte-deficient rats. J Periodontal Res. 1989;24:384–90.

Knobel H, Guelar A, Vallecillo G, Nogues X, Diez A. Osteopenia in HIV-infected patients: is it the disease or is it the treatment? AIDS. 2001;15:807–8.

Kobayashi K, Takahashi N, Jimi E, Udagawa N, Takami M, Kotake S, Nakagawa N, Kinosaki M, Yamaguchi K, Shima N, Yasuda H, Morinaga T, Higashio K, Martin TJ, Suda T. Tumor necrosis factor alpha stimulates osteoclast differentiation by a mechanism independent of the ODF/RANKL-RANK interaction. J Exp Med. 2000;191:275–86.

Kong YY, Yoshida H, Sarosi I, Tan HL, Timms E, Capparelli C, Morony S, Oliveira-dos-Santos AJ, Van G, Itie A, Khoo W, Wakeham A, Dunstan CR, Lacey DL, Mak TW, Boyle WJ, Penninger JM. OPGL is a key regulator of osteoclastogenesis, lymphocyte development and lymph-node organogenesis. Nature. 1999a;397:315–23.

Kong YY, Feige U, Sarosi I, Bolon B, Tafuri A, Morony S, Capparelli C, Li J, Elliott R, McCabe S, Wong T, Campagnuolo G, Moran E, Bogoch ER, Van G, Nguyen LT, Ohashi PS, Lacey DL, Fish E, Boyle WJ, Penninger JM. Activated T-cells regulate bone loss and joint destruction in adjuvant arthritis through osteoprotegerin ligand. Nature. 1999b;402:304–9.

Kuek A, Hazleman BL, Ostor AJ. Immune-mediated inflammatory diseases (IMIDs) and biologic therapy: a medical revolution. Postgrad Med J. 2007;83:251–60.

Lacey DL, Timms E, Tan HL, Kelley MJ, Dunstan CR, Burgess T, Elliott R, Colombero A, Elliott G, Scully S, Hsu H, Sullivan J, Hawkins N, Davy E, Capparelli C, Eli A, Qian YX, Kaufman S, Sarosi I, Shalhoub V, Senaldi G, Guo J, Delaney J, Boyle WJ. Osteoprotegerin ligand is a cytokine that regulates osteoclast differentiation and activation. Cell. 1998;93:165–76.

Lam J, Takeshita S, Barker JE, Kanagawa O, Ross FP, Teitelbaum SL. TNF-alpha induces osteoclastogenesis by direct stimulation of macrophages exposed to permissive levels of RANK ligand. J Clin Invest. 2000;106:1481–8.

Lee SK, Kalinowski JF, Jastrzebski SL, Puddington L, Lorenzo JA. Interleukin-7 is a direct inhibitor of in vitro osteoclastogenesis. Endocrinology. 2003;144:3524–31.

Lee SK, Kadono Y, Okada F, Jacquin C, Koczon-Jaremko B, Gronowicz G, Adams DJ, Aguila HL, Choi Y, Lorenzo JA. T lymphocyte-deficient mice lose trabecular bone mass with ovariectomy. J Bone Miner Res. 2006a;21:1704–12.

Lee SK, Kalinowski JF, Jacquin C, Adams DJ, Gronowicz G, Lorenzo JA. Interleukin-7 influences osteoclast function in vivo but is not a critical factor in ovariectomy-induced bone loss. J Bone Miner Res. 2006b;21:695–702.

Lewis JR, Hassan SK, Wenn RT, Moran CG. Mortality and serum urea and electrolytes on admission for hip fracture patients. Injury. 2006;37:698–704.

Li P, Schwarz EM. The TNF-alpha transgenic mouse model of inflammatory arthritis. Springer Semin Immunopathol. 2003;25:19–33.

Li Y, Toraldo G, Li A, Yang X, Zhang H, Qian WP, Weitzmann MN. B-cells and T-cells are critical for the preservation of bone homeostasis and attainment of peak bone mass in vivo. Blood. 2007a;109:3839–48.

Li Y, Li A, Yang X, Weitzmann MN. Ovariectomy-induced bone loss occurs independently of B-cells. J Cell Biochem. 2007b;100:1370–5.

Li JY, Chassaing B, Tyagi AM, Vaccaro C, Luo T, Adams J, Darby TM, Weitzmann MN, Mulle JG, Gewirtz AT, Jones RM, Pacifici R. Sex steroid deficiency-associated bone loss is microbiota dependent and prevented by probiotics. J Clin Invest. 2016;126:2049–63.

Lindberg MK, Svensson J, Venken K, Chavoshi T, Andersson N, Moverare Skrtic S, Isaksson O, Vanderschueren D, Carlsten H, Ohlsson C. Liver-derived IGF-I is permissive for ovariectomy-induced trabecular bone loss. Bone. 2006;38:85–92.

Lopez-Granados E, Temmerman ST, Wu L, Reynolds JC, Follmann D, Liu S, Nelson DL, Rauch F, Jain A. Osteopenia in X-linked hyper-IgM syndrome reveals a regulatory role for CD40 ligand in osteoclastogenesis. Proc Natl Acad Sci U S A. 2007;104:5056–61.

Lubberts E, van den Bersselaar L, Oppers-Walgreen B, Schwarzenberger P, Coenen-de Roo CJ, Kolls JK, Joosten LA, van den Berg WB. IL-17 promotes bone erosion in murine collagen-induced arthritis through loss of the receptor activator of NF-kappa B ligand/osteoprotegerin balance. J Immunol. 2003;170:2655–62.

Manolagas SC, Bellido T, Jilka RL. New insights into the cellular, biochemical, and molecular basis of postmenopausal and senile osteoporosis: roles of IL-6 and gp130. Int J Immunopharmacol. 1995;17:109–16.

Masuzawa T, Miyaura C, Onoe Y, Kusano K, Ohta H, Nozawa S, Suda T. Estrogen deficiency stimulates B lymphopoiesis in mouse bone marrow. J Clin Invest. 1994;94:1090–7.

Matsuzaki K, Udagawa N, Takahashi N, Yamaguchi K, Yasuda H, Shima N, Morinaga T, Toyama Y, Yabe Y, Higashio K, Suda T. Osteoclast differentiation factor (ODF) induces osteoclast-like cell formation in human peripheral blood mononuclear cell cultures. Biochem Biophys Res Commun. 1998;246:199–204.

McComsey GA, Huang JS, Woolley IJ, Young B, Sax PE, Gerber M, Swindells S, Bonilla H, Gopalakrishnan G. Fragility fractures in HIV-infected patients: need for better understanding of diagnosis and management. J Int Assoc Physicians AIDS Care (Chic). 2004;3:86–91.

McComsey GA, Tebas P, Shane E, Yin MT, Overton ET, Huang JS, Aldrovandi GM, Cardoso SW, Santana JL, Brown TT. Bone disease in HIV infection: a practical review and recommendations for HIV care providers. Clin Infect Dis. 2010;51:937–46.

Miyaura C, Onoe Y, Inada M, Maki K, Ikuta K, Ito M, Suda T. Increased B-lymphopoiesis by interleukin 7 induces bone loss in mice with intact ovarian function: similarity to estrogen deficiency. Proc Natl Acad Sci U S A. 1997;94:9360–5.

Mizuno A, Amizuka N, Irie K, Murakami A, Fujise N, Kanno T, Sato Y, Nakagawa N, Yasuda H, Mochizuki S, Gomibuchi T, Yano K, Shima N, Washida N, Tsuda E, Morinaga T, Higashio K, Ozawa H. Severe osteoporosis in mice lacking osteoclastogenesis inhibitory factor/osteoprotegerin. Biochem Biophys Res Commun. 1998;247:610–5.

Mondy K, Tebas P. Emerging bone problems in patients infected with human immunodeficiency virus. Clin Infect Dis. 2003;36:S101–5.

Mora S, Sala N, Bricalli D, Zuin G, Chiumello G, Vigano A. Bone mineral loss through increased bone turnover in HIV-infected children treated with highly active antiretroviral therapy. AIDS. 2001;15:1823–9.

Mundy GR, Luben RA, Raisz LG, Oppenheim JJ, Buell DN. Bone-resorbing activity in supernatants from lymphoid cell lines. N Engl J Med. 1974;290:867–71.

Najafian N, Sayegh MH. CTLA4-Ig: a novel immunosuppressive agent. Expert Opin Investig Drugs. 2000;9:2147–57.

Nakashima T, Hayashi M, Fukunaga T, Kurata K, Oh-Hora M, Feng JQ, Bonewald LF, Kodama T, Wutz A, Wagner EF, Penninger JM, Takayanagi H. Evidence for osteocyte regulation of bone homeostasis through RANKL expression. Nat Med. 2011;17:1231–4.

Niida S, Kaku M, Amano H, Yoshida H, Kataoka H, Nishikawa S, Tanne K, Maeda N, Kodama H. Vascular endothelial growth factor can substitute for macrophage colony- stimulating factor in the support of osteoclastic bone resorption. J Exp Med. 1999;190:293–8.

Ofotokun I, Weitzmann MN. HIV-1 infection and antiretroviral therapies: risk factors for osteoporosis and bone fracture. Curr Opin Endocrinol Diabetes Obes. 2010;17:523–9.

Ofotokun I, Weitzmann MN. HIV and bone metabolism. Discov Med. 2011;11:385–93.

Ofotokun I, Titanji K, Vikulina T, Roser-Page S, Yamaguchi M, Zayzafoon M, Williams IR, Weitzmann MN. Role of T-cell reconstitution in HIV-1 antiretroviral therapy-induced bone loss. Nat Commun. 2015;6:8282.

Ofotokun I, Titanji K, Vunnava A, Roser-Page S, Vikulina T, Villinger F, Rogers K, Sheth AN, Lahiri CD, Lennox JL, Weitzmann MN. Antiretroviral therapy induces a rapid increase in bone resorption that is positively associated with the magnitude of immune reconstitution in HIV infection. AIDS. 2016a;30:405–14.

Ofotokun I, Titanji K, Lahiri CD, Vunnava A, Foster A, Sanford SE, Sheth AN, Lennox JL, Knezevic A, Ward L, Easley KA, Powers P, Weitzmann MN. A single dose Zoledronic acid infusion prevents antiretroviral therapy-induced bone loss in treatment-naive HIV-infected patients: a phase IIb trial. Clin Infect Dis. 2016b;63:663–71.

Onal M, Xiong J, Chen X, Thostenson JD, Almeida M, Manolagas SC, O'Brien CA. Receptor activator of nuclear factor kappaB ligand (RANKL) protein expression by B lymphocytes contributes to ovariectomy-induced bone loss. J Biol Chem. 2012;287:29851–60.

Pacifici R. T-cells: critical bone regulators in health and disease. Bone. 2010;47:461–71.

Pacifici R. Role of T-cells in the modulation of PTH action: physiological and clinical significance. Endocrine. 2013;44:576–82.

Pacifici R, Brown C, Puscheck E, Friedrich E, Slatopolsky E, Maggio D, McCracken R, Avioli LV. Effect of surgical menopause and estrogen replacement on cytokine release from human blood mononuclear cells. Proc Natl Acad Sci U S A. 1991;88:5134–8.

Passeri G, Girasole G, Jilka RL, Manolagas SC. Increased interleukin-6 production by murine bone marrow and bone cells after estrogen withdrawal. Endocrinology. 1993;133:822–8.

Pineda B, Laporta P, Hermenegildo C, Cano A, Garcia-Perez MA. A C >T polymorphism located at position -1 of the Kozak sequence of CD40 gene is associated with low bone mass in Spanish postmenopausal women. Osteoporos Int. 2008;19:1147–52.

Piso RJ, Rothen M, Rothen JP, Stahl M. Markers of bone turnover are elevated in patients with antiretroviral treatment independent of the substance used. J Acquir Immune Defic Syndr. 2011;56:320–4.

Prieto-Alhambra D, Guerri-Fernandez R, De Vries F, Lalmohamed A, Bazelier M, Starup-Linde J, Diez-Perez A, Cooper C, Vestergaard P. HIV infection and its association with an excess risk of clinical fractures: a nationwide case-control study. J Acquir Immune Defic Syndr. 2014;66:90–5.

Prior J, Burdge D, Maan E, Milner R, Hankins C, Klein M, Walmsley S. Fragility fractures and bone mineral density in HIV positive women: a case-control population-based study. Osteoporos Int. 2007;18:1345–53.

Redlich K, Gortz B, Hayer S, Zwerina J, Doerr N, Kostenuik P, Bergmeister H, Kollias G, Steiner G, Smolen JS, Schett G. Repair of local bone erosions and reversal of systemic bone loss upon therapy with anti-tumor necrosis factor in combination with osteoprotegerin or parathyroid hormone in tumor necrosis factor-mediated arthritis. Am J Pathol. 2004;164:543–55.

Reid W, Sadowska M, Denaro F, Rao S, Foulke J Jr, Hayes N, Jones O, Doodnauth D, Davis H, Sill A, O'Driscoll P, Huso D, Fouts T, Lewis G, Hill M, Kamin-Lewis R, Wei C, Ray P, Gallo RC, Reitz M, Bryant J. An HIV-1 transgenic rat that develops HIV-related pathology and immunologic dysfunction. Proc Natl Acad Sci U S A. 2001;98:9271–6.

Rifas L, Weitzmann MN. A novel T-cell cytokine, secreted osteoclastogenic factor of activated T-cells, induces osteoclast formation in a RANKL-independent manner. Arthritis Rheumatol. 2009;60:3324–35.

Roggia C, Gao Y, Cenci S, Weitzmann MN, Toraldo G, Isaia G, Pacifici R. Up-regulation of TNF-producing T-cells in the bone marrow: a key mechanism by which estrogen deficiency induces bone loss in vivo. Proc Natl Acad Sci U S A. 2001;98:13960–5.

Roodman GD, Kurihara N, Ohsaki Y, Kukita A, Hosking D, Demulder A, Smith JF, Singer FR. Interleukin 6. A potential autocrine/paracrine factor in Paget's disease of bone. J Clin Invest. 1992;89:46–52.

Roser-Page S, Vikulina T, Zayzafoon M, Weitzmann MN. CTLA-4Ig-induced T-cell anergy promotes Wnt-10b production and bone formation in a mouse model. Arthritis Rheumatol. 2014;66:990–9.

Ruderman EM, Pope RM. The evolving clinical profile of abatacept (CTLA4-Ig): a novel co-stimulatory modulator for the treatment of rheumatoid arthritis. Arthritis Res Ther. 2005;7(Suppl 2):S21–5.

Ryan MR, Shepherd R, Leavey JK, Gao Y, Grassi F, Schnell FJ, Qian WP, Kersh GJ, Weitzmann MN, Pacifici R. An IL-7-dependent rebound in thymic T-cell output contributes to the bone loss induced by estrogen deficiency. Proc Natl Acad Sci U S A. 2005;102:16735–40.

Sarma U, Flanagan AM. Macrophage colony-stimulating factor induces substantial osteoclast generation and bone resorption in human bone marrow cultures. Blood. 1996;88:2531–40.

Sass DA, Liss T, Bowman AR, Rucinski B, Popoff SN, Pan Z, Ma YF, Epstein S. The role of the T lymphocyte in estrogen deficiency osteopenia. J Bone Miner Res. 1997;12:479–86.

Sato K, Suematsu A, Okamoto K, Yamaguchi A, Morishita Y, Kadono Y, Tanaka S, Kodama T, Akira S, Iwakura Y, Cua DJ, Takayanagi H. Th17 functions as an osteoclastogenic helper T-cell subset that links T-cell activation and bone destruction. J Exp Med. 2006;203:2673–82.

Sayegh MH. Finally, CTLA4Ig graduates to the clinic. J Clin Invest. 1999;103:1223–5.

Schett G, David JP. The multiple faces of autoimmune-mediated bone loss. Nat Rev Endocrinol. 2010;6:698–706.

Schett G, Redlich K, Hayer S, Zwerina J, Bolon B, Dunstan C, Gortz B, Schulz A, Bergmeister H, Kollias G, Steiner G, Smolen JS. Osteoprotegerin protects against generalized bone loss in tumor necrosis factor-transgenic mice. Arthritis Rheum. 2003;48:2042–51.

Sharma A, Flom PL, Weedon J, Klein RS. Prospective study of bone mineral density changes in aging men with or at risk for HIV infection. AIDS. 2010;24:2337–45.

Sharma A, Shi Q, Hoover DR, Anastos K, Tien PC, Young MA, Cohen MH, Golub ET, Gustafson D, Yin MT. Increased fracture incidence in middle-aged HIV-infected and HIV-uninfected women: updated results from the Women's Interagency HIV study. J Acquir Immune Defic Syndr. 2015;70:54–61.

Shinar DM, Rodan GA. Biphasic effects of transforming growth factor-beta on the production of osteoclast-like cells in mouse bone marrow cultures: the role of prostaglandins in the generation of these cells. Endocrinology. 1990;126:3153–8.

Simonet WS, Lacey DL, Dunstan CR, Kelley M, Chang MS, Luthy R, Nguyen HQ, Wooden S, Bennett L, Boone T, Shimamoto G, DeRose M, Elliott R, Colombero A, Tan HL, Trail G, Sullivan J, Davy E, Bucay N, Renshaw-Gegg L, Hughes TM, Hill D, Pattison W, Campbell P, Boyle WJ, et al. Osteoprotegerin: a novel secreted protein involved in the regulation of bone density. Cell. 1997;89:309–19.

Srivastava S, Weitzmann MN, Kimble RB, Rizzo M, Zahner M, Milbrandt J, Ross FP, Pacifici R. Estrogen blocks M-CSF gene expression and osteoclast formation by regulating phosphorylation of Egr-1 and its interaction with Sp-1. J Clin Invest. 1998;102:1850–9.
Suda T, Takahashi N, Udagawa N, Jimi E, Gillespie MT, Martin TJ. Modulation of osteoclast differentiation and function by the new members of the tumor necrosis factor receptor and ligand families. Endocr Rev. 1999;20:345–57.
Surh CD, Sprent J. Homeostasis of naive and memory T-cells. Immunity. 2008;29:848–62.
Takayanagi H, Ogasawara K, Hida S, Chiba T, Murata S, Sato K, Takaoka A, Yokochi T, Oda H, Tanaka K, Nakamura K, Taniguchi T. T-cell-mediated regulation of osteoclastogenesis by signalling cross-talk between RANKL and IFN-gamma. Nature. 2000;408:600–5.
Tanchot C, Lemonnier FA, Perarnau B, Freitas AA, Rocha B. Differential requirements for survival and proliferation of CD8 naive or memory T-cells. Science. 1997;276:2057–62.
Tawfeek H, Bedi B, Li JY, Adams J, Kobayashi T, Weitzmann MN, Kronenberg HM, Pacifici R. Disruption of PTH receptor 1 in T-cells protects against PTH-induced bone loss. PLoS One. 2010;5:e12290.
Tebas P, Powderly WG, Claxton S, Marin D, Tantisiriwat W, Teitelbaum SL, Yarasheski KE. Accelerated bone mineral loss in HIV-infected patients receiving potent antiretroviral therapy. AIDS. 2000;14:F63–7.
Teitelbaum SL. Bone resorption by osteoclasts. Science. 2000;289:1504–8.
Terauchi M, Li JY, Bedi B, Baek KH, Tawfeek H, Galley S, Gilbert L, Nanes MS, Zayzafoon M, Guldberg R, Lamar DL, Singer MA, Lane TF, Kronenberg HM, Weitzmann MN, Pacifici R. T lymphocytes amplify the anabolic activity of parathyroid hormone through Wnt10b signaling. Cell Metab. 2009;10:229–40.
Thirunavukkarasu K, Miles RR, Halladay DL, Yang X, Galvin RJ, Chandrasekhar S, Martin TJ, Onyia JE. Stimulation of osteoprotegerin (OPG) gene expression by transforming growth factor-beta (TGF-beta). Mapping of the OPG promoter region that mediates TGF-beta effects. J Biol Chem. 2001;276:36241–50.
Thomas J, Doherty SM. HIV infection – a risk factor for osteoporosis. J Acquir Immune Defic Syndr. 2003;33:281–91.
Titanji K, Vunnava A, Sheth AN, Delille C, Lennox JL, Sanford SE, Foster A, Knezevic A, Easley KA, Weitzmann MN, Ofotokun I. Dysregulated B-cell expression of RANKL and OPG correlates with loss of bone mineral density in HIV infection. PLoS Pathog. 2014;10:e1004497.
Toraldo G, Roggia C, Qian WP, Pacifici R, Weitzmann MN. IL-7 induces bone loss in vivo by induction of receptor activator of nuclear factor kappa B ligand and tumor necrosis factor alpha from T-cells. Proc Natl Acad Sci U S A. 2003;100:125–30.
Triant VA, Brown TT, Lee H, Grinspoon SK. Fracture prevalence among human immunodeficiency virus (HIV)-infected versus non-HIV-infected patients in a large U.S. healthcare system. J Clin Endocrinol Metab. 2008;93:3499–504.
Trummel CL, Mundy GR, Raisz LG. Release of osteoclast activating factor by normal human peripheral blood leukocytes. J Lab Clin Med. 1975;85:1001–7.
Tsuda E, Goto M, Mochizuki S, Yano K, Kobayashi F, Morinaga T, Higashio K. Isolation of a novel cytokine from human fibroblasts that specifically inhibits osteoclastogenesis. Biochem Biophys Res Commun. 1997;234:137–42.
Tyagi AM, Srivastava K, Mansoori MN, Trivedi R, Chattopadhyay N, Singh D. Estrogen deficiency induces the differentiation of IL-17 secreting Th17 cells: a new candidate in the pathogenesis of osteoporosis. PLoS One. 2012;7:e44552.
Valenzona HO, Pointer R, Ceredig R, Osmond DG. Prelymphomatous B-cell hyperplasia in the bone marrow of interleukin-7 transgenic mice: precursor B-cell dynamics, microenvironmental organization and osteolysis. Exp Hematol. 1996;24:1521–9.
Vikulina T, Fan X, Yamaguchi M, Roser-Page S, Zayzafoon M, Guidot DM, Ofotokun I, Weitzmann MN. Alterations in the immuno-skeletal interface drive bone destruction in HIV-1 transgenic rats. Proc Natl Acad Sci U S A. 2010;107:13848–53.

Vital EM, Emery P. Abatacept in the treatment of rheumatoid arthritis. Ther Clin Risk Manag. 2006;2:365–75.
Walker DG. Congenital osteopetrosis in mice cured by parabiotic union with normal siblings. Endocrinology. 1972;91:916–20.
Walker DG. Bone resorption restored in osteopetrotic mice by transplants of normal bone marrow and spleen cells. Science. 1975;190:784–5.
Wang MW, Wei S, Faccio R, Takeshita S, Tebas P, Powderly WG, Teitelbaum SL, Ross FP. The HIV protease inhibitor ritonavir blocks osteoclastogenesis and function by impairing RANKL-induced signaling. J Clin Invest. 2004;114:206–13.
Wei S, Kitaura H, Zhou P, Ross FP, Teitelbaum SL. IL-1 mediates TNF-induced osteoclastogenesis. J Clin Invest. 2005;115:282–90.
Weitzmann MN, Pacifici R. Estrogen deficiency and bone loss: an inflammatory tale. J Clin Invest. 2006;116:1186–94.
Weitzmann MN, Cenci S, Rifas L, Brown C, Pacifici R. Interleukin-7 stimulates osteoclast formation by up-regulating the T-cell production of soluble osteoclastogenic cytokines. Blood. 2000a;96:1873–8.
Weitzmann MN, Cenci S, Haug J, Brown C, DiPersio J, Pacifici R. B lymphocytes inhibit human osteoclastogenesis by secretion of TGFbeta. J Cell Biochem. 2000b;78:318–24.
Weitzmann MN, Cenci S, Rifas L, Haug J, Dipersio J, Pacifici R. T-cell activation induces human osteoclast formation via receptor activator of nuclear factor kappaB ligand-dependent and -independent mechanisms. J Bone Miner Res. 2001;16:328–37.
Weitzmann MN, Roggia C, Toraldo G, Weitzmann L, Pacifici R. Increased production of IL-7 uncouples bone formation from bone resorption during estrogen deficiency. J Clin Invest. 2002;110:1643–50.
Wohl D, Oka S, Clumeck N, Clarke A, Brinson C, Stephens J, Tashima K, Arribas JR, Rashbaum B, Cheret A, Brunetta J, Mussini C, Tebas P, Sax PE, Cheng A, Zhong L, Callebaut C, Das M, Fordyce M, Team GUS. A randomized, double-blind comparison of Tenofovir Alafenamide (TAF) vs. Tenofovir Disoproxil fumarate (TDF), each coformulated with Elvitegravir, Cobicistat, and Emtricitabine (E/C/F) for initial HIV-1 treatment: week 96 results. J Acquir Immune Defic Syndr. 2016;2(1):58–64.
Womack JA, Goulet JL, Gibert C, Brandt C, Chang CC, Gulanski B, Fraenkel L, Mattocks K, Rimland D, Rodriguez-Barradas MC, Tate J, Yin MT, Justice AC. Increased risk of fragility fractures among HIV infected compared to uninfected male veterans. PLoS One. 2011;6:e17217.
Wong BR, Josien R, Lee SY, Sauter B, Li HL, Steinman RM, Choi Y. TRANCE (tumor necrosis factor [TNF]-related activation-induced cytokine), a new TNF family member predominantly expressed in T-cells, is a dendritic cell-specific survival factor. J Exp Med. 1997a;186:2075–80.
Wong BR, Rho J, Arron J, Robinson E, Orlinick J, Chao M, Kalachikov S, Cayani E, Bartlett FS 3rd, Frankel WN, Lee SY, Choi Y. TRANCE is a novel ligand of the tumor necrosis factor receptor family that activates c-Jun N-terminal kinase in T-cells. J Biol Chem. 1997b;272:25190–4.
Xiong J, Onal M, Jilka RL, Weinstein RS, Manolagas SC, O'Brien CA. Matrix-embedded cells control osteoclast formation. Nat Med. 2011;17:1235–41.
Yamaza T, Miura Y, Bi Y, Liu Y, Akiyama K, Sonoyama W, Patel V, Gutkind S, Young M, Gronthos S, Le A, Wang CY, Chen W, Shi S. Pharmacologic stem cell based intervention as a new approach to osteoporosis treatment in rodents. PLoS One. 2008;3:e2615.
Yanaba K, Hamaguchi Y, Venturi GM, Steeber DA, St Clair EW, Tedder TF. B-cell depletion delays collagen-induced arthritis in mice: arthritis induction requires synergy between humoral and cell-mediated immunity. J Immunol. 2007;179:1369–80.
Yin MT, Shi Q, Hoover DR, Anastos K, Sharma A, Young M, Levine A, Cohen MH, Shane E, Golub ET, Tien PC. Fracture incidence in HIV-infected women: results from the Women's Interagency HIV study. AIDS. 2010a;24:2679–86.
Yin MT, Lu D, Cremers S, Tien PC, Cohen MH, Shi Q, Shane E, Golub ET, Anastos K. Short-term bone loss in HIV-infected premenopausal women. J Acquir Immune Defic Syndr. 2010b;53:202–8.

Young B, Dao CN, Buchacz K, Baker R, Brooks JT. Increased rates of bone fracture among HIV-infected persons in the HIV Outpatient Study (HOPS) compared with the US general population, 2000–2006. Clin Infect Dis. 2011;52:1061–8.

Yun TJ, Chaudhary PM, Shu GL, Frazer JK, Ewings MK, Schwartz SM, Pascual V, Hood LE, Clark EA. OPG/FDCR-1, a TNF receptor family member, is expressed in lymphoid cells and is up-regulated by ligating CD40. J Immunol. 1998;161:6113–21.

Zaiss MM, Axmann R, Zwerina J, Polzer K, Guckel E, Skapenko A, Schulze-Koops H, Horwood N, Cope A, Schett G. Treg cells suppress osteoclast formation: a new link between the immune system and bone. Arthritis Rheum. 2007;56:4104–12.

Zaiss MM, Frey B, Hess A, Zwerina J, Luther J, Nimmerjahn F, Engelke K, Kollias G, Hunig T, Schett G, David JP. Regulatory T-cells protect from local and systemic bone destruction in arthritis. J Immunol. 2010;184(12):7238–46. (PMID 20483756)

Zhang YH, Heulsmann A, Tondravi MM, Mukherjee A, Abu-Amer Y. Tumor necrosis factor-alpha (TNF) stimulates RANKL-induced osteoclastogenesis via coupling of TNF type 1 receptor and RANK signaling pathways. J Biol Chem. 2001;276:563–8.

Zwerina K, Koenders M, Hueber A, Marijnissen RJ, Baum W, Heiland GR, Zaiss M, McLnnes I, Joosten L, van den Berg W, Zwerina J, Schett G. Anti IL-17A therapy inhibits bone loss in TNF-alpha-mediated murine arthritis by modulation of the T-cell balance. Eur J Immunol. 2012;42:413–23.

Chapter 13
Bone and the Central Nervous System

Rishikesh N. Kulkarni and Paul A. Baldock

Abstract The human skeleton is a miracle of engineering, combining both strength and light weight to provide mechanical support to withstand the force of gravity and to transfer muscle forces during movement. The brain is well established as a master regulator of homeostasis in peripheral tissues. The discovery of bone regulation by central nervous system represents a growing area of study that is identifying novel regulatory axes between the nervous system and bone homeostasis, and revealing a far more complex, and interdependent bone biology than previously envisioned. This chapter examines the current understanding of the central regulation of bone homeostasis. Herein, we will discuss the contribution of central peptides, and other peptides such as leptin, and semaphorins and involvement of the brain in regulation of bone metabolism.

Keywords Central nervous system (CNS) • Sympathetic nervous system (SNS) • Hypothalamus • Adrenergic • Leptin • Neuropeptide Y (NPY) • Cannabinoid • Semaphorins • Melanocortin • ACTH • CART • Energy

R.N. Kulkarni
Bone and Mineral Research Program, Osteoporosis and Bone Biology Division, Garvan Institute of Medical Research, St Vincent's Hospital, 384 Victoria St., Darlinghurst, Sydney, NSW, Australia

P.A. Baldock (✉)
Bone and Mineral Research Program, Osteoporosis and Bone Biology Division, Garvan Institute of Medical Research, St Vincent's Hospital, 384 Victoria St., Darlinghurst, Sydney, NSW, Australia

Faculty of Medicine, University of New South Wales, Sydney, NSW, Australia

School of Medicine, The University of Notre Dame Australia, Sydney, NSW, Australia
e-mail: p.baldock@garvan.org.au; http://www.garvan.org.au

S.Y. Smith et al. (eds.), *Bone Toxicology*, Molecular and Integrative Toxicology,
DOI 10.1007/978-3-319-56192-9_13

13.1 Introduction

Bone architecture adapts to changes in the mechanical strain generated within it to ensure that it maintains sufficient strength to withstand the loads to which it is subjected. As part of this adaptation, bone is remodeled throughout life by means of basic multicellular units (BMUs), consisting of osteoclasts and osteoblasts acting in a coordinated fashion to resorb existing bone and form new bone, respectively (Burger et al. 1995; Burger and Klein-Nulend 1999). The prevalent and widely accepted view has been that bone remodeling is controlled in a predominantly endocrine manner while simultaneously responding to local mechanical stimuli. Of late, the discovery of bone regulation by central nervous system represents an emerging area of study that is identifying novel regulatory axes between the nervous system and bone homeostasis.

The brain is a powerful regulator of homeostatic processes in many peripheral tissues. There is now clear evidence for cross talk between the brain and bone via two distinct routes. The first comprises well-defined hormonal signals arising from neuroendocrine neurons of the hypothalamus and subsequently processed within the pituitary. The second, more recently appreciated pathway, consists of efferent neuronal discharges originating from the hypothalamus and processed through the brainstem. The hypothalamus, with its semipermeable blood brain barrier, is thus one of the most powerful regulatory regions within the body, integrating signals not only from peripheral tissues via the circulation but also from afferent neural return. A direct role of the nervous system in bone cells was strongly supported by immunocytochemistry studies which revealed the presence of innervation and neuropeptides receptors in bone cells (Elefteriou 2005). Furthermore, retrograde transsynaptic tracing has identified neural tracts from the femoral bone marrow linked direct to the central nervous system (Denes et al. 2005). These findings suggest the existence of neuronal connection between the brain and bone. These direct, neuronal pathways represent an emergent area of study identifying novel regulatory axes between the brain and the cells of bone. Moreover, this work is also providing insights into regulatory connections involving skeletal tissues that are proving unexpected, thereby outlining a level of interconnectedness that has been previously unappreciated. This chapter examines the expanded understanding of the central, neural outputs to bone metabolism and remodeling.

13.2 Leptin

Leptin is a small polypeptide hormone secreted into the peripheral circulation by adipocytes, and study of a mouse with nonfunctional leptin was the first indicator of the connection between the brain and bone. The leptin receptor in the hypothalamus senses the circulating levels of leptin which are positively correlated to total body adiposity and play a pivotal part in whole-body energy homeostasis. Mice with mutations in the leptin gene (*ob/ob*) or its receptor (*db/db*) display marked obesity accompanied by hyperglycemia, hyperinsulinemia, hypercorticism, and hypogonadism (Tartaglia et al. 1995; Halaas et al. 1995; Friedman and Halaas 1998). Interestingly, these mice also display marked bone phenotypes, which have been pivotal to understanding central control of bone but also link between energy and bone homeostasis. The hypogonadism and hypercorticism of *ob/ob* and *db/db* mice would be expected to result in reduced cancellous bone mass due to elevated bone loss, but, unexpectedly, it was found that cancellous bone volume was markedly increased in both models (Ducy et al. 2000). This increased bone mass was associated with a marked increase in bone formation, which surpassed the elevation of bone resorption. A key role for hypothalamic leptin activity in regulating bone mass was indicated following infusion of leptin into the cerebral ventricles, with no detectable change in serum leptin, which produced a decrease in cancellous bone volume in both *ob/ob* and wild-type mice (Ducy et al. 2000). The importance of central leptin signaling was further demonstrated by central infusion of leptin to one mouse of a pair of parabiosed *ob/ob* mice. With effective cross-circulation between the pair, correction of the *ob/ob* skeletal phenotype occurred only in the centrally infused recipient and not the contralateral mouse (Takeda et al. 2002). These findings provided strong evidence that, similar to the central regulation of obesity, the inhibition of osteoblast activity on cancellous bone surfaces is mediated by leptin within the hypothalamus.

The control of bone remodeling by leptin deficiency provides exciting new insights into neuronal mechanisms in skeletal physiology, but some reports in the literature suggest additional levels of complexity. Leptin action in cortical bone may be distinct from its effects in the cancellous compartment, as femoral length and total body bone mineral content (BMC) are decreased in *ob/ob* mice and peripheral leptin treatment of immature (4-week old) *ob/ob* mice partially restored both parameters (Steppan et al. 2000). In addition, similar to the *db/db* mouse, leptin-resistant Zucker (*fa/fa*) rats carry an inactivating Gln269Pro mutation in the extracellular domain of the Ob-Rb leptin receptor gene (Takaya et al. 1996) and exhibit reduced femoral bone mineral density (BMD) and calcium content (Mathey et al. 2002). Leptin can stimulate growth plate chondrocyte proliferation and differentiation in vitro (Nakajima et al. 2003); thus it is possible that the *ob/ob*-associated reductions in bone length and BMC may relate to growth plate rather than osteoblastic effects. Further complexity exists in *ob/ob* mice, between axial and appendicular skeletons. A subsequent study reported increased vertebral length and lumbar BMC in addition to reduced femur length mid-shaft cortical area and thickness and BMC (Hamrick et al. 2004). Whether these differences between effects of leptin deficiency on cortical and trabecular compartments or between appendicular and axial skeleton may relate to effects on muscle mass, marrow adipogenesis (Hamrick et al. 2004) or growth plate remains to be resolved; however, studies outlined below suggest a role for neuropeptide Y pathways in cortical bone control by leptin, district from the pathway to cancellous bone.

13.3 Leptin and Bone in Humans

Genetic studies in mice have convincingly demonstrated the association between leptin deficiency and altered skeletal homeostasis, but the limited genetic evidence in humans is mixed. Study of four individuals in a family carrying a missense mutation in the leptin gene showed only one exhibited low bone mass despite morbid obesity in all four; however, a high degree of consanguinity and multiple endocrine defects cloud interpretation of the observation (Ozata et al. 1999). Administration of leptin for a year to a young girl with a leptin mutation was associated with loss of body fat and an increase in whole-body bone mineral density (BMD), as would be expected from mouse studies (Farooqi et al. 1999). Indirect clinical evidence suggests that if there is an association between leptin deficiency and bone mass in humans, BMD is reduced after self-imposed calorie restriction in anorexia nervosa, which can lead to extremely low levels of the nutritionally dependent hormones leptin and IGF-1 and amenorrhea, consistent with the low bone mass, reduced fertility, and suppressed activity of the somatotropic axis in *ob/ob* mice (Miller et al. 2004; Soyka et al. 1999). Reinforcing the potential importance of IGF-1 in this context, leptin treatment in patients with hypothalamic amenorrhea due to strenuous exercise or low body weight resulted in increased levels of both IGF-1 and bone formation markers (Welt et al. 2004). Such studies involving reduced nutritional intake, excessive exercise, and disrupted menses are not, however, amenable to simple interpretation in terms of direct leptin effects on bone vs indirect effects acting through altered reproductive and other neuroendocrine function.

Before the discovery of leptin, initial correlations between bone mass and body weight or fat mass (Reid 2002) were attributed either to changes in mechanical load or estrogen production related to fat tissue mass. But after the discovery of leptin, this additional factor was also investigated. Serum leptin levels and BMD are directly correlated in pre- and postmenopausal females by some investigators (Thomas et al. 2001) but not others (Rauch et al. 1998). Furthermore, a positive association between leptin levels and BMC was observed in a group of healthy nonobese women (Pasco et al. 2001), and a weak correlation was observed in postmenopausal osteoporotic women but not the controls (Odabasi et al. 2000). This uncertain relationship between leptin and bone may be gender-specific, as studies in males have reported either no association (Thomas et al. 2001) or a negative association (Sato et al. 2001) between leptin and BMD. In this context, it is interesting that one study which detected a positive association between leptin and BMD found no association between leptin and biochemical markers of bone resorption or formation and concluded that leptin is not likely to play a significant direct role in regulating bone cell activity (Goulding and Taylor 1998). However, this interpretation is challenged by direct examinations of bone cell responses to leptin in vitro.

13.4 Bone Cell Responses to Leptin in Vitro

Evidence for presence of the signal-transducing Ob-Rb leptin receptor or leptin-binding sites has been detected on ossifying fetal cartilage (Hoggard et al. 1997), immortalized marrow stromal cells (Thomas et al. 1999), chondrocytes (Steppan et al. 2000; Nakajima et al. 2003; Maor et al. 2002; Cornish et al. 2002), primary osteoblasts (Steppan et al. 2000; Cornish et al. 2002; Enjuanes et al. 2002; Reseland et al. 2001; Lee et al. 2002), and some but not all osteosarcoma cell lines (Reseland et al. 2001). Adding further to the possibility of peripheral actions of leptin is evidence of leptin production by primary human and rat osteoblast cultures (Morroni ct al. 2004; Gordcladzc ct al. 2002).

A number of in vitro functional studies support direct leptin effects on chondrocytes and osteoblasts. Leptin has been observed to stimulate proliferation and differentiation of cultured growth plate chondrocytes (Nakajima et al. 2003) and to induce osteoblast differentiation and suppress adipogenesis in a human stromal cell line (Thomas et al. 1999). The hormone also has direct stimulatory effects on cell proliferation, differentiation, and function in human and rat osteoblastic cultures (Cornish et al. 2002; Gordeladze et al. 2002). In addition, leptin can inhibit osteoclastogenesis in vitro (Holloway et al. 2002; Burguera et al. 2001). One notable exception, the study which identified the central leptin pathway to bone, to these studies detected no osteoblastic expression of leptin or the leptin receptor long form and no effect of leptin treatment on formation of mineralized nodules in primary mouse osteoblastic cultures (Ducy et al. 2000). The majority of the in vitro evidence, however, supports peripheral actions of leptin in bone biology. It therefore remains possible that, particularly at high leptin concentrations direct, peripheral actions are prominent, and at low levels, central anti-actions may predominate.

13.5 Sympathetic Nervous System Regulation of Bone Mass

Although leptin deficiency is characterized by numerous endocrine changes, a humoral pathway was ruled out by parabiosis experiments in *ob/ob* mice (Takeda et al. 2002). This indicated that leptin signaling in the brain (supplied by brain-only infusion, intracerebroventricular icv) was responsible for the skeletal changes, without the necessity of a humoral signal. The ventromedial hypothalamus (VMH) was identified as the source of the leptin signaling and origin of the signal to cancellous bone. Thus implicated sympathetic activity, as sympathetic tone, is known to be altered in leptin deficiency and was likely the important downstream mediator of central leptin signaling. Takeda and colleagues investigated whether the central actions of leptin might be mediated via the sympathetic nervous system (SNS) and found that dopamine β-hydroxylase deficient mice, which are unable to produce epinephrine and norepinephrine, the catecholamine ligands for adrenergic receptors, also have high bone mass (Takeda et al. 2002). Importantly, icv infusion of leptin failed to reduce the bone mass of dopamine β-hydroxylase mice, indicating a requirement for a functional autonomic nervous system for leptin actions and identifying the sympathetic system as a neuronal mediator of leptin effects on bone (Takeda et al. 2002). This conclusion is supported by studies using β-adrenergic receptor (β-AR) agonists and antagonists. Administration of the β-agonist isoproterenol restored sympathetic activity in *ob/ob* mice and reduced cancellous bone mass in both wild-type and *ob/ob* mice without affecting body weight (Takeda et al. 2002). Conversely, administration of the nonselective β-adrenergic receptor antagonist propranolol increased bone mass in wild-type mice, demonstrating that modulation of SNS activity affects cancellous bone remodeling in a manner indicated by *ob/ob* mice (Takeda et al. 2002). Consistent with a role for SNS signaling in bone, mice deficient in ß2AR showed increased cancellous bone volume (Elefteriou et al. 2005), while deletion of a downstream mediator of ß2AR signaling, adenylyl cyclase 5, protects against age-related bone loss (Yan et al. 2007). Disruption of dopamine ß-hydroxylase, an enzyme generating adrenaline and noradrenaline, produced greater cancellous bone mass in wild-type mice. Importantly, reducing adrenergic signaling protected against cancellous bone loss following icv leptin, through loss of dopamine ß-hydroxylase in wild-type mice and propranolol treatment in *ob/ob* (Takeda et al. 2002). Unfortunately the effect of these agents on cortical bone was not systematically investigated. However, the novel role of ß-adrenergic signaling via SNS completed the central leptin signaling pathway to bone.

Consistent with the in vivo evidence for sympathetic control of bone formation, β-ARs have been identified on osteoblasts and in osteoblast-like cell lines. β1- and β2-adrenergic receptors are expressed at different levels in various human osteoblast-like cell lines (Kellenberger et al. 1998), and β2-adrenergic receptors have been identified on rat osteoblast-like cells, and in mouse primary osteoblast cultures (Takeda et al. 2002; Moore et al. 1993), suggesting that sympathetic activity might control bone formation through direct modulation of osteoblast function. Supporting this, administration of the specific β2-AR agonist, formoterol, induced cAMP and expression of the immediate-early gene *c-fos* in the SaOS-2 human osteosarcoma cell line (Kellenberger et al. 1998). Likewise, *c-fos* expression was inhibited by administration of a specific β2-AR antagonist, demonstrating that β2-ARs in osteoblasts are indeed functionally coupled to intracellular signaling pathways. Furthermore, in vitro studies have confirmed the ability of adrenergic signaling to alter bone remodeling. Administration of norepinephrine stimulated bone resorption in neonatal mouse calvariae in organ culture (Moore et al. 1993), and treatment of the MC3T3-E1 pre-osteoblastic mouse cell line with epinephrine stimulated expression of osteoclast differentiation factor (Takeuchi et al. 2000), suggesting that β-adrenergic stimulation of resorption may occur via an indirect osteoblast-mediated pathway.

13.6 Clinical Relevance of β-Adrenergic Control of Bone

Beta-blocker usage to treat cardiovascular diseases is prevalent, but effects on skeletal mass and strength had not been widely investigated prior to the evidence from mouse studies cited above. However, recent studies have assessed the effects of β-adrenergic antagonists on bone turnover, BMD, and fracture risk in human population-based studies. β-Blocker use was associated with a reduction in fracture risk and increased BMD at the hip and forearm in women over 50 years of age (Pasco et al. 2004), and reduced fracture risk in women and men between 30 and 79 years of age (Schlienger et al. 2004), consistent with the rodent model data. However, a third observational study contradicted these findings, with β-blocker usage associated with a threefold increase in fracture risk and reduced serum osteocalcin, an osteoblastic marker of bone formation, in premenopausal women (Rejnmark et al. 2004). The difference in this latter finding may relate to variations in study design. Thus, there is some epidemiological evidence to support the hypothesis that β-adrenergic signaling is involved in bone metabolism; however, placebo-controlled randomized clinical trials are necessary to more effectively assess potential therapeutic benefit from bone-specific β-adrenergic blockade in hypogonadal and age-related osteoporosis and other conditions that could benefit from stimulation of bone formation.

13.7 Neuropeptide Y and the Y Receptors

13.7.1 NPY System

The role of leptin in the hypothalamus was followed by the identification of a number of central pathways to bone. One neuronal system of particular importance to bone is the neuropeptide Y (NPY) system. The NPY system consists of three ligands: NPY, peptide YY (PYY), and pancreatic polypeptide (PP) mediating its effects via G protein-coupled receptors, of which five have been identified to date, Y1, Y2, Y4, Y5, and y6 (Blomqvist and Herzog 1997; Lin et al. 2005). NPY, a 36-amino acid peptide, is widely expressed in the central and peripheral nervous systems, and is present in both sympathetic and parasympathetic nerve fibers, often co-released with noradrenaline during nerve stimulation. It also circulates in the blood from neuronal and adrenal sources. NPY is a well-characterized vasoconstrictor in these neurons, in addition to enhancing the action of other pressor agents (Pernow et al. 1987; Parker and Herzog 1999; Morris 1994). NPY-ergic neurons are abundant in the brain, with high levels in the hypothalamic arcuate nucleus (ARC) and VMH (Hokfelt et al. 1998). Central NPY action is associated with the regulation of food intake, cardiac and respiratory activity, and the release of pituitary hormones (Hokfelt et al. 1998; Wettstein et al. 1995). PYY is produced primarily from the endocrine L cells of the gastrointestinal tract with some expression in the pancreatic islets, and their function is related to satiety control and gastrointestinal regulation (Lundberg et al. 1984). Moreover, PYY expression has also been reported in brainstem neurons, although the functional significance of this localization is unknown (Ekblad and Sundler 2002). PP is produced by endocrine islet cells of the pancreas and regulates endocrine functions of the pancreas as well as satiety centrally (Ueno et al. 1999).

13.7.2 NPY–Leptin Interaction

In keeping with NPY's role as a powerful modulator of energy homeostasis, both NPY and leptin have a close association within the hypothalamus. Indeed, a significant proportion of NPY-ergic neurons co-express the leptin receptor in the ARC (Mercer et al. 1996), and NPY expression is upregulated following the reduction in leptin due to starvation (Schwartz et al. 1995; Spanswick et al. 1997; Spiegelman and Flier 1996) and in leptin-deficient *ob/ob* mice (Wilding et al. 1993). Thus NPY is a critical downstream mediator of leptin-deficient starvation signaling in the hypothalamus and indices powerful increases in appetite and conservation of energy. Leptin receptors are also present in other nuclei in the hypothalamus, including the ventromedial hypothalamus, from which the leptin-mediated pathway to cancellous bone was shown to originate (Takeda et al. 2002). Thus the degree to which they function separately or can coordinate pathways is not known.

The first NPY model evaluated for skeletal activity was Y2 receptor null, $Y2^{-/-}$ mice, due to the known co-expression of Y2 and leptin receptors on neurons within the arcuate nucleus (Baskin et al. 1999; Broberger et al. 1997). Initial analysis of germline $Y2^{-/-}$ mice was similar to those following conditional deletion of Y2 receptors in the hypothalamus, demonstrating a role for central Y2 receptors in this pathway and the first specific gene deletion in the hypothalamus to alter bone homeostasis. Both models revealed a greater cortical and cancellous bone volume associated with a greater bone formation rate (Baldock et al. 2002, 2006). Bone resorption parameters were unchanged except for a modest increase in osteoclast number. Importantly, the skeletal changes observed in germline $Y2^{-/-}$ mice and hypothalamic $Y2^{-/-}$ mice occurred in the absence of any measurable changes in bone active endocrine factors. Thus these findings suggested that the bone anabolic effects after Y2 receptor deletion are mediated by a neuronal mechanism and not by endocrine effectors of bone turnover. The importance of the ARC to NPY–bone signaling was confirmed by viral-dependent overexpression of NPY restricted to the ARC resulting in marked bone loss, despite extreme weight gain (Baldock et al. 2005), consistent with stimulation of starvation pathways.

Several reports suggested that leptin- and NPY-mediated pathways to bone were similar. This was supported by findings in NPY, Y2 receptor, and leptin double mutant mice ($Y2^{-/-}$;*ob/ob*) in which no additive effect on cancellous bone volume or formation was evident (Baldock et al. 2006). Additionally, male $Y2^{-/-}Y4^{-/-}$ double NPY receptor knockout mice revealed a synergistic increase in cancellous bone volume compared with $Y2^{-/-}$ mice (Sainsbury et al. 2003). This gender-specific effect was coincident with a marked reduction in plasma leptin in male, but not in female *Y2*$^{-/-}$*Y4*$^{-/-}$ mice (Sainsbury et al. 2003), suggesting an additive effect of the Y2 receptor and leptin on bone.

However, when NPY was continuously administered into wild-type mice, mimicking the increase in *ob/ob*, the treatment reduced cancellous bone volume, suggesting that NPY and leptin might use different pathways to control bone mass (Ducy et al. 2000). Consistent with this notion, *NPY*$^{-/-}$ mice display a different bone phenotype to *ob/ob*, while both display anabolic effects in cancellous bone, and no additive effect as outlined above; however, they do differ in cortical bone. Leptin-deficient mice have markedly reduced cortical bone production, with shorter and smaller bones, particularly when corrected for the greater body weight of these very obese mice (Baldock et al. 2006), leading to their reduced whole-body BMC (Steppan et al. 2000). In contrast, *NPY*$^{-/-}$ mice display a generalized bone anabolic effect, with greater whole-body BMC and larger cortical bones, with greater bone formation similar to *Y2*$^{-/-}$ mice (Lee et al. 2011a). The nature of the relationship with leptin was examined in *NPY*$^{-/-}$*;ob/ob* double mutant mice (Wong et al. 2013). NPY levels are markedly elevated in *ob/ob* mice, as a direct effect of reduced activation of the leptin receptor in NPY-ergic neurons (Erickson et al. 1996). Loss of NPY in *ob/ob* mice had very specific effects upon bone metabolism, with a correction of the cortical deficit through a correction of bone formation; *NPY*$^{-/-}$*;ob/ob* cortical bone was returned to wild-type levels, indicating that the reduction in bone mass in *ob/ob* is driven by the increase in ARC NPY signaling following leptin withdrawal. However, there was no difference in cancellous bone between *ob/ob and NPY*$^{-/-}$*;ob/ob*, indicating that the VMH/SNS/β2 pathway is not altered by NPY. Thus this model identified two anatomically distinct pathways from two hypothalamic nuclei to bone, which respond to deficiency in the critical marker of energy homeostasis, leptin. In response to leptin deficiency, a condition associated with reduced energy storage (as fat), and thus starvation, NPY from the arcuate nucleus inhibits bone formation and bone accrual, thereby conserving energy. Simultaneously, the VMH alters SNS activity to maintain cancellous bone volume, perhaps to maintain some of the calcium and trace elements stored in bone tissue.

13.7.3 NPY in Bone Tissue

It is interesting to note that with leptin, the receptors and ligands present in the hypothalamus were also present in the cells of bone. There is also evidence that NPY-positive autonomic nerves are present in healthy bone tissue, particularly in association with blood vessel walls, suggestive of a role in modulation of vascular tone rather than direct effects on bone cells (Bjurholm et al. 1988; Ahmed et al. 1993; Lindblad et al. 1994; Sisask et al. 1996). In addition, sympathetic denervation has been observed to reduce NPY immunoreactive nerves in the periosteum (Hill and Elde 1991). Very early studies demonstrated that NPY treatment in osteoblastic cell lines inhibited the cAMP response to parathyroid hormone and norepinephrine (Bjurholm 1991; Bjurholm et al. 1992), suggesting the presence of functional Y receptors on bone cells and a possible regulatory role for NPY in bone.

Similar to leptin, direct NPY-mediated effects in bone were confirmed in mutant mouse models. Two NPY receptors, Y1 and Y2, have been connected with bone homeostasis. These receptors are abundant in the hypothalamus as well as in peripheral nerves (Kishi and Elmquist 2005; Kopp et al. 2002; Naveilhan et al. 1998). Similar to the phenotype resulting from Y2 receptor deletion, germline Y1 receptor deficiency resulted in a generalized anabolic phenotype with greater cortical bone and cancellous accrual, although with an additional increase in bone resorption (Baldock et al. 2007). However, unlike the anabolic effects in hypothalamus-specific Y2 deletion (Parker and Herzog 1999), loss of hypothalamic Y1 receptors had no effect on bone homeostasis, indicating a noncentral mechanism for Y1 action in bone. The existence of a direct Y1-mediated effect on bone anabolism was further suggested following identification of Y1 expression in osteoblastic cells in vivo (Baldock et al. 2007). NPY treatment of calvarial osteoblast cultures markedly decreased cell numbers, an effect absent in cultures from Y1 receptor knockout mice, indicating functional osteoblastic Y1 receptors. The direct regulation of osteoblasts by NPY was confirmed in osteoblast-specific Y1 receptor knockout mice, which displayed an increase in bone formation similar to germline Y1-null mice (Lee et al. 2011b). Loss of Y1 receptor has also been demonstrated to regulate mesenchymal stem cell activity and mineralization of osteoblastic cultures in vitro (Lee et al. 2010). In addition, NPY is produced by osteoblasts and elevation of NPY production in osteoblast-specific NPY transgenic mice shown an opposing phenotype to Y1 receptor deletion, with reduced bone volume and bone formation (Matic et al. 2012). The integration of NPY within the osteoblast lineage was confirmed by a study demonstrating the NPY and Y1 receptor expression is controlled in a differentiation-dependent manner and responds to loading, a fundamental aspect of bone physiology (Igwe et al. 2009). Adding a new dimension to neuronal/bone interactions, a recent study, which produced an osteoblast-specific deletion of p38a-Mapk14 demonstrated alterations in bone homeostasis, but also increased energy expenditure and reduced adiposity (Rodriguez-Carballo et al. 2015). Interestingly, these metabolic changes were associated with reduced NPY production by osteoblasts and blocked by i.p. NPY administration. Thus, taken together these models confirmed an NPY-mediated pathway from the arcuate nucleus of the hypothalamus to the osteoblast, as well as an active NPY loop within the osteoblast lineage. Moreover, they indicate that osteoblastic NPY can modulate whole-body energy homeostasis, as does central NPY in the hypothalamus, illustrating extremely novel interorgan communication from bone cells.

13.7.4 Cannabinoid Receptors

Endocannabinoid signaling has been identified as one of the central pathways to bone which mediates its actions via two cannabinoid receptors, CB1 and CB2 (Howlett et al. 2002). CB1 is primarily found within the CNS and accounts for most of the CNS actions of cannabinoid drugs and endocannabinoids (Mackie 2008), while CB2 is predominantly expressed in peripheral tissues (Tam et al. 2008). Endocannabinoids are generated as needed, whereas other neurotransmitters are released from vesicles. The main endogenous ligands of CB1 and CB2 receptors are 2-arachidonoylglycerol (2-AG) and *N*-arachidonoylethanolamine (AEA or anandamide) (Devane et al. 1992; Mechoulam et al. 1995).

It has been recently reported that endocannabinoids regulate bone homeostasis by modulating adrenergic signaling. The activation of CB1 receptor signaling on presynaptic nerve terminals in bone inhibits norepinephrine release by sympathetic neurons to balance the tonic sympathetic restrain of bone formation (Ishac et al. 1996; Niederhoffer et al. 2003). However, CB1 receptor inactivation not only increased bone mineral but also provided protection against ovariectomy-induced bone loss (Idris et al. 2005). Moreover, CB1 receptor inactivation inhibited osteoclastogenesis and bone resorption, while CB1 receptor-deficient mice were resistant to these effects suggesting cannabinoid signaling acts via the CB1 receptor to regulate osteoclasts (Idris et al. 2005).

CB2 receptor is abundantly expressed in osteoblasts, osteocytes, and osteoclasts. CB2 knockout mice have accelerated age-related cancellous bone loss and cortical expansion due to increased bone turnover (Ofek et al. 2006). These results corroborate with human genetic association studies linking CNR2 gene (encoding CB2) and reduced bone mass in women (Karsak et al. 2005; Yamada et al. 2007) and in vitro pharmacological studies demonstrating a direct activation of CB2 in osteoclasts. These in vitro studies indicate that CB2 signaling contributes to the maintenance of bone mass by stimulating stromal cells/osteoblasts directly and by inhibiting monocytes/osteoclasts. Also CB2 agonists, which are not psychoactive, attenuated ovariectomy-induced bone loss in mice suggesting that these compounds might be used for the treatment of low bone mass diseases (Ofek et al. 2006). Taken together, these data suggest that the cannabinoid system plays an important role in the regulation and maintenance of bone mass through the signaling of both the central and skeletal cannabinoid receptors. This system, along with the NPY system, highlights the importance of centrally expressed receptors but also the locally expressed receptors in bone.

13.7.5 Semaphorins

Semaphorins are a family of both secreted and membrane-associated proteins that can regulate cell–cell interactions as well as cell differentiation, morphology, and function. The best-characterized biological role of semaphorins is their ability to provide attractant or repellent cues for migrating cells and growing neurites, i.e., axons and dendrites (Derijck et al. 2010; Tran et al. 2007). Originally characterized as axon guidance molecules, recent studies have demonstrated that the semaphorin–plexin system has an important role in the crosstalk between osteoblasts and osteoclasts (Negishi-Koga and Takayanagi 2012; Kang and Kumanogoh 2013). Semaphorins are divided into eight subclasses, of which classes III–VII are vertebral semaphorins. Most of the effects of semaphorins are mediated by plexin and neuropilin receptors. Recent studies have shown that semaphorin 3A (SEMA3A) and semaphorin 4D (SEMA4D) are involved in bone homeostasis.

In mice, SEMA3A is produced by osteoblasts and inhibits osteoclast formation from precursor cells (Hayashi et al. 2012). SEMA3A-null mice have reduced bone density, and systemic administration of SEMA3A prevents bone loss in a mouse model of menopause, indicating that SEMA3A could be a potential therapeutic agent to reduce bone loss (Hayashi et al. 2012). SEMA3A not only has an inhibitory effect on osteoclast differentiation but it also repels osteoclast precursors. SEMA3A from autocrine or paracrine sources also promotes osteoblast formation and differentiation. Recent studies have shown that SEMA3A was shown to have a crucial role in the development of proper sensory innervation of bone tissue, and neuronal but not osteoblastic SEMA3A was found to be responsible for bone loss in SEMA3A-null mice (Fukuda et al. 2013). These data indicate that SEMA3A, through its modulation of sensory innervation during development, regulates bone metabolism.

SEMA4D is highly and selectively expressed by osteoclasts and it inhibits osteoblast formation and differentiation. Consistent with its role in inhibition of osteoblast formation, mice lacking SEMA4D or its receptor plexin-B1 have elevated bone mass (Negishi-Koga et al. 2011; Dacquin et al. 2011). Moreover, it has been reported that SEMA4D-specific antibody was able to reduce bone loss in ovariectomized mice by increasing bone formation (Negishi-Koga et al. 2011). In vitro studies have shown that plexin-B1-mediated, ERBB2-dependent RhoA activation is responsible for the inhibition of osteoblast differentiation. These data are supported by in vivo studies showing that the osteoblast-specific expression of RhoA mimics global SEMA4D and plexin-B1 knockout mouse phenotypes (Negishi-Koga et al. 2011).

13.7.6 POMC and Melanocortin System

The melanocortin system is involved in diverse physiological functions from coat color to body weight homoeostasis. Pro-opiomelanocortin (POMC)-derived peptides such as melanocortins, adrenocorticotropic hormone (ACTH), and β-endorphin exert their pleiotropic effects via binding to melanocortin receptors (MCRs) and opioid receptors. Of the five MCRs identified as G-protein-coupled receptors, MC1–MC5 (Beltramo et al. 2003; Nijenhuis et al. 2001), melanocortin 4 receptor (MC4R) is abundantly expressed in hypothalamic neurons. MC4R is a major regulator of bone homeostasis with patients lacking in MC4R exhibiting a high bone mineral density resulting from a decrease in bone resorption (Farooqi et al. 2000). Importantly, the greater BMD is still evident after correction of the obesity that is characteristic of MC4R deficiency (Farooqi et al. 2000). Most studies on POMC peptides and their receptors in bone have concentrated on effects of melanocortin peptides rather than on expression of POMC, POMC-derived peptides, and effects of endogenous opioids. In a mouse model, daily subcutaneous administration of 0.2 μg/kg ACTH (amino acids 1–24) protected against glucocorticoid-induced osteonecrosis of the femoral head (Isales et al. 2010). Moreover, ACTH stimulated vascular endothelial growth factor (VEGF) production, which supported the maturation and survival of osteoblasts. It has been observed that induction of VEGF expression and secretion from osteoblasts was mediated by MC2R. ACTH appeared to modulate osteogenic differentiation as well. In the human osteoblast-like Saos2 cells of an osteosarcoma origin, ACTH had a biphasic effect on transcripts of collagen type I, the major collagen in bone, for which expression is strongly elevated during osteoblast differentiation. ACTH at 10 nM increased collagen I mRNA and thus differentiation, whereas at lower concentrations, it opposed osteoblast differentiation (Isales et al. 2010). Interestingly, ACTH immunoreactivity and ACTH secretion was described in rat osteoclastic cells (Sun et al. 2006), but the significance of this finding remains to be shown. Contrary to ACTH, in vitro administration of 10^{-8} M α-MSH increased proliferation of fetal rat osteoblasts without affecting their differentiation. In cultures of mouse bone marrow, α-MSH also stimulated osteoclast formation from their precursors but has no effect on mature osteoclasts.

13.7.7 CART in Bone Homeostasis

Studies in mice have implicated the involvement of hypothalamic neuropeptide, CART in bone homeostasis. CART, a neuropeptide precursor protein involved in the regulation of food intake and energy expenditure, is broadly expressed in the hypothalamus and the peripheral organs such as the pancreas and adrenal glands (Elefteriou et al. 2005). Interestingly, the phenotype of leptin-deficient (*ob/ob*) mice suggested that leptin could be affecting bone resorption via central effects on CART. *ob/ob* mice, unlike the ß2-adrenergic receptor null mice, showed decrease in hypothalamic CART expression and increased bone resorption, thereby implicating CART as a potential regulator of bone resorption. Moreover, intraperitoneal treatment with leptin in *ob/ob* mice was able to restore the decreased CART expression (Kristensen et al. 1998). Consistent with this hypothesis, CART knockout mice are osteopenic due to an increase in bone resorption (Elefteriou et al. 2005). Interestingly, CART-deficient mice express higher levels of RANKL in bone than wild-type mice, with in vitro osteoclast differentiation experiments indicating that the effect of CART bone is not cell autonomous, suggesting a local mechanism for the central CART changes (Elefteriou et al. 2005). Moreover, hypothalamic CART expression is elevated in *MC4R*$^{-/-}$ mice, which display a high bone mass phenotype due to decreased osteoclast formation and activity(Elefteriou et al. 2005; Ahn et al. 2006), as evident in human studies. Further, MC4R mutant mice lacking one or two copies of CART exhibited a significantly lower bone mass (Elefteriou et al. 2005; Ahn et al. 2006), demonstrating increased CART signaling plays an important role in the low bone resorption/high bone mass phenotype observed in MC4R-null mice.

13.8 Conclusion

It has become clear in the last decade or so that direct neural connections between the brain and the cells of bone represent a major regulatory axis controlling bone homeostasis. However, signals from the major neuronal populations arising from hypothalamus are critical for sensing the current status of energy stores on a body-wide scale, and it is these nerves which appear to play a pivotal role in regulating bone mass. In this manner, integration of energy balance and nutritional state can be achieved not only through modulation of energy intake but also expenditure, through control of activity at a tissue level, in this case bone metabolism. Bone is a large and protein/ nutrient dense tissue and as such represents a significant energy store. In addition, osteoblasts are very active protein synthetic cells, requiring a substantial amount of energy to renew lost bone. Thus bone, being an energy store and requiring energy to function, is a vital element of the regulatory processes coordinated with the brain. This connection between energy and bone homeostasis renders many of these interactions context-dependent, with responses particularly evident in starvation-type context, such as low leptin, CART, or POMC or greater NPY. While this relationship may help shed light upon skeletal responses to altered nutrient balance, such as obesity or anorexia, it also means that care must be taken when comparing mouse results to population-based studies in humans. The parallels between bone regulatory pathways in the hypothalamus and within the cells of bone themselves present potential therapeutic opportunities to effect bone directly, without the complexity of central hypothalamic responses. Given the power of some of these pathways to control bone cell activity, such agents may offer exciting avenues for therapeutic development in the years to come.

References

Ahmed M, et al. Neuropeptide Y, tyrosine hydroxylase and vasoactive intestinal polypeptide-immunoreactive nerve fibers in the vertebral bodies, discs, dura mater, and spinal ligaments of the rat lumbar spine. Spine. 1993;18(2):268–73.

Ahn JD, et al. Cart overexpression is the only identifiable cause of high bone mass in melanocortin 4 receptor deficiency. Endocrinology. 2006;147(7):3196–202.

Baldock PA, et al. Hypothalamic Y2 receptors regulate bone formation. J Clin Invest. 2002;109(7):915–21.

Baldock PA, et al. Hypothalamic control of bone formation: distinct actions of leptin and y2 receptor pathways. J Bone Miner Res. 2005;20(10):1851–7.

Baldock PA, et al. Hypothalamic regulation of cortical bone mass: opposing activity of Y2 receptor and leptin pathways. J Bone Miner Res. 2006;21(10):1600–7.

Baldock PA, et al. Novel role of Y1 receptors in the coordinated regulation of bone and energy homeostasis. J Biol Chem. 2007;282(26):19092–102.

Baskin DG, Breininger JF, Schwartz MW. Leptin receptor mRNA identifies a subpopulation of neuropeptide Y neurons activated by fasting in rat hypothalamus. Diabetes. 1999;48(4):828–33.

Beltramo M, et al. Gene expression profiling of melanocortin system in neuropathic rats supports a role in nociception. Brain Res Mol Brain Res. 2003;118(1–2):111–8.

Bjurholm A. Neuroendocrine peptides in bone. Int Orthop. 1991;15(4):325–9.
Bjurholm A, et al. Neuropeptide Y-, tyrosine hydroxylase- and vasoactive intestinal polypeptide-immunoreactive nerves in bone and surrounding tissues. J Auton Nerv Syst. 1988;25(2–3):119–25.
Bjurholm A, et al. Neuroendocrine regulation of cyclic AMP formation in osteoblastic cell lines (UMR-106-01, ROS 17/2.8, MC3T3-E1, and Saos-2) and primary bone cells. J Bone Miner Res. 1992;7(9):1011–9.
Blomqvist AG, Herzog H. Y-receptor subtypes – how many more? Trends Neurosci. 1997;20(7):294–8.
Broberger C, et al. Subtypes Y1 and Y2 of the neuropeptide Y receptor are respectively expressed in pro-opiomelanocortin- and neuropeptide-Y-containing neurons of the rat hypothalamic arcuate nucleus. Neuroendocrinology. 1997;66(6):393–408.
Burger EH, Klein-Nulend J. Mechanotransduction in bone – role of the lacuno-canalicular network. FASEB J. 1999;13(12):S101–12.
Burger EH, et al. Function of osteocytes in bone – their role in mechanotransduction. J Nutr. 1995;125(7 Suppl):2020S–3S.
Burguera B, et al. Leptin reduces ovariectomy-induced bone loss in rats. Endocrinology. 2001;142(8):3546–53.
Cornish J, et al. Leptin directly regulates bone cell function in vitro and reduces bone fragility in vivo. J Endocrinol. 2002;175(2):405–15.
Dacquin R, et al. Control of bone resorption by semaphorin 4D is dependent on ovarian function. PLoS One. 2011;6(10):26.
Denes A, et al. Central autonomic control of the bone marrow: multisynaptic tract tracing by recombinant pseudorabies virus. Neuroscience. 2005;134(3):947–63.
Derijck AA, Van Erp S, Pasterkamp RJ. Semaphorin signaling: molecular switches at the midline. Trends Cell Biol. 2010;20(9):568–76.
Devane WA, et al. Isolation and structure of a brain constituent that binds to the cannabinoid receptor. Science. 1992;258(5090):1946–9.
Ducy P, et al. Leptin inhibits bone formation through a hypothalamic relay: a central control of bone mass. Cell. 2000;100(2):197–207.
Ekblad E, Sundler F. Distribution of pancreatic polypeptide and peptide YY. Peptides. 2002;23(2):251–61.
Elefteriou F. Neuronal signaling and the regulation of bone remodeling. Cell Mol Life Sci. 2005;62(19–20):2339–49.
Elefteriou F, et al. Leptin regulation of bone resorption by the sympathetic nervous system and CART. Nature. 2005;434(7032):514–20.
Enjuanes A, et al. Leptin receptor (OB-R) gene expression in human primary osteoblasts: confirmation. J Bone Miner Res. 2002;17(6):1135.
Erickson JC, Hollopeter G, Palmiter RD. Attenuation of the obesity syndrome of ob/ob mice by the loss of neuropeptide Y. Science. 1996;274(5293):1704–7.
Farooqi IS, et al. Effects of recombinant leptin therapy in a child with congenital leptin deficiency. New Engl J Med. 1999;341(12):879–84.
Farooqi IS, et al. Dominant and recessive inheritance of morbid obesity associated with melanocortin 4 receptor deficiency. J Clin Invest. 2000;106(2):271–9.
Friedman JM, Halaas JL. Leptin and the regulation of body weight in mammals. Nature. 1998;395(6704):763–70.
Fukuda T, et al. Sema3A regulates bone-mass accrual through sensory innervations. Nature. 2013;497(7450):490–3.
Gordeladze JO, et al. Leptin stimulates human osteoblastic cell proliferation, de novo collagen synthesis, and mineralization: impact on differentiation markers, apoptosis, and osteoclastic signaling. J Cell Biochem. 2002;85(4):825–36.
Goulding A, Taylor RW. Plasma leptin values in relation to bone mass and density and to dynamic biochemical markers of bone resorption and formation in postmenopausal women. Calcif Tissue Int. 1998;63(6):456–8.

Halaas JL, et al. Weight-reducing effects of the plasma protein encoded by the obese gene. Science. 1995;269(5223):543–6.
Hamrick MW, et al. Leptin deficiency produces contrasting phenotypes in bones of the limb and spine. Bone. 2004;34(3):376–83.
Hayashi M, et al. Osteoprotection by semaphorin 3A. Nature. 2012;485(7396):69–74.
Hill EL, Elde R. Distribution of CGRP-, VIP-, D beta H-, SP-, and NPY-immunoreactive nerves in the periosteum of the rat. Cell Tissue Res. 1991;264(3):469–80.
Hoggard N, et al. Leptin and leptin receptor mRNA and protein expression in the murine fetus and placenta. Proc Natl Acad Sci U S A. 1997;94(20):11073–8.
Hokfelt T, et al. Neuropeptide Y: some viewpoints on a multifaceted peptide in the normal and diseased nervous system. Brain Res Brain Res Rev. 1998;26(2–3):154–66.
Holloway WR, et al. Leptin inhibits osteoclast generation. J Bone Miner Res. 2002;17(2):200–9.
Howlett AC, et al. International Union of Pharmacology. XXVII. Classification of cannabinoid receptors. Pharmacol Rev. 2002;54(2):161–202.
Idris AI, et al. Regulation of bone mass, bone loss and osteoclast activity by cannabinoid receptors. Nat Med. 2005;11(7):774–9.
Igwe JC, et al. Neuropeptide Y is expressed by osteocytes and can inhibit osteoblastic activity. J Cell Biochem. 2009;108(3):621–30.
Isales CM, Zaidi M, Blair HC. ACTH is a novel regulator of bone mass. Ann N Y Acad Sci. 2010;1192:110–6.
Ishac EJ, et al. Inhibition of exocytotic noradrenaline release by presynaptic cannabinoid CB1 receptors on peripheral sympathetic nerves. Br J Pharmacol. 1996;118(8):2023–8.
Kang S, Kumanogoh A. Semaphorins in bone development, homeostasis, and disease. Semin Cell Dev Biol. 2013;24(3):163–71.
Karsak M, et al. Cannabinoid receptor type 2 gene is associated with human osteoporosis. Hum Mol Genet. 2005;14(22):3389–96.
Kellenberger S, et al. Formoterol and isoproterenol induce c-fos gene expression in osteoblast-like cells by activating beta2-adrenergic receptors. Bone. 1998;22(5):471–8.
Kishi T, Elmquist JK. Body weight is regulated by the brain: a link between feeding and emotion. Mol Psychiatry. 2005;10(2):132–46.
Kopp J, et al. Expression of the neuropeptide Y Y1 receptor in the CNS of rat and of wild-type and Y1 receptor knock-out mice. Focus on immunohistochemical localization. Neuroscience. 2002;111(3):443–532.
Kristensen P, et al. Hypothalamic CART is a new anorectic peptide regulated by leptin. Nature. 1998;393(6680):72–6.
Lee YJ, et al. Leptin receptor isoform expression in rat osteoblasts and their functional analysis. FEBS Lett. 2002;528(1–3):43–7.
Lee NJ, et al. Critical role for Y1 receptors in mesenchymal progenitor cell differentiation and osteoblast activity. J Bone Miner Res. 2010;25(8):1736–47.
Lee NJ, et al. Y2 and Y4 receptor signalling attenuates the skeletal response of central NPY. J Mol Neurosci. 2011a;43(2):123–31.
Lee NJ, et al. Osteoblast specific Y1 receptor deletion enhances bone mass. Bone. 2011b;48(3):461–7.
Lin S, et al. Compensatory changes in [125I]-PYY binding in Y receptor knockout mice suggest the potential existence of further Y receptor(s). Neuropeptides. 2005;39(1):21–8.
Lindblad BE, et al. Vasoconstrictive action of neuropeptide Y in bone. The porcine tibia perfused in vivo. Acta Orthop Scand. 1994;65(6):629–34.
Lundberg JM, et al. Comparative immunohistochemical and biochemical analysis of pancreatic polypeptide-like peptides with special reference to presence of neuropeptide Y in central and peripheral neurons. J Neurosci. 1984;4(9):2376–86.
Mackie K. Signaling via CNS cannabinoid receptors. Mol Cell Endocrinol. 2008;286:S60–5.
Maor G, et al. Leptin acts as a growth factor on the chondrocytes of skeletal growth centers. J Bone Miner Res. 2002;17(6):1034–43.
Mathey J, et al. Bone mass in obese diabetic Zucker rats: influence of treadmill running. Calcif Tissue Int. 2002;70(4):305–11.

Matic I, et al. Bone-specific overexpression of NPY modulates osteogenesis. J Musculoskelet Neuronal Interact. 2012;12(4):209–18.
Mechoulam R, et al. Identification of an endogenous 2-monoglyceride, present in canine gut, that binds to cannabinoid receptors. Biochem Pharmacol. 1995;50(1):83–90.
Mercer JG, et al. Coexpression of leptin receptor and preproneuropeptide Y mRNA in arcuate nucleus of mouse hypothalamus. J Neuroendocrinol. 1996;8(10):733–5.
Miller KK, et al. Preservation of neuroendocrine control of reproductive function despite severe undernutrition. J Clin Endocrinol Metab. 2004;89(9):4434–8.
Moore RE, et al. Characterization of beta-adrenergic receptors on rat and human osteoblast-like cells and demonstration that beta-receptor agonists can stimulate bone resorption in organ culture. Bone Miner. 1993;23(3):301–15.
Morris JL. Selective constriction of small cutaneous arteries by NPY matches distribution of NPY in sympathetic axons. Regul Pept. 1994;49(3):225–36.
Morroni M, et al. In vivo leptin expression in cartilage and bone cells of growing rats and adult humans. J Anat. 2004;205(4):291–6.
Nakajima R, et al. Effects of leptin to cultured growth plate chondrocytes. Horm Res. 2003;60(2):91–8.
Naveilhan P, et al. Complementary and overlapping expression of Y1, Y2 and Y5 receptors in the developing and adult mouse nervous system. Neuroscience. 1998;87(1):289–302.
Negishi-Koga T, Takayanagi H. Bone cell communication factors and Semaphorins. Bonekey Rep. 2012;1(183):183.
Negishi-Koga T, et al. Suppression of bone formation by osteoclastic expression of semaphorin 4D. Nat Med. 2011;17(11):1473–80.
Niederhoffer N, Schmid K, Szabo B. The peripheral sympathetic nervous system is the major target of cannabinoids in eliciting cardiovascular depression. Naunyn Schmiedeberg's Arch Pharmacol. 2003;367(5):434–43.
Nijenhuis WA, Oosterom J, Adan RA. AgRP(83-132) acts as an inverse agonist on the human-melanocortin-4 receptor. Mol Endocrinol. 2001;15(1):164–71.
Odabasi E, et al. Plasma leptin concentrations in postmenopausal women with osteoporosis. Eur J Endocrinol. 2000;142(2):170–3.
Ofek O, et al. Peripheral cannabinoid receptor, CB2, regulates bone mass. Proc Natl Acad Sci U S A. 2006;103(3):696–701.
Ozata M, Ozdemir IC, Licinio J. Human leptin deficiency caused by a missense mutation: multiple endocrine defects, decreased sympathetic tone, and immune system dysfunction indicate new targets for leptin action, greater central than peripheral resistance to the effects of leptin, and spontaneous correction of leptin-mediated defects. J Clin Endocrinol Metab. 1999;84(10):3686–95.
Parker RM, Herzog H. Regional distribution of Y-receptor subtype mRNAs in rat brain. Eur J Neurosci. 1999;11(4):1431–48.
Pasco JA, et al. Serum leptin levels are associated with bone mass in nonobese women. J Clin Endocrinol Metab. 2001;86(5):1884–7.
Pasco JA, et al. Beta-adrenergic blockers reduce the risk of fracture partly by increasing bone mineral density: Geelong Osteoporosis Study. J Bone Miner Res. 2004;19(1):19–24.
Pernow J, et al. Neuropeptide Y: presence in perivascular noradrenergic neurons and vasoconstrictor effects on skeletal muscle blood vessels in experimental animals and man. Regul Pept. 1987;19(5–6):313–24.
Rauch F, et al. Does leptin have an effect on bone in adult women? Calcif Tissue Int. 1998;63(6):453–5.
Reid IR. Relationships among body mass, its components, and bone. Bone. 2002;31(5):547–55.
Rejnmark L, et al. Fracture risk in perimenopausal women treated with beta-blockers. Calcif Tissue Int. 2004;75(5):365–72.
Reseland JE, et al. Leptin is expressed in and secreted from primary cultures of human osteoblasts and promotes bone mineralization. J Bone Miner Res. 2001;16(8):1426–33.

Rodriguez-Carballo E, et al. p38alpha function in osteoblasts influences adipose tissue homeostasis. In: FASEB J Faseb. United States; 2015. p. 1414–25.
Sainsbury A, et al. Synergistic effects of Y2 and Y4 receptors on adiposity and bone mass revealed in double knockout mice. Mol Cell Biol. 2003;23(15):5225–33.
Sato M, et al. Association between serum leptin concentrations and bone mineral density, and biochemical markers of bone turnover in adult men. J Clin Endocrinol Metab. 2001;86(11):5273–6.
Schlienger RG, et al. Use of beta-blockers and risk of fractures. JAMA. 2004;292(11):1326–32.
Schwartz MW, Dallman MF, Woods SC. Hypothalamic response to starvation: implications for the study of wasting disorders. Am J Phys. 1995;269(5 Pt 2):R949–57.
Sisask G, et al. The development of autonomic innervation in bone and joints of the rat. J Auton Nerv Syst. 1996;59(1–2):27–33.
Soyka LA, et al. The effects of anorexia nervosa on bone metabolism in female adolescents. J Clin Endocrinol Metab. 1999;84(12):4489–96.
Spanswick D, et al. Leptin inhibits hypothalamic neurons by activation of ATP-sensitive potassium channels. Nature. 1997;390(6659):521–5.
Spiegelman BM, Flier JS. Adipogenesis and obesity: rounding out the big picture. Cell. 1996;87(3):377–89.
Steppan CM, et al. Leptin is a potent stimulator of bone growth in ob/ob mice. Regul Pept. 2000;92(1–3):73–8.
Sun L, et al. FSH directly regulates bone mass. Cell. 2006;125(2):247–60.
Takaya K, et al. Molecular cloning of rat leptin receptor isoform complementary DNAs – identification of a missense mutation in Zucker fatty (fa/fa) rats. Biochem Biophys Res Commun. 1996;225(1):75–83.
Takeda S, et al. Leptin regulates bone formation via the sympathetic nervous system. Cell. 2002;111(3):305–17.
Takeuchi T, et al. Adrenergic stimulation of osteoclastogenesis mediated by expression of osteoclast differentiation factor in MC3T3-E1 osteoblast-like cells. Biochem Pharmacol. 2000;61:5.
Tam J, et al. The cannabinoid CB1 receptor regulates bone formation by modulating adrenergic signaling. FASEB J. 2008;22(1):285–94.
Tartaglia LA, et al. Identification and expression cloning of a leptin receptor, OB-R. Cell. 1995;83(7):1263–71.
Thomas T, et al. Leptin acts on human marrow stromal cells to enhance differentiation to osteoblasts and to inhibit differentiation to adipocytes. Endocrinology. 1999;140(4):1360–8.
Thomas T, et al. Role of serum leptin, insulin, and estrogen levels as potential mediators of the relationship between fat mass and bone mineral density in men versus women. Bone. 2001;29(2):114–20.
Tran TS, Kolodkin AL, Bharadwaj R. Semaphorin regulation of cellular morphology. Annu Rev Cell Dev Biol. 2007;23:263–92.
Ueno N, et al. Decreased food intake and body weight in pancreatic polypeptide-overexpressing mice. Gastroenterology. 1999;117(6):1427–32.
Welt CK, et al. Recombinant human leptin in women with hypothalamic amenorrhea. New Engl J Med. 2004;351(10):987–97.
Wettstein JG, Earley B, Junien JL. Central nervous system pharmacology of neuropeptide Y. Pharmacol Ther. 1995;65(3):397–414.
Wilding JP, et al. Increased neuropeptide-Y messenger ribonucleic acid (mRNA) and decreased neurotensin mRNA in the hypothalamus of the obese (ob/ob) mouse. Endocrinology. 1993;132(5):1939–44.
Wong IP, et al. Neuropeptide Y is a critical modulator of leptin's regulation of cortical bone. J Bone Miner Res. 2013;28(4):886–98.
Yamada Y, Ando F, Shimokata H. Association of candidate gene polymorphisms with bone mineral density in community-dwelling Japanese women and men. Int J Mol Med. 2007;19(5):791–801.
Yan L, et al. Type 5 adenylyl cyclase disruption increases longevity and protects against stress. Cell. 2007;130(2):247–58.

Chapter 14
Intestinal Microbiota and Bone Health: The Role of Prebiotics, Probiotics, and Diet

Fraser L. Collins, Soon Mi Kim, Laura R. McCabe, and Connie M. Weaver

Abstract In recent years, the intestinal microbiota has emerged as a crucial regulator of health with dysbiosis linked to a number of pathological states such as inflammatory bowel disease. In addition to a local intestinal effect, emerging evidence has demonstrated the potential for the microbiota to modulate systemic bone health via a gut-bone axis. In the present chapter, we discuss how diet can affect the composition of the intestinal microbiota, through the intake of prebiotics, and how these are utilized by the bacteria to influence the immune system and bone. In addition, we detail the recent murine studies that investigate how probiotic supplementation can increase bone mineral density in "healthy" individuals and protect against the pathological bone loss associated with menopausal estrogen deficiency. Finally, we highlight the advances made in unearthing the mechanisms that potentially lead to these observed beneficial effects.

Keywords Microbiota • Probiotic • Prebiotic • Bone • Osteoporosis

---

F.L. Collins
Department of Physiology, Michigan State University, East Lansing, MI 48824, USA

S.M. Kim
Department of Food and Nutrition, Gachon University, 461-701 Seongnam, South Korea

L.R. McCabe (✉)
Department of Physiology, Michigan State University, East Lansing, MI 48824, USA

Department of Radiology, Michigan State University, East Lansing, MI 48824, USA

Biomedical Imaging Research Center, Michigan State University, East Lansing, MI 48824, USA
e-mail: mccabel@msu.edu

C.M. Weaver (✉)
Department of Food and Nutrition, Purdue University, West Layfayette, IN 47907, USA
e-mail: weavercm@purdue.edu

S.Y. Smith et al. (eds.), *Bone Toxicology*, Molecular and Integrative Toxicology,
DOI 10.1007/978-3-319-56192-9_14

14.1 Gut Microbiota, Diet, and Health

14.1.1 Gut Microbiota and Diet

In the normal fetus, the gastrointestinal tract is sterile until the time of birth at which point the number of gut bacteria, the microbiota, increases rapidly through the normal delivery process and by oral and environmental factors such as suckling, feeding, and touching. In fact, the microbiota reaches 10^8–10^{10} colony-forming units (cfu)/g feces in just a few days after birth and continues to expand to reach 10^{13}–10^{14} microorganisms in the adult human intestine (Savage 1977; Oozeer et al. 2010). The gut microbiome, which is the genome of the gut microbiota, is at least a hundred times the size of the human genome (Gill et al. 2006). Gut microbiota, dietary proteins, and pathogens are all recognized as nonself by the human immune system and, therefore, can elicit an immune response. Consequently, the gut has to protect the host from continuously invading pathogens while maintaining immune tolerance with beneficial dietary protein and harmless commensal microbes (Tenorio et al. 2010; Sjögren et al. 2012; Kelsall and Leon 2005; Alpan 2001). For this to occur, the local immune system of the host gut is well developed and is known as gut-associated lymphoid tissue (GALT). GALT is the largest lymphoid organ in the body and contains Peyer's patches, lamina propria lymphocytes, and intraepithelial lymphocytes (IELs), which are specialized for the gut. Peyer's patches, which are mainly found in the terminal ileum contain B cells, T cells, dendritic cells, and specialized epithelial cells known as M cells. Dysbiosis of the gut microbiota can lead to pathological results (Chandran et al. 2003) such as obesity, diabetes, metabolic syndrome, and inflammatory bowel disease (Robles Alonso and Guarner 2013).

Bacteria secrete a variety of enzymes that are not secreted by the human alimentary systems. These enzymes are used by the bacteria to ferment nondigestible polycarbohydrates (mainly dietary fiber). This fermentation results in the production of energy, used by the bacteria, and short chain fatty acids (SCFA) used by the intestinal epithelium for proliferation (Abreu 2010; Nam et al. 2011; Sjögren et al. 2012). Endocrine cells in the intestinal epithelium can also be affected by the gut microbiota. Hormones or cytokines secreted from the endocrine cells, in response to bacterial stimulation, can communicate with immune cells or the autonomic nervous system to elicit local or systemic effects (Sjögren et al. 2012; Bonaz and Bernstein 2013).

The composition of the gut microbiome can be affected by three major factors (1) diet, (2) age, and (3) genetics. Of these factors, manipulation of diet has been demonstrated to be the easiest way to affect change with numerous studies reporting that composition, richness, and diversity of gut microbiota were all influenced by diet and dietary patterns in ways that influenced human health (Wu et al. 2011; Cotillard et al. 2013; Le Chatelier et al. 2013; Nam et al. 2011; Claesson et al. 2012; Gill et al. 2006). A suggested model for the impact of diet on the microbiome and health is shown in Fig. 14.1. Wu et al. (2011) conducted two studies to investigate the relationship between gut microbiota and dietary variables. First, microbiota

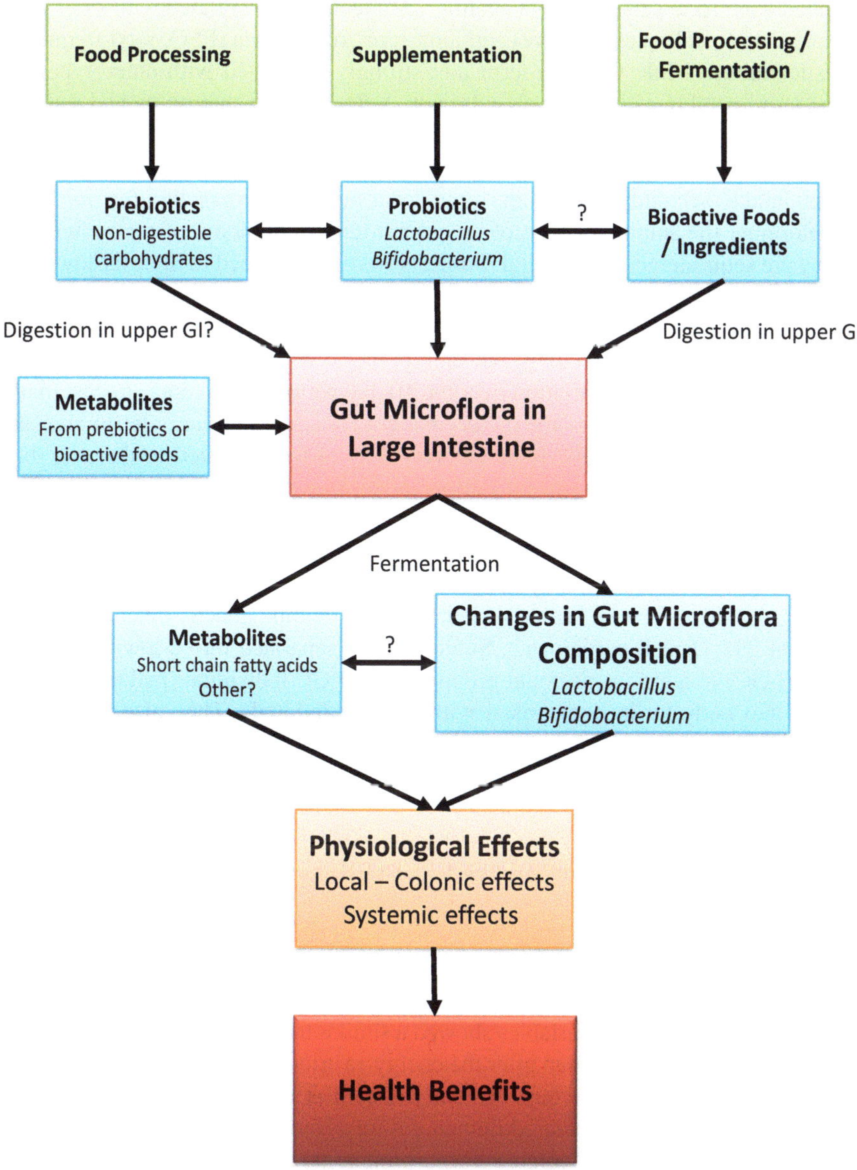

Fig. 14.1 Health effects ofprebiotics, probiotics, and bioactive foods through changing gut microbiota. Schematic representation of the interaction between the intestinal microbiota and prebiotics, probiotics, and bioactive foods and how they possibly lead to changes in health

isolated from fecal samples of 98 healthy volunteers was correlated with short-term dietary data (recall questionnaire) and long-term dietary data (FFQ, food frequency questionnaire). While the long-term diet strongly correlated with enterotype, the short-term recall questionnaire data did not. A *Bacteroides* enterotype had a strong relationship with animal protein, amino acids, and saturated fat consumption, while a *Prevotella* enterotype had a strong relationship with carbohydrates and simple sugars intake. Secondly, this group conducted a short-term controlled-feeding study to determine the stability of the observed nutrient microbiome of ten individuals who were sequestered in a hospital. Subjects were randomized to high-fat/low-fiber or low-fat/high-fiber diets for 10 days. Stool and rectal biopsies were sampled and analyzed. Microbiome composition clearly changed within 24 h following the beginning of controlled feeding, but enterotype identity remained stable without overcoming intersubject variation over the 10 day period. These data suggest that it may be necessary to change the dietary pattern for some time in order to alter the gut microbiome beneficially. If an enterotype is causally related to disease, then long-term dietary interventions may allow modulation of an individual's enterotype to improve health.

Recently, studies have presented evidence to suggest that the gut microbiome has a more important role in obesity than the human genome. Le Chatelier et al. (2013) investigated the effects of bacterial richness (microbiome genetic diversity) on the health of 292 Danish participants. Subjects were separated into a low gene count group (LGC) or a high gene count group (HGC). LGC individuals had higher body weight and body fat in comparison with HGC individuals. They also had higher concentrations of the inflammatory indicators serum insulin, triglycerides, free fatty acids, highly sensitive C-reactive protein (hsCRP), and white blood cells. Moreover, LGC individuals had a higher abundance of peroxidase, catalase, and TCA modules which suggest an increased capacity to handle exposure to oxygen/oxidative stress. In a dietary intervention study, Cotillard et al. (2013) showed the importance of diet to gut microbial richness in 49 overweight/obese individuals through an energy-restricted high-protein diet followed by a 6-week weight-maintenance diet. Similar to the Le Chatelier et al. study (2013), the LGC group had significantly higher insulin resistance and fasting serum triglyceride levels than the HGC group. LDL cholesterol and inflammation variables showed a similar trend. Gene richness increased significantly in the LGC group after the energy-restricted diet, but it did not change significantly in the HGC group. Recently, David et al. (2014) also showed that short-term dietary change altered human gut microbiota rapidly and reproducibly. An animal-based diet with high-fat/high-protein/no dietary fiber enhanced bile-tolerant microorganisms such as *Alistipes*, *Bilophila*, and *Bacteroides*. On the other hand, a plant-based diet composed of high-fiber/reduced fat and protein showed higher levels of *Firmicutes* that metabolize dietary fiber. Furthermore, the plant-based diet showed higher concentrations of carbohydrate-fermented short-chain fatty acids (SCFAs) such as acetate and butyrate. In turn, the animal-based diet showed increased concentration of amino acids from fermented products such as isovalerate and isobutyrate.

In summary, current studies have shown that the gut microbiota can be changed by transient dietary changes as well as long-term dietary patterns. Specific diet-induced changes of the gut microbiota can lead to the improvement of body weight and metabolic disease.

14.1.2 SCFAs as Bacterial Fermentation Products

The Cotillard et al. (2013) and Le Chatelier et al. (2013) studies showed that high GC individuals ate more fruits, vegetables, and fish products than low GC subjects. Fifty-eight species that were significantly different between LGC and HGC individuals were associated with changes in body mass index (BMI). Eight species that were more abundant in HGC than in LGC individuals were butyrate producers, likely due to high dietary fiber intake from greater plant consumption. This implies that characteristics of gut microbiota and their ability to produce SCFAs are affected by dietary patterns of each individual as well as their food culture. The Gordon group confirmed this study by comparing the gut microbiome of 531 individuals residing in either the Amazonas of Venezuela, rural Malawi, or US metropolitan areas. Individual gut microbiomes were clustered according to geographical location (Yatsunenko et al. 2012). Studies have also examined the microbiome of other cultures. Nam et al. (2011) characterized the gut microbiota in 20 Koreans. As many as 15 species level, phylotypes were related to butyrate-producing bacteria among 43 core gut microbiota distributed broadly in Korean individuals. They compared these results with those of individuals from the USA, China, and Japan. The major phyla of gut microbial community were *Firmicutes*, *Bacteroidetes*, and *Actinobacteria*. Differences were observed due to country of residence. Americans had higher *Firmicutes* than individuals from other countries. Japanese had higher *Actinobacteria* than those from other countries, and Koreans and Chinese had more *Bacteroidetes* than individuals from the other countries.

In the mammal intestine, SCFAs, such as acetate, propionate, and butyrate, are produced by groups of gut microbiota, mainly from dietary fibers (Cook and Sellin 1998). Total SCFAs concentration ranged from 80 mmol/kg in the descending colon to 131 mmol/kg in the caecum. Acetate was the most predominant SCFA with molar ratios of acetate:propionate:butyrate of 65:28:8 in mature pigs (Cummings et al. 1987). In a human study, subjects who consumed a low-protein diet (LP), a high-protein diet (HP), and high-protein diet added with dietary fiber of 21.8 g (HP+F) revealed that the concentration of total SCFAs ranged from 58 to 66 mmol/l, respectively (Cummings et al. 1979). SCFAs that arise from gut bacterial metabolism are rapidly absorbed into the large intestine and used as energy source for mucosal epithelial cells. Propionate and acetate are transported to the liver and metabolized after absorption, whereas butyrate is mostly (70–90%) metabolized by the colonocytes to support the growth of mucosal cells (Cook and Sellin 1998). Frankel et al. (1994) investigated whether the trophic effects of the SCFAs are mediated by autonomic nervous and/or enterotropic hormones. For this, SCFAs were infused into the cecum of rats that had undergone cecal denervation. The results showed that the jejunal tropic effects such as increased jejunal DNA, villous height, surface area, and crypt depth were mediated by autonomic nervous system and increased jejunal gastrin. In contrast, SCFAs were found to have local trophic effects on the colon. Interestingly, it was recently reported that propionate, a major bacterial fermentation metabolite, can also have a potentially systemic effect by reducing allergic airway inflammation (Trompette et al. 2014). In a murine model of allergic airway inflammation, mice were fed a control diet (dietary fiber 4%), a low-fiber diet (<0.3%), or a high-fiber diet (30% cellulose; a non-fermentable dietary fiber, or pectin; a soluble dietary fiber). The low-fiber diet greatly reduced the richness and diversity of gut microbiota and led microbiota to be dominated by *Firmicutes*. On the other hand, the high-fiber diet changed the composition of the gut and lung microbiota leading to a decreased ratio of *Firmicutes* to *Bacteroidetes* and increased local and systemic levels of SCFAs. The authors next determined whether the protective effects elicited by a high-fiber diet could be attributed to the increased levels of the SCFAs acetate and propionate. Administration of propionate was found to lessen allergic airway inflammation by signaling through GPR 41, the SCFA receptor. Interestingly, SCFAs were not detected directly in the lung. The authors suggested a mechanism for an intestinal-bone marrow-lung axis where increased levels of circulating SCFAs such as propionate enhance dendritic cell hematopoiesis in the bone marrow. These dendritic cells have an impaired ability to activate T_H2 effector cells in the lung, consequently modulating and resolving allergic airway inflammation.

In conclusion, studies have shown that dietary fiber has the potential to change gut microbiota richness, diversity, and composition. This has the subsequent effect of increasing the production of SCFAs that can have various health effects through not only the local immune system but also systemically.

14.2 Bone Metabolism and Gut Microbiota

The bone is a multifunctional organ performing very complicated functions. It supports the human body and serves as a reservoir for calcium and phosphorus. The bone marrow is also the site of hematopoiesis; hematopoietic stem cells differentiate into both blood and immune cells. The bone maintains its structural integrity via a continuous remodeling process. In humans, more than 10% of total bone mass is changed newly every year (Theill et al. 2002). Bone remodeling is accomplished by the coupled activities of a number of key cells. The osteoblast, a cell derived from pluripotent mesenchymal stromal cells that produce the organic bone matrix and aids in its mineralization (Karsenty et al. 2009); the osteoclast, a multinucleated giant cell of monocyte/macrophage lineage derived from hematopoietic stem cells responsible for the degradation of the bone and extracellular matrix (Teitelbaum 2007); the osteocyte, an osteoblast-derived cell that lies within the bone matrix which acts as a mechanosensor and an endocrine cell (Bonewald and Johnson

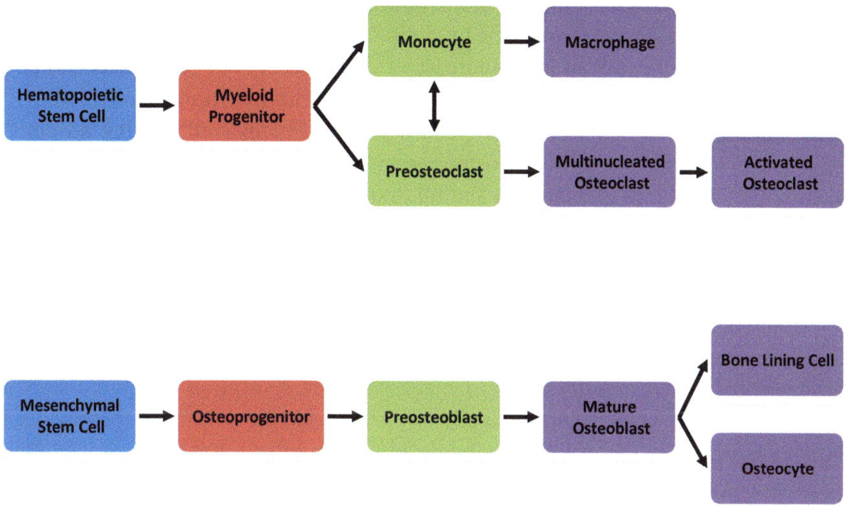

Fig. 14.2 Stages of osteoclast and osteoblast differentiation. Simplified schematic representation of osteoclast and osteoblast differentiation. Osteoclasts arise from hematopoietic stem cells and are of the monocyte/macrophage lineage. Under the influence of MCSF and RANKL, preosteoclasts fuse together and undergo terminal differentiation into the multinucleated osteoclasts which activate to resorb the bone. Osteoblasts arise from mesenchymal stem cells. Under the influence of external factors, mesenchymal stem cells differentiate into osteoprogenitors and preosteoblasts and differentiate into mature mineralizing osteoblasts. These mature osteoblasts will then either undergo apoptosis or terminally differentiate into osteocytes or bone lining cells

2008); and the bone lining cell, a second osteoblast-derived cell that helps couple bone formation to resorption (Fig. 14.2) (Andersen et al. 2009).

Cross talk between the osteoclast and the osteoblast is a key component of bone remodeling, with the RANKL-RANK-OPG axis a major regulator. RANKL (receptor activator of nuclear factor κB (NF-κB) ligand) is expressed extensively on the surface of osteoblast/stromal cells and plays a critical role in osteoclast differentiation along with the cytokine macrophage colony-stimulating factor (M-CSF). RANK (receptor activator of NF-κB), the receptor for RANKL, is expressed on the surface of hematopoietic osteoclast progenitors and mature osteoclasts. The binding of RANKL and RANK leads to activation of osteoclast differentiation through a signal transduction cascade and consequently stimulates bone resorbing activity. Osteoprotegerin (OPG) known as the protector of the bone is a potent inhibitor of osteoclastogenesis and is produced by osteoblasts/stromal cells, B cells, and dendritic cells as a soluble form. OPG acts as a decoy receptor for RANKL and competes with RANK for RANKL (Theill et al. 2002; Lorenzo et al. 2008; Yasui et al. 2011; Yun et al. 1998). Bone resorbing signals such as glucocorticoids, vitamin D_3, IL-1, IL-6, IL-11, IL-17, TNFα, PGE2, or PTH can upregulate the expression of RANKL on osteoblasts whereas estrogen can increase the production of OPG (Theill et al. 2002).

Like any other organ of the body, the bone is susceptible to disease. Diseases of the bone are typically characterized by a direct or indirect effect on the balance of remodeling, which can lead to an increase or decrease in bone resorption and bone formation. The etiology of these diseases is varied with nutrition, physical activity, inflammation, genetic abnormalities, and hormonal changes all playing a role. Of these adverse bone conditions, osteoporosis is the most common and costly. Approximately one in two women and up to one in four men age 50 and older will break a bone due to osteoporosis costing $18 billion a year in America (U.S. Department of Health and Human Services 2004; National Osteoporosis Foundation). Osteoporosis is characterized by low bone mineral density (BMD), deteriorated bone microarchitecture, and an increased risk of fracture (Harvey et al. 2010). A major cause of osteoporosis is imbalanced bone remodeling, where increased osteoclast activity leads to excessive bone resorption compared to bone formation (Bucay et al. 1998; Jilka et al. 1992). In females, onset of menopause is the major factor that contributes to the development of postmenopausal osteoporosis. Estrogen deficiency is linked to an increase in the expression of pro-inflammatory and osteogenic cytokines RANKL, TNF, IL-1, and IL-6 (Jilka et al. 1992; D'Amelio et al. 2008; D'Amelio et al. 2005; Pacifici et al. 1991). Treatments for osteoporosis include bisphosphonates, intermittent PTH, hormone therapy, and denosumab (a fully humanized monoclonal antibody directed against RANKL) (Papapoulos 2008; Rosen and Bilezikian 2001; Schwarz and Ritchlin 2007). Treatment with these drugs, however, can have adverse side effects. Long-term treatment with bisphosphonates can lead to nonunion fractures and osteonecrosis of the jaw, while denosumab can increase the risk of infection (Ing-Lorenzini et al. 2009; Shannon et al. 2011; Kendler et al. 2010).

As stated earlier, the gut microbiota plays an important role in maintaining host immune homeostasis, and the bone marrow is the key site for the development and maturation of immune cells (Kelsall and Leon 2005). It is, therefore, feasible to suggest that the gut microbiota may indirectly modulate bone turnover. Few studies however, have evaluated the relationship between the gut microbiota and bone metabolism. To examine the relationship between the gut microbiome and bone metabolism and whether modulation of the intestinal microbiome can be used to treat pathological bone loss animal models are commonly utilized. Non-pathologic, "healthy," animal models are routinely used to study baseline and pathologic microbial composition. Healthy animal models can also be used to determine the safety and efficacy of treatments as well as the mechanisms of action. Pathological bone loss is associated with the progression of a number of diseases including menopause, arthritis, periodontitis, and hypertension (Gough et al. 1994; U.S. Department of Health and Human Services 2004; Cochran 2008; Cappuccio et al. 1999). Animal models are routinely utilized to investigate the different mechanisms of disease progression, bone loss, and potential treatments. For example, the ovariectomized (OVX) rat/mouse model is routinely used to induce estrogen deficiency and is therefore a gold standard model to assess whether the use of treatments can protect against bone loss associated with menopause (Kalu 1991).

A key piece of evidence linking the intestinal microbiome and bone metabolism was provided by Sjögren et al. (2012). In their studies, Sjögren et al. used germ-free (GF) mice, conventional (CONV-R) mice, and colonized GF mice with normal gut microbiota to investigate the importance of the gut microbiota in affecting bone mass. Bone mass of GF mice was higher compared to that of CONV-R mice with no significant differences in testosterone levels, serum calcium, PTH, and 25-hydroxy vitamin D_3 levels. GF mice were also found to have reduced numbers of osteoclasts per bone surface and a decreased frequency of $CD4^+$ T cells and osteoclast precursor cells in their bone marrow compared with CONV-R mice. Increased bone mass and decreased $CD4^+$ T cells of GF mice were normalized with colonization of gut microbiota from CONV-R mice. Additionally, expression of TNFα, an inflammatory cytokine and a potent stimulator of bone resorption (Lorenzo et al. 2008), was reduced in the bone and bone marrow of GF mice. From these results, the authors suggested that the gut microbiota regulates bone mass in mice through modulating osteoclastogenesis via $CD4^+$ T cells and expression of inflammatory cytokines.

14.3 Bioactive Foods, Gut Microbiota, and Bone Health

14.3.1 Milk and Dairy Products

Milk and dairy foods contain essential nutrients required for bone formation including calcium, phosphorus, magnesium, vitamin D, and other trace nutrients, as well as good quality proteins such as whey protein and casein (Zhao et al. 2005). Milk or dairy products are also very important sources of bioactive peptides. These peptides have been shown to elucidate various health-promoting effects including immunomodulatory, antihypertensive, antioxidative, antithrombotic, antimicrobial activity, and mineral binding activity (Kamau et al. 2010; Nagpal et al. 2011; Beermann and Hartung 2013). The bioactive peptides encrypted in milk proteins can be released by proteolytic digestive enzymes, proteolysis using a starter culture or by naturally occurring fermentable microbes during fermentation (Narva et al. 2004a; Caplice and Fitzgerald 1999). Studies into the potential beneficial effects of casein for the bone have shown that when digested, it is broken down into phosphopeptides. These caseinophosphopeptides (CPP) are resistant to digestive proteolytic enzymes and help enhance calcium absorption by binding soluble calcium and inhibiting formation of insoluble calcium salts (Lee et al. 1980).

Lactose-derived galactooligosaccharides (GOS), a prebiotic that promotes the growth of beneficial gut microbiota including *bifidobacteria*, as well as whey or casein-derived active peptides may also be linked to the improved bone health associated with the use of fermented dairy products (Whisner et al. 2013). Human milk contains GOS though the abundant concentration in colostrum is rapidly decreased over time. GOS is also produced through transgalactosylation of lactose by β-galactosidase in cow's milk. Some microbes used frequently in food fermentation such as *Aspergillus oryzae*, *Lactobacillus reuteri*, and *Bacillus stearothermophilus* can convert lactose to GOS (Macfarlane et al. 2008). Weaver et al. (2011) reported that GOS supplementation derived from lactose improved mineral balance and bone properties in growing rats through gut fermentation. Furthermore, Whisner et al. (2013) determined that daily supplementation of 5 g GOS increased calcium absorption, changed gut microbiota, specifically *bifidobacteria*, and improved peak bone mass of adolescent girls.

14.3.2 Soy and Soy Products

Many East Asians traditionally consume a large amount of soybean foods including whole soybean, soy milk, and soy powder. These foods are both fermented and non-fermented and use the whole bean (Reinwald and Weaver 2010; Wu et al. 1998). This contrasts with the type of soy foods consumed in the west which are typically not fermented and not based on the whole bean (Reinwald and Weaver 2010). The natural calcium content of soy milk is very low (Zhao et al. 2005); however, they are known to contain other molecules such as phytoestrogens that may have a beneficial effect on the bone.

Isoflavones are a major class of phytoestrogen and are analogous to mammalian estrogens. Soy food naturally contains 1.2–1.3 mg isoflavone/g dry weight. The major isoflavones found in soy are the glycoside-type daidzin and genistin and their corresponding aglycone forms, daidzein and genistein (Chiang and Pan 2013). Soybean and unfermented soybean foods contain the glycoside daidzin and genistin. After consumption, the glycosides are hydrolyzed to their aglycone form by β-glucosidases. These enzyme activities are found in intestinal bacteria such as *Lactobacillus* spp. or *Bifidobacterium* spp. as well as on the intestinal brush border. The permeability of the aglycone form is higher compared to that of glycoside type (Steer et al. 2003; Chen et al. 2010; Chiang and Pan 2013; Cassidy 2006). Daidzein, the most abundant isoflavone in soybeans, can be metabolized to equol by the intestinal microflora and exhibits a stronger estrogenic effect than daidzein (Fonseca and Ward 2004; Fujioka et al. 2004). In murine studies, equol was shown to suppress OVX-induced bone loss though it was not able to affect the associated uterine atrophy (Fujioka et al. 2004). In contrast, in a rat OVX model, equol exhibited some uterine activity, but the weights were still lower than the sham controls (Legette et al. 2009). Unlike rodents, not all humans have gut microbiota that can convert daidzein to equol; levels of dietary fat can also affect equol excretion (Rowland et al. 2000). However, equol-producing status did not significantly affect bone calcium retention in postmenopausal women (Weaver and Legette 2010).

Soy food consumption has been shown to have many benefits and has been negatively associated with risk of coronary heart disease (Zhang et al. 2003), type 2 diabetes (Villegas et al. 2008), breast cancer (Wu et al. 1998), and prostate cancer (Yan and Spitznagel 2009). However, a positive association with bone health is unclear. Epidemiological studies in Asia have demonstrated a positive association with bone mineral density and fracture protection and whole soy foods (Reinwald and Weaver 2010). However, randomized controlled trials with isolated soy isoflavones have not typically been associated with improved BMD in humans (Alekel et al. 2010; Weaver et al. 2012). In a prospective cohort study of approximately 75,000 Chinese women, Zhang et al. (2005) reported that soy food consumption was associated with decreased risk of fracture in postmenopausal women, specifically in those in the early phase of menopause. Estimated soy isoflavone consumption was similarly related to decreased fracture risk. In a review by Chiang and Pan (2013), phytoestrogens and their metabolites produced by gut microbiota were shown to increase the rate of bone formation by enhancing the OPG/RANKL ratio and increasing levels of alkaline phosphatase, osteocalcin, osteopontin, and type 1 collagen. Phytoestrogens were also shown to reduce the rate of bone resorption by inhibiting differentiation and activation of osteoclasts and by decreasing expression of the osteoclast bone-degrading enzyme tartrate-resistant acid phosphatase (TRAP). In a study by Kim et al. (2009a), administration of 60 mg/kg soy isoflavones was shown to elicit similar protective effects on femoral and lumber vertebral bone loss in middle-aged OVX mice as the sex hormone 17β-estridiol.

In addition to the consumption of soy-derived phytoestrogens, ingestion of fermented soy foods or fermentation of soy foods by gut microbiota may have bone benefits. Tenorio et al. (2010) showed that soybean whey, an underused by-product of tofu preparation, contains nondigestible oligosaccharides (NDO) such as GOS stachyose and traces of inulin which act as prebiotics. The consumption of this soybean whey was demonstrated to improve mineral balance (especially calcium and magnesium) and enhanced the production of SCFA, particularly acetic acid and butyric acid. In an analogous study Haron et al. (2010) studied absorption of calcium from tempeh, a traditional fermented soy product, compared to milk in 20 postmenopausal Malay women using a randomized, crossover design. Calcium absorption was not significantly different between milk and tempeh when analyzed in the urine, but lower gut calcium absorption was not assessed. Cheung et al. (2011) tested the hypothesis that the beneficial outcomes of fermented soy milk on calcium bioavailability and bone health may be obtained by long-term consumption. They recruited 12 Australian osteopenic postmenopausal women and undertook a randomized crossover pilot study to determine the effects of calcium availability by fermentation of soy milk. The subjects consumed Ca-fortified soy milk or fermented and calcium-fortified soy milk for the acute pilot study. Soy milk was fortified by radiolabeled $^{45}CaCl_2$. Ca-fortified soy milk was fermented with *Lactobacillus acidophilus* ATCC 4962. However, the mean fractional calcium absorption between the two groups was not significantly different. In a 1-year prospective study of 73 healthy premenopausal women, Katsuyama et al. (2004) investigated the effects of Natto, a traditional Japanese food made from soybeans fermented with *Bacillus subtilis* (*Bacillus natto*), on bone turnover markers. Natto is known to contain high levels of vitamin K_2 which plays a role in maintaining bone strength by increasing osteocalcin levels and decreasing serum undercarboxylated osteocalcin (ucOC) levels (Kasukawa et al. 2014). High levels of Natto consumption (three times a week) was shown to be positively associated with an increase in the bone formation marker bone specific alkaline phosphatase. Levels of ucOC were also observed to be significantly lower after 6 months.

As well as soy-based isoflavones and fermented soy products, soy protein has been studied for its potential beneficial effects on bone. Chen et al. (2013) showed that consumption of a high-fat diet with soy protein isolate (SPI) prevented the high-fat-induced bone impairment in a casein-based diet (HF-Cas). HF-Cas fed rats had increased bone marrow adiposity, non-esterified free fatty acid (NEFA) levels, and insulin resistance compared to HF-SPI fed rats. The mechanism for the HF-SPI bone protection is not clear but is thought to relate to an increase in ucOC enhancing insulin sensitivity and insulin secretion from pancreatic β-cells (Clemens and Karsenty 2011) and changes in insulin signaling in osteoblasts.

Current studies have shown that the consumption of prebiotics such as GOS (Weaver et al. 2011; Whisner et al. 2013) and plant fibers (Whisner et al. 2014) have some beneficial effects on bone health. While the exact mechanisms through which these benefits are exerted are not clear, increased levels of SCFAs and alterations to gut microbiota composition are thought to play a key role. New light is also being shed on the potential role of fermented food products in enhancing bone health though the efficacy of these products still requires further investigation. In addition to the use of prebiotics to maintain a healthy microbiota, research into probiotic supplementation has also gained considerable attention over recent years.

14.4 Probiotic Bacteria and Bone Health

Probiotic bacteria are defined as "live microorganisms which when administered in adequate amounts confer a health benefit on the host" (FAO 2002). The use of orally administered probiotic bacteria to improve health by suppressing harmful bacteria is not a new one. As early as 1906, Tissier suggested the administration of *bifidobacteria* to infants suffering from diarrhea to suppress the disease-causing putrefying bacteria (Tissier 1906). In addition, Elie Metchnikoff stated in his 1907 book, "The prolongation of life: optimistic studies," that consumption of lactobacilli-containing yogurt resulted in a reduction of toxin-producing bacteria in the gut and that this was associated with increased longevity of the host (Metchnikoff 1907; de Vrese and Schrezenmeir 2008). Interest in probiotic bacteria is now at an all-time high (Sanders 2008) with many new products being released to the market and increasing evidence of their therapeutic value. Reported beneficial effects of probiotics include treatment of gastrointestinal ailments (Sartor 2004), increased bone health (Britton et al. 2014; McCabe et al. 2013; Ohlsson et al. 2014), suppression of intestinal inflammation and modulation of immune response (McCabe et al. 2013; Thomas et al. 2012; Ouwehand et al. 2002), and potentially influencing human mood and behavioral disorders (Bested et al. 2013).

The majority of the currently used probiotic bacteria belong to the *Lactobacillus* and *Bifidobacterium* species, though other bacteria and some yeasts have also been demonstrated to have probiotic properties (Table 14.1) (de Vrese and Schrezenmeir

Table 14.1 Probiotic microorganisms

Genus	Species	Genus	Species	Other
Lactobacillus	*acidophilus*	*Bifidobacterium*	*adolescentis*	*Bacillus cereus*
	casei		*animalis*	*Bacillus subtilis*
	crispatus		*bifidum*	*Enterococcus faecalis*
	fermentum		*breve*	*Enterococcus faecium*
	gasseri		*infantis*	*Escherichia coli* (strain Nissle 1917)
	johnsonii		*lactis*	*Lactococcus lactis*
	paracasei		*longum*	*Propionibacterium freudenreichii*
	plantarum			*Saccharomyces cerevisiae*
	reuteri			*Streptococcus thermophilus*
	rhamnosus			
	salivarius			

Adapted from Ouwehand et al. (2002) and de Vrese and Schrezenmeir (2008)

2008). A number of key criteria, as detailed below, are currently employed to determine whether a microbe is suitable for use as a probiotic (FAO 2002). Determination of these criteria is routinely done through in vitro assays. However, results from these assays may not be fully predictive of the in vivo effect (Ibnou-Zekri et al. 2003), and so in vivo validation is required.

The microbe in question must:

- Be safe for humans (free from pathogenic and toxic effects).
- Be able to survive travel through the gastrointestinal tract – tolerant to low pH, bile, and pancreatic enzyme.
- Be able to adhere to the intestinal mucosa and/or human epithelial cells.
- Be of human origin (though some exceptions exist such as *S. cerevisiae* (*boulardii*)).
- Have antimicrobial activity against potentially pathogenic bacteria.
- Have the ability to reduce pathogen adhesion to surfaces.
- Comply with technological requirements for production on a large scale.

In recent years, a number of studies have been published indicating the potential effects of probiotic dietary supplementation on increasing BMD in healthy and pathological states. These beneficial effects have been observed with different strains of bacteria and in different models of disease (Mutuş et al. 2006; Britton et al. 2014; McCabe et al. 2013; Ohlsson et al. 2014; Rodrigues et al. 2012). To try and elucidate the benefits and mechanism of probiotic supplementation on bone health, a number of experimental models have been used (Table 14.2). These studies can be divided into subsets that examine the effect of probiotics on (1) healthy rodents, (2) diseased rodent models, and (3) poultry.

Table 14.2 Effect of probiotics on the bone (animal studies)

Probiotic strain	Animal model	Duration	Analysis method	Bone effects	Ref
Bacillus licheniformis and *Bacillus subtilis*	Broiler	6 weeks	Measuring calipers	↑ Tibia lateral and medial wall thickness	Mutuş et al. (2006)
Brewer's yeast	Broiler	9 weeks	Visual	↓ Tibial dyschondroplasia	Plavnik and Scott (1980)
Lactobacillus	Broiler	28 days	Atomic absorption spectrophotometry Phosphomolybdic acid method	↑ Ca and P retention	Nahashon et al. (1994)
Lactobacillus reuteri (ATCC 6475)	Male mice	4 weeks	μCT	↑ Bone volume fraction ↑ Tb. N ↑ Tb. Th. ↑ BMC ↑ BMD ↑ Osteocalcin ↑ BFR	McCabe et al. (2013)
	Female mice (OVX)	4 weeks	μCT	↑ Bone volume fraction ↑ BMC ↑ BMD ↓ RANKL gene expression ↓ TRAP5 expression	Britton et al. (2014)
Lactobacillus casei, *L. reuteri*, and *L. gasseri*	Rats			↑ BMC ↑ Ca absorption	Ghanem et al. (2004)
Bifidobacterium longum-fermented broccoli	Male Wistar rats	12 weeks	Histology	↓ $TRAP^{+}$ osteoclasts	Tomofuji et al. (2012)
Bifidobacterium longum (ATCC 15707)	Male Wistar rats	28 days	Texture analyzer Plasma emission spectrophotometry	↑ Tibial Ca, P, and Mg content ↑ Fracture strength	Rodrigues et al. (2012)

Lactobacillus rhamnosus (HN001)	Male Sprague-Dawlry rats	3 weeks	DEXA	↑ Ca and Mg retention	Kruger et al. (2009)
	Female Sprague-Dawlry rats (OVX)	3 months	DEXA	↓ Bone loss	Kruger et al. (2009)
L. paracasei (NTU 101)- *or L. plantarum* (NTU 102)-fermented soy milk	Female mice (OVX)	8 weeks	μCT	↑ Bone volume fraction	Chiang and Pan (2011)
			SEM	↑ Tb. N	
L. paracasei or *L. paracasei* and *L. plantarum*	Female mice (OVX)	6 weeks	μCT	↑ Cortical BMC ↑ Cortical area ↑ OPG expression	Ohlsson et al. (2014)
L. casei 393-fermented milk	Female Sprague-Dawlry rats (OVX)	6 weeks	DEXA Texture analyzer Plasma emission spectrophotometry	↑ BMD ↑ Fracture strength ↑ Ca content ↓ TRAP activity	Kim et al. (2009a, b)
L. helveticus-fermented milk	Spontaneously hypertensive male rats	14 weeks	DEXA Plasma emission spectrophotometry	↑ BMD ↑ BMC	Narva et al. (2004a)
L. casei and *L. acidophilus*	Female Wistar rat (adjuvant-induced arthritis)	12 days	X-ray	↓ Bone loss	Amdekar et al. (2012)
Enterococcus faecium (with methotrexate)	Male Lewis rat (adjuvant-induced arthritis)	50 days	DEXA X-ray	↑ Anti-inflammatory effects ↑ Antiarthritic effects	Rovenský et al. (2004)

14.4.1 Probiotics and Non-Pathological Animal Models

As stated earlier in this chapter, non-pathologic "healthy" animal models are used to investigate the safety, efficacy, and mechanisms of probiotic treatment.

In a study using *Lactobacillus reuteri* (ATCC 6475) (McCabe et al. 2013), male mice housed under specific pathogen-free conditions were treated at 14 weeks of age for a period of 4 weeks under conventional housing conditions. At the end of this period, the femora and vertebrae were analyzed by μCT. Treatment with *L. reuteri* was reported to significantly increase femoral and vertebral trabecular bone volume/total volume (BV/TV), trabecular number (Tb. N.), trabecular thickness (Tb. Th.), bone mineral content (BMC), and BMD compared to non-treated males. Levels of the osteoblast marker osteocalcin and bone formation rate (BFR) were also significantly elevated in the *L. reuteri*-treated mice though no difference was observed in serum TRAP levels. Interestingly in this study, the effects of *L. reuteri* treatment were found to be gender specific with females showing no response to *L. reuteri* supplementation with regard to bone health (McCabe et al. 2013).

Analogous to the mouse studies, treatment of rats with *Lactobacillus casei*, *L. reuteri*, and *L. gasseri* containing probiotic yogurt increased BMC and calcium absorption compared to the control group (Ghanem et al. 2004). *Bifidobacterium longum* has also been reported to have a significant beneficial effect on bone health. In a study by Tomofuji et al. (2012), rats fed a high-cholesterol diet containing *B. longum*-fermented broccoli for 12 weeks were found to have a significantly lower number of TRAP-positive osteoclasts on the alveolar surface. In a separate study, treatment of male rats with *Bifidobacterium longum* (ATCC 15707) resulted in tibia with significantly increased calcium, phosphorous, and magnesium mineral content and increased fracture strength compared to rats fed a control diet (Rodrigues et al. 2012). Similarly, supplementation of growing male rats with *Lactobacillus rhamnosus* (HN001) was also found to improve calcium and magnesium retention (Kruger et al. 2009).

14.4.2 Probiotics and Animal Models of Bone Disease

A number of studies have utilized the OVX model to demonstrate the potential protective effects of different strains of probiotics, namely, *Lactobacillus* and *Bifidobacterium*, on OVX-induced bone loss (Britton et al. 2014; Ohlsson et al. 2014; McCabe et al. 2013). In a study by Britton et al. (2014), the bacterium *Lactobacillus reuteri* ATCC 6475 was used in the OVX mouse menopause model and the femora and vertebrae analyzed by μCT. Mice were provided with constant access to *L. reuteri* in their drinking water (1.5×10^8 cfu/mL) and gavaged three times a week (1×10^9 cfu/mL) over a period of 4 weeks following surgery. OVX mice were reported to lose approximately 50% of the trabeculae in the distal femur and 25% in the vertebrae. In sharp contrast, however, mice treated with the *L. reuteri* were completely protected from trabecular bone loss, resulting in a BV/TV comparable to the control mice. In addition, significant increases in trabecular BMD and BMC were observed in the OVX *L. reuteri*-treated mice compared to the OVX controls. Analysis of osteoclast markers showed a significant reduction in bone RANKL gene expression and a decrease in TRAP5 expression in *L. reuteri*-treated OVX mice compared to the OVX controls. A significant decrease was also observed in the osteoclastogenic potential of the OVX *L. reuteri*-treated bone marrow when cultured ex vivo, bringing it in-line with the level of osteoclastogenesis seen with control bone marrow. This study supports the results from a previous study where Chiang and Pan et al. (Chiang and Pan 2011) investigated the effect of treating OVX mice with either *L. paracasei* (NTU 101)- or *L. plantarum* (NTU 102)-fermented soy milk. The effect on both femoral cortical and trabecular bone was analyzed using μCT and scanning electron microscopy (SEM). OVX mice treated with the NTU 101- and NTU 102-fermented soy milk exhibited significantly increased BV/TV and Tb. N. compared to the OVX controls.

In a similar study by Ohlsson et al. (2014), mice were supplemented 2 weeks prior to surgery, and for 4 weeks after, with either a single *L. paracasei* strain (DSM13434) or a mixture of three strains (*L. paracasei* DSM13434, *L. plantarum* DSM15312 and DSM15313). Analysis of the cortical bone at the mid-diaphyseal region of the femur by μCT demonstrated that both the single strain and mixed strains protected the mice against OVX-induced cortical bone loss. Cortical BMC and cortical area were both elevated in OVX mice treated with the single and mixed strains compared to the vehicle-treated OVX cohort. Analysis of bone-related mRNA levels in the cortical bone revealed that the probiotic treatment decreased the RANKL/OPG ratio (a major determinant of osteoclastogenesis) by increasing OPG expression.

In a 2009 study, Kim et al. (2009b) demonstrated that providing ovariectomized rats with *L. casei*-fermented milk increased BMD and bone breaking force when compared to the control OVX mice. These effects were related to an increase in calcium content within the bones as well as a significant reduction in TRAP activity. In a different study, OVX rats fed with the bacterial strain *L. rhamnosus* (HN001) were demonstrated to have a reduced rate of bone loss and higher final whole bone density in the spine and femur after 12 weeks compared to the OVX controls (Kruger et al. 2009).

Studies have also investigated the potential therapeutic effect of probiotic supplementation on adverse bone pathology associated with arthritis and periodontitis (Amdekar et al. 2012; Rovenský et al. 2004; Messora et al. 2013; Foureaux et al. 2013). In the rat adjuvant-induced arthritis (AA) model, *Lactobacillus casei* and *L. acidophilus* were reported to protect against the associated bone loss (Amdekar et al. 2012). The bacterium *Enterococcus faecium* was also found to potentially increase the effectiveness of methotrexate treatment, enhancing its anti-inflammatory and antiarthritic effects (Rovenský et al. 2004). In a model of periodontitis, probiotic treatment has been observed to significantly reduce alveolar bone loss associated with ligature-induced periodontitis (Messora et al. 2013; Foureaux et al. 2013).

Interestingly, in a less well-known bone pathology model, spontaneously hypertensive male rats, Narva et al. (2004a) investigated the effect of *L. helveticus*-fermented milk on bone health. In this model, the rats develop osteoporosis with age, so at 6 weeks of age, the rats were divided into groups and treated for 14 weeks. Analysis of the femoral BMD and BMC by dual-energy X-ray absorptiometry (DEXA) revealed significantly increased BMD and BMC levels in the *L. helveticus*-treated group compared to controls.

14.4.3 Probiotics and Poultry

Low bone density is not only costly to humans, but each year, the agricultural sector also loses millions of dollars due to livestock developing skeletal abnormalities affecting quality and output (Payne 1977). This is a critical issue since the poultry industry is an important economic factor in many countries. Bone pathologies are thought to have developed because of the pressure to produce large, fast-growing, and affordable broilers in large-scale rearing facilities. This method of farming leads to increased stress for the birds resulting in elevated infection rates, increased muscle mass which the bone cannot support, and the development of skeletal problems, which is estimated to cost the industry $80–120 million annually (Sullivan 1994).

Traditionally, these complications were overcome with the use of growth factors, antibiotics, and other veterinary medicines. Continued use of subtherapeutic antibiotics in animal feed can result in trace amounts making its way into the human food chain and subsequently the evolution of drug-resistant microorganisms that can infect humans (Jin et al. 1997). As such, public opinion and government regulations toward the use of pharmacological intervention in the treatment of animals for consumption have shifted, resulting in investigation of alternatives such as probiotics.

The addition of probiotic supplementation to chicken feed has been demonstrated to provide a number of benefits. Probiotics have been shown to reduce mortality, improve weight gain and feed conversion, increase egg size, and decrease the severity and incidence of salmonella infections in chickens (Jin et al. 1997; Nava et al. 2005; Khan and Naz 2013; Nahashon et al. 1994). Several studies have also demonstrated the positive effects that probiotic supplementation can have on improving poultry bones (Mutuş et al. 2006; Plavnik and Scott 1980; Nahashon et al. 1994).

In a study by Mutus et al. (2006), dietary supplementation with *Bacillus licheniformis* and *Bacillus subtilis* resulted in significantly greater medial and lateral wall thickness of the tibia compared to the birds on the control diet. This supported an earlier study by Plavnik and Scott (1980) who reported an increase in bone strength and lower incidence of tibial dyschondroplasia in chickens receiving brewer's yeast. The exact mechanism how the supplemental probiotics lead to an increase in chicken bone strength is not yet fully elucidated. A positive correlation between *Lactobacillus* supplementation and phosphorus and calcium retention has been observed, suggesting the increased retained phosphorous and calcium might be deposited in the skeleton (Nahashon et al. 1994).

Table 14.3 In vitro effects of probiotic strains

Probiotic strain	Cell type	Effect	Ref
L. reuteri	RAW 264.7	↓ OC differentiation	Britton et al. (2014)
	THP-1	↓TNF production	Thomas et al. (2012)
L. casei-fermented milk	MC3T3-E1	↑ Cell proliferation	Kim et al. (2009a, b)
L. helveticus-fermented milk	Primary osteoblasts	↑ Ca accumulation	Narva et al. (2004b)
	Primary osteoclasts	No effect	

14.4.4 Probiotic In Vitro Studies

To determine if probiotic secretory products can directly affect bone cells, in vitro studies have been utilized (Table 14.3). Osteoclast differentiation from RAW 264.7 cells was significantly inhibited when the cells were cultured with *L. reuteri* conditioned media, suggesting that the probiotic releases a soluble factor that is important for reducing bone turnover (Britton et al. 2014). This is supported by an earlier study which demonstrated that *L. reuteri* secretes histamine which is capable of suppressing TNF production from human monocytoid cells (Thomas et al. 2012). The positive effects of *L. helveticus* and *L. casei* on BMD have also been confirmed in in vitro studies. Cultures of *L. casei*-fermented milk were demonstrated to increase the osteoblast precursor MC3T3-E1 cell proliferation in a dose-dependent manner (Kim et al. 2009a, b). The addition of *L. helveticus*-fermented milk products to cultures of primary bone marrow (BM) cells isolated from 8 to 12 week female mice resulted in a 1.3–1.4 increase in calcium accumulation in the osteoblast cultures compared to the controls suggesting their potential to increase osteoblast differentiation. Interestingly, the products of *L. helveticus*-fermented milk had no significant effect on osteoclast differentiation (Narva et al. 2004b).

14.4.5 Possible Mechanisms of Probiotic Action

While the exact mechanisms by which probiotics exert a beneficial effect on the bone are not fully known, progress has been made. Studies in GF mice have shown that the absence of the gut microbiota leads to increased bone mass, which is associated with a decrease in osteoclast precursors and $CD4^+$ T cells in the bone marrow. Upon colonizing the GF mice with a normal gut microbiota, the bone mass returns to the level of conventional mice as does the numbers of osteoclast precursors and $CD4^+$ T cells (Sjögren et al. 2012). This study demonstrated that the gut microbiota plays an important role in shaping the systemic immune system and subsequently bone health.

The mechanism by which probiotic bacteria exert a positive influence on the bone is most likely multifaceted. Studies have shown that the bacteria have the ability to synthesize vitamins and enzymes required for matrix formation and bone growth, including vitamins D, K, C, and folate (Crittenden et al. 2003; Arunachalam 1999). In addition, *bifidobacteria* are able to produce SCFAs that can reduce the pH of the intestinal tract and increase the solubility and absorption of minerals (Campbell et al. 1997).

The studies with *L. reuteri* 6475 demonstrated that the probiotic is capable of reducing expression of pro-inflammatory cytokines and osteoclastogenic cytokines systemically, in the bone marrow and in the ileum and jejunum (McCabe et al. 2013; Britton et al. 2014). Reducing epithelial cell inflammation in the gastrointestinal tract may directly enhance the transport of calcium and other ions across the gut barrier; alternatively, the *L. reuteri* could impact bone health by increasing calcium solubility (McCabe et al. 2013). Inhibition of TNF expression by *L. reuteri* secreted factors (Thomas et al. 2012) may also play a big role in the probiotic modulation of bone health, as TNF can directly and indirectly stimulate osteoclastogenesis and elevation in its levels are associated with number of bone diseases including osteoporosis (Bismar et al. 1995; Pacifici et al. 1991; Kobayashi et al. 2000; Azuma et al. 2000).

It is becoming apparent that different probiotics can have different modes of action. *L. helveticus* has been shown to produce the bioactive peptides isoleucyl-prolyl-proline (IPP) and valyl-prolyl-proline (VPP). These peptides are known to inhibit the angiotensin-converting enzyme (ACE), subsequently preventing the formation of angiotensin II (Ang II), a stimulator of osteoclasts resorption, from angiotensin I (Ang I) (Narva et al. 2004a, b). Other mechanisms through which *L. helveticus* may enhance bone density include increased calcium intake and better mineral absorption (Narva et al. 2004a).

Estrogen deficiency, as observed in menopause, is associated with increased oxidative stress in the bone. This increase in oxidative stress can inhibit osteoblastic differentiation and participates in the upregulation of osteoclast differentiation (Bai et al. 2004; Suda et al. 1993; Manolagas 2010). Excess reactive oxygen species connected with oxidative stress also induce the expression of inducible nitric oxide synthase (iNOS), an inflammatory mediator of bone loss (Cuzzocrea et al. 2003). *Bifidobacterium longum* reduces periodontal oxidative stress by decreasing NF-κB gene expression (Tomofuji et al. 2012), potentially inhibiting osteoclasts differentiation while stimulating osteoblasts differentiation. In addition, it has also been shown to downregulate iNOS gene expression in the periodontal tissue.

14.5 Conclusion and Future Research

Gut microbiota have a symbiotic relationship with the host. Dysbiosis of gut microbiota can lead to pathologies such as obesity, diabetes, metabolic syndrome, and inflammatory bowel disease. Recently, many studies report that the composition, richness, and diversity of gut microbiota are influenced by diet and dietary patterns, which in turn influence human health. The gut microbiota can secrete various enzymes which are not secreted in the digestive system of the host and so dietary constituents which did not get absorbed in the small intestine can undergo subsequent break down by fermentation in the large intestine. This may lead to relatively higher growth of the gut microbiota, which use the metabolites as an energy source.

Therefore, foods that promote the growth of beneficial gut microbiota may be the very ones that promote health benefits (Fig. 14.1).

The bone may be one tissue that benefits from this symbiotic relationship between the gut microbiome and the host though it is far less studied than gut health. Plant-based foods like fruits, vegetables, and grains contain many kinds of oligosaccharides as well as dietary fibers (Jovanovic-Malinovska et al. 2014; Costabile et al. 2008). These prebiotics enhance absorption of minerals required for bone formation including calcium through the enlarged absorptive surface area of small intestine and reduced pH by SCFA production. SCFAs, a by-product of gut fermentation, may also influence the bone through their effect on the systemic immune system or as signaling molecules that regulate bone metabolism. Besides prebiotics found naturally in plant foods, commercial prebiotics produced by enzymatic processes are also available, such as those found in fermented foods. Few investigators have studied the relationship between consumption of prebiotics or fermented foods, the changes in the gut microbiota community, and the subsequent effects on the bone (Weaver et al. 2011; Whisner et al. 2013). However, new research is highlighting the potential importance of the gut-bone axis though more research is needed to understand the mechanisms involved.

In addition to prebiotics, research into the supplementation of the gut microbiota with beneficial probiotic bacteria has recently gained traction. While past studies have concluded that probiotics can have a beneficial effect on BMD and BMC, countering the adverse effects observed in conditions such as osteoporosis, much work still remains to be done. Whether all probiotics have a positive effect on the bone or whether it is only specific strains remains to be seen. While current animal studies have highlighted potential mechanisms, further work is required to determine whether these translate into humans.

Acknowledgments The writing of this review and research in the author's labsis supported by funding from the National Institutes of Health Grants: DK101050 (LRM), AT007695 (LRM) and Berries and Bones R01AT008754 (CMW).

References

Abreu MT. Toll-like receptor signalling in the intestinal epithelium: how bacterial recognition shapes intestinal function. Nat Rev Immunol. 2010;10(2):131–44. Available at: http://dx.doi.org/10.1038/nri2707.

Alekel DL, et al. The Soy Isoflavones for Reducing Bone Loss (SIRBL) study: a 3-y randomized controlled trial in postmenopausal women. Am J Clin Nutr. 2010;91(1):218–30.

Alpan O. Oral tolerance and gut-oriented immune response to dietary proteins. Curr Allergy Asthma Rep. 2001;1(6):572–7.

Amdekar S, et al. Lactobacillus protected bone damage and maintained the antioxidant status of liver and kidney homogenates in female wistar rats. Mol Cell Biochem. 2012;368:155–65.

Andersen TL, et al. A physical mechanism for coupling bone resorption and formation in adult human bone. Am J Pathol. 2009;174(1):239–47. Available at: http://dx.doi.org/10.2353/ajpath.2009.080627. [Accessed 28 Oct 2013].

Arunachalam KD. Role of bifidobacteria in nutrition, medicine and technology. Nutr Res. 1999;19(10):1559–97.
Azuma Y, et al. Tumor necrosis factor-alpha induces differentiation of and bone resorption by osteoclasts. J Biol Chem. 2000;275(7):4858–64. Available at: http://www.ncbi.nlm.nih.gov/pubmed/10671521. Accessed 28 Oct 2013.
Bai XC, et al. Oxidative stress inhibits osteoblastic differentiation of bone cells by ERK and NF-??B. Biochem Biophys Res Commun. 2004;314:197–207.
Beermann C, Hartung J. Physiological properties of milk ingredients released by fermentation. Food Funct. 2013;4:185–99.
Bested AC, Logan AC, Selhub EM. Intestinal microbiota, probiotics and mental health: from Metchnikoff to modern advances: part II – contemporary contextual research. Gut Pathog. 2013;5(1):3. Available at: http://www.pubmedcentral.nih.gov/articlerender.fcgi?artid=3601973&tool=pmcentrez&rendertype=abstract. Accessed 16 Nov 2014.
Bismar H, et al. Increased cytokine secretion by human bone marrow cells after menopause or discontinuation of estrogen replacement. J Clin Endocrinol Metab. 1995;80(11):3351–5. Available at: http://www.ncbi.nlm.nih.gov/pubmed/7593450. Accessed 20 Nov 2013.
Bonaz BL, Bernstein CN. Brain-gut interactions in inflammatory bowel disease. Gastroenterology. 2013;144(1):36–49.
Bonewald LF, Johnson ML. Osteocytes, mechanosensing and Wnt signaling. Bone. 2008;42(4):606–15. Available at: http://www.pubmedcentral.nih.gov/articlerender.fcgi?artid=2349095&tool=pmcentrez&rendertype=abstract. Accessed 31 Oct 2013.
Britton RA, et al. Probiotic L. reuteri treatment prevents bone loss in a menopausal ovariectomized mouse model. J Cell Physiol. 2014;229(11):1822–30. Available at" http://www.ncbi.nlm.nih.gov/pubmed/24677054. Accessed 29 Dec 2014.
Bucay N, et al. Osteoprotegerin-deficient mice develop early onset osteoporosis and arterial calcification. Genes Dev. 1998;12(9):1260–8. Available at: http://www.pubmedcentral.nih.gov/articlerender.fcgi?artid=316769&tool=pmcentrez&rendertype=abstract. Accessed 31 Oct 2013.
Campbell JM, Fahey GC, Wolf BW. Selected indigestible oligosaccharides affect large bowel mass, cecal and fecal short-chain fatty acids, pH and microflora in rats. J Nutr. 1997;127:130–6.
Caplice E, Fitzgerald GF. Food fermentations: role of microorganisms in food production and preservation. Int J Food Microbiol. 1999;50(1–2):131–49.
Cappuccio FP, et al. High blood pressure and bone-mineral loss in elderly white women: a prospective study. Study of Osteoporotic Fractures Research Group. Lancet. 1999;354(9183):971–5.
Cassidy A. Factors affecting the bioavailability of soy isoflavones in humans. J AOAC Int. 2006;89(4):1182–8.
Chandran P, et al. Inflammatory bowel disease: dysfunction of GALT and gut bacterial flora (I). Surg J R Coll Surg Edinb Irel. 2003;1(2):63–75.
Chen TR, Su RQ, Wei QK. Hydrolysis of isoflavone phytoestrogens in soymilk fermented by lactobacillus and bifidobacterium cocultures. J Food Biochem. 2010;34(1):1–12.
Chen JR, et al. Soy protein isolates prevent loss of bone quantity associated with obesity in rats through regulation of insulin signaling in osteoblasts. FASEB J. 2013;27(9):3514–23.
Cheung ALTF, et al. Fermentation of calcium-fortified soya milk does not appear to enhance acute calcium absorption in osteopenic post-menopausal women. Br J Nutr. 2011;105(2):282–6.
Chiang SS, Pan TM. Antiosteoporotic effects of lactobacillus-fermented soy skim milk on bone mineral density and the microstructure of femoral bone in ovariectomized mice. J Agric Food Chem. 2011;59:7734–42.
Chiang SS, Pan TM. Beneficial effects of phytoestrogens and their metabolites produced by intestinal microflora on bone health. Appl Microbiol Biotechnol. 2013;97(4):1489–500.
Claesson MJ, et al. Gut microbiota composition correlates with diet and health in the elderly. 2012. Available at: http://dx.doi.org/10.1038/nature11319.
Clemens TL, Karsenty G. The osteoblast: an insulin target cell controlling glucose homeostasis. J Bone Miner Res. 2011;26(4):677–80.
Cochran DL. Inflammation and bone loss in periodontal disease. J Periodontol. 2008;79(8 Suppl):1569–76.

Cook SI, Sellin JH. Review article: short chain fatty acids in health and disease. Aliment Pharmacol Ther. 1998;12(6):499–507.
Costabile A, et al. Whole-grain wheat breakfast cereal has a prebiotic effect on the human gut microbiota: a double-blind, placebo-controlled, crossover study. Br J Nutr. 2008;99(1):110–20.
Cotillard A, et al. Dietary intervention impact on gut microbial gene richness. Nature. 2013;500(7464):585–8. Available at: http://www.ncbi.nlm.nih.gov/pubmed/23985875.
Crittenden RG, Martinez NR, Playne MJ. Synthesis and utilisation of folate by yoghurt starter cultures and probiotic bacteria. Int J Food Microbiol. 2003;80:217–22.
Cummings JH, et al. The effect of meat protein and dietary fiber on colonic function and metabolism. II. Bacterial metabolites in feces and urine. Am J Clin Nutr. 1979;32(10):2094–101.
Cummings JH, et al. Short chain fatty acids in human large intestine, portal, hepatic and venous blood. Gut. 1987;28(10):1221–7.
Cuzzocrea S, et al. Inducible nitric oxide synthase mediates bone loss in ovariectomized mice. Endocrinology. 2003;144:1098–107.
D'Amelio P, Grimaldi A, Pescarmona GP, Tamone C, Roato I, Isaia G. Spontaneous osteoclast formation from peripheral blood mononuclear cells in postmenopausal osteoporosis. FASEB J. 2005;19(3):410–2.
D'Amelio P, et al. Estrogen deficiency increases osteoclastogenesis up-regulating T cells activity: a key mechanism in osteoporosis. Bone. 2008;43(1):92–100. Available at: http://www.ncbi.nlm.nih.gov/pubmed/18407820. Accessed 28 Oct 2013.
David LA, et al. Diet rapidly and reproducibly alters the human gut microbiome. Nature. 2014;505(7484):559–63. Available at: http://www.pubmedcentral.nih.gov/articlerender.fcgi?artid=3957428&tool=pmcentrez&rendertype=abstract.
De Vrese M, Schrezenmeir J. Probiotics, prebiotics, and synbiotics. Adv Biochem Eng Biotechnol. 2008;111:1–66. Available at: http://www.ncbi.nlm.nih.gov/pubmed/18461293.
Food and Agriculture Organization of the United Nations (FAO). Report of a joint FAO/WHO working group on drafting guide- lines for the evaluation of probiotics in food. London, Ontario, Canada, 30 April–1 May 2002. http://www.who.int/foodsafety/fs_management/en/probiotic_guidelines.pdf.
Fonseca D, Ward WE. Daidzein together with high calcium preserve bone mass and biomechanical strength at multiple sites in ovariectomized mice. Bone. 2004;35(2):489–97.
Foureaux RDC, et al. Effects of probiotic therapy on metabolic and inflammatory parameters of rats with ligature-induced periodontitis associated with restraint stress. J Periodontol. 2013;0:1–15.
Frankel WL, et al. Mediation of the trophic effects of short-chain fatty acids on the rat jejunum and colon. Gastroenterology. 1994;106(2):375–80.
Fujioka M, et al. Equol, a metabolite of daidzein, inhibits bone loss in ovariectomized mice. J Nutr. 2004;134(10):2623–7.
Ghanem KZ, Badawy IH, Abdel-Salam AM. Influence of yoghurt and probiotic yoghurt on the absorption of calcium, magnesium, iron and bone mineralization in rats. Milchwissenschaft. 2004;59(9–10):472–5. Available at: http://cat.inist.fr/?aModele=afficheN&cpsidt=16082591. Accessed 20 Jan 2015.
Gill SR, et al. Metagenomic analysis of the human distal gut microbiome. Science (New York, NY). 2006;312(5778):1355–9.
Gough AK, et al. Generalised bone loss in patients with early rheumatoid arthritis. Lancet. 1994;344(8914):23–7.
Haron H, et al. Absorption of calcium from milk and tempeh consumed by postmenopausal Malay women using the dual stable isotope technique. Int J Food Sci Nutr. 2010;61(2):125–37.
Harvey N, Dennison E, Cooper C. Osteoporosis: impact on health and economics. Nat Rev Rheumatol. 2010;6(2):99–105. Available at: http://dx.doi.org/10.1038/nrrheum.2009.260.
Ibnou-Zekri N, et al. Divergent patterns of colonization and immune response elicited from two intestinal lactobacillus strains that display similar properties in vitro. Infect Immun. 2003;71(1):428–36. Available at: http://iai.asm.org/cgi/doi/10.1128/IAI.71.1.428-436.2003.

Ing-Lorenzini K, et al. Low-energy femoral fractures associated with the long-term use of bisphosphonates: a case series from a Swiss university hospital. Drug Saf Int J Med Toxicol Drug Experience. 2009;32(9):775–85.

Jilka RL, et al. Increased osteoclast development after estrogen loss: mediation by interleukin-6. Science (New York, NY). 1992;257(5066):88–91. Available at: http://www.ncbi.nlm.nih.gov/pubmed/1621100. Accessed 9 Jan 2014.

Jin LZ, et al. Probiotics in poultry: modes of action. World Poult Sci Assoc. 1997;53:351–63.

Jovanovic-Malinovska R, Kuzmanova S, Winkelhausen E. Oligosaccharide profile in fruits and vegetables as sources of prebiotics and functional foods. Int J Food Prop. 2014;17(5):949–65. Available at: http://www.tandfonline.com/doi/abs/10.1080/10942912.2012.680221.

Kalu DN. The ovariectomized rat model of postmenopausal bone loss. Bone Miner. 1991;15(3):175–91. Available at: http://www.ncbi.nlm.nih.gov/pubmed/11502475.

Kamau SM, et al. Functional significance of bioactive peptides derived from milk proteins. Food Rev Intl. 2010;26(4):386–401.

Karsenty G, Kronenberg HM, Settembre C. Genetic control of bone formation. Annu Rev Cell Dev Biol. 2009;25:629–48. Available at: http://www.ncbi.nlm.nih.gov/pubmed/19575648. Accessed 29 Jan 2014.

Kasukawa Y, et al. Effects of risedronate alone or combined with vitamin K2on serum undercarboxylated osteocalcin and osteocalcin levels in postmenopausal osteoporosis. J Bone Miner Metab. 2014;32(3):290–7.

Katsuyama H, et al. Promotion of bone formation by fermented soybean (Natto) intake in premenopausal women. J Nutr Sci Vitaminol. 2004;50(2):114–20.

Kelsall BL, Leon F. Involvement of intestinal dendritic cells in oral tolerance, immunity to pathogens, and inflammatory bowel disease. Immunol Rev. 2005;206:132–48.

Kendler DL, et al. Effects of denosumab on bone mineral density and bone turnover in postmenopausal women transitioning from alendronate therapy. J Bone Miner Res Off J Am Soc Bone Miner Res. 2010;25(1):72–81.

Khan RU, Naz S. The applications of probiotics in poultry production. Worlds Poult Sci J. 2013;69(03):621–32. Available at: http://www.journals.cambridge.org/abstract_S0043933913000627. Accessed 13 Jan 2015.

Kim DW, et al. Soy isoflavones mitigate long-term femoral and lumbar vertebral bone loss in middle-aged ovariectomized mice. J Med Food. 2009a;12(3):536–41.

Kim JG, et al. Effects of a Lactobacillus casei 393 fermented milk product on bone metabolism in ovariectomised rats. Int Dairy J. 2009b;19(11):690–5. Available at: http://dx.doi.org/10.1016/j.idairyj.2009.06.009.

Kobayashi K, et al. Tumor necrosis factor alpha stimulates osteoclast differentiation by a mechanism independent of the ODF/RANKL-RANK interaction. J Exp Med. 2000;191(2):275–86. Available at: http://www.pubmedcentral.nih.gov/articlerender.fcgi?artid=2195746&tool=pmcentrez&rendertype=abstract. Accessed 29 Jan 2014.

Kruger MC, et al. The effect of Lactobacillus rhamnosus HN001 on mineral absorption and bone health in growing male and ovariectomised female rats. Dairy Sci Technol. 2009;89:219–31.

Le Chatelier E, et al. Richness of human gut microbiome correlates with metabolic markers. Nature. 2013;500(7464):541–6. Available at: http://www.ncbi.nlm.nih.gov/pubmed/23985870.

Lee YS, Noguchi T, Naito H. Phosphopeptides and soluble calcium in the small intestine of rats given a casein diet. Br J Nutr. 1980;43(3):457–67.

Legette LL, et al. Supplemental dietary racemic equol has modest benefits to bone but has mild uterotropic activity in ovariectomized rats. J Nutr. 2009;139(10):1908–13.

Lorenzo J, Horowitz M, Choi Y. Osteoimmunology: interactions of the bone and immune system. Endocr Rev. 2008;29(4):403–40.

Macfarlane GT, Steed H, Macfarlane S. Bacterial metabolism and health-related effects of galacto-oligosaccharides and other prebiotics. J Appl Microbiol. 2008;104(2):305–44.

Manolagas SC. From estrogen-centric to aging and oxidative stress: a revised perspective of the pathogenesis of osteoporosis. Endocr Rev. 2010;31:266–300.

McCabe LR, et al. Probiotic use decreases intestinal inflammation and increases bone density in healthy male but not female mice. J Cell Physiol. 2013;228(8):1793–8. Available at: http://www.ncbi.nlm.nih.gov/pubmed/23389860. Accessed 6 Nov 2013.

Messora MR, et al. Probiotic therapy reduces periodontal tissue destruction and improves the intestinal morphology in rats with ligature-induced periodontitis. J Periodontol. 2013;84:1818–26. Available at: http://www.ncbi.nlm.nih.gov/pubmed/23327675.

Metchnikoff E. Essais optimistes. In: The prolongation of life. Optimistic studies. Translated and edited by P. Chalmers Mitchell. London: Heinemann; 1907.

Mutuş R, et al. The effect of dietary probiotic supplementation on tibial bone characteristics and strength in broilers. Poult Sci. 2006;85(9):1621–5. Available at: http://www.ncbi.nlm.nih.gov/pubmed/16977848. Accessed 6 Jan 2015.

Nagpal R, et al. Bioactive peptides derived from milk proteins and their health beneficial potentials: an update. Food Funct. 2011;2(1):18–27.

Nahashon SN, Nakaue HS, Mirosh LW. Production variables and nutrient retention in single comb White Leghorn laying pullets fed diets supplemented with direct-fed microbials. Poult Sci. 1994;73(11):1699–711. Available at: http://www.ncbi.nlm.nih.gov/pubmed/7862610. Accessed 12 Jan 2015.

Nam YD, et al. Comparative analysis of korean human gut microbiota by barcoded pyrosequencing. PLoS ONE. 2011;6(7):e22109.

Narva M, Collin M, et al. Effects of long-term intervention with lactobacillus helveticus-fermented milk on bone mineral density and bone mineral content in growing rats. Ann Nutr Metab. 2004a;48:228–34.

Narva M, Halleen J, et al. Effects of Lactobacillus helveticus fermented milk on bone cells in vitro. Life Sci. 2004b;75:1727–34.

National Osteoporosis Foundation. What is osteoporosis and what causes it? [Internet]. National Osteoporosis Foundation Website. Available from: https://www.nof.org/patients/what-isosteoporosis/.

Nava GM, et al. Probiotic alternatives to reduce gastrointestinal infections: the poultry experience. Anim Health Res Rev Conf Res Workers Anim Dis. 2005;6(1):105–18. Available at: http://www.journals.cambridge.org/abstract_S1466252305000071. Accessed 12 Jan 2015.

Ohlsson C, et al. Probiotics protect mice from ovariectomy-induced cortical bone loss. PLoS One. 2014;9(3):e92368. Available at: http://www.pubmedcentral.nih.gov/articlerender.fcgi?artid=3956931&tool=pmcentrez&rendertype=abstract. Accessed 18 Apr 2014.

Oozeer R, et al. Gut health: predictive biomarkers for preventive medicine and development of functional foods. Br J Nutr. 2010;103(10):1539–44.

Ouwehand AC, Salminen S, Isolauri E. Probiotics: an overview of beneficial effects. Antonie Van Leeuwenhoek. 2002;82(1–4):279–89. Available at: http://www.ncbi.nlm.nih.gov/pubmed/12369194. Accessed 29 Dec 2014.

Pacifici R, et al. Effect of surgical menopause and estrogen replacement on cytokine release from human blood mononuclear cells. Proc Natl Acad Sci U S A. 1991;88(12):5134–8. Available at: http://www.pubmedcentral.nih.gov/articlerender.fcgi?artid=51826&tool=pmcentrez&rendertype=abstract. Accessed 28 Oct 2013.

Papapoulos SE. Bisphosphonates: how do they work? Best Pract Res Clin Endocrinol Metab. 2008;22(5):831–47. Available at: http://www.ncbi.nlm.nih.gov/pubmed/19028359. Accessed 6 Oct 2013.

Payne JM. Metabolic diseases in farm animals. London: Butterworth-Heinemann; 1977.

Plavnik I, Scott ML. Effects of additional vitamins, minerals, or brewer's yeast upon leg weaknesses in broiler chickens. Poult Sci. 1980;59(2):459–67. Available at: http://www.ncbi.nlm.nih.gov/pubmed/7413573. Accessed 12 Jan 2015.

Reinwald S, Weaver CM. Soy components vs. whole soy: are we betting our bones on a long shot? J Nutr. 2010;140(12):2312S–7S.

Robles Alonso V, Guarner F. Linking the gut microbiota to human health. Br J Nutr. 2013;109(Suppl):S21–6. Available at: http://www.ncbi.nlm.nih.gov/pubmed/23360877.

Rodrigues FC, et al. Yacon flour and Bifidobacterium longum modulate bone health in rats. J Med Food. 2012;15(7):664–70. Available at: http://www.ncbi.nlm.nih.gov/pubmed/22510044. Accessed 6 Jan 2015.

Rosen CJ, Bilezikian JP. Clinical review 123: anabolic therapy for osteoporosis. J Clin Endocrinol Metab. 2001;86(3):957–64. Available at: http://www.ncbi.nlm.nih.gov/pubmed/11238469. Accessed 8 Nov 2013

Rovenský J, et al. The effects of Enterococcus faecium and selenium on methotrexate treatment in rat adjuvant-induced arthritis. Clin Dev Immunol. 2004;11:267–73.

Rowland IR, Wiseman H, Sanders TA, Adlercreutz H, Bowey EA. Interindividual variation in metabolism of soy isoflavones and lignans: influence of habitual diet on equol production by the gut microflora. Nutr Cancer. 2000;36(1):27–32.

Sanders ME. Probiotics: definition, sources, selection, and uses. Clin Infect Dis Off Publ Infect Dis Soc Am. 2008;46(Suppl 2):S58–S61; discussion S144–51. Available at: http://www.ncbi.nlm.nih.gov/pubmed/18181724. Accessed 29 Dec 2014.

Sartor RB. Therapeutic manipulation of the enteric microflora in inflammatory bowel diseases: antibiotics, probiotics, and prebiotics. Gastroenterology. 2004;126(6):1620–33. Available at: http://linkinghub.elsevier.com/retrieve/pii/S0016508504004561. Accessed 24 Nov 2014.

Savage DC. Microbial ecology of the gastrointestinal tract. Annu Rev Microbiol. 1977;31:107–33.

Schwarz EM, Ritchlin CT. Clinical development of anti-RANKL therapy. Arthritis Res Ther. 2007;6:S7. Available at: http://www.pubmedcentral.nih.gov/articlerender.fcgi?artid=1924522&tool=pmcentrez&rendertype=abstract. Accessed 6 Oct 2013.

Shannon J, et al. Bisphosphonates and osteonecrosis of the jaw. J Am Geriatr Soc. 2011;59:2350–5.

Sjögren K, et al. The gut microbiota regulates bone mass in mice. J Bone Miner Res. 2012;27(6):1357–67.

Steer TE, et al. Metabolism of the soybean isoflavone glycoside genistin in vitro by human gut bacteria and the effect of prebiotics. Br J Nutr. 2003;90(3):635–42.

Suda N, et al. Participation of oxidative stress in the process of osteoclast differentiation. Biochim Biophys Acta. 1993;1157:318–23.

Sullivan TW. Skeletal problems in poultry: estimated annual cost and descriptions. Poult Sci. 1994;73(6):879–82. Available at: http://www.ncbi.nlm.nih.gov/pubmed/8072932. Accessed 6 Jan 2015.

Teitelbaum SL. Osteoclasts: what do they do and how do they do it? Am J Pathol. 2007;170(2):427–35. Available at: http://www.pubmedcentral.nih.gov/articlerender.fcgi?artid=1851862&tool=pmcentrez&rendertype=abstract. Accessed 28 Oct 2013.

Tenorio MD, et al. Soybean whey enhance mineral balance and caecal fermentation in rats. Eur J Nutr. 2010;49(3):155–63.

Theill LE, Boyle WJ, Penninger JM. RANK-L and RANK: T cells, bone loss, and mammalian evolution. Annu Rev Immunol. 2002;20:795–823.

Thomas CM, et al. Histamine derived from probiotic Lactobacillus reuteri suppresses TNF via modulation of PKA and ERK signaling. PLoS One. 2012;7(2):e31951. Available at: http://www.pubmedcentral.nih.gov/articlerender.fcgi?artid=3285189&tool=pmcentrez&rendertype=abstract. Accessed 6 Nov 2013.

Tissier H. Traitement des infections intestinales par la méthode de la flore bactérienne de l'intestin. CR Soc Biol. 1906;60:359–61.

Tomofuji T, et al. Supplementation of broccoli or Bifidobacterium longum-fermented broccoli suppresses serum lipid peroxidation and osteoclast differentiation on alveolar bone surface in rats fed a high-cholesterol diet. Nutr Res. 2012;32(4):301–7. Available at: http://dx.doi.org/10.1016/j.nutres.2012.03.006.

Trompette A, et al. Gut microbiota metabolism of dietary fiber influences allergic airway disease and hematopoiesis. Nat Med. 2014;20(2):159–66. Available at: http://www.nature.com.proxy.lib.uiowa.edu/nm/journal/v20/n2/full/nm.3444.html.

U.S. Department of Health and Human Services. Bone health and osteoporosis: a report of the surgeon general. Rockville: U.S. Department of Health and Human Services/Office of the Surgeon General; 2004.

Villegas R, et al. Legume and soy food intake and the incidence of type 2 diabetes in the Shanghai Women's Health Study. Am J Clin Nutr. 2008;87(1):162–7.

Weaver CM, Legette LL. Equol, via dietary sources or intestinal production, may ameliorate estrogen deficiency-induced bone loss. J Nutr. 2010;140(7):1377S–9S.

Weaver CM, et al. Galactooligosaccharides improve mineral absorption and bone properties in growing rats through gut fermentation. J Agric Food Chem. 2011;59(12):6501–10.

Weaver CM, et al. Flavonoid intake and bone health. J Nutr Gerontol Geriatr. 2012;31(3):239–53.

Whisner CM, et al. Galacto-oligosaccharides increase calcium absorption and gut bifidobacteria in young girls: a double-blind cross-over trial. Br J Nutr. 2013;110(7):1292–303. Available at: http://www.ncbi.nlm.nih.gov/pubmed/23507173.

Whisner CM, et al. Soluble maize fibre affects short-term calcium absorption in adolescent boys and girls: a randomised controlled trial using dual stable isotopic tracers. Br J Nutr. 2014;112(3):446–56. Available at: http://www.ncbi.nlm.nih.gov/pubmed/24848974.

Wu AH, et al. Soy intake and risk of breast cancer in Asians and Asian Americans. Am J Clin Nutr. 1998;68(6 Suppl):1437S–43S.

Wu GD, et al. Linking long-term dietary patterns with gut microbial enterotypes. Science. 2011;334(6052):105–8.

Yan L, Spitznagel EL. Soy consumption and prostate cancer risk in men: a revisit of a meta-analysis. Am J Clin Nutr. 2009;89(4):1155–63.

Yasui T, et al. Epigenetic regulation of osteoclast differentiation. Ann N Y Acad Sci. 2011;1240(1):7–13.

Yatsunenko T, et al. Human gut microbiome viewed across age and geography. Nature. 2012;486:222–7.

Yun TJ, et al. OPG/FDCR-1, a TNF receptor family member, is expressed in lymphoid cells and is up-regulated by ligating CD40. J Immunol. 1998;161(11):6113–21.

Zhang X, et al. Soy food consumption is associated with lower risk of coronary heart disease in Chinese women. J Nutr. 2003;133(9):2874–8.

Zhang X, et al. Prospective cohort study of soy food consumption and risk of bone fracture among postmenopausal women. Arch Intern Med. 2005;165(16):1890–5.

Zhao Y, Martin BR, Weaver CM. Calcium bioavailability of calcium carbonate fortified soymilk is equivalent to cow's milk in young women. J Nutr. 2005;135(10):2379–82.

Chapter 15
Bone and Energy Metabolism

Clifford J. Rosen

Abstract There is an intimate relationship between fat and bone metabolism that has only recently been appreciated. Adipose depots regulate energy utilization in coordination with the liver and skeletal muscle through an intricate series of local, systemic, and neuronal mediators. Although adipose tissue is one of the largest organs in the body, its functions vary by location and origin. Adipocytes can act in an autocrine manner to regulate energy balance by sequestering triglycerides and subsequently, depending on demand, releasing fatty acids through lipolysis for energy utilization, and in some cases through uncoupling protein 1 for generating heat. Adipose tissue can also act in an endocrine or paracrine manner by releasing adipokines that modulate the function of other organs. Bone is one of those target tissues, although the skeleton can also reciprocate by releasing its own factors that modulate adipose tissue function and pancreatic islet cells production of insulin in the pancreas. The central nervous system is the final common pathway modulating these signals, both in the skeleton and in adipose tissue. Disruption in this complex regulatory circuit and its downstream tissues is manifested in a wide range of metabolic disorders, including type 2 diabetes mellitus (T2D). The aim of this chapter is to summarize our knowledge of common determinants in bone and adipose function.

Keywords Adipose • Marrow • Metabolic disease • PPAR • FGF • Leptin • Adiponectin • Serotonin • Glucocorticoids • Adrenergic • Osteocalcin • ATP

Supported by Funding NIDDK: R24DK092759-05 (CJR) and NIAMS R21 AR066120-01 (CJR)

C.J. Rosen, MD (✉)
Maine Medical Center Research Institute, Scarborough, ME, USA

Tufts University School of Medicine, Boston, MA, USA
e-mail: cjrofen@gmail.com

S.Y. Smith et al. (eds.), *Bone Toxicology*, Molecular and Integrative Toxicology,
DOI 10.1007/978-3-319-56192-9_15

15.1 Introduction

Energy homeostasis is a vital component of mammalian life and has evolved over millions of years. Traditional perspectives have focused on the role of the adipocyte in managing energy demands through the "yin and yang" of lipid storage and lipolysis. Tissues that have traditionally been recognized for high energy needs have been cardiac and skeletal muscle as well as brown adipose tissue. More recently, attention has turned to the skeleton in part because of the huge energy "sump" required during bone remodeling and because of the presence of marrow adipocytes. Moreover, skeletal complications have recently been recognized as another of the several comorbidities associated with diseases of energy utilization such as type 1 (T1D) and type 2 (T2D) diabetes mellitus (Schwartz 2015). Clinical studies suggest that disordered glucose and lipid metabolism have a profound effect on bone, and this had led investigators to explore the relationship between energy requirements in bone and adipose tissue. However, the regulation of energy metabolism relative to substrate utilization and the consequences of disordered utilization in respect to the skeleton are only now beginning to be elucidated. Understanding the complexity of these homeostatic systems begins by addressing the biology of both the adipocyte and the osteoblast.

15.2 Adipocyte Biology and Its Relationship to Bone

Adipocytes and osteoblasts arise from the same MSC progenitor but are distinct in many ways beyond the specific transcription factors that determine fate (see Fig. 15.1). Adipose tissue consists of distinct morphological characteristics defined by local, systemic, and environmental signals. The major building blocks of adipose tissue are adipocytes, highly specialized cells that control fuel management through a continuous process of lipogenesis, lipid storage, and lipolysis. Adipose tissue also consists of other cell types, with macrophages playing an important role in managing the inflammatory environment that impacts adipocyte development, function, and insulin sensitivity. Unlike other cells, the adipocyte is particularly well suited for energy storage through lipid droplets which can expand or contract depending on the energy demands of the organism. However, during adipocyte expansion, cytokines and adipokines are released that can influence other tissues as well as the immune response. This is particularly true for type I macrophages which are attracted to the stromal tissue surround the adipocyte by the release of pro-inflammatory peptides like IL-1, IL-6, and TNF. Furthermore, lipids released during expansion of adipocytes such as lysophosphatidylcholine (LPC) can induce an innate immune response by binding to receptors on T-helper cells (Chang et al. 2008). It is not clear whether these factors could have a chronic effect on the skeleton as well as the cardiovascular system and the vasculature, resulting in higher atherosclerotic risk.

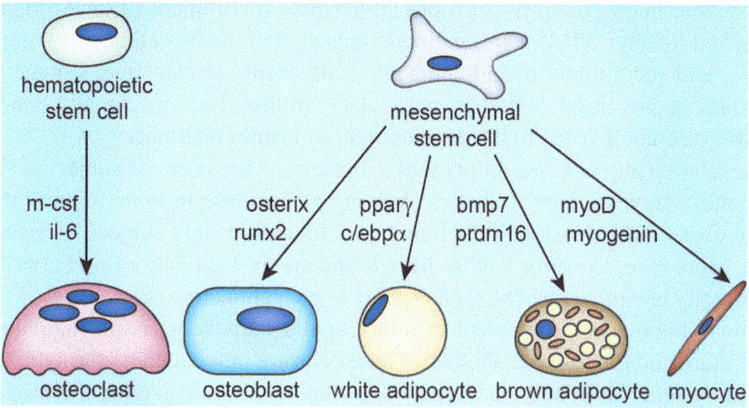

Fig. 15.1 The MSC progenitor can differentiate into several cell types depending on transcription factors and bioenergetics

Another unique feature of the adipose organ is that its distribution is directly associated with its function (Giralt and Villarroya 2013). Thus, subcutaneous fat provides insulation for internal organs and is more insulin sensitive, while adipose tissue surrounding the heart has thermogenic characteristics, and visceral fat is more pro-inflammatory and insulin resistant. Adipose tissue is also present in the bone marrow and our understanding of its function is just emerging. In that context, we juxtaposed bone and fat because the regulation of mesenchymal stem cell (MSC) differentiation toward adipocytes vs. osteoblasts was considered as mutually exclusive and because of the clinical finding of an inverse correlation between bone mass and marrow fat content. However, there is emerging evidence that bone marrow adipose tissue (MAT) is characterized by a high degree of plasticity. The regulation of MAT is still being explored but in some circumstances, marrow adiposity is influenced by the same modulators (e.g., sympathetic tone) as extramedullary adipose tissue, while in other circumstances, MAT is independent of peripheral adipose tissue regulators (e.g., anorexia nervosa) (Lecka-Czernik 2012; Rosen and Klibanski 2009).

There are several types of adipocytes categorized by their morphology and metabolic function. Classically these have been considered either white (e.g., classic, yellow) or brown. More recently, a third type has emerged, "beige" adipocytes, cells that can be recruited from white progenitors, or trans-differentiate in response to cold or β-adrenergic stimuli (Wu et al. 2012). White adipose tissue (WAT) consists of adipocytes that expand in response to nutrient availability by incorporating fatty acids to store as triglycerides for later use. Classically, visceral adipocytes, particularly during expansion, have been associated with the elaboration of inflammatory cytokines such as TNFα, IL-1, IL-6, and resistin, as well as several adipokines (Fain 2010; De Nardo and Latz 2011; Vandanmagsar et al. 2011; Skurk et al. 2007; Dandona et al. 2004). These peptides may have variable effects on the skeleton depending on time, site, and relative concentration. Generation of new animal models supports the precept that during high fat feeding, most of the increase in adipose volume is due to hypertrophy rather than recruitment of new progenitors (Wang and Scherer 2014). Once cellular expansion of the fat cell occurs, apoptosis, fibrosis, and recruitment of inflammatory cells occur. At that time, storage of triglycerides begins in other tissues, particularly in the liver, muscle, and bone marrow; this ultimately leads to the development of insulin resistance.

The relationship of white adipocytes to bone mass has been the subject of significant controversy, with most studies showing an increase in bone mineral density among obese individuals and low bone mass in underweight subjects (Evans et al. 2014). More recently, some studies have found an inverse relationship between visceral obesity and microstructural changes in bone architecture (Bredella et al. 2012).

Brown adipocytes are present in interscapular adipose tissue to regulate body temperature and glucose metabolism. These cells are innervated by the sympathetic nervous system (SNS) and are the major regulators of non-shivering thermogenesis in virtually all neonates (Carobbio et al. 2013). With the advent of positron emission tomography using (18)F-fluorodeoxyglucose ((18)FDG-PET), metabolically active brown adipose tissue (BAT) has been detected in both neonates and adults as discrete loci located in the neck and supraclavicular region (Nedergaard and Cannon 2013; Cronin et al. 2012). Brown adipocytes are derived from a Myf-5-positive muscle-like cellular lineage and have a specific metabolic program (Farmer 2008) with very high numbers of mitochondria. Uncoupling protein 1 (UCP1) and ppar gamma co-activator 1 alpha (Pgc1α), expressed in BAT, are necessary for fatty acid oxidation and uncoupled heat generation (Fedorenko et al. 2012; Uldry et al. 2006). That metabolic machinery also makes BAT a target for insulin-mediated glucose uptake. Recent work suggests that the volume of BAT in adults may be positively related to BMD (Ponrartana et al. 2012; Lee et al. 2013; Bredella et al. 2013). This may in part be due to endocrine secretory factors or confounding from the strong positive relationship of brown fat to muscle mass, likely from the shared transcriptional determinant, Myf5 (Farmer 2008). Impairment in BAT function or ablation of beige fat in mice leads to development of insulin resistance and low bone mass (Stanford et al. 2013; Cohen et al. 2014; Nguyen et al. 2015). In humans, the activity of BAT is attenuated in diabetes and during aging, both conditions associated with increased fractures (Ouellet et al. 2011).

Beige adipocytes are found in some white adipocyte depots. In rodents, beige adipocytes are present predominantly in the inguinal fat depots (Waldén et al. 2012). In humans, they are identified in the classical BAT depots but they may also develop as discrete loci in subcutaneous fat in response to distinct hormonal and environmental stimuli (Wu et al. 2012). These include chronic cold exposure, adrenergic signaling, and pharmacologic and nutritional factors which increase mitochondrial number and enhance expression of brown adipose-specific proteins to uncouple respiration (Bonet et al. 1831). It remains controversial as to whether white adipocytes can transdifferentiate into beige cells or whether their progenitor is distinct (Waldén et al. 2012). Beige adipocytes are myf 5 negative implying their progenitor is different from "brown" adipocytes. Nevertheless, beige cells are controlled by similar transcription factors found in brown adipocytes (Prdm16, Tbx15, FoxC2, others) as well as SNS tone. But yet they can be distinguished from brown and white adipocytes by expression of unique sets of gene biomarkers (Waldén et al. 2012; Ussar et al. 2014;

Boström et al. 2012). The involvement of the SNS in beige fat development and bone remodeling is complex. In states in which SNS tone is high (e.g., cold exposure or FGF-21 excess), beige cells are found in the subcutaneous or inguinal depot (Bornstein et al. 2014). On the other hand, bone mass is significantly reduced during high SNS tone, principally due to activation of the β2 receptor on osteoblasts. Activation of this receptor suppresses the key transcription factor, ATF4, and enhances RANKL production (Karsenty 2006; Motyl et al. 2013; Ducy et al. 2000; Elefteriou et al. 2005; Elefteriou 2008; Ma et al. 2011; Kode et al. 2012).

Besides their role in fatty acid oxidation and increased respiration, beige adipocytes secrete factors that may have anabolic effects on bone. Mice with adipocyte-specific expression of the FoxC2 transcription factor, which increases mitochondrial biogenesis and promotes "browning" of white adipocytes, has high bone mass (Rahman et al. 2013). A secretome of beige adipocytes isolated from an epididymal fat depot or from bone marrow of these mice includes Wnt10b and IGFBP2 proteins, which increase osteoblast differentiation and function (Rahman et al. 2013).

Although identified in humans, the thermogenic functions of beige adipocytes are still unclear. Moreover, it is not known whether recruitment of human beige adipocytes by environmental or hormonal factors can enhance their thermogenic capacity. Recent efforts toward characterization of beige adipocytes led to the identification of specific biomarkers, which will undoubtedly constitute a very promising step toward determining their contribution to energy homeostasis and provide a basis for development of new therapies to treat metabolic diseases and perhaps osteoporosis (Xue et al. 2015).

Marrow adipose tissue (MAT) represents a unique fat depot. It is an obvious candidate for a regulatory effect on bone due to its proximity and juxtaposition to skeletal surfaces. Indeed, MAT likely modulates hematopoietic and skeletal turnover in several different ways (Lecka-Czernik 2012; Cawthorn et al. 2014). In adult humans, MAT may constitute up to 1 kg of tissue, and its volume may increase in conditions of impaired energy metabolism including obesity, diabetes, aging, lipodystrophy, and anorexia nervosa (reviewed in (Elefteriou et al. 2005)). However, unlike in WAT, this increase is due to de novo differentiation rather than an increase in volume of existing adipocytes (Cawthorn et al. 2014).

In some clinical scenarios (e.g., aging, radiation, drugs, and diabetes), particularly in humans, increased total adipocyte volume per marrow volume correlates with a decrease in bone mass/quality and increase in fractures (reviewed in Lecka-Czernik and Rosen 2015). But because the origin and function of marrow adipocytes are not known, our understanding of bone-fat interactions, particularly in the niche, is lacking. It has been hypothesized that some marrow adipocytes develop from progenitors delivered during the process of vascularization of developing bone (Bianco et al. 2013). However, and unlike other marrow components, either mesenchymal or hematopoietic, these progenitors appear to be dormant during early bone growth. In humans, adipocytes start to populate marrow in long bone after the first decade of life, whereas in mice marrow adipocytes start to appear at skeletal maturation between 1 and 3 months (Moore and Dawson 1990; Lazarenko et al. 2007). It is unclear which skeletal or systemic compartments signal to adipocytic progenitors to induce their differentiation. Also it should be noted that marrow adipocytes are not surrounded by fibrous or inflammatory components per se, unlike adipose depots in other sites. This may influence their secretory properties in vivo.

It is also unclear whether marrow adipocytes are directly related to osteoblasts. Several models suggest that both adipocytes and osteoblasts originate from a common Myf5-negative progenitor, and determination of their fate is under control of retinoblastoma protein (pRB) (Calo et al. 2010; Krause et al. 2010). However, in contrast to the periphery, a significant fraction of marrow adipocytes may express the osteoblast-specific lineage marker osterix (Liu et al. 2013). This observation suggests that some adipocytes are in a closer developmental association with osteoblasts than with peripheral adipocytes. Whether bone marrow adipocytes arise from bone-lining cells also remains a matter of debate since these cells also express osterix but appear to be more fibroblastic than pre-osteoblastic. Interestingly, it has been recently suggested that adipocytes do not share the same musculoskeletal ancestor as osteoblasts, chondrocytes, and muscle cells by demonstrating that Gremlin 1 positive mesenchymal progenitors can differentiate into the above lineages but not adipocytes (Worthley et al. 2015).

It is conceivable that MAT does not constitute a homogeneous population of adipocytes, but rather a mixture of lipid-accumulating cells, of different origins, and of different functions and in distinct locations within the marrow. MAT has both WAT- and BAT-like characteristics with respect to the expression of gene markers (Krings et al. 2012). Whether that translates into functional activity remains to be determined since various depots can have a mix of metabolic expression signatures relative to brown or white adipocytes. In mice, BAT-like or beige fat features of MAT are compromised with diabetes and aging, suggesting a positive correlation between the beige metabolic profile of MAT and bone health (Krings et al. 2012). An analysis of BAT gene markers in the bone marrow of either old (24 months) mice or yellow agouti A^{vy}/a mice, a murine model of human type 2 diabetes, showed remarkable reduction in basal expression of Prdm16, FoxC2, β3-ADR, and Dio2 transcripts despite the fact that the volume of MAT in old and in diabetic animals was almost twofold higher than in controls (Krings et al. 2012). However, it remains to be established whether MAT's decreased expression of beige markers affects bone remodeling.

MAT may also impact the hematopoietic niche, which in turn is regulated by endosteal osteoblasts (Sacchetti et al. 2007; Omatsu et al. 2010). Some have proposed that MAT serves as a clearing site for circulating lipids (e.g., obesity, diabetes, anorexia nervosa), while other groups support its role as an endocrine organ (Cawthorn et al. 2014) or a paracrine tissue providing energy for emergency situations requiring new bone formation (Elefteriou et al. 2005). Interestingly, MAT volume decreases substantially during lactation when skeletal calcium stores are mobilized and there is increased bone resorption (Bornstein et al. 2014). Lineage tracing studies from several groups have now begun to address these various hypotheses, and it is anticipated that further work will uncover both the origin and function of marrow adipocytes. This is especially relevant for the question of whether increased marrow adipose tissue, present in patients with diabetes, affects osteoblast function and particularly skeletal fragility.

15.3 Molecular and Metabolic Determinants of Adipogenesis and Osteogenesis

15.3.1 PPARγ: A Master Regulator of Insulin Sensitivity and Adipogenesis

Differentiation toward WAT-, BAT-, MAT-, or beige-type adipocytes is controlled by the transcription factor, PPARγ. For its activation, PPARγ requires heterodimeric formation with an RXR nuclear receptor and binding of a specific ligand, of which there are several both natural and artificial. Natural PPARγ ligands consist of polyunsaturated fatty acids (PUFA) and their oxidized derivatives, certain phospholipids, and oxidized forms of prostaglandin J2. Our understanding of PPARγ's biological function was expanded in the late 1990s with the discovery of artificial high-affinity agonists, thiazolidinediones (TZDs), which are potent insulin sensitizers. Two of them, rosiglitazone and pioglitazone, have been used successfully for the last decade in clinical settings to control hyperglycemia in diabetic patients. TZDs are unique and their effect is most robust in combination with other antidiabetic therapies, as they comprise PPARγ-mediated activity to sensitize adipocytes to insulin. Despite the beneficial antidiabetic effect of TZDs, this class of drugs possesses off target side effects, including weight gain, edema, cardiovascular events, and fractures (Cariou et al. 2012).

Upon ligand binding, PPARγ undergoes posttranslational modifications (PTMs) specifying its activities. These PTMs determine assembly of specific co-activators on the PPARγ/RXR dimer. The functionally important PTMs include (1) dephosphorylation of Ser273 by inhibiting Cdk5 activity and the de-summoylation mediated by FGF-21, which induce insulin-sensitizing activity (Choi et al. 2011; Dutchak et al. 2012); (2) dephosphorylation of Ser112 by PP5 phosphatase, which induces adipogenic activity (Hinds et al. 2011); and (3) lysine deacetylation by SirT1 which induces PPARγ transcriptional switch to activate BAT- instead of WAT-specific gene expression (Qiang et al. 2012).

The oft-cited "mutually exclusive" hypothesis that MSCs can either go into one lineage or the other is questionable when considering the actions of the TZDs. Rosiglitazone forces MSCs toward the adipogenic lineage and blocks osteogenesis, in part by increasing β-catenin degradation, leading to enhanced marrow adiposity and bone loss (Rzonca et al. 2004; Ackert-Bicknell et al. 2009; Rahman et al. 2012). This has led to the tenet that there must be exclusive commitment toward one cell type over another. However, other PPARγ agonists that have pro-adipogenic effects also have a neutral or even positive effect on bone forming cells, raising provocative questions about the role of PPARγ in osteogenesis (Lee et al. 2012). Notwithstanding, rosiglitazone also activates Pgc1β, an important cofactor for energy utilization by the osteoclast, and ultimately this pathway leads to bone resorption from cell autonomous actions on the differentiation of monocytic cells (Wei et al. 2010).

PPARγ is one of two major transcription factors (the other being β-catenin) that occupy a central role in determining the fate of bone marrow MSCs (Siersbaek et al.

2010). Both these proteins are modified not only by posttranscriptional events but also by binding to other transcription factors such as FoxO1, as well as through the non-cell autonomous actions of reactive oxygen species (ROS) (Ambrogini et al. 2010; Iyer et al. 2013; Rached et al. 2010). FoxO1 exerts multiple actions by regulating insulin sensitivity (Accili and Arden 2004) and, by binding to Runx2 and ATF4, modulating oxidative stress. It also reduces osteocalcin activity through upregulation of the Esp1 phosphatase, thereby altering glucose homeostasis and promoting glucose intolerance (Siersbaek et al. 2010). FoxO1 is also important for the early differentiation of adipocytes, although downregulation occurs during later stages, particularly after the incorporation of free fatty acids (Subauste and Burant 2007; Munekata and Sakamoto 2009). Importantly, FoxO1 has been shown to regulate dopamine beta hydroxylase activity in neurons of the locus coeruleus and sympathetic ganglion, supporting its central role in modulating energy metabolism (Kajimura et al. 2014).

Rosiglitazone treatment favors activation of PPARγ at the expense of β-catenin, and this is accompanied by upregulation of FoxO1 (Rahman et al. 2012). However, PPARγ is also expressed in early osteoblasts and may play some role in defining the ultimate fate of later stage progenitors (Sun et al. 2013). In that same vein, conditional deletion of PPARγ in osteoblasts results in high bone mass and greater bone formation via activation of mTOR, supporting the tenet proposed by Krebsbach that there are likely to be both cell autonomous and cell non-autonomous actions of PPARγ on bone cells (Sun et al. 2013).

15.3.2 Metabolic Determinants of Adipocytes

Two PPARγ-regulated adipokines, leptin and adiponectin, have non-cell autonomous actions on adipose tissue and bone. Leptin is an adipokine secreted by adipocytes in response to storage of triglyceride and cell expansion. It circulates in relatively high concentrations, crosses the blood brain barrier, and has a multitude of effects on the brain, principally targeting the hypothalamus. As fat cells expand and release leptin, circulating concentrations increase, (although transport across the blood brain barrier may slow with higher circulating levels) to suppress orexigenic factors, alter hypothalamic reproductive factors, and enhance sympathetic tone (Ducy et al. 2000). Once sympathetic activity is increased, there is activation of the β2 receptor on osteoblasts. This results in suppressed bone formation and enhanced bone resorption. On the other hand, there is some evidence, albeit preliminary, that peripheral leptin has an effect on osteoblasts, although it is not clear whether it acts directly through the leptin receptor or indirectly by regulating the myeloid lineage and the bone marrow milieu (Turner et al. 2013; Scheller et al. 2012; Zhou et al. 2014). Interestingly however, in MAT, leptin expression is very low, highlighting the distinct phenotype of marrow adipocytes and the modest direct effect on osteoblasts and osteoclasts of locally generated leptin (Krings et al. 2012). T2D is often associated with hyperlipidemia and development of leptin resistance.

Although it is not known how this might affect bone, it can be speculated that high leptin levels in type 2 diabetes will increase SNS tone and β-adrenergic signaling leading to bone loss.

Adiponectin, in contrast, has both direct and indirect effects on bone, first by directly suppressing bone formation and later by blocking SNS outflow from the CNS (Kajimura et al. 2013). Hence, the global deletion of adiponectin results in a context specific skeletal phenotype; i.e., at a young age, the animals have high bone mass due to the direct loss of adiponectin suppression on osteoblasts; at an older age (36 weeks) adipo−/− mice have low vertebral trabecular bone volume due to the loss of antagonism of the SNS. This "yin and yang" model proposed and validated by the Karsenty laboratory provides strong evidence that metabolic status, integrated in the brain but originating from the adipocyte, has a major impact on the state of the skeleton (Kajimura et al. 2013). In contrast to leptin, adiponectin is relatively highly expressed in MAT and marrow adiponectin may contribute to the overall pool in circulation especially in states of WAT wasting and MAT expansion such as calorie restriction or anorexia nervosa (Cawthorn et al. 2014). On the other hand, in states of WAT and MAT expansion associated with development of insulin resistance such as diabetes and aging, circulating levels of adiponectin decrease suggesting that MAT is subjected to the same changes as WAT in respect to decrease in endocrine function (Cartwright et al. 2007).

Serotonin is a neurotransmitter and gut secretory peptide that can affect bone and fat in two distinct but opposing ways. Brain-derived serotonin, catalyzed by *Tph2*, binds to the serotonin receptor (5-HTR) in the ventral medial hypothalamus (VMH), inactivates tone from the SNS, and enhances skeletal acquisition. Leptin suppresses *Tph2* expression in the brain stem, resulting in decreased serotonin in the brain, and bone loss due to suppressed serotonin signaling in the VMH. Gut-derived serotonin, catalyzed by *Tph1*, is taken up by platelets through serotonin transporter (5-HTT) and has been shown to suppress skeletal accrual. Osteoblasts and osteoclasts express 5-HT receptors and 5-HTT, and thus the serotonin network is likely to be operative in the skeletal environment (Kawai and Rosen 2010). Moreover it has been demonstrated that Lrp5 in the gut can inhibit serotonin synthesis which in turn has been shown in several studies to block osteoblast proliferation (Karsenty and Yadav 2011).

Serum serotonin levels are elevated in diabetes probably due to hyperreactivity of platelets, which are storage cells for peripheral serotonin (Malyszko et al. 1994). In contrast, prolonged use of selective serotonin reuptake inhibitors (SSRIs) is often associated with increased body weight and obesity (Raeder et al. 2006). It has been shown recently that peripheral serotonin plays an important role in regulation of energy homeostasis by acting directly on adipose tissue (Oh 2015). In mice, pharmacologic inhibition of serotonin synthesis or its production in adipocytes inhibits lipogenesis in WAT, induces browning in inguinal WAT and increases thermogenic activity in BAT. In humans, pharmacologic inhibition of serotonin receptor by the drug sarpogrelate increases insulin sensitivity and adiponectin production (Nomura et al. 2005). Serotonin is implicated in the degradation of insulin substrate 1 (IRS-1) in adipocytes, a factor essential for insulin signaling (Li et al. 2013). Thus, serotonin may stimulate lipogenesis in WAT implicating its contribution to WAT expansion.

Hence, theoretically, serotonin levels in diabetics may increase MAT expansion and suppress insulin signaling in both adipocytes and osteoblasts, the latter resulting in decreased bone turnover (Oh 2015; Wei et al. 2014).

Glucocorticoids (GCs) are factors that contribute to the development of diabetes and have profound effects on fat and bone metabolism (Anagnostis et al. 2009). Excessive pharmacologic exposure to GCs or elevated levels of endogenous cortisol (due to either hyperactivity of hypothalamic-pituitary-adrenal axes or activity of 11beta-hydroxysteroid dehydrogenase-1 at the tissue level) is recognized as a risk for development of T2D and osteoporosis (Weinstein et al. 2010; Brennan-Speranza et al. 2012). There is a significant dose response such that higher levels of GCs for a prolonged duration produce more bone loss and a shift in adipocyte distribution toward the centripetal regions of the body. This predisposes to the development of central obesity and metabolic syndrome. GCs regulate adrenergic responsiveness by regulating the expression of all three forms of the β-adrenergic receptors. GCs inhibit the transcriptional response of the UCP1 gene to adrenergic stimulation in brown adipocytes by inhibiting the expression of β1 and β3 adrenergic receptors and suppress brown or beige adipocyte activity in the periphery and perhaps in bone marrow. In epididymal WAT, increased levels of endogenous cortisol correlate with downregulation of β1 and β3 adrenergic receptor expression and upregulation of β2 adrenergic receptor expression (Farias-Silva et al. 2004). Taken together, glucocorticoids have a dual effect on adrenergic signaling. They decrease adrenergic receptor-mediated metabolic activity of brown fat and increase catabolic activity in bone. Thus, activation of glucocorticoid signaling in marrow adipocytes may lead to a decrease in β3 adrenergic receptor expression and loss of brown-like potential of marrow adipocytes, which is seen with aging and diabetes. Alternatively, GCs' effects on energy metabolism may be mediated through the skeleton. As demonstrated in rodents, GCs suppress osteoblast function and osteocalcin synthesis in a dose-dependent manner and enhance marrow adipogenesis (Brennan-Speranza et al. 2012; Ma et al. 2011).

15.4 Osteoblast Biology and Energy Utilization

Osteoblasts have two major roles in the skeletal remodeling unit: synthesis of collagen and mineralization of osteoid. Both require energy and both have been shown to be linked to metabolic homeostasis through a complex sequence of events. Osteocalcin was the first bone-specific protein to be identified as a modulator of peripheral adipose sensitivity and insulin secretion. Osteocalcin is produced by osteoblasts and circulates both in its native form and after release from the skeletal matrix during bone resorption. Beyond its recognition as a marker of differentiated osteoblasts, its functional significance for the bone remodeling unit has been difficult to clarify. However, it is established that undercarboxylated osteocalcin is the protein released during active bone resorption. Undercarboxylated osteocalcin (unOC) was shown to enhance glucose uptake in peripheral fat depots and increase insulin secretion in mice. In addition, it was found that insulin promotes the release of unOC by stimulating resorption through the elaboration of RANKL to enhance osteoclastogenesis; thus, there is a fast-forward system whereby release of osteocalcin from bone promotes glucose sensitivity through increased β cell production of insulin and higher insulin sensitivity in peripheral fat and muscle tissue. Further proof of that tenet comes from studies of Esp1, a phosphatase that is responsible for the degradation of osteocalcin in murine bone cells. The genetic absence of Esp1 in mice results in significant hypoglycemia, due to excess undercarboxylated osteocalcin, enhanced insulin sensitivity with low glucose, and dephosphorylation of the insulin receptor (Ferron et al. 2010). In contrast, deletion of the insulin receptor in osteoblasts leads to an insulin resistance state due to lower levels of unOC and increased free fatty acids resulting in insulin receptor ubiquitination (Wei et al. 2014). More recent work confirms the feedback circuit between bone and the β-cell of the pancreas. Abdallah et al. reported that unOC stimulates DLK1 production in the beta cells of the pancreas, which in turn circulates and inhibits further osteoblast production of unOC (Abdallah et al. 2015).

Despite the vast support for this critical metabolic regulatory loop in mice, studies in humans are conflicting with some showing a positive correlation between unOC and glucose control and others showing no effect. Other matrix bound proteins may also influence whole-body homeostasis. IGF-I is another peptide synthesized and stored in bone, although the vast majority of circulating IGF-I is derived from the liver. IGF in bone is bound to a series of IGF binding proteins and can be released during bone resorption. IGF-I can regulate glucose and fatty acid synthesis through the type I IGFR as well as hybrid insulin/IGF receptors. Recent evidence in humans shows a strong correlation between low levels of IGF-I and high circulating fatty acids as well as reduced GLUT1 expression in skeletal muscles (Thankamony et al. 2014). It is also likely that there are other bone proteins that influence glucose homeostasis.

More recently there has been interest in the osteocyte specific production of FGF-23, which belongs to a family of FGF proteins (FGF-19, FGF-21, FGF-23) with insulin-sensitizing properties. Several papers have suggested that glucose tolerance is indirectly related to circulating FGF-23, particularly in renal insufficiency (Wojcik et al. 2012). Whether this effect is predominant remains to be clarified, although it is very evident that FGF-23 is the phosphaturic hormone of the body, regulating directly or indirectly 1,25-dihydroxyvitamin D synthesis in the kidney.

15.5 Bioenergetic Determinants of Adipogenesis and Osteogenesis

Recent attention has focused on mechanisms related to mesenchymal cell fate and, in particular, the metabolic program of MSC progenitors (Figs. 15.1 and 15.2). There is now preliminary evidence that adipocytes and osteoblasts have distinct energy utilization pathways as they move down their respective lineages during

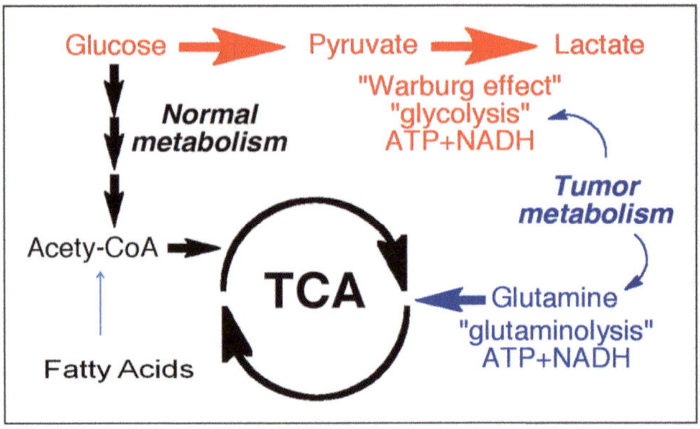

Fig. 15.2 ATP generation for energy demands in a cell can occur by glycolysis (*red*), through the citric acid cycle (*black*) and from glutaminolysis (*blue*). The most ATP is generated per mol of glucose from TCA generation, but glycolysis is a rapid process

differentiation (Fig. 15.2). During these stages, ATP demand drives the type of substrate utilization that is operative and thus is very context specific. Early progenitors and pluripotent stem cells utilize glucose as their primary fuel, even in aerobic states, in a manner analogous to cancer cells (i.e., the Warburg effect) (Ng and Daley George 2015). But pre-adipocytes and newly differentiated adipocytes primarily use mitochondrial respiration to supply energy for their metabolic needs (Wang et al. 2010; Tormos et al. 2011). The process of glucose entry and fatty acid oxidation through the Krebs cycle generates more molecules of ATP per mole of glucose (36:1) than glycolysis (2:1) but it comes at a cost, as mitochondrial respiration leads to the generation of reactive oxygen species (ROS) from the electron transport chain (ETC). ROS compounds (e.g., H_2O_2, superoxides, etc.) can further suppress mitochondrial respiration and promote an adipogenic program that is associated with more insulin resistance and less lipolysis (Wang et al. 2010; Tormos et al. 2011). Excess ROS in adipocytes may also cause mitochondrial DNA damage or further changes to complex I in the ETC leading to metabolic dysfunction.

In contrast, pre-osteoblasts utilize a distinct metabolic program that features oxidative phosphorylation and glycolysis, although the latter predominates especially during differentiation. Pre-osteoblasts differentiate under the influence of various ligands, particularly the Wnts, TGF-beta, and IGFs. Long and colleagues first demonstrated that glycolysis is a major feature of Wnt3a-induced osteoblast differentiation (Esen et al. 2013). But remarkably, much older ex vivo studies from Neuman and colleagues demonstrated that PTH produced lactic acid in calvarial osteoblasts supporting the tenet that osteoblasts utilize glycolysis to generate lactate during collagen synthesis and mineralization (Nichols and Neuman 1987). Guntur et al. also showed that glycolysis is essential for terminal differentiation of osteoblasts and that oxidative phosphorylation is more important early in the differentiation scheme (Guntur et al. 2014). The Long group reported that HIF1α is a critical transcriptional regulator of glycolysis, triggered in part, by relative hypoxia in the bone marrow niche (Regan et al. 2014). More recently, Karsenty and colleagues demonstrated the importance of the Glut1 transporter in Runx2-mediated early osteoblast differentiation (Wei et al. 2015). In that same vein, Karner et al. demonstrated that glutaminolysis also is essential for osteoblast differentiation through the Wnt signaling system (Karner et al. 2014).

In sum, there are at least three substrates for ATP generation and differentiation of MSCs: Glutamine, which then enters the Krebs cycle through alpha ketoglutarate; glucose, which via glycolysis can generate lactate as well as ribose nucleotides via the pentose phosphate shunt; and fatty acids, which are metabolized via acetyl CoA in the mitochondria. Osteoblasts have the machinery to use all three substrates but their relative utilization likely depends on their availability. Notwithstanding, the bioenergetics of the niche must be defined in order to fully understand lineage allocation, and the role transcription factors play in defining the ultimate fate of MSCs in the marrow (Burkewitz et al. 2014). ATP demand is at the forefront of what biochemical program is activated and that in turn is determined by the differentiative function of that cell. It is remarkable, however, that a picture is emerging of a metabolic program in osteoblasts that resembles tumor cell bioenergetics. The relative hypoxic environment, the additive effects of glutaminolysis with glycolysis on ATP generation, and the evolutionary importance of substrate availability that is intimately tied to glycolysis in single cell organisms provide us with potential clues about progenitor cells and their ultimate fate either to store energy (i.e., adipocytes) or to use sufficient energy to build and mineralize matrix (i.e., osteoblasts). What is still not known is how osteoblasts handle a high glucose load in vivo and what bioenergetic pathways are utilized particularly in states of chronic hyperglycemia.

15.6 Summary

Bone and fat arise from progenitors with distinct metabolic programs. Their physiologic interactions are critical for metabolic homeostasis. Hence, alterations starting at the molecular level and proceeding to the brain (e.g., hypothalamus and SNS) can result in metabolic syndromes. Notwithstanding, we still know little about the intricate cell-cell communication that must take place, whether in the marrow niche, in the circulation or in bone, between adipocytes and osteoblasts. Furthermore, to better understand the skeletal defects in T2D, we must define fuel preferences by differentiating osteoblasts, the role of the insulin receptor in mediating substrate utilization in humans, and the importance of endocrine (e.g., adiponectin, leptin, etc.) and paracrine mediators during chronic states of hyperglycemia. Newer therapies for T2D aimed at enhancing glucose utilization, such as agents that can brown white adipocytes or enhance the function of preformed brown adipose tissue, may have off target effects on the skeleton requiring careful phenotyping in both humans and animal models. Notwithstanding, in this era of personalized medicine, the development of new drugs to treat chronic conditions will demand an even greater understanding of the molecular and cellular underpinnings of the disease itself.

References

Abdallah BM, Ditzel N, Laborda J, Karsenty G, Kassem M. DLK1 regulates whole body glucose metabolism: a negative feedback regulation of the osteocalcin-insulin loop. Diabetes. 2015; doi:10.2337/db14-1642.

Accili D, Arden KC. FoxOs at the crossroads of cellular metabolism, differentiation, and transformation. Cell. 2004;117:421–6.

Ackert-Bicknell CL, Shockley KR, Horton LG, Lecka-Czernik B, Churchill GA, Rosen CJ. Strain-specific effects of rosiglitazone on bone mass, body composition, and serum insulin-like growth factor-I. Endocrinology. 2009;150:1330–40. doi:10.1210/en.2008-0936.

Ambrogini E, Almeida M, Martin-Millan M, Paik JH, DePinho RA, Han L, et al. FoxO-mediated defense against oxidative stress in osteoblasts is indispensable for skeletal homeostasis in mice. Cell Metab. 2010;11:136–46.

Anagnostis P, Athyros VG, Tziomalos K, Karagiannis A, Mikhailidis DP. Clinical review: the pathogenetic role of cortisol in the metabolic syndrome: a hypothesis. J Clin Endocrinol Metab. 2009;94:2692–701. doi:10.1210/jc.2009-0370.

Bianco P, Cao X, Frenette PS, Mao JJ, Robey PG, Simmons PJ, et al. The meaning, the sense and the significance: translating the science of mesenchymal stem cells into medicine. Nat Med. 2013;19:35–42.

Bonet ML, Oliver P, Palou A. Pharmacological and nutritional agents promoting browning of white adipose tissue. Biochim Biophys Acta. 1831;2013:969–85. doi:10.1016/j.bbalip.2012.12.002.

Bornstein S, Brown SA, Le PT, Wang X, Demambro V, Horowitz MC, et al. FGF-21 and skeletal remodeling during and after lactation in C57BL6 mice. Endocrinology. 2014. en20141083; doi:10.1210/en.2014-1083.

Boström P, Wu J, Jedrychowski MP, Korde A, Ye L, Lo JC, et al. A PGC1-α-dependent myokine that drives brown-fat-like development of white fat and thermogenesis. Nature. 2012;481:463–8. doi:10.1038/nature10777.

Bredella MA, Lin E, Gerweck AV, Landa MG, Thomas BJ, Torriani M, et al. Determinants of bone microarchitecture and mechanical properties in obese men. J Clin Endocrinol Metab. 2012;97:4115–22.

Bredella MA, Gill CM, Rosen CJ, Klibanski A, Torriani M. Positive effects of brown adipose tissue on femoral bone structure. Bone. 2013;58C:55–8. doi:10.1016/j.bone.2013.10.007.

Brennan-Speranza TC, Henneicke H, Gasparini SJ, Blankenstein KI, Heinevetter U, Cogger VC, et al. Osteoblasts mediate the adverse effects of glucocorticoids on fuel metabolism. J Clin Invest. 2012;122:4172–89. doi:10.1172/JCI63377.

Burkewitz K, Zhang Y, Mair WB. AMPK at the nexus of energetics and aging. Cell Metab. 2014;20:10–25. doi:10.1016/j.cmet.2014.03.002.

Calo E, Quintero-Estades JA, Danielian PS, Nedelcu S, Berman SD, Lees JA. Rb regulates fate choice and lineage commitment in vivo. Nature. 2010;466:1110–4. doi:10.1038/nature09264.

Cariou B, Charbonnel B, Staels B. Thiazolidinediones and PPARγ agonists: time for a reassessment. Trends Endocrinol Metab. 2012;23:205–15. doi:10.1016/j.tem.2012.03.001.

Carobbio S, Rosen B, Vidal-Puig A. Adipogenesis: new insights into brown adipose tissue differentiation. J Mol Endocrinol. 2013;51:T75–85.

Cartwright MJ, Tchkonia T, Kirkland JL. Aging in adipocytes: potential impact of inherent, depot-specific mechanisms. Exp Gerontol. 2007;42:463–71. doi:10.1016/j.exger.2007.03.003.

Cawthorn WP, Scheller EL, Learman BS, Parlee SD, Simon BR, Mori H, et al. Bone marrow adipose tissue is an endocrine organ that contributes to increased circulating adiponectin during caloric restriction. Cell Metab. 2014:1–8. doi:10.1016/j.cmet.2014.06.003

Chang DH, Deng H, Matthews P, Krasovsky J, Ragupathi G, Spisek R, et al. Inflammation-associated lysophospholipids as ligands for CD1d-restricted T cells in human cancer. Blood. 2008;112:1308–16. doi:10.1182/blood-2008-04-149831.

Choi JH, Banks AS, Kamenecka TM, Busby SA, Chalmers MJ, Kumar N, et al. Antidiabetic actions of a non-agonist PPARγ ligand blocking Cdk5-mediated phosphorylation. Nature. 2011;477:477–81. doi:10.1038/nature10383.

Cohen P, Levy JD, Zhang Y, Frontini A, Kolodin DP, Svensson KJ, et al. Ablation of PRDM16 and beige adipose causes metabolic dysfunction and a subcutaneous to visceral fat switch. Cell. 2014;156:304–16. doi:10.1016/j.cell.2013.12.021.

Cronin CG, Prakash P, Daniels GH, Boland GW, Kalra MK, Halpern EF, et al. Brown fat at PET/CT: correlation with patient characteristics. Radiology. 2012;263:836–42. doi:10.1148/radiol.12100683.

Dandona P, Aljada A, Bandyopadhyay A. Inflammation: the link between insulin resistance, obesity and diabetes. Trends Immunol. 2004;25:4–7.

De Nardo D, Latz E. NLRP3 inflammasomes link inflammation and metabolic disease. Trends Immunol. 2011;32:373–9. doi:10.1016/j.it.2011.05.004.

Ducy P, Amling M, Takeda S, Priemel M, Schilling AF, Beil FT, et al. Leptin inhibits bone formation through a hypothalamic relay: a central control of bone mass. Cell. 2000;100:197–207. doi:10.1016/S0092-8674(00)81558-5.

Dutchak PA, Katafuchi T, Bookout AL, Choi JH, Yu RT, Mangelsdorf DJ, et al. Fibroblast growth factor-21 regulates PPARγ activity and the antidiabetic actions of thiazolidinediones. Cell. 2012;148:556–67. doi:10.1016/j.cell.2011.11.062.

Elefteriou F. Regulation of bone remodeling by the central and peripheral nervous system. Arch Biochem Biophys. 2008;473:231–6. doi:10.1016/j.abb.2008.03.016.

Elefteriou F, Ahn JD, Takeda S, Starbuck M, Yang X, Liu X, et al. Leptin regulation of bone resorption by the sympathetic nervous system and CART. Nature. 2005;434:514–20. doi:10.1038/nature03398.

Esen E, Chen J, Karner CM, Okunade AL, Patterson BW, Long F. WNT-LRP5 signaling induces Warburg effect through mTORC2 activation during osteoblast differentiation. Cell Metab. 2013;17:745–55. doi:10.1016/j.cmet.2013.03.017.

Evans AL, Paggiosi MA, Eastell R, Walsh JS. Bone density, microstructure and strength in obese and normal weight men and women in younger and older adulthood. J Bone Miner Res. 2014; doi:10.1002/jbmr.2407.

Fain JN. Release of inflammatory mediators by human adipose tissue is enhanced in obesity and primarily by the nonfat cells: a review. Mediat Inflamm. 2010;2010:513948. doi:10.1155/2010/513948.

Farias-Silva E, dos Santos IN, Corezola do Amaral ME, Grassi-Kassisse DM, Spadari-Bratfisch RC. Glucocorticoid receptor and Beta-adrenoceptor expression in epididymal adipose tissue from stressed rats. Ann N Y Acad Sci. 2004;1018:328–32. doi:10.1196/annals.1296.040.

Farmer SR. Brown fat and skeletal muscle: unlikely cousins? Cell. 2008;134:726–7. doi:10.1016/j.cell.2008.08.018.

Fedorenko A, Lishko PV, Kirichok Y. Mechanism of fatty-acid-dependent UCP1 uncoupling in brown fat mitochondria. Cell. 2012;151:400–13. doi:10.1016/j.cell.2012.09.010.

Ferron M, Wei J, Yoshizawa T, Del Fattore A, DePinho RA, Teti A, et al. Insulin signaling in osteoblasts integrates bone remodeling and energy metabolism. Cell. 2010;142:296–308.

Giralt M, Villarroya F. White, brown, beige/brite: different adipose cells for different functions? Endocrinology. 2013;154:2992–3000.

Guntur AR, Le PT, Farber CR, Rosen CJ. Bioenergetics during calvarial osteoblast differentiation reflect strain differences in bone mass. Endocrinology. 2014;155:1589–95. doi:10.1210/en.2013-1974.

Hinds TD, Stechschulte LA, Cash HA, Whisler D, Banerjee A, Yong W, et al. Protein phosphatase 5 mediates lipid metabolism through reciprocal control of glucocorticoid receptor and peroxisome proliferator-activated receptor-γ (PPARγ). J Biol Chem. 2011;286:42911–22. doi:10.1074/jbc.M111.311662.

Iyer S, Ambrogini E, Bartell SM, Han L, Roberson PK, De Cabo R, et al. FOXOs attenuate bone formation by suppressing Wnt signaling. J Clin Invest. 2013;123:3409–19.

Kajimura D, Lee HW, Riley KJ, Arteaga-Solis E, Ferron M, Zhou B, et al. Adiponectin regulates bone mass via opposite central and peripheral mechanisms through foxo1. Cell Metab. 2013;17:901–15.

Kajimura D, Paone R, Mann JJ, Karsenty G. Foxo1 regulates Dbh expression and the activity of the sympathetic nervous system in vivo. Mol Metab. 2014;3:770–7. doi:10.1016/j.molmet.2014.07.006.

Karner CM, Esen E, Okunade AL, Patterson BW, Long F. Increased glutamine catabolism mediates bone anabolism in response to WNT signaling. J Clin Invest. 2014;125:551–62. doi:10.1172/JCI78470.

Karsenty G. Convergence between bone and energy homeostases: leptin regulation of bone mass. Cell Metab. 2006;4:341–8.

Karsenty G, Yadav VK. Regulation of bone mass by serotonin: molecular biology and therapeutic implications. Annu Rev Med. 2011;62:323–31. doi:10.1146/annurev-med-090710-133426.

Kawai M, Rosen CJ. Minireview: a skeleton in serotonin's closet? Endocrinology. 2010;151:4103–8. doi:10.1210/en.2010-0499.

Kode A, Mosialou I, Silva BC, Joshi S, Ferron M, Rached MT, et al. FoxO1 protein cooperates with ATF4 protein in osteoblasts to control glucose homeostasis. J Biol Chem. 2012;287:8757–68.

Krause U, Harris S, Green A, Ylostalo J, Zeitouni S, Lee N, et al. Pharmaceutical modulation of canonical Wnt signaling in multipotent stromal cells for improved osteoinductive therapy. Proc Natl Acad Sci U S A. 2010;107:4147–52. doi:10.1073/pnas.0914360107.

Krings A, Rahman S, Huang S, Lu Y, Czernik PJ, Lecka-Czernik B. Bone marrow fat has brown adipose tissue characteristics, which are attenuated with aging and diabetes. Bone. 2012;50:546–52. doi:10.1016/j.bone.2011.06.016.

Lazarenko OP, Rzonca SO, Hogue WR, Swain FL, Suva LJ, Lecka-Czernik B. Rosiglitazone induces decreases in bone mass and strength that are reminiscent of aged bone. Endocrinology. 2007;148:2669–80. doi:10.1210/en.2006-1587.

Lecka-Czernik B. Marrow fat metabolism is linked to the systemic energy metabolism. Bone. 2012;50:534–9. doi:10.1016/j.bone.2011.06.032.

Lecka-Czernik B, Rosen CJ. Energy excess, glucose utilization, and skeletal remodeling: new insights. J Bone Miner Res. 2015; doi:10.1002/jbmr.2574.

Lee DH, Huang H, Choi K, Mantzoros C, Kim Y-B. Selective PPARγ modulator INT131 normalizes insulin signaling defects and improves bone mass in diet-induced obese mice. Am J Physiol Endocrinol Metab. 2012;302:E552–60. doi:10.1152/ajpendo.00569.2011.

Lee P, Brychta RJ, Collins MT, Linderman J, Smith S, Herscovitch P, et al. Cold-activated brown adipose tissue is an independent predictor of higher bone mineral density in women. Osteoporos Int. 2013;24:1513–8. doi:10.1007/s00198-012-2110-y.

Li Q, Hosaka T, Harada N, Nakaya Y, Funaki M. Activation of Akt through 5-HT2A receptor ameliorates serotonin-induced degradation of insulin receptor substrate-1 in adipocytes. Mol Cell Endocrinol. 2013;365:25–35. doi:10.1016/j.mce.2012.08.022.

Liu Y, Strecker S, Wang L, Kronenberg MS, Wang W, Rowe DW, et al. Osterix-Cre labeled progenitor cells contribute to the formation and maintenance of the bone marrow stroma. PLoS One. 2013;8:e71318.

Ma Y, Nyman JS, Tao H, Moss HH, Yang X, Elefteriou F. β2-adrenergic receptor signaling in osteoblasts contributes to the catabolic effect of glucocorticoids on bone. Endocrinology. 2011;152:1412–22. doi:10.1210/en.2010-0881.

Malyszko J, Urano T, Knofler R, Taminato A, Yoshimi T, Takada Y, et al. Daily variations of platelet aggregation in relation to blood and plasma serotonin in diabetes. Thromb Res. 1994;75:569–76.

Moore SG, Dawson KL. Red and yellow marrow in the femur: age-related changes in appearance at MR imaging. Radiology. 1990;175:219–23. doi:10.1148/radiology.175.1.2315484.

Motyl KJ, Bishop KA, Demambro VE, Bornstein SA, Le P, Kawai M, et al. Altered thermogenesis and impaired bone remodeling in Misty mice. J Bone Miner Res. 2013;28:1885–97. doi:10.1002/jbmr.1943.

Munekata K, Sakamoto K. Forkhead transcription factor Foxo1 is essential for adipocyte differentiation. In Vitro Cell Dev Biol Anim. 2009;45:642–51.

Nedergaard J, Cannon B. How brown is brown fat? It depends where you look. Nat Med. 2013;19:540–1. doi:10.1038/nm.3187.

Ng S-C, Daley George Q. Metabolic switches linked to pluipotency and embryonic stem cell differentiation. Cell Metab. 2015;21:349–50.

Nguyen AD, Lee NJ, Wee NKY, Zhang L, Enriquez RF, Khor EC, et al. Uncoupling protein-1 is protective of bone mass under mild cold stress conditions. Bone. 2015; doi:10.1016/j.bone.2015.05.037.

Nichols FC, Neuman WF. Lactic acid production in mouse calvaria in vitro with and without parathyroid hormone stimulation: lack of acetazolamide effects. Bone. 1987;8:105–9.

Nomura S, Shouzu A, Omoto S, Nishikawa M, Iwasaka T. 5-HT2A receptor antagonist increases circulating adiponectin in patients with type 2 diabetes. Blood Coagul Fibrinolysis. 2005;16:423–8.

Oh C-M, Namkung J, Go Y, Shong KE, Kim K, Kim H, et al. Regulation of systemic energy homeostasis by serotonin in adipose tissues. Nat Commun. 2015;6:6794. doi:10.1038/ncomms7794.

Omatsu Y, Sugiyama T, Kohara H, Kondoh G, Fujii N, Kohno K, et al. The essential functions of adipo-osteogenic progenitors as the hematopoietic stem and progenitor cell niche. Immunity. 2010;33:387–99. doi:10.1016/j.immuni.2010.08.017.

Ouellet V, Routhier-Labadie A, Bellemare W, Lakhal-Chaieb L, Turcotte E, Carpentier AC, et al. Outdoor temperature, age, sex, body mass index, and diabetic status determine the prevalence, mass, and glucose-uptake activity of 18F-FDG-detected BAT in humans. J Clin Endocrinol Metab. 2011;96:192–9. doi:10.1210/jc.2010-0989.

Ponrartana S, Aggabao PC, Hu HH, Aldrovandi GM, Wren TAL, Gilsanz V. Brown adipose tissue and its relationship to bone structure in pediatric patients. J Clin Endocrinol Metab. 2012;97:2693–8. doi:10.1210/jc.2012-1589.

Qiang L, Wang L, Kon N, Zhao W, Lee S, Zhang Y, et al. Brown remodeling of white adipose tissue by SirT1-dependent deacetylation of Pparγ. Cell. 2012;150:620–32. doi:10.1016/j.cell.2012.06.027.

Rached MT, Kode A, Silva BC, Jung DY, Gray S, Ong H, et al. FoxO1 expression in osteoblasts regulates glucose homeostasis through regulation of osteocalcin in mice. J Clin Invest. 2010;120:357–68.

Raeder MB, Bjelland I, Emil Vollset S, Steen VM. Obesity, dyslipidemia, and diabetes with selective serotonin reuptake inhibitors: the Hordaland Health Study. J Clin Psychiatry. 2006;67:1974–82.

Rahman S, Czernik PJ, Lu Y, Lecka-Czernik B. β-catenin directly sequesters adipocytic and insulin sensitizing activities but not osteoblastic activity of PPARγ2 in marrow mesenchymal stem cells. PLoS One. 2012;7:e51746. doi:10.1371/journal.pone.0051746.

Rahman S, Lu Y, Czernik PJ, Rosen CJ, Enerback S, Lecka-Czernik B. Inducible brown adipose tissue, or beige fat, is anabolic for the skeleton. Endocrinology. 2013;154:2687–701. doi:10.1210/en.2012-2162.

Regan JN, Lim J, Shi Y, Joeng KS, Arbeit JM, Shohet RV, et al. Up-regulation of glycolytic metabolism is required for HIF1α-driven bone formation. Proc Natl Acad Sci U S A. 2014;111:8673–8. doi:10.1073/pnas.1324290111.

Rosen CJ, Klibanski A. Bone, fat, and body composition: evolving concepts in the pathogenesis of osteoporosis. Am J Med. 2009;122:409–14. doi:10.1016/j.amjmed.2008.11.027.

Rzonca SO, Suva LJ, Gaddy D, Montague DC, Lecka-Czernik B. Bone is a target for the antidiabetic compound rosiglitazone. Endocrinology. 2004;145:401–6. doi:10.1210/en.2003-0746.

Sacchetti B, Funari A, Michienzi S, Di Cesare S, Piersanti S, Saggio I, et al. Self-renewing osteoprogenitors in bone marrow sinusoids can organize a hematopoietic microenvironment. Cell. 2007;131:324–36. doi:10.1016/j.cell.2007.08.025.

Scheller EL, Song J, Dishowitz MI, Hankenson KD, Krebsbach PH. A potential role for the myeloid lineage in leptin-regulated bone metabolism. Horm Metab Res. 2012;44:1–5. doi:10.1055/s-0031-1297971.

Schwartz AV. Epidemiology of fractures in type 2 diabetes. Bone. 2015; doi:10.1016/j.bone.2015.05.032.

Siersbaek R, Nielsen R, Mandrup S. PPARgamma in adipocyte differentiation and metabolism – novel insights from genome-wide studies. FEBS Lett. 2010;584:3242–9.

Skurk T, Alberti-Huber C, Herder C, Hauner H. Relationship between adipocyte size and adipokine expression and secretion. J Clin Endocrinol Metab. 2007;92:1023–33. doi:10.1210/jc.2006-1055.

Stanford KI, Middelbeek RJW, Townsend KL, An D, Nygaard EB, Hitchcox KM, et al. Brown adipose tissue regulates glucose homeostasis and insulin sensitivity. J Clin Invest. 2013;123:215–23. doi:10.1172/JCI62308.

Subauste AR, Burant CF. Role of FoxO1 in FFA-induced oxidative stress in adipocytes. Am J Physiol Endocrinol Metab. 2007;293:E159–64.

Sun H, Kim JK, Mortensen R, Mutyaba LP, Hankenson KD, Krebsbach PH. Osteoblast-targeted suppression of PPARc increases osteogenesis through activation of mTOR signaling. Stem Cells. 2013;31:2183–92.

Thankamony A, Capalbo D, Marcovecchio ML, Sleigh A, Jørgensen SW, Hill NR, et al. Low circulating levels of IGF-1 in healthy adults are associated with reduced β-cell function, increased intramyocellular lipid, and enhanced fat utilization during fasting. J Clin Endocrinol Metab. 2014;99:2198–207. doi:10.1210/jc.2013-4542.

Tormos KV, Anso E, Hamanaka RB, Eisenbart J, Joseph J, Kalyanaraman B, et al. Mitochondrial complex III ROS regulate adipocyte differentiation. Cell Metab. 2011;14:537–44.

Turner RT, Kalra SP, Wong CP, Philbrick KA, Lindenmaier LB, Boghossian S, et al. Peripheral leptin regulates bone formation. J Bone Miner Res. 2013;28:22–34. doi:10.1002/jbmr.1734.

Uldry M, Yang W, St-Pierre J, Lin J, Seale P, Spiegelman BM. Complementary action of the PGC-1 coactivators in mitochondrial biogenesis and brown fat differentiation. Cell Metab. 2006;3:333–41. doi:10.1016/j.cmet.2006.04.002.

Ussar S, Lee KY, Dankel SN, Boucher J, Haering M-F, Kleinridders A, et al. ASC-1, PAT2, and P2RX5 are cell surface markers for white, beige, and brown adipocytes. Sci Transl Med. 2014;6:247ra103. doi:10.1126/scitranslmed.3008490.

Vandanmagsar B, Youm Y-H, Ravussin A, Galgani JE, Stadler K, Mynatt RL, et al. The NLRP3 inflammasome instigates obesity-induced inflammation and insulin resistance. Nat Med. 2011;17:179–88. doi:10.1038/nm.2279.

Waldén TB, Hansen IR, Timmons JA, Cannon B, Nedergaard J. Recruited vs. nonrecruited molecular signatures of brown, "brite," and white adipose tissues. Am J Physiol Endocrinol Metab. 2012;302:E19–31. doi:10.1152/ajpendo.00249.2011.

Wang QA, Scherer PE. The AdipoChaser mouse: a model tracking adipogenesis in vivo. Adipocytes. 2014;3:146–50. doi:10.4161/adip.27656.

Wang T, Si Y, Shirihai OS, Si H, Schultz V, Corkey RF, et al. Respiration in adipocytes is inhibited by reactive oxygen species. Obesity (Silver Spring). 2010;18:1493–502.

Wei W, Wang X, Yang M, Smith LC, Dechow PC, Wan Y. PGC1?? Mediates PPAR?? Activation of osteoclastogenesis and rosiglitazone-induced bone loss. Cell Metab. 2010;11:503–16. doi:10.1016/j.cmet.2010.04.015.

Wei J, Ferron M, Clarke CJ, Hannun YA, Jiang H, Blaner WS, et al. Bone-specific insulin resistance disrupts whole-body glucose homeostasis via decreased osteocalcin activation. J Clin Invest. 2014;124:1–13. doi:10.1172/JCI72323.

Wei J, Shimazu J, Makinistoglu MP, Maurizi A, Kajimura D, Zong H, et al. Glucose uptake and Runx2 synergize to orchestrate osteoblast differentiation and bone formation. Cell. 2015;161:1576–91. doi:10.1016/j.cell.2015.05.029.

Weinstein RS, Wan C, Liu Q, Wang Y, Almeida M, O'Brien CA, et al. Endogenous glucocorticoids decrease skeletal angiogenesis, vascularity, hydration, and strength in aged mice. Aging Cell. 2010;9:147–61. doi:10.1111/j.1474-9726.2009.00545.x.

Wojcik M, Janus D, Dolezal-Oltarzewska K, Drozdz D, Sztefko K, Starzyk JB. The association of FGF23 levels in obese adolescents with insulin sensitivity. J Pediatr Endocrinol Metab. 2012;25:687–90. doi:10.1515/jpem-2012-0064.

Worthley DL, Churchill M, Compton JT, Tailor Y, Rao M, Si Y, et al. Gremlin 1 identifies a skeletal stem cell with bone, cartilage, and reticular stromal potential. Cell. 2015;160:269–84. doi:10.1016/j.cell.2014.11.042.

Wu J, Boström P, Sparks LM, Ye L, Choi JH, Giang AH, et al. Beige adipocytes are a distinct type of thermogenic fat cell in mouse and human. Cell. 2012;150:366–76. doi:10.1016/j.cell.2012.05.016.

Xue R, Lynes MD, Dreyfuss JM, Shamsi F, Schulz TJ, Zhang H, et al. Clonal analyses and gene profiling identify genetic biomarkers of the thermogenic potential of human brown and white preadipocytes. Nat Med. 2015;21:760–8. doi:10.1038/nm.3881.

Zhou BO, Yue R, Murphy MM, Peyer JG, Morrison SJ. Leptin-receptor-expressing mesenchymal stromal cells represent the main source of bone formed by adult bone marrow. Cell Stem Cell. 2014;15:154–68. doi:10.1016/j.stem.2014.06.008.

Index

S.Y. Smith et al. (eds.), *Bone Toxicology*, Molecular and Integrative Toxicology,
DOI 10.1007/978-3-319-56192-9

MIX
Papier aus verantwortungsvollen Quellen
Paper from responsible sources
FSC® C105338

If you have any concerns about our products,
you can contact us on
ProductSafety@springernature.com

In case Publisher is established outside the EU,
the EU authorized representative is:
Springer Nature Customer Service Center GmbH
Europaplatz 3, 69115 Heidelberg, Germany

Printed by Libri Plureos GmbH
in Hamburg, Germany